Differentiation Formulas

$$\frac{dc}{dx} = 0, \text{ if } c \text{ is a constant}$$

$$\frac{d}{dx}(u \pm v) = \frac{du}{dx} \pm \frac{dv}{dx}$$

$$\frac{d(cu)}{dx} = c\frac{du}{dx}, \text{ if } c \text{ is a constant}$$

Product Rule: $\dfrac{d}{dx}(uv) = u\dfrac{dv}{dx} + v\dfrac{du}{dx}$

Quotient Rule: $\dfrac{d}{dx}\left(\dfrac{u}{v}\right) = \dfrac{v(du/dx) - u(dv/dx)}{v^2}$

Chain Rule: $\dfrac{dy}{dx} = \dfrac{dy}{du}\dfrac{du}{dx}$

Power Rule: $\dfrac{du^n}{dx} = nu^{n-1}\dfrac{du}{dx}$

$$\frac{d}{dx}(\ln u) = \frac{1}{u}\frac{du}{dx}$$

$$\frac{d}{dx}(\log_a u) = \frac{\log_a e}{u}\frac{du}{dx}$$

$$\frac{d}{dx}e^u = e^u\frac{du}{dx}$$

$$\frac{d}{dx}a^u = a^u \ln a\frac{du}{dx}$$

$$\frac{d}{dx}\cos u = -\sin u\frac{du}{dx}$$

$$\frac{d}{dx}\sin u = \cos u\frac{du}{dx}$$

$$\frac{d}{dx}\cot u = -\csc^2 u\frac{du}{dx}$$

$$\frac{d}{dx}\tan u = \sec^2 u\frac{du}{dx}$$

$$\frac{d}{dx}\csc u = -\csc u \cot u\frac{du}{dx}$$

$$\frac{d}{dx}\sec u = \sec u \tan u\frac{du}{dx}$$

$$\frac{d}{dx}\text{Arccos } u = -\frac{1}{\sqrt{1-u^2}}\frac{du}{dx}$$

$$\frac{d}{dx}\text{Arcsin } u = \frac{1}{\sqrt{1-u^2}}\frac{du}{dx}$$

$$\frac{d}{dx}\text{Arccot } u = -\frac{1}{u^2+1}\frac{du}{dx}$$

$$\frac{d}{dx}\text{Arctan } u = \frac{1}{u^2+1}\frac{du}{dx}$$

Technical Calculus
with Analytic Geometry
Fourth Edition

Peter Kuhfittig
Milwaukee School of Engineering

THOMSON
™
BROOKS/COLE

Australia • Brazil • Canada • Mexico • Singapore
Spain • United Kingdom • United States

Technical Calculus with Analytic Geometry
Fourth Edition
Peter Kuhfittig

Acquisitions Editor: *John-Paul Ramin*
Assistant Editor: *Katherine Brayton*
Editorial Assistant: *Leata Holloway*
Technology Project Manager: *Earl Perry*
Marketing Manager: *Karin Sandberg*
Project Manager, Editorial Production: *Belinda Krohmer*
Creative Director: *Rob Hugel*
Art Director: *Vernon Boes*
Print Buyer: *Karen Hunt*
Permissions Editor: *Kiely Sisk*

Production Service: *Interactive Composition Corporation*
Text Designer: *Jeanne Calabrese*
Copy Editor: *AmyLyn Reynolds*
Cover Designer: *Jeanne Calabrese*
Cover Image: *Getty Images*
Cover Printer: *Phoenix Color Corp*
Compositor: *Interactive Composition Corporation*
Printer: *RR Donnelly Crawfordsville*

For more information about our products, contact us at:
Thomson Learning Academic Resource Center
1-800-423-0563

For permission to use material from this text or product, submit a request online at
http://www.thomsonrights.com.
Any additional questions about permissions can be submitted by e-mail to
thomsonrights@thomson.com.

Thomson Higher Education
10 Davis Drive
Belmont, CA 94002-3098
USA

Asia (including India)
Thomson Learning
5 Shenton Way
#01-01 UIC Building
Singapore 068808

Australia/New Zealand
Thomson Learning Australia
102 Dodds Street
Southbank, Victoria 3006
Australia

Canada
Thomson Nelson
1120 Birchmount Road
Toronto, Ontario M1K 5G4
Canada

UK/Europe/Middle East/Africa
Thomson Learning
High Holborn House
50–51 Bedford Row
London WC1R 4LR
United Kingdom

Library of Congress Control Number: 2005925199

ISBN 0-495-01876-7

Preface

The main purpose of the fourth edition of *Technical Calculus with Analytic Geometry* is to continue and enhance the student-oriented features that have contributed to the success of the first three editions.

The most important enhancement is the use of a functional second color to help explain steps. A second color is often more effective than a verbal description. A similar enhancement is the use of graphing calculators. Although graphing calculators had already been introduced in the third edition (Section 1.5, for example), in this edition more emphasis is placed on amplifying and clarifying certain concepts such as graphing tangent and normal lines, graphing parametric and polar curves, and graphing solutions of differential equations. Without graphing calculators, such graphs might never be seen. Occasional use of a computer algebra system or the integration and root-finding capabilities of a graphing utility are also intended to enhance rather than replace traditional topics. Examples are integration by parts, partial fraction expansions, and calculation of arc length. These tools are recommended in some of the exercises.

Many of the exercise sets have been overhauled or at least revised. More applied problems have been added throughout. Additional details have been supplied in many of the examples.

Two new topics were introduced in this edition: cylindrical coordinates (Section 9.8) and numerical solutions of differential equations (Section 11.5).

The fourth edition, like the first three, has been written for today's technology student: the approach is nonrigorous and intuitive with emphasis on applications of calculus to technology. An important aspect of this is the use of letters such as t and V and occasional subscripts, particularly in Chapters 6, 7, and 11. The purpose is to acquaint the technical student with notations frequently encountered in technology. (Students seem to have extra difficulties with problems that use an unfamiliar notation.)

Again with the student in mind, explanations and suitable motivations are provided for new concepts as they arise. For example, the base to natural logarithms is introduced only after the derivative of the general logarithmic function has been discussed. Integrals are defined by means of an informal discussion of Riemann sums to provide a smoother transition from areas to volumes of revolution and to work and fluid pressure.

Other student-oriented features are included in this edition: the important concepts in each section are boxed and labeled for easy reference. Many other concepts are identified by marginal labels. Of particular importance is the use of marginal notes to help explain steps, particularly in the examples. This feature complements the functional second color. The examples are designed to illustrate the procedures in the exercises; special care has been taken to provide sufficient detail, especially in the fourth edition, to make the examples easier to follow and emulate.

While emphasis on applications helps provide the calculus skills needed in technology courses, a good deal of practice in the basic operations is equally important. Consequently, a large number of drill exercises has been included.

The basic plan of the book is to introduce differential and integral calculus in the first five chapters, with the two-year technology programs in mind. The later chapters treat various more advanced topics and may be selected according to requirements of individual programs. On the other hand, the topics have been covered thoroughly enough to meet the needs of four-year technology programs designed around a complete calculus sequence through differential equations.

The prerequisites for the topics in this book are courses in algebra and trigonometry. However, review sections on trigonometric, inverse trigonometric, logarithmic, and exponential functions are duly presented.

Chapter 1 covers the traditional topics of analytic geometry. Although intended mainly for use in calculus, a number of applications of conic sections are also discussed. The derivative is introduced in Chapter 2 and applications of the derivative in Chapter 3. Chapter 4 on integration begins with antiderivatives and continues with areas and the fundamental theorem of calculus. As noted at the end of Section 4.1, Section 4.5 may be taken up first, if desired. Chapter 5 continues with applications of integration. The emphasis throughout is on setting up integrals by means of a shortcut using a single typical element ("sloppy Riemann sum"). The purpose is to enable the student to set up integrals in many different situations, that is, to *apply* the integral concept, rather than to rely on memorized formulas. For this approach to make sense, some discussion, however informal, of the integral as a limit of a sum is unavoidable.

Chapter 6 covers transcendental functions and their applications to various technical fields. Also included is a brief discussion of L'Hospital's rule. Chapter 7 develops different integration techniques. These techniques are covered in considerable detail and may be selected according to individual needs. For example, if time constraints do not allow a detailed discussion of trigonometric substitution, integrals of this form can be obtained by use of tables, discussed in Section 7.9. (An obvious alternative is to use a computer algebra system, although this approach is not emphasized.) The discussion of partial fractions (Section 7.8) can be postponed to Chapter 13.

Chapter 8 introduces parametric equations in conjunction with the vector concept and continues with a discussion of arc length and polar coordinates. Chapter 9 begins with a discussion of quadric surfaces, followed by partial derivatives and double integrals. Triple integrals are treated briefly. An optional section on curve fitting is also included. The chapter ends with a section on cylindrical coordinates. Chapter 10 emphasises power-series expansions but contains an optional section on tests of convergence. A section on Fourier series is included for use in electrical technology curricula.

Chapters 11, 12, and 13 are all devoted to differential equations since these provide particularly interesting and powerful applications to numerous technical fields. Chapter 13 on Laplace transforms includes a section on partial fractions. Although time-consuming, partial fractions are essential to the Laplace transform technique.

The topics in Chapters 8 through 13 do not have to be covered in the order presented, and numerous omissions, even of entire chapters, are possible. The topics in Chapter 1 can be treated in less detail and, as already noted, so can the integration techniques in Chapter 7.

Each chapter ends with a carefully selected set of review exercises that cover all the topics in the chapter.

The answers to the odd-numbered exercises are given in the answer section; however, for the review exercises in each chapter all answers are given. A separate Student Solutions Manual contains detailed solutions to all the odd-numbered exercises. An Instructor's Manual with all exercises worked out is also available.

I am indebted to the students of the Milwaukee School of Engineering for providing the stimulus for writing this book. I am especially grateful to Professors Stanley Guberud and Dorothy Johnson for reading the entire manuscript for the first edition and checking all the answers to the exercises; a special note of thanks goes to Professor Edward Griggs for many helpful discussions. For the fourth edition I would like to thank Professors Edward Griggs and James Carr for their invaluable technical assistance and Professor Robert Strangeway for a number of helpful suggestions. I would also like to thank the staff of Brooks/Cole Publishing Company and the following reviewers for their cooperation and help in the preparation of all four editions.

For the first edition: William Brower, New Jersey Institute of Technology; Joseph Colla, Milwaukee Area Technical College; Gerald Flynn, State University of New York at Farmingdale; J. Richard Garnham, Rochester Institute of Technology; Frank Lopez, Eastfield College; Gary M. Simundza, Wentworth Institute of Technology; V. Merriline Smith, California Polytechnic State University; Ara B. Sullenberger, Tarrant County Junior College—South; Roman Voronka, New Jersey Institute of Technology; Harry Wilson, California Polytechnic State University; and James Wolfe, Hocking Technical College. *For the second edition:* William Brower, New Jersey Institute of Technology; Molly Fails, Terra Technical College; John Monroe, University of Akron; Catherine Murphy, Purdue University—Calumet Campus; James Schmeidler; Al Swimmer, Arizona State University; and Jan Wynn, Brigham Young University. *For the third edition:* Linda J. S. Allen, Texas Tech University; Cecil Coone, Tennessee State Technical Institute at Memphis; Hsin Fan, California State Polytechnic University; Henry Hosek, Purdue University—Calumet; Jim Jones, California State University-Chico; Robert J. Radin, University of Hartford; and James A. Runyon, Rochester Institute of Technology. *For the fourth edition:* Albert Bronstein, Purdue University; Bruno Wichnoski, University of North Carolina, Charlotte; Irving Tang, Oklahoma State University; Randy Norton, Idaho State University.

Peter Kuhfittig

Brief Contents

Contents

7   Integration Techniques   252

8   Parametric Equations, Vectors, and Polar Coordinates   292

9   Three-Dimensional Space; Partial Derivatives; Multiple Integrals   318

10   Infinite Series   362

11   First-Order Differential Equations   399

12   Higher-Order Linear Differential Equations   425

13   The Laplace Transform   451

Appendix A: Tables   466

Appendix B: Answers to Selected Exercises   471

Index   514

1 Introduction to Analytic Geometry

Analytic geometry may be defined as the study of classical geometry by means of algebra. Credit for the discovery of this method is usually given to René Descartes (1596–1650), the famous French mathematician and philosopher, but the same method was discovered independently and somewhat earlier by Pierre de Fermat (1601–1665), a French lawyer.

Prior to the development of analytic geometry, algebra and geometry were largely independent of each other—lines belonged to geometry and equations belonged to algebra. The fusion of these two branches of mathematics led to entirely new scientific applications, as we will see later in this chapter. Analytic geometry also paved the way for the discovery of calculus a few decades later.

1.1 The Cartesian Coordinate System

In this section we will study basic concepts from analytic geometry: a coordinate system for plotting points, a formula for determining the distance between two points, and the midpoint formula.

Numbers

Rational number

Recall from algebra that real numbers are either rational or irrational. A **rational number** can be written as a ratio of two integers. For example, $-2.5 = -5/2$. Rational numbers have either terminating decimals (such as $3/5 = 0.6$) or infinite repeating decimals (such as $1/3 = 0.3333\ldots$).

Irrational number

A number that is not rational is called **irrational.** Irrational numbers have non-repeating decimal expansions. For example, $\sqrt{2} \approx 1.4142136$; there are no repeating cycles.

Complex numbers

Numbers of the form $a + bj$, where $j = \sqrt{-1}$, are called **complex numbers.** Analytic geometry deals mostly with real numbers.

The Cartesian Coordinate System

The set of real numbers may be exhibited on a line: we assume that to every number there corresponds a point and to every point a number (Figure 1.1). By choosing an arbitrary point for the number zero, called the **origin,** we can place the positive numbers to the right of zero and the negative numbers to the left in such a way that $-a$ is the "mirror image" of $+a$.

$$-\tfrac{5}{2} \qquad \sqrt{2}$$

$$-5 \; -4 \; -3 \; -2 \; -1 \;\; 0 \;\; 1 \;\; 2 \;\; 3 \;\; 4 \;\; 5$$

Figure 1.1

x-axis
y-axis
Origin

To locate points in a plane, construct a reference system consisting of two perpendicular lines, where the horizontal line is called the ***x*-axis,** the vertical line the ***y*-axis,** and their intersection the **origin** (Figure 1.2). Each line is a copy of the line in Figure 1.1. The position of the point can now be described by giving its perpendicular distance and direction from each axis. Distances measured to the right of the *y*-axis or up from the *x*-axis are regarded as positive; those to the left and down, respectively, are considered negative. The distance from the *y*-axis, with proper sign, is called the ***x*-coordinate** or **abscissa,** and the distance from the *x*-axis is called the ***y*-coordinate** or **ordinate.** Together they form the **coordinates** of the point.

x-coordinate
y-coordinate
Axes
Quadrants

The *x*- and *y*-axes, also called the **coordinate axes,** divide the plane into four **quadrants,** numbered I, II, III, and IV, as in Figure 1.2. We see that both coordinates are positive in the first quadrant, that the abscissa is negative and the ordinate positive in the second quadrant, and so on. In naming a point, it is customary to write the coordinates in parentheses with the *x*-coordinate first. For example, in Figure 1.2 the point P is $(-2, 4)$ and point Q is $(-3, -1)$. This system of locating points is called the **Cartesian** or **rectangular coordinate system.**

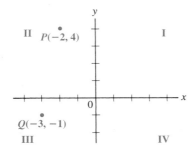

Figure 1.2

The Distance Formula

In the Cartesian coordinate system, fixed points are usually distinguished by subscripts such as P_1, P_2, P_3 or by (x_1, y_1) or (x_2, y_2). Using this notation, let us find a useful formula for the distance between two arbitrary points $P_1(x_1, y_1)$ and $P_2(x_2, y_2)$, shown in Figure 1.3.

First join P_1 and P_2 by a straight line. Then construct a right triangle by drawing a line through P_1 parallel to the *x*-axis and a line through P_2 parallel to the *y*-axis. Denote the point of intersection by P_3. The coordinates of P_3 are (x_2, y_1).

The length of the segment $P_1 P_3$ is $|x_2 - x_1|$ and the length of $P_2 P_3$ is $|y_2 - y_1|$. (See Exercise 12 at the end of the section.) Since $P_1 P_2$ is the hypotenuse, from the Pythagorean theorem we get

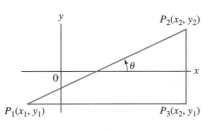

Figure 1.3

$$(P_1 P_2)^2 = (P_1 P_3)^2 + (P_2 P_3)^2 \qquad \text{or} \qquad P_1 P_2 = \sqrt{(x_2 - x_1)^2 + (y_2 - y_1)^2}$$

Distance Formula

The distance d from (x_1, y_1) to (x_2, y_2) is given by

$$d = \sqrt{(x_2 - x_1)^2 + (y_2 - y_1)^2} \qquad\qquad (1.1)$$

Example 1 Find the distance between $(-2, -4)$ and $(6, -2)$.

Solution. Let $(x_2, y_2) = (-2, -4)$ and $(x_1, y_1) = (\mathbf{6, -2})$. Then

$$d = \sqrt{(x_2 - x_1)^2 + (y_2 - y_1)^2} = \sqrt{(-2 - 6)^2 + [-4 - (-2)]^2}$$
$$= \sqrt{64 + 4} = \sqrt{68} = \sqrt{4 \cdot 17} = 2\sqrt{17} \approx 8.25$$

The same result is obtained by letting $(x_2, y_2) = (6, -2)$ and $(x_1, y_1) = (-2, -4)$. ■

We can easily find the midpoint of the line segment with endpoints $P_1(x_1, y_1)$ and $P_2(x_2, y_2)$, as shown in Figure 1.4.

Midpoint Formula

The midpoint of the line segment joining (x_1, y_1) and (x_2, y_2) is

$$\left(\frac{x_1 + x_2}{2}, \frac{y_1 + y_2}{2} \right) \tag{1.2}$$

To derive this formula, let $P(x, y)$ be the midpoint of the line segment $P_1 P_2$. Draw the triangle shown in Figure 1.4 by constructing lines perpendicular to the x-axis from $P_1(x_1, y_1)$, $P(x, y)$, and $P_2(x_2, y_2)$. It follows that $2(x - x_1) = x_2 - x_1$ and $2(y - y_1) = y_2 - y_1$. Solving, we get $x = (1/2)(x_1 + x_2)$ and $y = (1/2)(y_1 + y_2)$, which are the coordinates of the midpoint of the line segment.

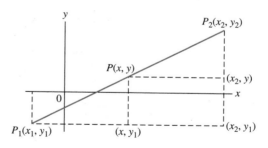

Figure 1.4

■ Exercises / Section 1.1

In Exercises 1–9, plot each pair of points and find the distance between them.

1. $(2, 4)$ and $(5, 2)$

2. $(-3, 2)$ and $(5, -4)$

3. $(-3, -6)$ and $(5, -2)$

4. $(0, 0)$ and $(-\sqrt{5}, 2)$

5. $(0, 2)$ and $(\sqrt{3}, 4)$

6. $(\sqrt{2}, \sqrt{5})$ and $(\sqrt{2}, 0)$

7. $(1, -\sqrt{2})$ and $(-1, 0)$

8. $(1, 6)$ and $(-3, 2)$

9. $(-10, -2)$ and $(-12, 0)$

10. Draw the triangle whose vertices are $(0, 0)$, $(4, 3)$, and $(6, 0)$.

11. If (x, y) is a point, determine in which quadrant x/y is
(a) positive; (b) negative.

12. In the derivation of the distance formula show that the length of P_1P_3 is $|x_2 - x_1|$ and the length of P_2P_3 is $|y_2 - y_1|$, no matter in which quadrants the points lie. The distance formula is therefore valid for all points. (See Figure 1.3.)

13. Find the set of all points whose **(a)** abscissas are zero; **(b)** ordinates are zero.

14. Find the perimeter of the triangle in Exercise 10.

15. Show that $(-2, 5)$, $(3, 5)$, and $(-2, 2)$ are vertices of a right triangle by using the Pythagorean theorem.

16. Repeat Exercise 15 for $(1, 7)$, $(-2, 1)$, and $(10, -5)$.

17. Show that $(12, 0)$, $(-4, 8)$, and $(-1, -13)$ lie on a circle whose center is $(1, -2)$.

18. Show that $(-2, 10)$, $(3, -2)$, and $(15, 3)$ are vertices of an isosceles triangle.

19. Show that $(-1, -1)$, $(2, 8)$, and $(5, 17)$ are collinear, that is, lie on the same line.

20. Find the relation between x and y so that (x, y) is 3 units from $(-1, 2)$.

21. Find the relation between x and y so that (x, y) is equidistant from the y-axis and the point $(2, 0)$.

In Exercises 22–28, find the midpoint of the line segment joining the given points.

22. $(-3, -5)$ and $(-1, 7)$
23. $(-2, 6)$ and $(2, -4)$
24. $(-3, 5)$ and $(-2, 9)$
25. $(5, 0)$ and $(9, 4)$
26. $(-4, 3)$ and $(-1, -7)$
27. $(-3, 10)$ and $(-6, 0)$
28. $(-4, -5)$ and $(-6, 5)$

1.2 The Slope

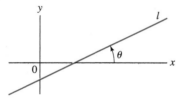

Figure 1.5

We saw in Section 1.1 that every point has a definite location with respect to the coordinate axes. In a similar way, a line has a definite orientation with respect to these axes. For example, the angle θ in Figure 1.5 is a measure of steepness, just as the steepness of a road is measured with respect to the horizontal. This angle $\theta(0° \leq \theta < 180°)$ is called the **angle of inclination.** It turns out to be more convenient, however, to define steepness by the formula $m = \tan \theta$, called the **slope** of the line.

Definition of Slope
$m = \tan \theta$ (where θ = angle of inclination)   (1.3)

By definition of the tangent function, m can be viewed as the vertical displacement divided by the corresponding horizontal displacement or by the change in the y-direction with respect to the change in the x-direction, sometimes referred to as *rise over run*. (The importance of this idea will be seen in Chapter 2.)

We can obtain a formula for the slope of a line determined by two points $P_1(x_1, y_1)$ and $P_2(x_2, y_2)$, as shown in Figures 1.6 and 1.7. Consider the *directed line segment* P_1P_3, defined by $P_1P_3 = x_2 - x_1$. A directed line segment can be positive or negative. For example, in Figure 1.6, P_1P_3 is positive (since $x_2 > x_1$) and in Figure 1.7, P_1P_3 is negative (since $x_2 < x_1$). $P_2P_3 = y_2 - y_1$ can also be positive or negative depending on the relative positions of P_2 and P_3. Since $\alpha = \theta$ (so that

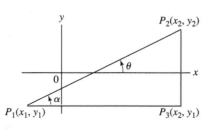

Figure 1.6

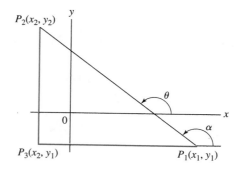

Figure 1.7

$\tan \alpha = \tan \theta$), we now have

$$m = \tan \theta = \tan \alpha = \frac{y_2 - y_1}{x_2 - x_1}$$

> **Slope of a Line**
>
> The slope m of the line passing through the points (x_1, y_1) and (x_2, y_2) is
>
> $$m = \frac{y_2 - y_1}{x_2 - x_1} = \frac{y_1 - y_2}{x_1 - x_2} \qquad (1.4)$$

　　The angle θ in Figure 1.6 is acute ($0° < \theta < 90°$), so that the line rises to the right. As a result, $x_2 - x_1 > 0$ and $y_2 - y_1 > 0$, so that m is positive. This also follows from the fact that $m = \tan \theta$ is positive for any angle between $0°$ and $90°$.

　　In Figure 1.7, θ is obtuse ($90° < \theta < 180°$), so that the line rises to the left. It follows that m is negative since $x_2 - x_1 < 0$ and $y_2 - y_1 > 0$. This agrees with the fact that $m = \tan \theta$ is negative for any angle between $90°$ and $180°$.

　　When $\theta = 0°$ (when the line is horizontal) $m = 0$. If the line is vertical, then $\theta = 90°$. Since $\tan 90°$ is undefined, m is undefined.

　　Summary: $m > 0$, or positive, when θ is acute (line rising to the right), $m < 0$, or negative, when θ is obtuse (line rising to the left), $m = 0$ when $\theta = 0°$ (horizontal line), and m is undefined when $\theta = 90°$ (vertical line).

Example 1　Find the slope of the line passing through $(-6, -2)$ and $(2, 4)$.

Solution. Let $(x_1, y_1) = (-6, -2)$ and $(x_2, y_2) = (\mathbf{2}, \mathbf{4})$. Then

$$m = \frac{y_2 - y_1}{x_2 - x_1} = \frac{4 - (-2)}{\mathbf{2} - (-6)} = \frac{6}{8} = \frac{3}{4}$$

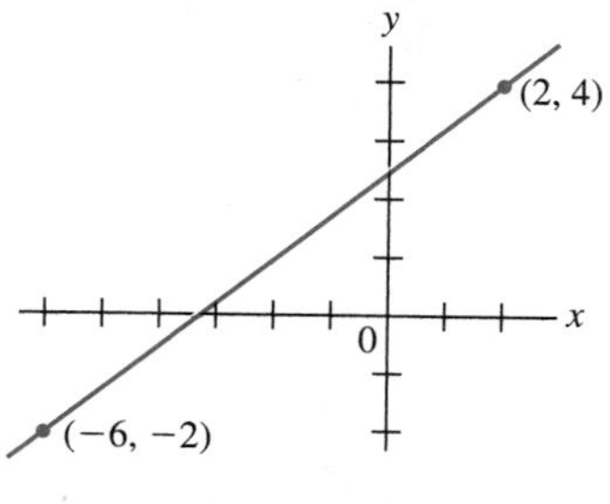

(See Figure 1.8.) Note that m is positive since the line rises to the right (θ acute). We get the same result by letting $(x_1, y_1) = (2, 4)$ and $(x_2, y_2) = (-6, -2)$:

$$m = \frac{y_2 - y_1}{x_2 - x_1} = \frac{-2 - 4}{-6 - 2} = \frac{-6}{-8} = \frac{3}{4} \qquad ■$$

Figure 1.8

Example 2   Find the slope of the line determined by $(-2, 5)$ and $(3, -1)$.

Solution. Let $(x_1, y_1) = (3, -1)$ and $(x_2, y_2) = (-2, 5)$. Then

$$m = \frac{y_2 - y_1}{x_2 - x_1} = \frac{5 - (-1)}{-2 - 3} = \frac{6}{-5} = -\frac{6}{5}$$

(See Figure 1.9.) Here m is negative since the line rises to the left (θ obtuse).   ■

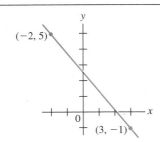

Figure 1.9

Example 3   Determine the slope of the line passing through $(-4, -2)$ and $(5, -2)$.

Solution.

$$m = \frac{-2 - (-2)}{5 - (-4)} = \frac{0}{9} = 0$$

Note that $m = 0$ because the line is horizontal. (See Figure 1.10.)   ■

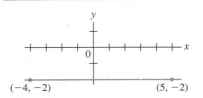

Figure 1.10

If the angle of inclination is close to $90°$, then m is numerically large. For example, for the line in Figure 1.11,

$$m = \frac{12 - (-2)}{2 - 1} = \frac{14}{1} = 14$$

As noted earlier, if the line is vertical the slope is undefined (infinitely large). If we calculate the slope of the line in Figure 1.12, for example, we get

$$m = \frac{4 - (-3)}{-1 - (-1)} = \frac{7}{0} \quad \text{(undefined)}$$

In other words, when a line is vertical, $x_1 = x_2$, so that $x_2 - x_1 = 0$. Since division by zero is not allowed, the slope is undefined.

Figure 1.11

Parallel and Perpendicular Lines

The slope can be used to determine whether two lines are parallel or perpendicular. If two lines are parallel, they have equal angles of inclination and therefore equal slopes. If two lines are perpendicular, their angles of inclination differ by $90°$, that is, $\theta_2 = \theta_1 + 90°$. From Figure 1.13, we have

$$\tan \theta_1 = \frac{b}{a}$$

and

$$\tan \theta_2 = \tan(\theta_1 + 90°) = \frac{a}{-b} = -\frac{1}{b/a} = -\frac{1}{\tan \theta_1}$$

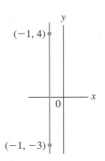

Figure 1.12

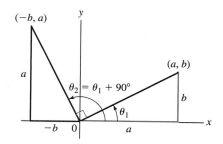

Figure 1.13

Thus $m_2 = -1/m_1$. These ideas are summarized next.

Parallel and Perpendicular Lines

Two lines are **parallel** if

$$m_1 = m_2 \qquad\qquad (1.5)$$

Two lines are **perpendicular** if

$$m_2 = -\frac{1}{m_1} \qquad \text{or} \qquad m_1 m_2 = -1 \qquad\qquad (1.6)$$

Example 4   Show that the line determined by $(-1, 4)$ and $(5, 2)$ is parallel to the line determined by $(-2, -1)$ and $(1, -2)$.

Solution. The lines are shown in Figure 1.14.

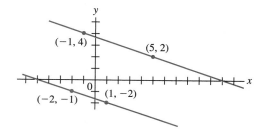

Figure 1.14

The respective slopes are

$$m_1 = \frac{2 - 4}{5 - (-1)} = \frac{-2}{6} = -\frac{1}{3}$$

$$m_2 = \frac{-2 - (-1)}{1 - (-2)} = \frac{-1}{3} = -\frac{1}{3} \qquad m_1 = m_2$$

Since the slopes are equal, the lines are parallel. ∎

Example 5   Show that the line through $(7, -1)$ and $(4, -6)$ is perpendicular to the line through $(2, 0)$ and $(-3, 3)$.

Solution. The respective slopes are

$$m_1 = \frac{-1 - (-6)}{7 - 4} = \frac{5}{3} \quad \text{and} \quad m_2 = \frac{0 - 3}{2 - (-3)} = -\frac{3}{5} \qquad m_1 = -\frac{1}{m_2}$$

Since the slopes are negative reciprocals, the lines are perpendicular. (See Figure 1.15.) ■

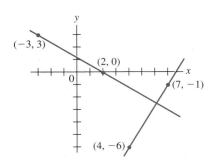

Figure 1.15

■ Exercises / Section 1.2

In Exercises 1–13, find the slope of the line through the given points.

1. $(2, 6)$ and $(1, 7)$

2. $(-3, -10)$ and $(-5, 2)$

3. $(0, 2)$ and $(-4, -4)$

4. $(6, -3)$ and $(4, 0)$

5. $(7, 8)$ and $(-3, -4)$

6. $(0, 0)$ and $(8, -4)$

7. $(0, 0)$ and $(-4, -8)$

8. $(1, 3)$ and $(-2, 0)$

9. $(3, 4)$ and $(3, -5)$

10. $(-4, -3)$ and $(0, 6)$

11. $(5, -3)$ and $(9, -3)$

12. $(1, 4)$ and $(5, -7)$

13. $(0, -6)$ and $(8, 0)$

14. Find the slope of the line whose angle of inclination is:

a. $0°$ **b.** $30°$ **c.** $150°$ **d.** $90°$ **e.** $45°$ **f.** $135°$

15. Draw the line passing through $(0, 2)$ and having a slope of:

a. 2 **b.** $\frac{1}{2}$ **c.** $-\frac{1}{2}$ **d.** 0 **e.** 3 **f.** $-\frac{2}{3}$

16. The vertices of a triangle are $A(-2, 0)$, $B(-1, 2)$, and $C(2, -3)$. Find the slope of each of its sides.

17. Find the slope of a line perpendicular to the line through $(-7, 1)$ and $(6, -5)$.

18. Show that the points $(-2, 8)$, $(1, -1)$, and $(3, -7)$ are collinear.

19. Without using the distance formula show that the points $(-4, 6)$, $(6, 10)$, and $(10, 0)$ are vertices of a right triangle.

20. Show that $(-4, 2)$, $(-1, 8)$, $(9, 4)$, and $(6, -2)$ are vertices of a parallelogram.

21. Show that $(0, -3)$, $(-2, 3)$, $(7, 6)$, and $(9, 0)$ are vertices of a rectangle.

22. Find the slope of the line through $(6, -4)$ and the midpoint of the line segment from $(-2, -4)$ and $(8, 8)$.

23. Find the slope of the line through $(5, 6)$ and the midpoint of the line segment from $(-3, 0)$ to $(9, -2)$.

24. Show that $(0, 1)$, $(3, 3)$, $(8, 2)$, and $(-1, -4)$ are vertices of an isosceles trapezoid.

25. Find the slopes of the medians in the triangle in Exercise 16.

26. If the center of a circle is $(1, 2)$ and one end of the diameter is $(4, -3)$, find the other end.

27. Find the value of x so that the line through $(x, 2)$ and $(4, -6)$ is parallel to the line through $(-1, -1)$ and $(3, -5)$.

28. Find x so that the line through $(x, -2)$ and $(4, -7)$ is perpendicular to the line through $(2, -1)$ and $(-3, 2)$.

1.3 The Straight Line

Equations and Graphs

As noted earlier, analytic geometry combines algebra and geometry. Let us consider a simple example. The equation $x + 2y = 2$ has many solutions. If we let $x = 0$, for example, we get $0 + 2y = 2$, or $y = 1$. Thus $x = 0$ and $y = 1$ is a solution. A few other pairs of solutions are given in the following chart:

x:	-1	0	1	2
y:	$\frac{3}{2}$	1	$\frac{1}{2}$	0

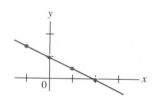

Figure 1.16

If we plot the points $(-1, 3/2)$, $(0, 1)$, $(1, 1/2)$, and $(2, 0)$ and connect them by a smooth curve, we get the graph shown in Figure 1.16. The graph looks like a straight line. It is. This example shows that the line in Figure 1.16 is the *solution set* of the equation $x + 2y = 2$.

Much of analytic geometry is concerned with geometric figures and their corresponding equations. In this section we will study the following two fundamental problems of analytic geometry for the straight line:

---

1. Given the geometric figure, find the corresponding equation.
2. Given the equation, draw the figure.

The two problems will be considered separately.

Moving from the Figure to the Equation

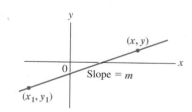

Figure 1.17

Suppose a line has slope m and passes through the point (x_1, y_1). To derive the equation, let (x, y) be any point on the line. Then the slope of the line segment joining (x_1, y_1) and (x, y) is the same regardless of which point is chosen. (See Figure 1.17.) So if m is the given slope, then

$$m = \frac{y - y_1}{x - x_1} \qquad \text{or} \qquad y - y_1 = m(x - x_1)$$

which is called the **point-slope form** of the line.

Point-Slope Form of a Line

The equation of a line with slope m and passing through the point (x_1, y_1) is given by

$$y - y_1 = m(x - x_1) \tag{1.7}$$

The point-slope form is used to write the equation of a given line, as shown in the next example.

Example 1   Find the equation of the line through $(-3, 5)$ and having a slope of 2.

Solution. From Equation (1.7) we get

$$y - 5 = 2[x - (-3)] \qquad \text{or} \qquad 2x - y + 11 = 0$$

Example 2   Find the equation of the line determined by the points $(-5, -1)$ and $(5, 4)$.

Solution. The slope of the line is found to be $1/2$. The point-slope form can now be obtained by using either point, say $(-5, -1)$. Thus

$$y - (-1) = \frac{1}{2}[x - (-5)] \qquad \text{or} \qquad x - 2y + 3 = 0$$

(See Figure 1.18.) We get the same equation by using the point $(5, 4)$.

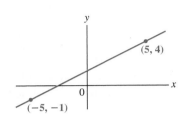

Figure 1.18

If the line is horizontal, then $m = 0$ and Equation (1.7) reduces to

Horizontal line

> **Horizontal Line**
>
> $$y = y_1 \tag{1.8}$$

If the line is vertical, the slope is undefined. However, all points on a vertical line have a constant abscissa, here called x_1. So the equation is

Vertical line

> **Vertical Line**
>
> $$x = x_1 \tag{1.9}$$

Example 3   The horizontal line $y = 3$ is shown in Figure 1.19. The vertical line $x = -2$ is shown in Figure 1.20.

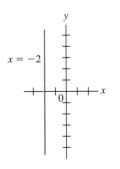

Figure 1.20

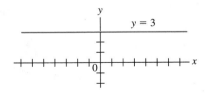

Figure 1.19

Moving from the Equation to the Figure

To see how we can obtain the line from its equation, let us consider the following special case of the point-slope form: Suppose that a line has slope m and y-intercept

b (that is, the line crosses the y-axis at $(0, b)$). Then, from Equation (1.7),

$$y - b = m(x - 0) \quad \text{or} \quad y = mx + b$$

which is called the **slope-intercept form** of the line.

Slope-Intercept Form of the Line

The equation of the line with slope m and y-intercept $(0, b)$ is given by

$$y = mx + b \tag{1.10}$$

Given the equation of a line, we can use the slope-intercept form to obtain the slope and y-intercept, and hence the graph, of the line. In other words, we use the slope-intercept form to move from the equation to the figure.

Example 4   Show that the line $x + 2y + 1 = 0$ is perpendicular to the line $2x - y + 3 = 0$.

Solution.   Solving each equation for y, we get

$$y = -\frac{1}{2}x - \frac{1}{2} \qquad \textbf{slope} = -\frac{1}{2}; \textbf{\textit{y}-intercept} = \left(\mathbf{0}, \ -\frac{1}{2}\right)$$

and

$$y = 2x + 3 \qquad \textbf{slope = 2; \textit{y}-intercept = (0, 3)}$$

respectively. Since the slopes 2 and $-1/2$ are negative reciprocals, the lines are perpendicular to each other. The lines are shown in Figure 1.21. ∎

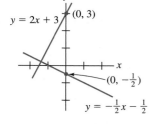

Figure 1.21

These examples show that every straight line has an equation that can be written as

---

Straight-Line Equation

$$Ax + By + C = 0 \qquad (A \text{ and } B \text{ not both zero})$$

Either A or B could be zero, but not both. If $B \neq 0$, we can solve the equation for y and obtain the slope-intercept form. If $B = 0$ (and $A \neq 0$), then $x = -C/A$, which is a vertical line. If $A = 0$ (and $B \neq 0$), then $y = -C/B$, which is a horizontal line.

As in algebra, we normally use x and y for the variables. The purpose of this convention is to generalize: since the variables can stand for any physical quantity, no restriction is placed on the problems to which the equations can be applied. However, it is sometimes convenient to use letters that suggest the physical quantities more directly.

Example 5   If E is the voltage drop across a resistor R, then the current I through the resistor is $I = E/R$ by Ohm's law. If $R = 2\,\Omega$, then $E = 2I$ is the line in Figure 1.22. (For a summary of SI units, see Table 1 in Appendix A.)   ■

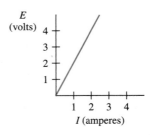

Figure 1.22

■ Exercises / Section 1.3

In Exercises 1–18, write the equation of the line satisfying the given conditions.

1. Passing through $(-7, 2)$ and having a slope of $1/2$
2. Passing through $(0, 3)$ and having a slope of -4
3. Passing through $(3, -4)$ and having a slope of 3
4. Passing through $(1, 2)$ and having a slope of 0
5. Passing through $(0, 0)$ and having a slope of $-1/3$
6. Passing through $(-3, -10)$ and parallel to the y-axis
7. Passing through $(-4, 0)$ and parallel to the line $y = 1$
8. Passing through $(-3, 2)$ and $(7, 6)$
9. Passing through $(-3, 4)$ and $(3, -6)$
10. Passing through $(2, 3)$ and $(-4, 6)$
11. Passing through $(5, 0)$ and $(9, -4)$
12. Passing through $(-3, 5)$ and $(1, 7)$
13. Passing through $(2, 3)$ and $(-6, 4)$
14. Passing through $(1, 2)$ and having an angle of inclination of $135°$
15. Passing through $(0, 10)$ and having an angle of inclination of $45°$
16. Slope of 3 and y-intercept 4
17. Slope of $-1/3$ and y-intercept -2
18. $b = 2$ and $m = 1$

In Exercises 19–24, change each of the straight-line equations to slope-intercept form and draw the lines. (See Example 4.)

19. $6x + 2y = 5$
20. $x - y = 1$
21. $2x = 3y$

22. $\dfrac{1}{3}x + \dfrac{1}{12}y = 1$
23. $2y - 7 = 0$
24. $\dfrac{1}{2}x - 2y = 3$

In Exercises 25–30, state whether the lines are parallel, perpendicular, or neither.

25. $2x - 3y = 1$ and $4x - 6y + 3 = 0$
26. $2x + 4y + 3 = 0$ and $y - 2x = 2$
27. $3x - 4y = 1$ and $3y - 4x = 3$
28. $7x - 10y = 6$ and $y - 4 = 0$
29. $x + 3y = 5$ and $y - 3x - 2 = 0$
30. $2x + 5y = 2$ and $6x + 15y = 1$
31. Find the equation of the line through $(-1, 1)$ and parallel to the line $3x - 4y = 7$.
32. Find the equation of the line through $(-1, 1)$ and perpendicular to the line $3x - 4y = 7$.
33. Find the equation of the line passing through the intersection of the lines $2x - 4y = 1$ and $3x + 4y = 4$ and parallel to $5x + 7y + 3 = 0$.
34. Show that the following lines form the sides of a right triangle: $x - 3y + 3 = 0$, $12x + 4y + 25 = 0$, and $2y + x + 8 = 0$.
35. Sketch the graph of $v = 1 + 10t$, $t \geq 0$, where v is the velocity of an object and t is time, letting the t-axis be horizontal. (Note that the graph is a straight line and gives a pictorial representation of the velocity at any time.)
36. If a spring is stretched x units, it pulls back with a force $F = kx$ by Hooke's law; k is a constant called the *force*

constant of the spring. If $k = 3$, draw the graph of the equation.

37. A force of 3 lb is required to stretch a spring $1/2$ ft. Find k and draw the graph. (Refer to Exercise 36.)

38. The value (in dollars) of a piece of equipment is $y = 1200 - 200t$ (where $t \geq 0$ is measured in years). Sketch the graph. What is the significance of the y-intercept? The t-intercept?

39. The relationship between temperature measured in degrees Celsius (C) and degrees Fahrenheit (F) is known to be linear—that is, it has the form $F = mC + b$. Water boils at $100°C$ and $212°F$ and freezes at $0°C$ and $32°F$. Find the relationship. (For a table of SI units, see Appendix A.)

40. A consultant charges a flat fee of $1800 plus $500/d. Find the equation relating the cost C (in dollars) and the time t (in days).

41. The resistance R in a certain wire has the form $R = aT + b$, where T is the temperature. A lab technician determines that $R = 51\,\Omega$ when $T = 100°C$ and $R = 54\,\Omega$ when $T = 400°C$. Find the relationship.

1.4 Curve Sketching

We saw in the previous section that a graph gives a revealing geometric picture of the relationship between two quantities. Thus an industrial firm may use graphs to show product and market statistics. A scientist may show graphically how much of 100 g of the radioactive element polonium is left after a certain number of days (Figure 1.23), as we will see in Chapter 11.

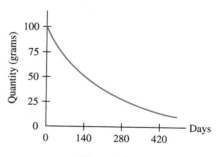

Figure 1.23

In this section we will develop some techniques for sketching the graph of a given equation. The purpose of these techniques is to obtain as much information as possible about the graph from the equation and to use this information to sketch the graph rapidly.

To get an overview of curve sketching, let us first consider a typical example, the equation $y = (1/9)x^3$. As an algebraic equation it is satisfied by infinitely many pairs of values. A few of these pairs are given in the following table:

x:	-3	-2	-1	0	1	2	3
y:	-3	$-\frac{8}{9}$	$-\frac{1}{9}$	0	$\frac{1}{9}$	$\frac{8}{9}$	3

We can plot these pairs as in Figure 1.24, and draw a smooth curve through them. The more points we plot, the more accurate the graph becomes. In general, a graph of an

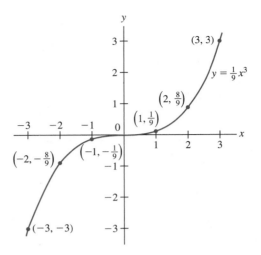

Figure 1.24

equation in two dimensions is the set of all possible solutions to that equation. *The graph contains all points, and only those points, whose coordinates satisfy the equation.* Since a graph is a geometric figure, we have established a basic connection between algebra and geometry.

The Graph of an Equation

The graph of an equation, which is a set of points called a **locus,** is the solution set of the equation.

Plotting the points of an equation one by one can be a lengthy procedure. What is worse, the curve may have many interesting properties not easily detected by the point-plotting method. As noted earlier, much of this information can be extracted directly from the equation. In fact, the whole idea is to obtain as much information as possible from the equation, thereby reducing point plotting to a minimum. This procedure is referred to as the **discussion of a curve.**

Since the discussion of a curve consists of several parts, let us outline the procedure at this point and then examine each part in detail.

Discussion of a Curve

Step 1. Find the **intercepts.**
Step 2. Determine **symmetry** with respect to the axes.
Step 3. Find the **asymptotes.**
Step 4. Determine the **extent** of the curve.
Step 5. Sketch the curve.

Intercepts

y-intercepts
x-intercepts

The **intercepts** are the points where the graph crosses the coordinate axes. The points where the graph crosses the *y*-axis are called the **y-intercepts** and the points where the graph crosses the *x*-axis are called the **x-intercepts.** To obtain the *y*-intercepts, we let $x = 0$ and solve for *y*; similarly, to determine the *x*-intercepts, we let $y = 0$ and solve for *x*.

Consider, for example, the curve $y = 4 - x^2$. If $x = 0$, then

$$y = 4 - 0^2 = 4 \qquad \textbf{y = 4 − x}^2$$

Thus (0, 4) is the *y*-intercept. Letting $y = 0$, we get

$$0 = 4 - x^2 \qquad \text{or} \qquad x = \pm 2 \qquad \textbf{Don't forget the negative root.}$$

Thus the *x*-intercepts are (2, 0) and (−2, 0), usually written (±2, 0). The intercepts can tell us a great deal about the graph. In particular, if there are no intercepts, we know that the graph cannot cross the coordinate axes.

Symmetry

We will use the term **symmetry** to denote symmetry with respect to the coordinate axes. Taking the usual meaning of the term, a graph is symmetric with respect to the *y*-axis if the left half of the graph is the "mirror image" of the right half. More precisely, if the point (a, b) lies on the graph, so does the point $(-a, b)$. To check this condition for all points simultaneously, we replace *x* by −*x* in the equation, and if the equation reduces to the given equation, then the graph is symmetric with respect to the *y*-axis. For example, let $y = x^2 + 1$. Upon replacing *x* by −*x*, we obtain $y = (-x)^2 + 1$, which reduces to the original equation $y = x^2 + 1$. Hence the graph of this equation is symmetric with respect to the *y*-axis.

By a similar procedure we can check the equation $y^2 + y^4 = x$ for symmetry with respect to the *x*-axis. Replacing *y* by −*y*, we get $(-y)^2 + (-y)^4 = x$, or $y^2 + y^4 = x$. The graph is therefore symmetric with respect to the *x*-axis.

Finally, a curve is symmetric with respect to the origin if the equation remains unchanged after substituting −*x* for *x* and −*y* for *y* at the same time. This statement implies that whenever the point (a, b) lies on the graph, so does the point $(-a, -b)$. Symmetry with respect to both axes implies symmetry with respect to the origin, but not conversely. For example, the graph in Figure 1.24 is symmetric with respect to the origin but not symmetric with respect to either axis. Thus

$$y = \frac{1}{9}x^3 \qquad \textbf{given equation}$$

$$-y = \frac{1}{9}(-x)^3 \qquad \textbf{replacing x by −x and y by −y}$$

$$-y = \frac{1}{9}(-x^3) \qquad \textbf{(−x)}^3 \textbf{ = −x}^3$$

$$y = \frac{1}{9}x^3 \qquad \textbf{reduces to the given equation}$$

The curve is therefore symmetric with respect to the origin.

Asymptotes

Asymptotes can best be understood from an example. Consider the equation

$$y = \frac{1}{x-1}$$

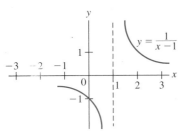

Figure 1.25

We see at once that y is undefined if $x = 1$, because substituting the number 1 for x would lead to division by zero. The real question is: what happens in the vicinity of $x = 1$? Since the denominator is close to zero near this value, the absolute value of y is very large and keeps increasing as x approaches 1. Consequently, the graph will approach the vertical line $x = 1$ as x approaches 1, but without ever reaching it. (See Figure 1.25.) In general, we define an **asymptote** to be a line approached by the graph in the sense that the distance between the line and a point on the graph approaches zero as the point moves to infinity along the graph. For the curves discussed in this section:

> To obtain the vertical asymptotes, solve the equation for y in terms of x. If the solution is a quotient (fraction) with the variable x in the denominator, set the denominator equal to zero and solve for x. At these x values the graph will have vertical asymptotes.

If the asymptote is horizontal, the procedure is a little more complicated. Keeping in mind the definition of asymptote, however, it is clear that we need to study the behavior of the graph as x gets large. It is possible to reverse the foregoing procedure and solve the equation for x in terms of y, but in a case such as

$$y = \frac{3x^2 + 2x - 1}{2x^2 + 2}$$

it would be better to work with the equation directly. Suppose we divide numerator and denominator by the highest power of x found in the denominator, in this case x^2. Then

$$y = \frac{\dfrac{3x^2}{x^2} + \dfrac{2x}{x^2} - \dfrac{1}{x^2}}{\dfrac{2x^2}{x^2} + \dfrac{2}{x^2}} \qquad \text{or} \qquad y = \frac{3 + \dfrac{2}{x} - \dfrac{1}{x^2}}{2 + \dfrac{2}{x^2}}$$

Since $1/x^n (n > 0)$ approaches zero as x gets large, it follows that

$$\frac{2}{x}, \ -\frac{1}{x^2}, \text{ and } \frac{2}{x^2}$$

all approach zero. As a result,

$$y = \frac{3 + \dfrac{2}{x} - \dfrac{1}{x^2}}{2 + \dfrac{2}{x^2}} \quad \text{approaches } \frac{3}{2} \text{ when } x \text{ gets large}$$

Thus $y = 3/2$ is a horizontal asymptote.

In summary, to find a horizontal asymptote, divide numerator and denominator by the highest power of x contained in the denominator and let x get large. If y approaches a constant as x gets large, then the value approached by y is the horizontal asymptote. In other words, if y approaches a, then $y = a$ is the horizontal asymptote.

Extent (Along the x-Axis)

The extent of a curve along the x-axis consists of those values of x for which y is defined and real. For example, the extent of the curve $y = 1/x$ consists of all x except $x = 0$ (since division by zero is undefined).

The extent of the curve $y = \sqrt{x}$ is the set of all x such that $x \geq 0$. (For $x < 0$, $\sqrt{x}$ is complex; if $x = -4$, for example, then $y = \sqrt{-4} = \sqrt{4(-1)} = \sqrt{4}\sqrt{-1} = 2j$, an imaginary number.)

To determine the extent of a curve, we solve the given equation for y in terms of x. If the resulting expression contains radicals of even index (such as square roots or fourth roots), then the values of x that make the expression under the radical sign negative must be excluded (to avoid complex values).

Consider, for example, the equation

$$x^2 + 4y^2 = 1$$

Solving for y, we obtain

$$y^2 = \frac{1}{4}(1 - x^2) \qquad \text{and} \qquad y = \pm\frac{1}{2}\sqrt{1 - x^2}$$

So if x exceeds unity in absolute value, then the y-values become imaginary. We would now say that the extent of the curve is from -1 to 1; that is, $-1 \leq x \leq 1$. As noted earlier, we also exclude values of x that lead to division by zero. We will illustrate the ideas in this section by several examples.

Example 1   Discuss and sketch the curve $y = x^2 - 1$.

Solution.

Step 1. Intercepts: if $x = 0$, then

$$y = 0^2 - 1 = -1$$

If $y = 0$, then

$$0 = x^2 - 1$$
$$x^2 = 1$$
$$x = \pm 1 \qquad \textbf{x = 1 and x = -1}$$

Thus the intercepts are $(\pm 1, 0)$ and $(0, -1)$.

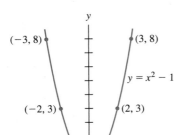

Step 2. Symmetry: replacing x by $-x$, we get

$$y = (-x)^2 - 1$$
$$y = x^2 - 1 \qquad \text{note the even power on } x$$

Since the equation reduces to the given equation, the curve is symmetric with respect to the y-axis.

Step 3. Asymptotes: none. (The equation is not in the form of a quotient with a variable in the denominator.)

Step 4. Extent: all x (no radicals).

We now plot the intercepts and four more points and then connect these by a smooth curve. The graph is shown in Figure 1.26. Since the curve is symmetric with respect to the y-axis, it is sufficient to draw the curve through the points on the right, $(0, -1)$, $(1, 0)$, $(2, 3)$, and $(3, 8)$, and then draw the "mirror image" on the left side. ∎

Figure 1.26

Example 2   Discuss and sketch the curve $y = (x + 1)(x - 2)(x - 3)$

Solution.

Step 1. Intercepts: if $x = 0$, then $y = 6$. If $y = 0$, then

$$(x + 1)(x - 2)(x - 3) = 0$$

and

$$x = -1, 2, 3$$

Step 2. Symmetry: none. (Replacing x by $-x$ or y by $-y$ changes the equation.)

Step 3. Asymptotes: none (no x in the denominator).

Step 4. Extent: all x.

We now plot the intercepts and two more points and connect these by a smooth curve. The graph is shown in Figure 1.27. Two additional points (not shown) are $(4, 10)$ and $(-2, -20)$.

One final comment: y changes signs at $x = -1, 2,$ and 3. As a result, the graph crosses the x-axis at the intercepts. ∎

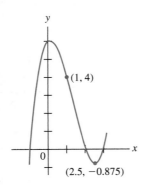

Figure 1.27

The next example illustrates a convenient method for determining the extent of a curve by means of a chart.

Example 3   Discuss and sketch the curve $y^2 = (x + 1)(x - 2)(x - 3)$.

Solution.

Step 1. Intercepts: if $x = 0$, $y = \pm\sqrt{6}$, and if $y = 0$, $x = -1, 2, 3$.

Step 2. Symmetry: x-axis. (Replacing y by $-y$ leaves the equation unchanged.)

Step 3. Asymptotes: none.
Step 4. Extent: Solving for y, we find that

$$y = \pm\sqrt{(x+1)(x-2)(x-3)}$$

As noted earlier, we must exclude all values of x for which the radicand is negative. We observe that $(x+1)(x-2)(x-3)$ changes signs only at $x = 3, 2$, and -1. If $x > 3$, all the factors are positive, so the radicand is positive. If x is between 2 and 3, then $x - 3 < 0$, so the entire radicand is negative. Hence the interval $2 < x < 3$ must be excluded from the graph (see Figure 1.28). At $x = 2$, the sign of the radicand changes again to become positive. Finally, at $x = -1$ the sign of the radicand changes back to negative, so that all values of x less than -1 must also be excluded. We conclude that the extent of the curve consists of the intervals $-1 \le x \le 2$ and $x \ge 3$. (The graph can now be drawn with the help of a few additional points which are shown in the table.)

x	y
-0.5	± 2.1
0	± 2.45
0.4	± 2.41
1	± 2
1.5	± 1.4
3.5	± 1.8
4	± 3.2

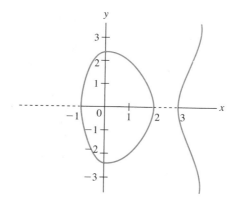

Figure 1.28

Summary: In determining the extent of the curve, note that the radicand is zero only at $x = -1, 2$, and 3 and must therefore be different from zero at all other points. Consequently, it is sufficient to substitute arbitrary "test values" to check signs. When $x = 0$, for example, the radicand is 6 and hence positive in the interval $-1 < x < 2$. These observations are summarized in the following chart:

	Test values	$x+1$	$x-2$	$x-3$	$(x+1)(x-2)(x-3)$
$x > 3$	4	+	+	+	+
$2 < x < 3$	2.5	+	+	−	−
$-1 < x < 2$	0	+	−	−	+
$x < -1$	−2	−	−	−	−

■

Example 4   Discuss and sketch the curve $y = x^2/(x^2 - 1)$.

Solution.

Step 1. Intercepts: if $x = 0$, $y = 0$, and if $y = 0$, $x = 0$.
Step 2. Symmetry: y-axis. (Note the even power on x.)

x	y
0.5	-0.33
0.75	-1.3
1.5	1.8
2	1.3
3	1.1

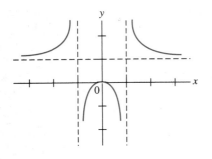

Figure 1.29

Step 3. Vertical asymptotes: setting the denominator equal to zero, we get $x^2 - 1 = 0$, so $x = \pm 1$. Horizontal asymptote: if numerator and denominator are divided by x^2, the equation becomes

$$y = \frac{\dfrac{x^2}{x^2}}{\dfrac{x^2}{x^2} - \dfrac{1}{x^2}} \qquad \text{or} \qquad y = \frac{1}{1 - \dfrac{1}{x^2}}$$

Since $-1/x^2$ approaches zero as x gets large, the right side approaches 1 as x gets large. Thus the line $y = 1$ is a horizontal asymptote.

Step 4. Extent: all x except $x = \pm 1$ (to avoid division by zero).

The graph is shown in Figure 1.29. Note that it follows from the discussion that the left and right branches cannot approach the vertical asymptotes in the downward direction without crossing the x-axis—but the only intercept is the origin. For the middle branch we find that $y \leq 0$ whenever $-1 < x < 1$. ∎

Example 5   Discuss and sketch the curve $y^2 = x/(x^2 - 4)$.

Solution.

Step 1. Intercepts: if $x = 0$, $y = 0$ and if $y = 0$, $x = 0$.
Step 2. Symmetry: x-axis. (Note the even power on y.)
Step 3. Vertical asymptotes: $x = \pm 2$. Horizontal asymptotes: dividing numerator and denominator of the fraction by x^2, we get

$$y^2 = \frac{\dfrac{1}{x}}{1 - \dfrac{4}{x^2}} \qquad \textbf{dividing by } x^2$$

As x gets large, y^2 approaches zero. Hence the asymptote is $y = 0$.

Step 4. Extent: solving for y, we find that

$$y = \pm \sqrt{\frac{x}{(x-2)(x+2)}} \qquad \textbf{taking square roots}$$

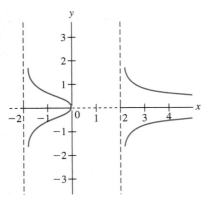

Figure 1.30

The radicand $x/[(x-2)(x+2)]$ changes signs only at $x = -2$, 0, and 2. If $x > 2$, all the factors are positive. If x is between 0 and 2, then $x - 2 < 0$, so that the radicand is negative. Hence the interval $0 < x < 2$ must be excluded (see Figure 1.30). At $x = 0$ the sign of the radicand changes back to positive. Finally, if $x < -2$ the radicand is negative; so the extent is given by $-2 < x \leq 0$ and $x > 2$.

By using "test values," these results can be summarized as follows:

	Test values	$x + 2$	x	$x - 2$	$\dfrac{x}{(x-2)(x+2)}$
$x > 2$	3	+	+	+	+
$0 < x < 2$	1	+	+	−	−
$-2 < x < 0$	−1	+	−	−	−
$x < -2$	−3	−	−	−	+

∎

■ Exercises / Section 1.4

In Exercises 1–37, use the format in Examples 1–5 to discuss and sketch the graphs of the equations.

1. $y = 2x - 1$

2. $y = 2 - 4x$

3. $y = x^2 - 9$

4. $y = x^2 + 1$

5. $y = 1 - x^2$

6. $y = 2x^2 - 1$

7. $y = 4 - 2x^2$

8. $y = 5 - x^2$

9. $y^2 = x$

10. $y^2 = 4x$

11. $y^2 = x + 1$

12. $y^2 = 2 - x$

13. $y = (x - 3)(x + 5)$

14. $y = (x - 4)(x + 6)$

15. $y = x(x + 3)(x - 2)$

16. $y = x(x - 1)(x - 4)$

17. $y = x(x - 1)(x - 2)^2$

18. $y = x^2(x + 2)(x - 3)$

19. $y = x(x - 1)^2(x - 2)$

20. $y = x^2(x + 2)(x - 3)^2$

21. $y = \dfrac{2}{x + 2}$

22. $y = \dfrac{3}{x - 3}$

23. $y = \dfrac{2}{(x - 1)^2}$

24. $y = \dfrac{x}{x + 2}$

25. $y = \dfrac{x^2}{x - 1}$

26. $y = \dfrac{x}{(x - 3)(x - 2)}$

27. $y = \dfrac{x + 1}{(x - 1)(x + 2)}$

28. $y = \dfrac{x(x - 1)}{(x - 2)(x + 1)}$

29. $y = x\sqrt{1 - x^2}$

30. $y = \sqrt{4 - x^2}$

31. $y = \dfrac{x^2 - 4}{x^2 - 1}$

32. $y^2 = \dfrac{x^2 - 1}{x^2 - 4}$

33. $y^2 = (x - 3)(x + 5)$

34. $y^2 = \dfrac{x}{x + 2}$

35. $y^2 = \dfrac{x}{(x - 3)(x - 2)}$

36. $y^2 = \dfrac{x + 1}{(x - 1)(x + 2)}$

37. $y^2 = \dfrac{x^2 - 4}{x^2 - 1}$

38. Recall that by Ohm's law $I = E/R$. Suppose that $E = 3$ volts and that R is a variable resistance. Discuss and sketch the resulting equation.

39. Suppose that two capacitors C_1 and C_2 are connected in series. The equivalent capacitance C is given by the equation

$$C = \frac{C_1 C_2}{C_1 + C_2}$$

If $C_2 = 10^{-2}$ farad, discuss and sketch the resulting graph for $C_1 \geq 0$.

40. The force of gravitational attraction between two point masses m_1 and m_2 is given by

$$F = G\frac{m_1 m_2}{x^2}$$

where x is the distance between the masses and G is a constant. Discuss the curve. What is the significance of the asymptotes?

41. A projectile shot directly upward with a speed of 60 m/s moves according to the law

$$S = 60t - 5t^2 \qquad (t \geq 0)$$

Sketch the curve.

42. The charge on a certain capacitor is $q = 1/(t + 1)^2$. Discuss and sketch the curve for $t \geq 0$.

43. The period P of a pendulum is $P = 1.1\sqrt{L}$, where P is measured in seconds and L in feet. Discuss and sketch the curve.

44. A manufacturer determines that the profit P (in dollars) on a certain item is

$$P = 800{,}000\frac{x}{(x + 2)^2}$$

where x (in thousands) is the number of units sold. Sketch the graph and estimate the number of units for which the profit is a maximum.

1.5   Discussion of Curves with Graphing Utilities (Optional)

The difficult task of curve sketching can be facilitated by the use of graphing calculators or computers with appropriate software. In this chapter it is assumed that graphing calculators are the primary tool. (Many of the graphs in this section were plotted using a graphing utility called *TEMATH*.)

Graphing utilities are usually limited to *functions,* which will be discussed in detail in Section 2.1. For now it is sufficient to say that y is a function of x if for every value of x in some set (called the **domain**) there corresponds one, and only one, value of y. The usual notation for a function is $y = f(x)$, read "f of x." Thus $y = x^2 - 1$ and $y = \sqrt{x + 4}$ are functions. The function $y = x^2 - 1$ has for its domain the set of all real numbers, while the function $y = \sqrt{x + 4}$ has domain $x \geq -4$ (to avoid imaginary values of y). Observe that the domain is just the extent of the curve along the x-axis.

When viewing a graph on a screen, we see only a portion of the plane. This portion will be denoted by the **viewing rectangle [L, R] by [B, T]**, where $L \leq x \leq R$ and $B \leq y \leq T$. The letters L, R, B, and T stand for left, right, bottom, and top, respectively.

Example 1   Draw the graph of the function $y = x^2 + 2$.

Solution. Using the standard viewing rectangle $[-10, 10]$ by $[-10, 10]$, we get the graph in Figure 1.31. This graph appears to be **complete** in the sense that it exhibits the basic behavior, a bowl-shaped curve symmetric with respect to the y-axis. We can check this behavior by using a large viewing rectangle, say $[-50, 50]$ by $[-100, 100]$ (Figure 1.32). The graph has remained essentially the same.

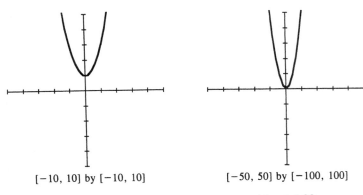

$[-10, 10]$ by $[-10, 10]$ $[-50, 50]$ by $[-100, 100]$

Figure 1.31   **Figure 1.32**

To see a smaller portion of a graph more closely, we use the **zoom** feature or a smaller viewing rectangle. For example, the graph in the viewing rectangle $[-2, 2]$ by $[-6, 6]$ is shown in Figure 1.33.

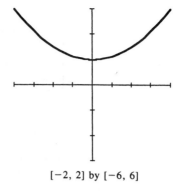

$[-2, 2]$ by $[-6, 6]$

Figure 1.33

Observe that in Figure 1.33 the scale marks on the two axes are equally spaced but that the distances between consecutive scale marks on these axes are different. The shape of a graph therefore depends on the viewing rectangle. In the next example the coefficients are so large that drastically different scales are needed.

Example 2   Draw the graph of

$$y = \frac{100x^2}{\sqrt{x - 1}}$$

Find the vertical asymptote and determine the domain.

Solution. It may require several tries to get an effective picture. For example, the standard rectangle $[-10, 10]$ by $[-10, 10]$ is too small to show anything. The viewing rectangle $[0, 5]$ by $[0, 2000]$ yields the graph shown in Figure 1.34.

The graph in Figure 1.34 suggests that $x = 1$ is indeed a vertical asymptote and that the domain is the set of all $x > 1$.

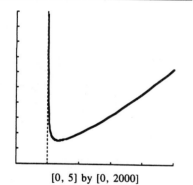

$[0, 5]$ by $[0, 2000]$

Figure 1.34

Example 3   Draw the graph of

$$y = \frac{120x}{(x + 1)^2}$$

Find the positive x-value for which y attains the maximum (largest) value.

Solution. The graph in the viewing rectangle $[0, 5]$ by $[0, 75]$ is shown in Figure 1.35.

The graph appears to peak around $x = 1$. When $x = 1$, then $y = 30$. So let us consider a small viewing rectangle around the point $(1, 30)$: say $[0.5, 1.5]$ by $[29, 31]$ (Figure 1.36). The graph clearly shows a peak near $x = 1$. So the highest point on the graph appears to be $(1, 30)$.

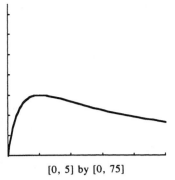

[0, 5] by [0, 75]

Figure 1.35

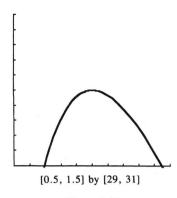

[0.5, 1.5] by [29, 31]

Figure 1.36

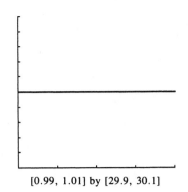

[0.99, 1.01] by [29.9, 30.1]

Figure 1.37

To confirm this behavior, let us take the small viewing rectangle [0.99, 1.01] by [29.9, 30.1] around the apparent maximum point (1, 30), shown in Figure 1.37. The graph has evidently leveled off.

If your calculator has a "trace" feature for moving the cursor along a graph, you will see the y-coordinate attain a maximum value when x is about 1. (The value is not likely to be exactly 30, however; graphing calculators can only approximate results.)   ■

■ Exercises / Section 1.5

In Exercises 1–8, graph each function and determine the x-intercepts.

1. $y = x^2(x - 1)(x - 2)$
2. $y = x^2(x + 2)(x - 1)$
3. $y = x^2(x - 1)^2(x - 2)$
4. $y = \dfrac{1}{4}x^2(x + 1)^2(x - 1)(x + 2)$
5. $y = x^4 - 2x^3$
6. $y = 30x^3 - 20x^5$
7. $y = 10x^2 - 5x^5$
8. $y = 10x^4 - 5x^3$

In Exercises 9–16, graph each function. Determine the vertical asymptotes and the domain in each case.

9. $y = \dfrac{\sqrt{x}}{1 + \sqrt{x}}$
10. $y = \dfrac{x}{\sqrt{x + 1}}$
11. $y = \sqrt{x} + \dfrac{1}{x}$
12. $y = \dfrac{\sqrt{x}}{x^2 - 1}$
13. $y = \dfrac{10x^2}{2x^2 - 3}$
14. $y = \dfrac{20x}{2x^2 - 5}$
15. $y = \dfrac{x^2}{2 - \sqrt{x}}$
16. $y = \dfrac{1.4\sqrt{x + 1}}{x}$

17. Estimate the coordinates of the highest point on the graph of
$$y = \frac{x^2}{50(2 - \sqrt{x})}$$
for $x > 4$.

18. Estimate the coordinates of the lowest point on the graph of $y = 3.4x^2 - 1.9/x$ for $x < 0$.

19. The charge (in coulombs) on a certain capacitor as a function of time (in seconds) is given by
$$q(t) = \frac{4.1t}{t^2 + 1.3}$$
Determine the maximum charge.

20. A projectile is shot directly upward with an initial velocity of 95 m/s. The distance s (in meters) from the ground as a function of time t (in seconds) is given by $s = 95t - 5t^2$. Find the maximum altitude attained by the projectile.

21. The power P (in watts) delivered to a resistor with variable resistance R (in ohms) is given by
$$P = \frac{100R}{(R + 10)^2}$$
Determine the maximum power.

22. The efficiency E of a certain screw as a function of the pitch angle T (in radians) is given by

$$E = \frac{T(1 - 0.26\,T)}{T + 0.26}$$

Find the value of T $(T < \pi/2)$ for which E is a maximum.

23. The deflection (in feet) of a certain beam as a function of the distance from its left end $(x = 0)$ is given by

$$d(x) = 3.4 \times 10^{-5} x^2 (10 - x)^2$$

Find the maximum deflection.

1.6 The Conics

In the next few sections we are going to return to the two fundamental problems mentioned in Section 1.3:

1. Given the figure, find the corresponding equation.
2. Given the equation, draw the figure.

In Section 1.4 we developed a number of techniques for handling the second problem, and we are going to study other techniques in Chapter 3. Problem 1 often involves finding a best-fitting curve for a given set of data, a technique usually studied in numerical analysis. (A brief treatment can be found in Chapter 9.) On the other hand, when dealing with a special group of figures such as the conics, it is often possible to go directly from the figure to the equation and vice versa.

The term **conic,** or **conic section,** comes from the fact that the curves for which equations are sought correspond to sections cut from a conic surface by planes. A conic surface consists of two funnel-shaped sheets, called **nappes,** extending indefinitely in both directions from a point (vertex) where the nappes meet. If a plane cuts entirely across one nappe, the intersection of plane and nappe is called an *ellipse* (see Figure 1.38). A special case, the *circle,* is obtained if the plane is perpendicular to the axis of the cone. If the plane is parallel to one line on the surface of the cone, the intersection of plane and nappe is called a *parabola* (see Figure 1.39). If the plane intersects both nappes, the intersection is called a *hyperbola* (see Figure 1.40). Certain special cases occur if the plane passes through the vertex (a point), through the axis (two intersecting lines), or through a single line on the surface.

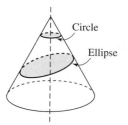

Figure 1.38

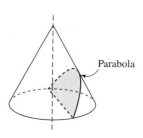

Figure 1.39

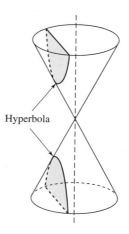

Figure 1.40

The conics were well known to the ancient Greeks. Many of the geometric properties of conic sections, including the ones to be used in our later definitions, were studied in antiquity. Although Euclid's own contributions to that study were later lost, they appear to have been incorporated in the work of Apollonius of Perga (circa 230 B.C.), to whom the names of the curves have been attributed. Archimedes (287–212 B.C.) was the first to find the area under a parabolic arch. In the remainder of this chapter we will study a number of scientific applications of the conics.

1.7   The Circle

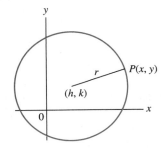

Figure 1.41

We recall from geometry that a **circle** is the locus of all points equidistant from a given point called the *center*. Let (h, k) in Figure 1.41 be the center and let $P(x, y)$ be any point on the circle. If r is the radius, then from the distance Formula (1.1) we get

$$\sqrt{(x - h)^2 + (y - k)^2} = r$$

Squaring both sides of this equation, we get the **standard form of the equation of a circle.**

Standard Form of the Equation of a Circle

$$(x - h)^2 + (y - k)^2 = r^2 \qquad (1.11)$$

where (h, k) is the center and r is the radius.

Example 1   Find the equation of the circle with center at $(-2, 1)$ and radius 3.

Solution. Since $h = -2$, $k = 1$, and $r = 3$, we have

$$(x - h)^2 + (y - k)^2 = r^2$$
$$[x - (-2)]^2 + (y - 1)^2 = 3^2$$

and

$$(x + 2)^2 + (y - 1)^2 = 9 \qquad \blacksquare$$

If the circle is centered at the origin, then $h = 0$ and $k = 0$, and the equation has the form given next.

Equation of a Circle Centered at the Origin

$$x^2 + y^2 = r^2 \qquad (1.12)$$

If we multiply the terms of Equation (1.11), we may conclude that any circle has the following form:

General Form of the Equation of a Circle

$$x^2 + y^2 + ax + by + c = 0 \qquad (1.13)$$

Example 2   Put $x^2 + y^2 + 6x - 4y + 9 = 0$ into standard form (1.11) and find the center and radius.

Solution.   To convert the equation to standard form we complete the square on each quadratic expression. Transposing the 9 and rearranging terms, we get

$$(x^2 + 6x \quad) + (y^2 - 4y \quad) = -9$$

Since the coefficient of x is 6, we must add the square of one-half of 6 to both sides. In other words, since

$$\left[\frac{1}{2} \cdot 6\right]^2 = 9$$

we add 9 to both sides.
Similarly, since the coefficient of y is -4, we add

$$\left[\frac{1}{2}(-4)\right]^2 = 4$$

to both sides. Thus

$$(x^2 + 6x + 9) + (y^2 - 4y + 4) = -9 + 9 + 4$$

Since $x^2 + 6x + 9 = (x + 3)^2$ and $y^2 - 4y + 4 = (y - 2)^2$, we get

$$(x + 3)^2 + (y - 2)^2 = 4$$

From the form

$$(x - h)^2 + (y - k)^2 = r^2$$

we conclude that the center is $(-3, 2)$ and the radius $\sqrt{4} = 2$. (See Figure 1.42.)

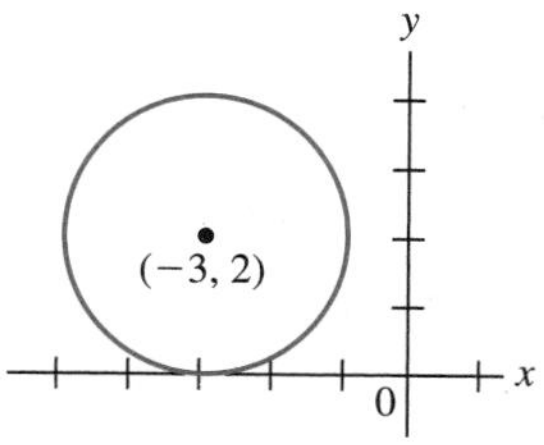

Figure 1.42

Example 3   Write $x^2 + y^2 - 2x + 4y + 5 = 0$ in standard form.

Solution.   Before completing the square on each quadratic expression, we rearrange the terms to group the variables together:

$$(x^2 - 2x \quad) + (y^2 + 4y \quad) = -5$$

One-half of the coefficients of x and y are -1 and 2, respectively. We add the square of each to both sides; hence

$$(x^2 - 2x + 1) + (y^2 + 4y + 4) = -5 + 1 + 4 \qquad (-1)^2 = 1, \ 2^2 = 4$$

or

$$(x - 1)^2 + (y + 2)^2 = 0$$

Point circle

This equation is satisfied by the coordinates of only one real point $(1, -2)$. The locus is sometimes called a **point circle.**

The equation

$$(x - 1)^2 + (y + 2)^2 = -1$$

is not satisfied by any real point: the squared terms on the left are positive or zero and the right side is strictly negative. This case will be called an *imaginary circle.* ∎

■ Exercises / Section 1.7

In Exercises 1–10, find the equation of each of the circles.

1. Center at the origin, radius 5
2. Center at the origin, radius 7
3. Center at the origin, passing through $(-6, 8)$
4. Center at the origin, passing through $(1, -4)$
5. Center at $(-2, 5)$, radius 1
6. Center at $(2, -3)$, radius $\sqrt{2}$
7. Center at $(-1, -4)$, passing through origin
8. Center at $(3, 4)$, passing through $(5, 10)$
9. Ends of diameter at $(-2, -6)$ and $(1, 5)$
10. Center on the line $2y = 3x$, tangent to the y-axis, radius 2

In Exercises 11–28, write each equation in standard form and determine the center and radius.

11. $x^2 + y^2 - 2x - 2y - 2 = 0$
12. $x^2 + y^2 - 2x - 4y + 4 = 0$
13. $x^2 + y^2 + 4x - 8y + 4 = 0$
14. $x^2 + y^2 + 2x + 6y + 3 = 0$
15. $x^2 + y^2 + 4x + 2y + 2 = 0$
16. $x^2 + y^2 - 8x + 6y + 20 = 0$
17. $x^2 + y^2 - 4x + y + \dfrac{9}{4} = 0$
18. $x^2 + y^2 + x - 4y + \dfrac{5}{4} = 0$

19. $4x^2 + 4y^2 - 8x - 12y + 9 = 0$
20. $4x^2 + 4y^2 - 20x - 8y + 25 = 0$
21. $x^2 + y^2 + 4x - 2y - 4 = 0$
22. $x^2 + y^2 + 2x + 8y + 1 = 0$
23. $x^2 + y^2 - x - 2y + \dfrac{1}{4} = 0$
24. $x^2 + y^2 - 6x - 8y + 19 = 0$
25. $x^2 + y^2 - 4x + y + \dfrac{9}{4} = 0$
26. $x^2 + y^2 + x - y - \dfrac{1}{2} = 0$
27. $4x^2 + 4y^2 + 12x + 16y + 5 = 0$
28. $36x^2 + 36y^2 - 144x - 120y + 219 = 0$

In Exercises 29–34, determine which is a point circle and which is an imaginary circle.

29. $4x^2 + 4y^2 - 20x - 4y + 26 = 0$
30. $x^2 + y^2 + 4x - 2y + 7 = 0$
31. $x^2 + y^2 - 6x + 8y + 25 = 0$
32. $x^2 + y^2 + 2x + 4y + 5 = 0$
33. $x^2 + y^2 - 6x - 8y + 30 = 0$
34. $x^2 + y^2 - 6x + 4y + 13 = 0$

35. A washer has inner radius 2.00 cm and outer radius 3.40 cm. Find the equations of the two circles, using the center of the washer as the origin.

36. Find the equation of the balancing hole in Figure 1.43.

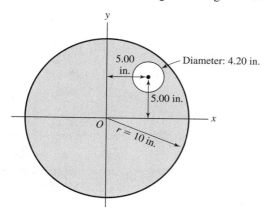

Figure 1.43

37. Write the equation of the path of a communications satellite that remains in a circular orbit 22,300 mi above a point on the equator. Assume that the radius of the earth is 4000 mi. (Let the center of the earth be the origin.)

38. Use a graphing utility to graph the circle $y^2 + (x - 1)^2 = 4$ by solving the equation for y in terms of x and graphing $y = \sqrt{4 - (x - 1)^2}$ and $y = -\sqrt{4 - (x - 1)^2}$ separately.

1.8   The Parabola

Focus
Directrix
Vertex
Axis

Let F be a fixed point and l be a fixed line. Then the locus of all points equidistant from F and l is called a **parabola** (Figure 1.44). F is called the **focus** of the parabola and l the **directrix**. If we draw a line from F perpendicular to l, intersecting l at Q, then the midpoint V of FQ is called the **vertex,** which is the point on the parabola closest to the line l. The line through F and Q is called the **axis** of the parabola.

   To get the equation of a parabola, we select a convenient coordinate system and apply the definition. Let the vertex be at the origin and denote the focus by $(p, 0)$, as in Figure 1.45. If p is positive, then the focus lies to the right of the origin; if p is negative, it lies to the left. The directrix is the line $x = -p$ in each case. Now let $P(x, y)$ be any point on the parabola (Figure 1.45). It follows that the distance to the directrix is $x + p$. By the definition of the parabola, the distance d_1 from $P(x, y)$ to the focus is equal to the distance d_2 from $P(x, y)$ to the directrix:

$$d_1 = d_2$$
$$\sqrt{(x - p)^2 + y^2} = x + p \qquad \textbf{by Distance Formula (1.1)}$$

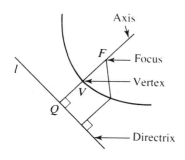

Figure 1.44

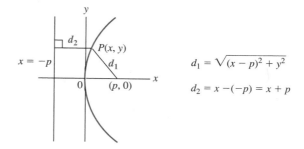

$$d_1 = \sqrt{(x - p)^2 + y^2}$$
$$d_2 = x - (-p) = x + p$$

Figure 1.45

Squaring both sides, we get

$$(x - p)^2 + y^2 = (x + p)^2$$

or

$$x^2 - 2px + p^2 + y^2 = x^2 + 2px + p^2 \qquad \textbf{(a ± b)}^2 = \textbf{a}^2 ± \textbf{2ab} + \textbf{b}^2$$

Combining like terms, the last expression reduces to

$$y^2 = 4px$$

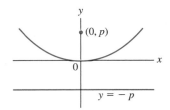

Figure 1.46

Conversely, any point whose coordinates satisfy this equation lies on the parabola. We owe the simplicity of this equation to the manner in which the coordinate system was chosen. It will be seen later that the equation can be modified to study a parabola whose vertex is not at the origin.

If the focus is at $(0, p)$ and the directrix is the line $y = -p$ (Figure 1.46), then the equation becomes

$$x^2 = 4py$$

These two cases are summarized next.

Equation of a Parabola

The equation of a parabola with vertex at the origin is

$$y^2 = 4px \qquad \text{(axis horizontal)} \tag{1.14}$$

or

$$x^2 = 4py \qquad \text{(axis vertical)} \tag{1.15}$$

To remember the form (and position) of a parabola, note that $y^2 = 4px$ is symmetric with respect to the x-axis, while $x^2 = 4py$ is symmetric with respect to the y-axis.

Example 1 Find the focus and directrix of the parabola $y^2 = -9x$ and sketch the curve.

Solution. To find the focus and directrix, we need to write the equation in the form $y^2 = 4px$. Thus

$$y^2 = -9x = 4\left(-\frac{9}{4}\right)x \qquad \textbf{y}^2 = \textbf{4px}$$

Since $p = -9/4$, the focus is at $(-9/4, 0)$ and the directrix is

$$x = -p = -\left(-\frac{9}{4}\right) = \frac{9}{4}$$

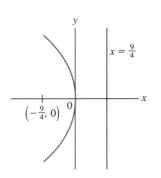

Figure 1.47

(See Figure 1.47.) ∎

Example 2   Find the focus and directrix of the parabola $x^2 = 14y$ and sketch the curve.

Solution.

$$x^2 = 14y = 4\left(\frac{14}{4}\right)y$$

or

$$x^2 = 4\left(\frac{7}{2}\right)y \qquad \textbf{\textit{x}}^2 = \textbf{4}\textbf{\textit{py}}$$

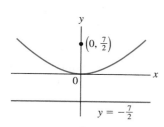

Figure 1.48

Since $p = 7/2$, we conclude from Equation (1.15) that the focus is at $(0, 7/2)$ and the directrix is the line $y = -p = -7/2$ (Figure 1.48). ∎

Example 3   Find the equation of the parabola with focus at $(0, 3)$ and the directrix $y = -3$.

Solution. Since the focus is on the y-axis, the form is $x^2 = 4py$. Since $p = 3$, we get

$$x^2 = 4(3)y \qquad \textbf{\textit{x}}^2 = \textbf{4}\textbf{\textit{py}}$$

or

$$x^2 = 12y$$

∎

Example 4   Find the equation of the parabola with focus at $(-5, 0)$ and vertex at the origin.

Solution. Since the focus is on the x-axis, the form of the equation is $y^2 = 4px$ by Equation (1.14). Since $p = -5$, we get

$$y^2 = 4(-5)x \qquad \textbf{\textit{y}}^2 = \textbf{4}\textbf{\textit{px}}$$

or

$$y^2 = -20x$$

∎

Figure 1.49

Applications of the Parabola

1. The cable of a suspension bridge whose weight is uniformly distributed over the entire length of the bridge is in the form of a parabola (Figure 1.49).
2. The path of a projectile is a parabola if air resistance is neglected (Figure 1.50). (If the object is heavy and the initial velocity low, the air resistance is negligible.)

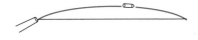

Figure 1.50

3. If a parabola is rotated about its axis, it forms a parabolic surface. If a reflecting surface is parabolic with a light source placed at the focus, the rays will be reflected parallel to the axis. Conversely, light rays coming in parallel to the axis are reflected so that they pass through the focus. These principles are employed in the construction of searchlights and reflecting telescopes (Figure 1.51).
4. Another parabolic surface is obtained if a cylindrical vessel is partly filled with water and rotated about the axis of the cylinder: a plane through the axis will cut the surface of the water in a parabola.
5. Steel bridges are frequently built with parabolic arches (Figure 1.52).
6. The relationship between the period of a pendulum and its length can be described in the form of a parabolic equation.

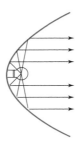

Figure 1.51

Figure 1.52

Example 5

The entrance to a formal garden is spanned by a parabolic arch 10 ft high and 10 ft across at its base. How wide should the walk through the center of the arch be if a minimum clearance of 7 ft is wanted above the walk?

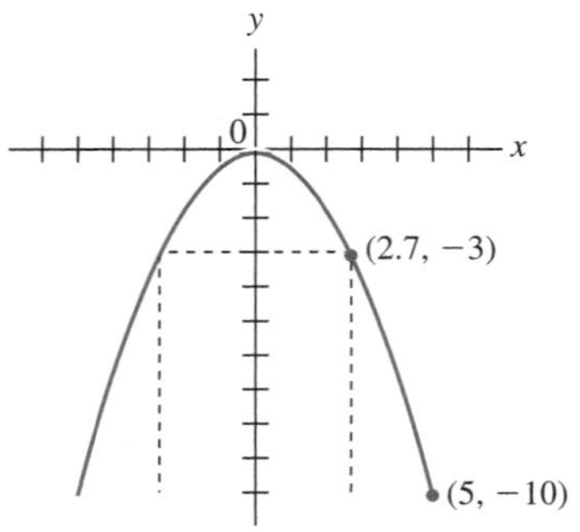

Solution. Let the vertex of the arch be at the origin (Figure 1.53). Since the arch is 10 ft high and 10 ft across at its base, we know that the point $(5, -10)$ lies on the curve and that the coordinates of this point must satisfy the equation of the parabola $x^2 = 4py$. We begin by substituting:

$$5^2 = 4p(-10) \qquad x^2 = 4py$$

Solving for $4p$, we get

$$4p = \frac{5^2}{-10} = \frac{25}{-10} = -\frac{5}{2}$$

So the equation becomes

$$x^2 = 4py$$
$$x^2 = -\frac{5}{2}y$$

Figure 1.53

Now, a clearance of 7 ft corresponds to a point for which $y = -3$ (see Figure 1.53); its x-coordinate tells us the width of the walk. Thus

$$x^2 = -\frac{5}{2}(-3) = \frac{15}{2} \qquad y = -3$$

and $x \approx 2.7$. So the walk must be $2(2.7) = 5.4$ ft wide. ∎

Example 6   An amateur astronomer constructs a reflecting telescope with a parabolic mirror at one end of a hollow tube. The image is formed at the other end, where it may be viewed through an eyepiece. If the mirror has a radius of 20.0 cm and a depth of 3.33 mm in the center, find the position of the image.

Solution. We may assume the parabolic cross-section to have the form

$$y^2 = 4px$$

We need to find the focus of this parabola. Since 3.33 mm = 0.333 cm, one point in the first quadrant is (0.333, 20.0). The coordinates of this point satisfy the equation. Hence

$$(20.0)^2 = 4p \cdot 0.333$$

and

$$p = \frac{(20.0)^2}{4 \cdot 0.333} = 300 \text{ cm}$$

Therefore, the image is 3 m from the center of the mirror. ∎

■ Exercises / Section 1.8

In Exercises 1–14, write the equations of the parabolas.

1. Vertex at the origin, focus at (3, 0)
2. Vertex at the origin, focus at (−3, 0)
3. Vertex at the origin, focus at (0, −5)
4. Vertex at the origin, focus at (0, 4)
5. Vertex at the origin, focus at (−4, 0)
6. Vertex at the origin, focus at (0, −6)
7. Vertex at the origin, directrix $x = -1$
8. Vertex at the origin, directrix $y = 2$
9. Directrix $x - 2 = 0$, vertex at the origin
10. Directrix $x + 2 = 0$, vertex at the origin
11. Vertex at the origin, axis along the x-axis, passing through (−2, −4)

12. Vertex at the origin, axis along the y-axis, passing through (−1, 1)
13. Vertex at the origin, passing through (1, 1)
14. Vertex at the origin, passing through (2, −1)

In Exercises 15–30, find the coordinates of the focus and the equation of the directrix in each exercise and sketch the curve.

15. $x^2 = 8y$
16. $x^2 = 16y$
17. $x^2 = -12y$
18. $x^2 = -16y$
19. $y^2 = 16x$
20. $y^2 = 8x$
21. $y^2 = -4x$
22. $y^2 = -12x$
23. $x^2 - 4y = 0$
24. $x^2 + 8y = 0$
25. $y^2 = 9x$
26. $y^2 = 10x$
27. $y^2 = -x$
28. $2x^2 - 3y = 0$

29. $3y^2 + 2x = 0$ **30.** $y^2 = 2ax$

31. The chord of a parabola that passes through the focus and is parallel to the directrix is called the *focal chord*. Find the equation of the circle having for a diameter the focal chord of the parabola $x^2 = 12y$.

32. Derive Equation (1.15).

33. Use the definition of the parabola to find the equation of the parabola whose focus is at (4, 1) and whose directrix is the y-axis.

34. Repeat Exercise 33 for the parabola having a focus at (4, 7) and directrix $y + 1 = 0$.

35. Find the equations of the two parabolas with vertices at the origin and which pass through $(-2, -4)$.

36. Repeat Exercise 35 for the point $(3, -5)$.

37. The support cables of a suspension bridge hang between towers in the form of a parabolic curve. The towers are 90.0 m high and 200.0 m apart. The roadway is suspended on other cables hung from the support cables. The shortest distance from the road surface to the supporting cable is 20.0 m. Determine the length of the vertical suspension cable which is 30.0 m from the center of the bridge. (Choose for the origin the lowest point on the supporting cable.)

38. The headlight on a car is 20.0 cm in diameter and has a parabolic surface; the headlight is 12.0 cm deep. Where should the light bulb be placed so that the reflected rays are parallel?

39. Determine the distance across the base of a parabolic arch that measures 25.0 ft at its highest point if the road through the center of the arch is 40.0 ft wide and must have a minimum clearance of 12.0 ft. (See Figure 1.54.)

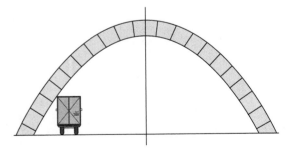

Figure 1.54

40. A stream is 60.0 m wide and is spanned by a parabolic arch. Boats pass through the 40.0-m-wide channel in the center of the stream. How high must the arch be so that the minimum clearance above the stream will be 10.0 m?

41. An entrance to a castle is in the form of a parabolic arch 6 m across at the base and 3 m high in the center. What is the length of a beam across the entrance, parallel to the base and 2 m above it?

42. The period T of a pendulum is directly proportional to the square root of its length L, provided that the arc of the swing is small (less than 15°). This can be described by $L = kT^2$. If $T = 6.0$ s when $L = 9.0$ m, find the relationship.

43. A solar collector for heating water has a cross-section in the shape of a parabola, as shown in Figure 1.55. Water is flowing through the pipe located at the focus. Find the distance from the pipe to the vertex.

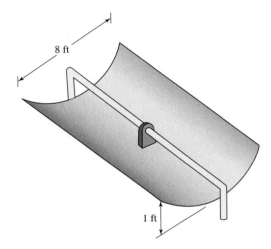

Figure 1.55

44. A satellite dish with a parabolic cross-section has a focus located 11.9 in. from the vertex. If the disk is 1.7 in. deep in the center (Figure 1.56), what is the diameter?

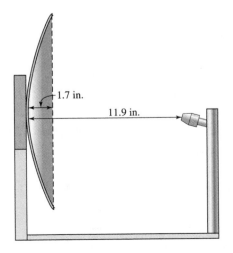

Figure 1.56

1.9   The Ellipse

Foci

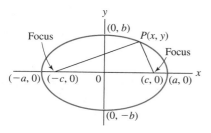

Figure 1.57

An **ellipse** is the locus of points such that the sum of the distances from two fixed points is constant. Suppose that the two fixed points, called the **foci** of the ellipse, are at $(c, 0)$ and $(-c, 0)$, respectively (Figure 1.57). The Distance Formula (1.1) can now be used to obtain the equation. A particularly simple form will result if we choose $2a$ for the constant, since the points $(a, 0)$ and $(-a, 0)$ will then lie on the ellipse. From the definition, the sum of the distances from $P(x, y)$ to the foci is

$$\sqrt{(x - c)^2 + y^2} + \sqrt{(x + c)^2 + y^2} = 2a$$

Conversely, any point whose coordinates satisfy this equation will lie on the ellipse. The equation can be simplified by transposing one of the radicals and squaring both sides:

$$[\sqrt{(x - c)^2 + y^2}]^2 = [2a - \sqrt{(x + c)^2 + y^2}]^2$$

or

$$(x - c)^2 + y^2 = 4a^2 - 4a\sqrt{(x + c)^2 + y^2} + (x + c)^2 + y^2$$

Since $(x \pm c)^2 = x^2 \pm 2xc + c^2$, we get

$$x^2 - 2cx + c^2 + y^2 = 4a^2 - 4a\sqrt{(x + c)^2 + y^2} + x^2 + 2cx + c^2 + y^2$$

If we now subtract x^2, c^2, and y^2 from both sides, we obtain

$$-2cx = 4a^2 - 4a\sqrt{(x + c)^2 + y^2} + 2cx$$

and

$$-4cx = 4a^2 - 4a\sqrt{(x + c)^2 + y^2} \qquad \text{subtracting } 2cx$$
$$cx = -a^2 + a\sqrt{(x + c)^2 + y^2} \qquad \text{dividing by } -4$$
$$a^2 + cx = a\sqrt{x^2 + 2cx + c^2 + y^2} \qquad \text{adding } a^2$$

Squaring both sides again, we get

$$a^4 + 2a^2cx + c^2x^2 = a^2(x^2 + 2cx + c^2 + y^2)$$

or

$$a^4 + c^2x^2 = a^2x^2 + a^2c^2 + a^2y^2 \qquad \text{subtracting } 2a^2\, cx$$

We now collect all x and y terms on one side of the equation and factor the resulting expressions:

$$a^2x^2 - c^2x^2 + a^2y^2 = a^4 - a^2c^2$$

or

$$(a^2 - c^2)x^2 + a^2y^2 = a^2(a^2 - c^2) \qquad \text{\textbf{common factors } } x^2 \textbf{ and } a^2$$

After dividing both sides by $a^2(a^2 - c^2)$, the equation reduces to

$$\frac{x^2}{a^2} + \frac{y^2}{a^2 - c^2} = 1 \qquad \text{\textbf{dividing by } } a^2 \, (a^2 - c^2)$$

A final simplification of the form can be obtained by letting

$$b^2 = a^2 - c^2$$

so that the equation becomes

$$\frac{x^2}{a^2} + \frac{y^2}{b^2} = 1$$

As a result, the intercepts are simply $x = \pm a$ and $y = \pm b$. (See Figure 1.57.) The curve itself is symmetric with respect to both axes. Note that $a > b$.

If the foci are at $(0, \pm c)$, we obtain the equation

$$\frac{x^2}{b^2} + \frac{y^2}{a^2} = 1$$

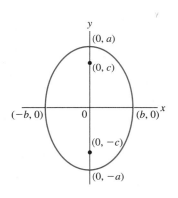

Figure 1.58

(See Figure 1.58.)

The two forms of the ellipse are easily distinguished by examining the intercepts. For the equation

$$\frac{x^2}{a^2} + \frac{y^2}{b^2} = 1$$

the x-intercepts are $x = \pm a$. Because $a > b$, the intercepts with the larger absolute values lie on the x-axis. For the other equation,

$$\frac{x^2}{b^2} + \frac{y^2}{a^2} = 1$$

the numerically larger intercepts, $y = \pm a$, lie on the y-axis.

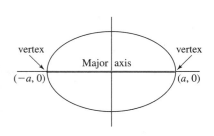

Figure 1.59

Major axis
Semimajor axis
Minor axis
Semiminor axis
Center
Vertex

The symmetric properties of the ellipse suggest some additional terminology: the line segment through the foci, extending between $(a, 0)$ and $(-a, 0)$ or between $(0, a)$ and $(0, -a)$, is called the **major axis;** its length is $2a$. The length a is called the **semimajor axis** (Figure 1.59). The segment of the other line of symmetry between $(b, 0)$ and $(-b, 0)$ or between $(0, b)$ and $(0, -b)$ is called the **minor axis;** its length is $2b$. The length b is called the **semiminor axis** (Figure 1.60). The intersection of the two axes is called the **center** of the ellipse. The ends of the major axis are called the **vertices** (Figure 1.59). These ideas are summarized next.

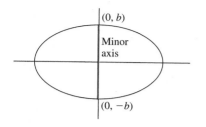

Figure 1.60

Equation of an Ellipse

The equation of an ellipse with center at the origin is

$$\frac{x^2}{a^2} + \frac{y^2}{b^2} = 1 \qquad \text{(major axis horizontal)} \tag{1.16}$$

or

$$\frac{x^2}{b^2} + \frac{y^2}{a^2} = 1 \qquad \text{(major axis vertical)} \tag{1.17}$$

The foci lie on the major axis, c units from the center, with

$$b^2 = a^2 - c^2 \tag{1.18}$$

Example 1   Sketch the ellipse and find the foci for the following equation:

$$\frac{x^2}{16} + \frac{y^2}{4} = 1$$

Solution. From the equation

$$\frac{x^2}{a^2} + \frac{y^2}{b^2} = 1$$

we have $a^2 = 16$ and $b^2 = 4$. It follows that $a = 4$ and $b = 2$.

The foci are found by using the equation $b^2 = a^2 - c^2$:

$$b^2 = a^2 - c^2$$
$$4 = 16 - c^2$$
$$c^2 = 12$$
$$c = \pm\sqrt{12} = \pm\sqrt{4 \cdot 3} = \pm\sqrt{4}\sqrt{3} = \pm 2\sqrt{3}$$

or

$$c = \pm 2\sqrt{3}$$

The foci are at $(\pm 2\sqrt{3}, 0)$. (See Figure 1.61.) ∎

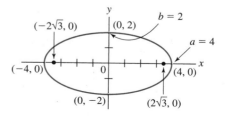

Figure 1.61

Example 2

Find the vertices, foci, and semiminor axis of the ellipse $2x^2 + y^2 = 4$ and sketch.

Solution. To get the equation into standard form, we divide both sides by 4:

$$\frac{x^2}{2} + \frac{y^2}{4} = 1$$

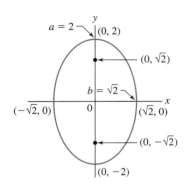

Figure 1.62

By Formula (1.17),

$$\frac{x^2}{b^2} + \frac{y^2}{a^2} = 1 \qquad \text{(major axis vertical)}$$

we have $a^2 = 4$ and $b^2 = 2$, so that $a = 2$ and $b = \sqrt{2}$. Since the major axis is vertical, we conclude that the vertices are at $(0, \pm 2)$. The length of the semiminor axis is $b = \sqrt{2}$. (Note that the x-intercepts are therefore $(\pm\sqrt{2}, 0)$.) From Equation (1.18), we have

$$b^2 = a^2 - c^2$$
$$2 = 4 - c^2 \qquad \text{or} \qquad c = \pm\sqrt{2}$$

Hence the foci are at $(0, \pm\sqrt{2})$ on the major axis. (See Figure 1.62.)

To graph the ellipse in the example with a graphing utility, we need to solve the equation $2x^2 + y^2 = 4$ for y in terms of x and then graph $y = \sqrt{4 - 2x^2}$ and $y = -\sqrt{4 - 2x^2}$ separately (Figure 1.63). ∎

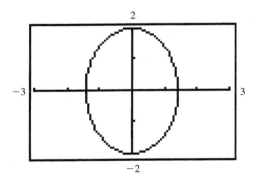

Figure 1.63

Example 3

Find the equation of the ellipse with center at the origin, foci at $(0, \pm 3)$, and major axis 8.

Solution. Since the length of the major axis is $2a = 8$, it follows that $a = 4$ and $a^2 = 16$. Since the foci are at $(0, \pm 3)$, $c = 3$. So by Equation (1.18) we have

$$b^2 = a^2 - c^2 = 4^2 - 3^2 = 7 \qquad \textbf{a = 4, c = 3}$$

Since the foci lie on the major axis, we use Form (1.17):

$$\frac{x^2}{b^2} + \frac{y^2}{a^2} = 1 \qquad \text{(major axis vertical)}$$

or

$$\frac{x^2}{7} + \frac{y^2}{16} = 1 \qquad \textbf{b² = 7, a² = 16}$$ ∎

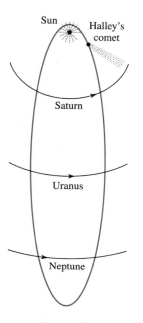

Figure 1.64

Applications of the Ellipse

1. The paths of the planets and comets as they move about the sun are ellipses with the sun at one of the foci (Figure 1.64). Similarly, artificial satellites move about the earth in elliptical orbits with the center of the earth at one of the foci.

2. If an ellipse is rotated about its major axis, we obtain a solid called an *ellipsoid*. If a sound wave originates at one of the foci within an ellipsoid enclosure or chamber, the reflected waves converge at the other focus. A room whose ceiling is a semi-ellipsoid is called a "whispering gallery." A person standing at one focus can hear even a slight noise made at the other focus, while a person standing between does not hear anything. (A famous example of a whispering gallery is the Statuary Hall in the United States Capitol in Washington. Guides will position visitors in the proper places to hear faint sounds from another part of the hall. The hall was originally constructed for use as the House of Representatives, but congressmen found the acoustics of the hall distressing since confidential conferences in one place could be overheard easily at the opposite end of the hall.)

3. Arches of stone bridges are frequently semiellipses (Figure 1.65).

4. Elliptical gears, which revolve on a shaft turning through a focus, are sometimes used in power punches since they yield powerful movements with quick returns.

Figure 1.65

Example 4

A certain whispering gallery has a ceiling with elliptical cross-sections. If the room is 30 ft long and 12 ft high in the center, where should two visitors place themselves to get the whispering effect?

Solution. If the center of the ellipse is placed at the origin, we get $a = 15$ and $b = 12$. Thus

$$c^2 = a^2 - b^2 = 225 - 144 = 81$$

and $c = 9$. So the visitors must place themselves on the two foci, each 9 ft from the center.   ∎

■ Exercises / Section 1.9

In Exercises 1–17, find the vertices, foci, and semiminor axis of each of the ellipses, and sketch each.

1. $\dfrac{x^2}{25} + \dfrac{y^2}{16} = 1$

2. $\dfrac{x^2}{16} + \dfrac{y^2}{9} = 1$

3. $\dfrac{x^2}{9} + \dfrac{y^2}{4} = 1$

4. $\dfrac{x^2}{4} + \dfrac{y^2}{9} = 1$

5. $\dfrac{x^2}{16} + y^2 = 1$

6. $\dfrac{x^2}{2} + \dfrac{y^2}{4} = 1$

7. $16x^2 + 9y^2 = 144$

8. $x^2 + 2y^2 = 4$

9. $5x^2 + 2y^2 = 20$

10. $5x^2 + 9y^2 = 45$

11. $5x^2 + y^2 = 5$

12. $x^2 + 4y^2 = 4$

13. $x^2 + 2y^2 = 6$

14. $9x^2 + 2y^2 = 18$

15. $15x^2 + 7y^2 = 105$

16. $9x^2 + y^2 = 27$

17. $2x^2 + 5y^2 = 50$

In Exercises 18–21, use a graphing utility to graph the given ellipses. (See Example 2.)

18. $6x^2 + 2y^2 = 5$

19. $5x^2 + y^2 = 6$

20. $\dfrac{1}{4}x^2 + y^2 = 1$

21. $x^2 + 4y^2 = 10$

In Exercises 22–32, find the equation of each of the ellipses.

22. Foci at $(\pm1, 0)$, vertices at $(\pm2, 0)$

23. Foci at $(\pm3, 0)$, vertices at $(\pm4, 0)$

24. Foci at $(0, \pm2)$, major axis 6

25. Foci at $(0, \pm2)$, major axis 8

26. Foci at $(0, \pm2)$, minor axis 4

27. Foci at $(0, \pm3)$, minor axis 6

28. Minor axis 5, vertices at $(\pm7, 0)$

29. Center at the origin, one vertex at $(0, 8)$, one focus at $(0, -5)$

30. Center at the origin, one focus at $(-3, 0)$, semiminor axis 4

31. Foci at $(\pm2\sqrt{3}, 0)$, minor axis 4

32. Foci at $(\pm\sqrt{5}, 0)$, vertices at $(\pm\sqrt{7}, 0)$

33. Find the locus of points such that the sum of the distances from $(\pm6, 0)$ is 16.

34. The shape of an ellipse depends on the values of c and a. The fraction $e = c/a$ is called the *eccentricity* of the ellipse. Find the equation of the ellipse with vertices at $(\pm4, 0)$ and $e = 1/2$.

35. Find the eccentricity of the ellipse $9x^2 + 5y^2 = 45$.

36. Find the equation of the ellipse centered at the origin, minor axis 6, and passing through the point $(1, 4)$.

37. Find the locus of points for which the distance from $(0, 0)$ is twice the distance from $(3, 0)$. What is the locus?

38. The area of an ellipse is given by $A = \pi ab$. Find the area of the floor of the whispering gallery in Example 4.

39. Find the equation of the ellipse to be inscribed in a rectangle 4 m long and 3 m wide if the center of the rectangle is placed at the origin with the long side horizontal.

40. A gateway is in the form of a semiellipse 10 ft across at the base and 15 ft high in the center. What is the length of the horizontal beam across the gateway 10 ft above the floor?

41. A road 8.0 m wide is spanned by a semielliptical arch 12.0 m across at its base. If the minimum clearance over the road is 4.0 m, what is the height of the arch?

42. Describe a mechanical method for drawing an ellipse inscribed in a rectangle. Refer to Exercise 39.

43. A satellite is in elliptical orbit about the earth. If the maximum altitude of the satellite is 120 mi and the minimum 80 mi, find the equation of the orbit. (The center of the earth is at a focus and the radius of the earth is approximately 4000 mi.)

44. The earth moves about the sun in an elliptical orbit with the center of the sun at one of the foci. The minimum distance to the center of the sun is 9.14×10^7 mi and the maximum distance 9.46×10^7 mi. Show that the eccentricity is only about $1/60$, that is, the path is nearly circular. Refer to Exercise 34.

45. Halley's comet moves in an elliptical orbit with the sun at one of the foci. (See Figure 1.64.) Its maximum distance from the sun is 3.285×10^9 mi and its minimum distance 5.48×10^7 mi. Find the eccentricity of the orbit.

1.10 The Hyperbola

Starting with two points, $(c, 0)$ and $(-c, 0)$, let us consider the locus of points for which the difference of the distances from the two fixed points is numerically equal to $2a$. The locus is called a **hyperbola** (Figure 1.66).

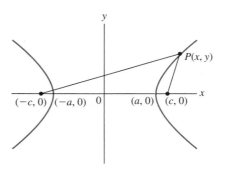

Figure 1.66

The derivation of the equation is similar to that of the ellipse. By the Distance Formula (1.1):

$$\sqrt{(x-c)^2 + y^2} - \sqrt{(x+c)^2 + y^2} = \pm 2a \tag{1.19}$$

which reduces to

$$\frac{x^2}{a^2} - \frac{y^2}{c^2 - a^2} = 1 \tag{1.20}$$

(The simplification will be left as an exercise.) This form suggests the substitution:

$$b^2 = c^2 - a^2$$

Hence

$$\frac{x^2}{a^2} - \frac{y^2}{b^2} = 1$$

Conversely, any point whose coordinates satisfy this equation lies on the hyperbola. The x-intercepts of this curve are seen to be $(\pm a, 0)$, but the y-intercepts are lacking. Therefore, there appears to be no obvious interpretation of the constant b—yet the opposite is true. To see why, let us solve the last equation for y in terms of x. Thus

$$\frac{y^2}{b^2} = \frac{x^2}{a^2} - 1$$

and

$$y = \pm b \sqrt{\frac{x^2}{a^2} - 1}$$

Now note that for very large values of x, the 1 in this equation becomes negligible. So y gets closer and closer to $y = \pm b\sqrt{x^2/a^2} = \pm b(x/a)$. In other words, the lines

$$y = \pm \frac{b}{a}x \tag{1.21}$$

are the **asymptotes** of the hyperbola (Figure 1.67).

    As in the case of the ellipse, the points $(\pm c, 0)$ are called the **foci** and $(\pm a, 0)$ the **vertices** of the hyperbola. The line segment of length $2a$ joining the vertices is called the **transverse axis** of the hyperbola, and the line segment of length $2b$ joining $(0, b)$ and $(0, -b)$ is called the **conjugate axis.** (See Figure 1.68.) Finally, the point of intersection of the two axes is the **center.**

Asymptotes
Foci
Vertices
Transverse axis
Conjugate axis
Center

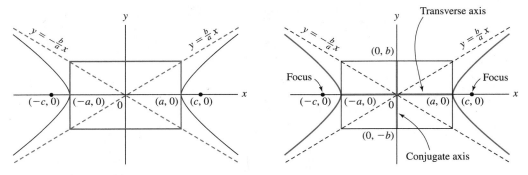

Figure 1.67   **Figure 1.68**

If the foci lie along the y-axis at $(0, \pm c)$, we obtain the form

$$\frac{y^2}{a^2} - \frac{x^2}{b^2} = 1$$

shown in Figure 1.69.

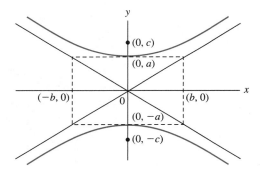

Figure 1.69

Caution: Even though the equations of the hyperbola resemble those of the ellipse, some important differences should be noted. It is clear from the graph of the hyperbola, as well as from the formula $b^2 = c^2 - a^2$, that $c > a$ (Figure 1.70)—just the opposite of the case of the ellipse (Figure 1.71). Moreover, a is not necessarily larger than b; rather, the constant a must be identified from the form of the equation.

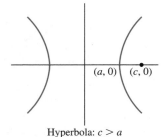

Hyperbola: $c > a$

Figure 1.70

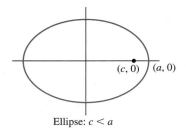

Ellipse: $c < a$

Figure 1.71

Equation of a Hyperbola

The equation of a hyperbola with center at the origin is

$$\frac{x^2}{a^2} - \frac{y^2}{b^2} = 1 \qquad \text{(transverse axis horizontal)} \tag{1.22}$$

or

$$\frac{y^2}{a^2} - \frac{x^2}{b^2} = 1 \qquad \text{(transverse axis vertical)} \tag{1.23}$$

The foci lie on the transverse axis, c units from the center, with

$$b^2 = c^2 - a^2 \tag{1.24}$$

To distinguish between the two forms of the hyperbola, note that for Equation (1.22) the only intercepts are $x = \pm a$. (If $x = 0$, then y is imaginary.) Similarly, the only intercepts for Equation (1.23) are $y = \pm a$.

Example 1   Sketch the hyperbola

$$\frac{x^2}{4} - \frac{y^2}{9} = 1$$

and find the foci, vertices, and asymptotes.

Solution. From Equation (1.22), the transverse axis is horizontal and $a = 2$ and $b = 3$. (Note that $a < b$ in this case.) These numbers can be used to construct the asymptotes, which should always be drawn first to facilitate the sketching. Plot $(\pm 2, 0)$ and $(0, \pm 3)$ and draw the rectangle determined by these points. (See Figure 1.72.) The dotted rectangle is called the **auxiliary rectangle.**

Auxiliary rectangle

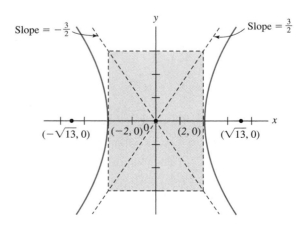

Figure 1.72

Now consider the lines through the vertices of the rectangle. The slope of the line through the upper right-hand corner is $b/a = 3/2$, and the slope of the other line is $-3/2$. Thus the lines are indeed the asymptotes. Their respective equations are

$$y = \frac{3}{2}x \qquad \text{and} \qquad y = -\frac{3}{2}x$$

From Equation (1.24),

$$b^2 = c^2 - a^2$$

we now get

$$9 = c^2 - 4 \qquad \text{and} \qquad c^2 = 13 \qquad \textbf{b = 3, a = 2}$$

so that the foci are at $(\pm\sqrt{13}, 0)$. Since $a = 2$, the vertices are at $(\pm 2, 0)$. (The vertices and foci lie on the transverse axis.) ∎

Example 2   Sketch the hyperbola $3y^2 - 2x^2 = 12$, including the asymptotes, and find the foci and vertices.

Solution. The equation can be written in the form

$$\frac{y^2}{4} - \frac{x^2}{6} = 1 \qquad \textbf{transverse axis vertical}$$

Therefore, $a = 2$ and $b = \sqrt{6}$ by Equation (1.23) and may be used to draw the auxiliary rectangle in Figure 1.73. The foci are found to be $(0, \pm\sqrt{10})$.
Since $a = 2$, the vertices are at $(0, \pm 2)$ on the transverse axis. ∎

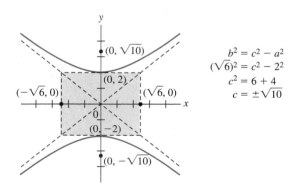

Figure 1.73

Example 3   Find the equation of the hyperbola satisfying the following conditions: the length of the transverse axis is 4 and the foci are at $(0, \pm 3)$.

Solution. Since the foci are on the y-axis, the form of the equation is

$$\frac{y^2}{a^2} - \frac{x^2}{b^2} = 1$$

The transverse axis, which is the line segment joining the vertices, has length 4, so that $a = 2$ and $a^2 = 4$. Since $c = 3$, from Equation (1.24) we get

$$b^2 = c^2 - a^2 = 3^2 - 2^2 = 5 \qquad c = 3, a = 2$$

The equation is therefore given by

$$\frac{y^2}{4} - \frac{x^2}{5} = 1 \qquad a^2 = 4, b^2 = 5 \qquad \blacksquare$$

Example 4 Find the equation of the hyperbola with asymptotes $3y = \pm 2x$, conjugate axis 8, and foci on the x-axis.

Solution. Since $y = \pm(2/3)x$, we have $b/a = 2/3$ by Equation (1.21). Since the length of the conjugate axis is $2b = 8$, we get $b = 4$. We can solve the equation $b/a = 2/3$ for a to obtain

$$\frac{4}{a} = \frac{2}{3} \qquad \text{or} \qquad a = 6$$

So by Equation (1.22),

$$\frac{x^2}{a^2} - \frac{y^2}{b^2} = 1$$

we get

$$\frac{x^2}{36} - \frac{y^2}{16} = 1 \qquad a = 6, b = 4 \qquad \blacksquare$$

Hyperbolas are used in long-range navigation as part of the LORAN system of navigation. A transmitter is located at each focus and radio signals are sent to the navigator simultaneously from each station. The difference in time at which the signals are received enables the navigator to determine his position. To be able to do so readily, however, the navigator uses charts already prepared on which hyperbolas are plotted from these differences in time. This way the navigator can find the hyperbola on which he is located. From a second pair of stations another curve of position is then found, and the navigator then determines his location by noting the point of intersection of the two hyperbolas.

■ Exercises / Section 1.10

In Exercises 1–12, find the vertices and foci of each hyperbola; draw each auxiliary rectangle and the asymptotes; and sketch the curves.

1. $\dfrac{x^2}{16} - \dfrac{y^2}{9} = 1$

2. $\dfrac{x^2}{9} - \dfrac{y^2}{4} = 1$

3. $\dfrac{x^2}{9} - \dfrac{y^2}{16} = 1$

4. $\dfrac{x^2}{16} - \dfrac{y^2}{4} = 1$

5. $\dfrac{y^2}{4} - \dfrac{x^2}{4} = 1$

6. $\dfrac{y^2}{4} - \dfrac{x^2}{8} = 1$

7. $x^2 - \dfrac{y^2}{5} = 1$

8. $9y^2 - 2x^2 = 18$

9. $2y^2 - 3x^2 = 24$

10. $x^2 - y^2 = 6$

11. $3y^2 - 2x^2 = 6$

12. $11x^2 - 7y^2 = 77$

In Exercises 13–24, determine the equation of each hyperbola.

13. Transverse axis 6, conjugate axis 4, foci on x-axis, center at origin

14. Transverse axis 6, foci at $(\pm 4, 0)$

15. Conjugate axis 8, foci at $(0, \pm 5)$

16. Vertices at $(0, \pm 5)$, conjugate axis 10

17. Transverse axis 12, foci at $(0, \pm 8)$

18. Transverse axis 8, foci at $(0, \pm 6)$

19. Vertices at $(\pm 4, 0)$, conjugate axis 8

20. Vertices at $(\pm 6, 0)$, conjugate axis 4

21. Vertices at $(\pm 3, 0)$, foci at $(\pm 6, 0)$

22. Vertices at $(\pm 2, 0)$, foci at $(\pm 4, 0)$

23. Asymptotes $y = \pm 2x$, vertices at $(\pm 1, 0)$

24. Asymptotes $y = \pm(1/2)x$, conjugate axis 3, transverse axis horizontal

25. Find the equation of the locus of points such that the difference of the distances from $(0, \pm 5)$ is ± 6.

26. Reduce Equation (1.19) to (1.20).

27. Find the equation of the locus of points such that the difference of the distances from $(1, 2)$ and $(-3, 2)$ is ± 2.

28. Derive the equation of the hyperbola whose asymptotes are $y = \pm(4/3)x$ and whose foci are at $(\pm 3, 0)$.

29. Write the equation of the hyperbola whose vertices are at $(0, \pm 12)$ and which passes through the point $(-1, 13)$.

30. Graph the equation $xy = 2$. It can be shown that the graph is a hyperbola.

31. Boyle's law states that for an ideal gas the product of the pressure and volume is always constant at a constant temperature. Mathematically, $pV = k$, where k is a constant. Suppose $V = 3.0\,\text{m}^3$ when $p = 12$ Pa (N/m^2) for some ideal gas. Find k and graph the equation. Refer to Exercise 30.

32. Some telescopes use a combination of parabolic and hyperbolic mirrors. A hyperbolic mirror has the property that a light directed at one focus will be reflected to the other focus. The mirror in Figure 1.74 has the equation

$$\frac{x^2}{16} - \frac{y^2}{33} = 1$$

Where will the reflected ray cross the x-axis?

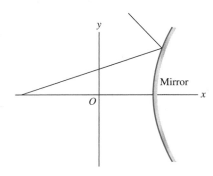

Figure 1.74

1.11 **Translation of Axes; Standard Equations of the Conics**

In this section we are going to obtain more general forms of the equations of the conics by means of *translation of axes*. A translation of axes gives a new set of axes parallel to, and oriented the same way as, the original axes.

Translation of Axes

If the origin, O, is translated to the point $O'(h, k)$, then each point has two sets of coordinates. Suppose (x, y) and (x', y') are the coordinates of a point P with respect to the old and new axes, respectively (Figure 1.75). We see from the diagram that

$$x = OB = OA + AB = h + x'$$

and

$$y = BP = BC + CP = k + y'$$

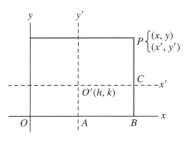

Figure 1.75

Thus $x' = x - h$ and $y' = y - k$.

Translation Formulas

$$x' = x - h \quad \text{and} \quad y' = y - k$$

Example 1   Consider the following equation of a circle:

$$(x + 1)^2 + (y - 3)^2 = 4$$

If the origin is translated to $(-1, 3)$, we get

$$x' = x + 1 \quad \text{and} \quad y' = y - 3$$

and the equation becomes

$$x'^2 + y'^2 = 4$$

Both forms are familiar from Section 1.7.   ■

Example 2   Consider the equation

$$\frac{(x - 2)^2}{9} + \frac{(y + 4)^2}{4} = 1$$

Selecting the point $(2, -4)$ for O', we have

$$x' = x - 2 \quad \text{and} \quad y' = y + 4$$

and we get

$$\frac{x'^2}{9} + \frac{y'^2}{4} = 1$$

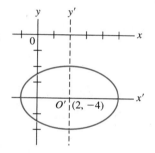

Figure 1.76

which is the equation of an ellipse (Figure 1.76).   ■

Standard Equations of the Conics

It follows from the preceding discussion that we can obtain a more general form of the conics. (For completeness the equation of a circle is included.)

Standard Equations of the Conics

$$\left.\begin{aligned} (y - k)^2 &= 4p(x - h) \\ (x - h)^2 &= 4p(y - k) \end{aligned}\right\} \quad \text{parabola; vertex at } (h, k)$$

$$\frac{(x - h)^2}{a^2} + \frac{(y - k)^2}{b^2} = 1 \quad \text{ellipse; center at } (h, k)$$

$$\frac{(x - h)^2}{a^2} - \frac{(y - k)^2}{b^2} = 1 \quad \text{hyperbola; center at } (h, k)$$

$$(x - h)^2 + (y - k)^2 = r^2 \quad \text{circle; center at } (h, k)$$

These forms are easily remembered since we are merely replacing x by $x - h$ and y by $y - k$. Geometrically, we have translated the original figure in such a way that the new center or vertex is at the point (h, k).

The translation formulas can be applied equally well to any graph: if h and k are positive, *replacing x by $x - h$ in an equation will move the corresponding graph h units to the right and replacing y by $y - k$ will move the graph k units up.* Similarly, replacing x by $x + h$ will move the graph h units to the left, and replacing y by $y + k$ will move the graph k units down.

Example 3   Write the following equation in standard form and sketch the curve:

$$y^2 - 4y - 5x - 1 = 0$$

Solution.

$$y^2 - 4y - 5x - 1 = 0 \qquad \textbf{given equation}$$
$$y^2 - 4y = 5x + 1 \qquad \textbf{adding 5x + 1}$$

We complete the square on the left by adding $[(1/2)(-4)]^2 = 4$ to both sides:

$$y^2 - 4y + 4 = 5x + 1 + 4 \qquad \textbf{adding 4 to both sides}$$
$$y^2 - 4y + 4 = 5x + 5$$
$$(y - 2)^2 = 5x + 5 \qquad \textbf{factoring the left side}$$

To obtain the standard form, we need to factor 5 on the right side:

$$(y - 2)^2 = 5(x + 1)$$

which is the equation of a parabola with axis horizontal. From

$$(y - 2)^2 = 4\left(\frac{5}{4}\right)(x + 1)$$

we see that the vertex is at **(−1, 2)** and that $\boldsymbol{p = 5/4}$. It follows that the focus is at

$$\left(-1 + \frac{5}{4}, 2\right) = \left(\frac{1}{4}, 2\right) \qquad \textbf{axis is horizontal}$$

(See Figure 1.77.) The directrix is the line

$$x = -1 - \frac{5}{4} = -\frac{9}{4}$$

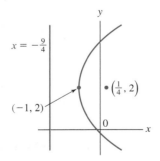

Figure 1.77

Example 4   Write the following equation in standard form and sketch the curve:

$$25x^2 - 9y^2 + 150x + 36y - 36 = 0$$

Solution. We proceed by rearranging the terms and completing the square as we did in Section 1.7:

$$25x^2 + 150x - 9y^2 + 36y = 36$$

At this point we need to factor out the coefficients of the squared terms, so that

$$25(x^2 + 6x \quad) - 9(y^2 - 4y \quad) = 36 \qquad \textbf{factoring 25 and } -9$$

Then complete the square inside the parentheses. Care must be taken when balancing the equation:

$$[(1/2) \cdot 6]^2 = 9 \text{ and } [(1/2)(-4)]^2 = 4; \text{ thus}$$

$$25(x^2 + 6x + 9) - 9(y^2 - 4y + 4) = 36 + 25 \cdot 9 - 9 \cdot 4$$

or

$$25(x + 3)^2 - 9(y - 2)^2 = 225$$

This equation simplifies to

$$\frac{(x + 3)^2}{9} - \frac{(y - 2)^2}{25} = 1 \qquad \textbf{dividing by 225}$$

This is the equation of a hyperbola with center at $(-3, 2)$ and transverse axis horizontal. Since $a = 3$ and $b = 5$, the vertices are at $(0, 2)$ and $(-6, 2)$, while the ends of the conjugate axis are at $(-3, 7)$ and $(-3, -3)$. From the formula $b^2 = c^2 - a^2$, we get $c = \sqrt{34}$, so that the foci are at $(-3 \pm \sqrt{34}, 2)$. Using the values of a and b, we now draw the auxiliary rectangle centered at $(-3, 2)$, and the two asymptotes (Figure 1.78).

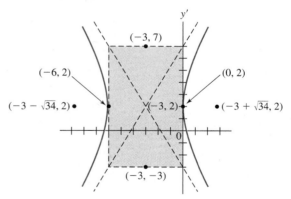

Figure 1.78   ■

Example 5
Find the equation of the following ellipse: center at $(-3, -2)$, one vertex at $(0, -2)$, and one focus at $(-5, -2)$.

Solution. The points are shown in Figure 1.79. Observe that

$$c = 2 \qquad \textbf{distance from center to focus}$$

and

$$a = 3 \qquad \textbf{distance from center to vertex}$$

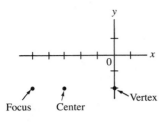

Figure 1.79

Also,

$$b^2 = a^2 - c^2 = 3^2 - 2^2 = 5$$

Since $(h, k) = (-3, -2)$, we get from the form

$$\frac{(x - h)^2}{a^2} + \frac{(y - k)^2}{b^2} = 1 \qquad \text{(major axis horizontal)}$$

the equation

$$\frac{(x + 3)^2}{9} + \frac{(y + 2)^2}{5} = 1$$

It follows from the preceding examples that any equation of the form

$$Ax^2 + By^2 + Cx + Dy + E = 0$$

represents a conic since completing the square will convert the equation to one of the standard forms. The types of loci may be summarized as follows:

Types of Loci

1. If $A = B$, the locus is a circle.
2. If $A = 0$ and $B \neq 0$, or if $B = 0$ and $A \neq 0$, then the locus is a parabola.
3. If $A \neq B$ and A and B have like signs, the locus is an ellipse.
4. If A and B have opposite signs, the locus is a hyperbola.

Certain degenerate cases may also occur:

5. The locus may be a point. We encountered this case in Section 1.7, although the possibility exists for type 3 also. Imaginary loci may occur in types 1 and 3.
6. The equation may factor into two linear factors, in which case the locus is two straight lines. For example, the equation $x^2 - 1 = 0$ can be written $(x - 1)(x + 1) = 0$, or $x = \pm 1$, which represents two parallel lines.

■ Exercises / Section 1.11

In Exercises 1–24, identify the conic, find the center (or vertex), and sketch.

1. $(x - 1)^2 + (y - 2)^2 = 3$

2. $\dfrac{(x - 1)^2}{9} + \dfrac{(y - 2)^2}{5} = 1$

3. $(y + 3)^2 = 8(x - 2)$

4. $\dfrac{(x - 3)^2}{4} - \dfrac{y^2}{9} = 1$

5. $2x^2 - 3y^2 + 8x - 12y + 14 = 0$

6. $4x^2 - 4x - 48y + 193 = 0$

7. $16x^2 + 4y^2 + 64x - 12y + 57 = 0$

8. $y^2 - 12y - 5x + 41 = 0$

9. $x^2 + y^2 + 2x - 2y + 2 = 0$

10. $x^2 + 2y^2 - 6x + 4y + 1 = 0$

11. $2x^2 - 12y^2 + 60y - 63 = 0$

12. $2x^2 + 3y^2 - 8x - 18y + 35 = 0$

13. $64x^2 + 64y^2 - 16x - 96y - 27 = 0$

14. $4x^2 - 4x - 16y + 5 = 0$

15. $3x^2 + y^2 - 18x + 2y + 29 = 0$

16. $100x^2 - 180x - 100y + 81 = 0$

17. $x^2 + 2x - 12y + 25 = 0$

18. $2x^2 - y^2 - 8x - 2y + 3 = 0$

19. $x^2 + 2y^2 + 6x - 4y + 9 = 0$

20. $y^2 + 2y - 12x + 49 = 0$

21. $x^2 + 4x + 4y + 16 = 0$

22. $3y^2 - 2x^2 - 18y - 8x + 7 = 0$

23. $x^2 + 2y^2 - 4x + 12y + 14 = 0$

24. $x^2 + 4y^2 + 6x + 24y + 41 = 0$

In Exercises 25–41, find the equations of the conics.

25. Parabola: vertex at $(-1, 2)$, focus at $(3, 2)$

26. Parabola: focus at $(2, 5)$, directrix $y + 1 = 0$

27. Ellipse: center at $(-3, 0)$, one vertex at the origin, passing through $(-3, -2)$

28. Ellipse: vertices at $(3, 1)$ and $(3, -5)$, one focus at $(3, 0)$

29. Hyperbola: vertices at $(1, 1)$ and $(-7, 1)$, one focus at $(3, 1)$

30. Hyperbola: center at $(-1, -2)$, one focus at $(2, -2)$, conjugate axis 4

31. Ellipse: center at $(2, 3)$, one vertex at $(-3, 3)$, minor axis 4

32. Parabola: vertex at $(5, -4)$, focus at $(8, -4)$

33. Hyperbola: center at $(1, 0)$, one vertex at $(3, 0)$, one asymptote $x - 2y = 1$

34. Parabola: vertex at $(-1, 3)$, passing through the origin

35. Ellipse: center at $(-3, 1)$, one focus at $(-3, -3)$, one vertex at $(-3, 6)$

36. Ellipse: center at $(4, 0)$, one focus at the origin, passing through $(4, -3)$

37. Parabola: vertex at $(4, -2)$, focus at $(1, -2)$

38. Hyperbola: center at $(3, 1)$, one vertex at $(0, 1)$, one focus at $(-1, 1)$

39. Parabola: focus at $(-2, -8)$, directrix x-axis

40. Ellipse: center at $(1, 1)$, one focus at $(1, 4)$, major axis 8

41. Hyperbola: center at $(-1, 1)$, one focus at $(-1, -2)$, one vertex at $(-1, 3)$

■ Review Exercises / Chapter 1*

1. Show that $(4, 2)$, $(7, 4)$, and $(3, 10)$ are the vertices of a right triangle.

2. Find the equation of the perpendicular bisector of the line segment joining $(-1, -2)$ and $(5, 4)$.

3. Fahrenheit and Celsius temperatures are related by $C = (5/9)(F - 32)$. What temperature is the same in both Fahrenheit and Celsius?

4. What is the physical significance of the F-intercept in Exercise 3?

5. Show that $(-1, 5)$, $(3, 1)$ $(3, 9)$, and $(7, 5)$ are the vertices of a square.

6. Find the equation of the line through $(4, 1)$ and perpendicular to the line $x + 2y - 5 = 0$.

7. Find the equation of the line through $(-1, 5)$ and parallel to the line $3x + y = 3$.

8. Show that the following lines form the sides of a right triangle: $4x - y - 7 = 0$, $x + 4y - 1 = 0$, and $y + 2x = 0$.

9. Find the equation of the circle with center at $(1, -2)$ and passing through the origin.

10. Find the equation of the circle centered at $(2, 5)$ and tangent to the x-axis.

11. Find the center and radius of the circle $x^2 + y^2 + 2x + 2y = 0$.

12. Show that the circle $x^2 + y^2 - 10x - 8y + 16 = 0$ is tangent to the y-axis.

In Exercises 13–18, identify the conic, find the vertices and foci, and sketch.

13. $\dfrac{x^2}{9} + \dfrac{y^2}{16} = 1$

14. $x^2 + 4y^2 = 1$

15. $\dfrac{y^2}{4} - \dfrac{x^2}{7} = 1$

*All the answers to the review exercises are given in the answer section (Appendix B).

16. $\dfrac{x^2}{9} - \dfrac{y^2}{16} = 1$

17. $y^2 = -3x$

18. $x^2 = 9y$

In Exercises 19–23, identify the conic, find the center (or vertex), and sketch.

19. $y^2 + 6y + 4x + 1 = 0$

20. $x^2 + y^2 - 8x + 10y - 4 = 0$

21. $16x^2 + 9y^2 - 64x + 18y = 71$

22. $x^2 + 4x + 8y - 20 = 0$

23. $x^2 - y^2 - 4x + 8y - 21 = 0$

In Exercises 24–33, find the equations of the curves indicated.

24. Parabola: vertex at origin, directrix $x = -2$

25. Parabola: vertex at $(1, 3)$, directrix $y = 0$

26. Ellipse: center at $(-2, -4)$, one vertex at $(-2, 0)$, one focus at $(-2, -2)$

27. Ellipse: center at origin, one vertex at $(0, 4)$, one focus at $(0, 3)$

28. Hyperbola: center at origin, one vertex at $(3, 0)$, asymptotes $y = \pm(3/4)x$

29. Hyperbola: vertices at $(0, 5)$ and $(0, -1)$, one focus at $(0, -2)$

30. Parabola: focus at $(-8, 2)$, directrix y-axis

31. Parabola: focus at $(0, 3)$, directrix $y = 1$

32. Hyperbola: center at $(2, -1)$, one focus at $(2, -4)$, one vertex at $(2, 1)$

33. Ellipse: center at $(4, -1)$, one focus at $(4, 1)$, one vertex at $(4, -5)$

In Exercises 34–40, discuss and sketch the given curves.

34. $y = x^3 + 1$

35. $y = (x + 1)^3$

36. $y = 2x - x^3$

37. $y^2 = x(x - 4)$

38. $y = \dfrac{x^2}{x^2 + 1}$

39. $y = \dfrac{x}{x^2 - 4}$

40. $y = x\sqrt{x + 1}$

41. A parabolic reflector has a diameter of 1.20 ft and a depth of 0.90 ft. Where should the light be placed to produce a beam of parallel rays?

42. A semielliptic arch measures 50.0 ft across the base and 15.0 ft high in the center. Find the heights of two vertical pillars each placed 15.0 ft from the center.

43. Equal squares of side x are cut from the corners of a piece of tin 6 in. square, and the edges are folded up to form a box with an open top. Express the volume V of the box in terms of x and draw the graph. Estimate the value of x for which the volume is a maximum.

44. If a stone is thrown upward from the earth's surface with an initial velocity of 96 ft/s, then the distance s above the ground is given by $s = 96t - 16t^2$, where s is measured in feet and t in seconds. Sketch the curve and estimate the greatest height reached by the stone.

45. Explorer 18 was launched on November 26, 1963. Its low and high points over the surface of the earth were 119 mi and 122,000 mi, respectively. Find the eccentricity of its elliptical orbit. (Assume that the radius of the earth is 4000 mi.)

Introduction to Calculus: The Derivative

Newton

Leibniz

Calculus was developed during the seventeenth century as a response to problems that were beyond the reach of elementary mathematics. Since mathematics and physics did not yet constitute different branches of knowledge, many of the motivations were based on physical problems. For example, the heliocentric (sun-centered) theory of Nicholas Copernicus (Polish astronomer, 1473–1543) and developed further by Johannes Kepler (German astronomer and physicist, 1571–1630) created an interest in problems of motion. However, instead of dealing only with average velocities and accelerations, calculus, once developed sufficiently, was able to handle problems dealing with instantaneous velocities and accelerations. Similarly, the concept of slope of a line was extended to slopes of arbitrary curves, and work done by a variable force could now be determined in addition to work done by a constant force. Equally remarkable were the advances in the study of areas and volumes. Classical geometry is confined to areas bounded by polygons; calculus deals with areas bounded by arbitrary curves. Basic to all these concepts is the notion of *limit*, to be discussed in detail in Section 2.2.

Credit for the discovery of calculus is usually given to Isaac Newton (English physicist and mathematician, 1642–1727) and to Gottfried Wilhelm Leibniz (German mathematician and philosopher, 1646–1716). Newton was frail as a child and, unable to participate in the usual games of boys his age, was forced to create his own diversions; he was especially fond of experimentation and showed considerable mechanical ability while still in his early teens. He attended Cambridge University as a young man; his education was interrupted by the great plague of 1664–1665, which he avoided by returning to his home in Lincolnshire for two years. In Lincolnshire, as a result of his meditations, he made his famous discoveries: the nature of colors, the calculus, and the universal law of gravitation; the last of these is probably the most far-reaching physical law ever discovered. He was twenty-three.

Leibniz possessed knowledge so comprehensive that it was said he achieved "universal knowledge" at a time when most scholars could no longer keep track of all new discoveries. He made significant contributions to mathematics, physical science, law, theology, statecraft, history, logic, and philosophy. Only in physics did he lag behind Newton. Trained in law, Leibniz made his living as a genealogist and family historian to the ancient aristocratic house of Brunswick-Lüneburg, later Hanover (some of whose descendants became kings of England). Since his work required extensive travel, Leibniz came into contact with many of the leading scholars of his day. He did not turn his attention to mathematics until age twenty-six, apparently inspired by discussions with Christian Huygens, the Dutch physicist and mathematician. Leibniz's main contributions to calculus were made around 1675, some time after Newton's. Because Newton had not published his results, an unfortunate and bitter controversy over priority resulted. It is now generally agreed that the discoveries were made independently. The notations dy/dx and $\int f(x)\, dx$ are due to Leibniz.

2.1   Functions and Intervals

In this section we are going to discuss intervals and functions. An **interval** is the set of all points on the number line between a and b. In some cases we may wish to include either or both endpoints in a designated interval, in other cases not. As you can see from the following chart, we use a square bracket to indicate that the endpoint is included and a parenthesis to indicate that it is not:

Interval	Meaning	Graph
(a, b)	$a < x < b$	
$[a, b]$	$a \leq x \leq b$	
$[a, b)$	$a \leq x < b$	
$(a, b]$	$a < x \leq b$	

These intervals have been given special names. The notation (a, b) denotes an *open* interval; $[a, b]$ is called a *closed* interval; and the remaining two are *half-open* intervals. The endpoints a and b need not be finite; for example, (a, ∞) is the set of all x such that $a < x < \infty$, or simply $x > a$. Because ∞ is not a number, in the present context $(a, \infty]$ has no meaning.

Functions

The concept of a function is closely connected to that of a curve. Consider, for example, the line $y = x - 2$. For any x we care to choose, we obtain one, and only one, value for y. The uniqueness of y makes the equation a *function of x*.

Definition of a Function

If for every value of the variable x there corresponds one and only one value of the variable y, then we call y a **function** of x, denoted by $y = f(x)$ (to be read "f of x").

The variable x is called the **independent** variable and y the **dependent** variable. (It is understood in this definition that we use only those values of x for which $f(x)$ is defined and real.)

To illustrate the definition of a function, consider the equation $y = 2x + 3$. Since for any value of x we obtain a unique value for y, the equation defines a function. The parabola $y = x^2$ also defines a function, since every value of x yields one value for y, and only one; the fact that $x = 2$ and $x = -2$ produce the same y value, namely $y = 4$, is of no consequence (Figure 2.1). Compare this case to that of the circle $x^2 + y^2 = 4$: for every x in the open interval $(-2, 2)$ we obtain two y values. For example, if $x = 1$, then $y = \pm\sqrt{3}$ (Figure 2.2). Hence the equation of the circle does not define a function.

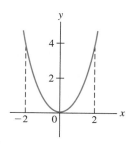

Figure 2.1   Figure 2.2

Example 1   The equation

$$y = \frac{x}{x^2 + 1}$$

represents a function: for every value of x we get one, and only one, value of y. A few such pairs are given in the following table:

x:	-1	0	1	2	3	4
y:	$-\frac{1}{2}$	0	$\frac{1}{2}$	$\frac{2}{5}$	$\frac{3}{10}$	$\frac{4}{17}$

The independent variable is x and the dependent variable is y.   ∎

Vertical line test

Geometrically, we can identify a function at a glance: a graph represents a function of x if *a vertical line intersects the graph at only one point*. More precisely, no vertical line intersects the graph at more than one point. Thus the graph in Figure 2.3 defines a function, but the graph in Figure 2.4 does not.

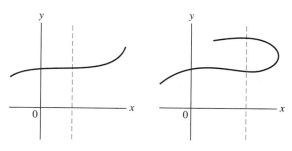

Figure 2.3   Figure 2.4

Domain

Range

Domain and Range

The set of all values of the independent variable x for which the dependent variable y is defined is called the **domain** of the function. (Note that the domain of the function is the same as the extent along the x-axis of the curve representing the function.) For example, the domain of the function $y = 1/x$ is the set of all x such that $x \neq 0$ (to avoid division by zero). The domain of the function $y = \sqrt{x}$ is the interval $[0, \infty)$. (Whenever $x < 0$, y is not a real number.)

The set of all y values that occur is called the **range** of the function. For example, the range of the function $y = x^2$ is $[0, \infty)$. (See Figure 2.1.)

Example 2   State the domain and range of the function $y = \sqrt{x^2 - 1}$.

Solution. Since y is real only if $x^2 - 1 \geq 0$, the domain consists of the intervals $(-\infty, -1]$ and $[1, \infty)$. Since y cannot be negative, the range is the interval $[0, \infty)$. ∎

Many formulas and physical laws are expressed as functions. For instance, if r is the radius of a sphere and V the volume, then $V = (4/3)\pi r^3$. Since every value of r yields a unique value of V, the relation is a function. (For physical reasons the domain is the interval $[0, \infty)$.) The formula for converting degrees Fahrenheit to degrees Celsius also defines a function:

$$C = \frac{5}{9}(F - 32)$$

Example 3   Write the area of a circle as a function of its radius.

Solution. The formula is $A = \pi r^2$. For every value of r we get a unique value of A. So the relation is indeed a function. ∎

Notation for Functions

When writing functions, we can use any letters we choose; the most common letters are x and y for the variables and f for the function. If more than one function is referred to in a discussion, we may have to use a different letter. The most commonly used letters are f, g, F, G, and ϕ. Thus $y = F(x)$, $y = g(x)$, and $y = \phi(x)$ could all represent functions, but so could $y = g(t)$ and $w = \phi(z)$. At times we will simply refer to the functions F, g, or ϕ. Finally, since the letters themselves are arbitrary,

$$F(x) = x^2 + 2x + 1 \qquad \text{and} \qquad g(t) = t^2 + 2t + 1$$

represent the same function.

We can see from the last two equations that a function can be viewed as a *rule* for performing certain operations. For example, the function $F(x) = x^2 + 2x + 1$ tells us to "square the value of the independent variable, add twice the value of the independent variable, and then add 1 to the result." Since $g(t) = t^2 + 2t + 1$ gives the same set of instructions, we see again that $F(x)$ and $g(t)$ represent the same function.

Example 4   **(a)** $\phi(z) = 1 - 2z^2$ and $h(w) = 1 - 2w^2$ represent the same function. **(b)** Since the general form of a function is $y = f(x)$, the function $y = 9x^2 - 4x$ can also be written $f(x) = 9x^2 - 4x$; in fact, functions are often written in this form. ■

This functional notation can also be used to specify individual function values. For the function $F(x) = x^2 + 2x + 1$, if $x = -3$, then $F(x) = 4$, or simply $F(-3) = 4$. Similarly, $F(0) = 1$ and $F(a) = a^2 + 2a + 1$. In fact, x may be replaced by any number or expression; for example,

$$F(a - 1) = (a - 1)^2 + 2(a - 1) + 1 = a^2 - 2a + 1 + 2a - 2 + 1 = a^2$$

The last case may seem like a useless exercise but is actually quite important, as we will see in Section 2.4. A simple way to make a substitution of this type is to leave a blank space for x and fill in the blanks:

$$F(\ \ ) = (\ \ )^2 + 2(\ \ ) + 1$$

We will illustrate the preceding ideas with some examples.

Example 5   If $f(x) = x^2 + 3x^3$, find **(a)** $f(1)$; **(b)** $f(a^2)$; **(c)** $f(x - 2)$; **(d)** $f(x + h)$.

Solution. $f(x) = x^2 + 3x^3$
   a. $f(1) = 1^2 + 3 \cdot 1^3 = 4$
   b. $f(a^2) = (a^2)^2 + 3(a^2)^3 = a^4 + 3a^6$
For parts **(c)** and **(d)** we leave a blank space for x:

$$f(\ \ ) = (\ \ )^2 + 3(\ \ )^3$$

   c. Placing $x - 2$ in the blank spaces, we get

$$f(x - 2) = (x - 2)^2 + 3(x - 2)^3 = x^2 - 4x + 4 + 3(x^3 - 6x^2 + 12x - 8)$$
$$= 3x^3 - 17x^2 + 32x - 20$$

   d. We fill the blanks with $x + h$ to get

$$f(x + h) = (x + h)^2 + 3(x + h)^3$$ ■

If $f(x)$ and $g(x)$ are two functions, then $f(g(x))$ is called a **composite function.**

Example 6   If $f(x) = \sqrt{x + 1}$ and $g(x) = x^2 - 1$, $x \geq 0$, find $f(g(x))$.

Solution. From $f(\ \ ) = \sqrt{(\ \ ) + 1}$, we get

$$f(g(x)) = f(x^2 - 1) = \sqrt{(x^2 - 1) + 1} = x$$ ■

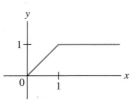

Figure 2.5

A function need not be defined by a single equation or even by an equation at all. The function

$$f(x) = \begin{cases} x & 0 \le x < 1 \\ 1 & x \ge 1 \end{cases}$$

is a perfectly good function whose domain is $[0, \infty)$ (Figure 2.5). Note that $f(0) = 0$, $f(1/2) = 1/2$, $f(1) = 1$, and $f(1000) = 1$.

Example 7   Suppose the input voltage in a DC electrical circuit is given by

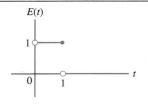

Figure 2.6

$$E(t) = \begin{cases} 1 & 0 < t \le 1 \\ 0 & t > 1 \end{cases}$$

Draw the graph.

Solution. The graph is shown in Figure 2.6. The domain is the interval $(0, \infty)$. ∎

■ Exercises / Section 2.1

1. Write the area of a square, A, as a function of its side, x.

2. Suppose that $E = 3$ volts in a DC circuit. If a rheostat is used to control the change in R, write I as a function of R.

3. Express the volume V of a cone of height 2 as a function of the radius r.

4. Express the area of the surface of a sphere as a function of the radius.

5. A variable resistor R (in ohms) is in series with a 10-ohm resistor. Write the resistance R_T of the combination as a function of R.

6. The cost of running a manufacturing plant is $10,000 per day plus $20 for each unit produced. Write the daily cost C as a function of x, the number of units produced.

7. Express the distance d traveled at 50 mi/h as a function of t (in hours).

8. A bar is 20 ft long. If x is the distance from the left end, write the distance y from the right end as a function of x.

9. A trouble-shooter charges $200/h for the first three hours and $100/h thereafter. Write the cost C as a function of time t, where $t \ge 3$ h.

10. Write the distance d from home plate as a function of the distance x from second base (Figure 2.7).

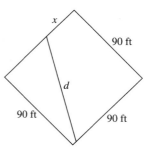

Figure 2.7

11. The resonance frequency f (in cycles per second) of a certain circuit as a function of the inductance L is

$$f(L) = \frac{1}{2\pi \sqrt{L}}$$

What is the domain of this function?

12. The period T (in seconds) of a pendulum is $T = 2\pi\sqrt{L/32}$, where L is measured in feet. If $2\,\text{ft} \le L \le 8\,\text{ft}$, what is the range in T?

In Exercises 13–26, find the domain and range of the given functions. Use a graphing utility to verify your answer.

13. $y = x + 2$

14. $y = x^2 + 2$

15. $y = \sqrt{x - 2}$

16. $y = \sqrt{1 - x^2}$

17. $y = \sqrt{4 - x^2}$

18. $y = \sqrt{5 - x}$

19. $y = x\sqrt{x - 3}$

20. $y = x^2\sqrt{x - 4}$

21. $y = (x - 1)\sqrt{x - 2}$

22. $y = x^2\sqrt{x^2 - 1}$

23. $y = \dfrac{1}{x - 1}$

24. $y = \dfrac{x}{x^2 - 1}$

25. $y = 1 + \sqrt{1 - x^2}$

26. $y = \sqrt{4 - x^2} - 1$

27. If $f(x) = x^2$, find $f(2)$ and $f(-2)$.

28. If $f(x) = 1 - x^2$, find $f(1)$ and $f(2)$.

29. If $f(x) = 2x$, find $f(0)$, $f(6)$.

30. If $g(x) = 1 - 3x$, find $g(0)$, $g(2)$.

31. If $h(x) = x^2 + 2x$, find $h(1)$, $h(3)$.

32. If $F(x) = -2x^3$, find $F(1)$, $F(-2)$.

33. If $f(x) = x^3 + 1$, find $f(0), f(1), f(-2)$.

34. If $g(x) = \sqrt{x}$, find $g(0)$, $g(4)$, $g(5)$.

35. If $\phi(x) = 1/x$, find $\phi(3)$, $\phi(a)$.

36. If $F(t) = 1/t^2$, find $F(x)$, and $F(a)$.

37. If $G(z) = \sqrt{z^2 - 1}$, find $G(a^2)$, $G(x - 1)$.

38. If $f(t) = 16t^2 - 2t$, find $f(x)$, $f(x + \Delta x)$, where Δx is a constant.

39. If $f(x) = 1 - x^2$, find $f(x + \Delta x)$, $f(x - \Delta x)$.

40. If $g(t) = 1/\sqrt[3]{t}$, find $g(1 - t)$, $g(x - 1)$.

41. If $f(x) = x^2$ and $g(x) = x + 1$, find:
    a. $f(g(x))$ **b.** $g(f(x))$ **c.** $f(f(x))$

42. If $f_1(x) = 1/x$ and $f_2(x) = x^3$, find:
    a. $f_1(f_2(x))$ **b.** $f_2(f_1(x))$ **c.** $f_1(f_1(x))$

43. Graph the function

$$f(x) = \begin{cases} -1 & -\infty < x < 0 \\ 1 & 0 \le x < \infty \end{cases}$$

and find $f(-1)$, $f(0)$, $f(2)$, $f(10)$.

44. Graph the function

$$f(x) = \begin{cases} x & 0 \le x \le 1 \\ x^2 & x > 1 \end{cases}$$

and find $f(0)$, $f(1)$, $f(3)$.

2.2 Limits

The notion of **limit** is one of the most important in calculus, because many concepts are based on it. Fortunately, we have already touched on limits in our discussion of asymptotes. For example, to find the horizontal asymptote of the function $y = 1/x$, we note that y approaches zero as x gets large, so the x-axis is an asymptote. This statement can be expressed more simply in symbolic form:

$$\lim_{x \to \infty} \frac{1}{x} = 0$$

Notice that we are not asserting that $1/x = 0$ when $x = \infty$ since ∞ is not a number. All we are claiming is that $1/x$ *approaches* zero as x increases. (See Figure 2.8.)
    Now consider the expression

$$\lim_{x \to 4} \frac{x^2}{x + 4}$$

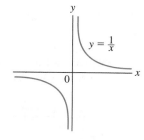

Figure 2.8

which means: what happens to the values of the function $f(x) = x^2/(x + 4)$ as x approaches 4? Suppose we study the behavior of the function by letting x get closer and closer to 4, as shown in the following table:

x:	4.1	4.01	4.001	4.0001	3.9	3.99	3.999	3.9999
$\dfrac{x^2}{x+4}$:	2.1	2.01	2.001	2.0001	1.9	1.99	1.999	1.9999

Since the values are getting closer and closer to 2, it is reasonable to conclude that

$$\lim_{x \to 4} \frac{x^2}{x+4} = 2$$

It seems that the limit could have been obtained more simply by letting $x = 4$ in the function $f(x) = x^2/(x+4)$, or

$$f(4) = \frac{4^2}{4+4} = 2$$

However, the equality $f(4) = 2$ and the statement $\lim_{x \to 4} x^2/(x+4) = 2$ do not say the same thing. If we say that $f(4) = 2$, we are merely giving the function value at $x = 4$. But if we say that

$$\lim_{x \to 4} \frac{x^2}{x+4} = 2$$

we are not interested in the value of $f(x)$ at $x = 4$, *but only in the value approached.* (See Figure 2.9.)

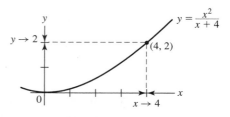

Figure 2.9

An interesting contrast to the limit just discussed is

$$\lim_{x \to 0} \frac{\sin x}{x}$$

Now the question is: what happens to the values of $f(x) = (\sin x)/x$ as x approaches zero? As before, we are not interested in the value of $f(x)$ at $x = 0$, *but only in the value approached.* Moreover, substituting 0 for x is useless anyway since $(\sin 0)/0$ is undefined (because of the division by zero). We will study the behavior of the function by means of a table. (To obtain the values, your calculator must be set in the radian mode.)

x:	0.5	0.25	0.2	0.1	0.05	0.01	0.001
$\dfrac{\sin x}{x}$:	0.9589	0.9896	0.9933	0.9983	0.9996	0.99998	0.9999998

The trend is clear: as x approaches zero, $(\sin x)/x$ approaches 1, or

$$\lim_{x \to 0} \frac{\sin x}{x} = 1$$

(We will see in Chapter 6 that this is indeed the case; merely substituting values does not, of course, constitute a proof.) We see from the table that even though the function $f(x) = (\sin x)/x$ is not defined at $x = 0$, the limit as $x \to 0$ exists just the same. The graph of $y = (\sin x)/x$ (Figure 2.10) tells us why: there is a hole where one would expect to find the point $(0, 1)$!

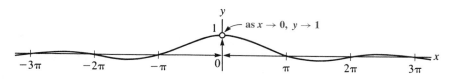

Figure 2.10

We are now ready to define the limit of a function in general.

> **Definition of Limit**
>
> Suppose that $f(x)$ becomes arbitrarily close to the number $L (f(x) \to L)$ as x approaches $a (x \to a)$. Then we say that the limit of $f(x)$ as x approaches a is L and write
>
> $$\lim_{x \to a} f(x) = L$$
>
> The number a may be replaced by ∞ or $-\infty$.

Example 1   Evaluate

$$\lim_{x \to -1} (x^2 - 3)$$

Solution. This limit is similar to $\lim_{x \to 4} x^2/(x + 4)$ discussed earlier. Since $f(x) = x^2 - 3$ is defined at $x = -1$, we can obtain the limit by inspection: as x gets closer and closer to -1, $f(x) = x^2 - 3$ gets closer and closer to -2. Thus

$$\lim_{x \to -1} (x^2 - 3) = -2 \qquad \blacksquare$$

Example 2   Evaluate

$$\lim_{x \to -2} \frac{x^2 - 4}{x + 2}$$

Solution. This limit is similar to $\lim_{x \to 0}(\sin x)/x$ discussed earlier. Direct substitution of -2 for x in $f(x) = (x^2 - 4)/(x + 2)$ yields $0/0$, which is undefined. We can obtain the limit by simplifying the expression for $f(x)$. Note that

$$\frac{x^2 - 4}{x + 2} = \frac{(x - 2)(x + 2)}{x + 2}$$

This fraction reduces to $x - 2$, provided that x is *not* equal to -2 (to avoid division by zero). But in the statement

$$\lim_{x \to -2} \frac{x^2 - 4}{x + 2}$$

x is never equal to -2; x only *approaches* -2. Since the limit concept avoids division by zero, we may say that

$$\lim_{x \to -2} \frac{x^2 - 4}{x + 2} = \lim_{x \to -2} \frac{(x - 2)(x + 2)}{x + 2}$$

$$= \lim_{x \to -2} (x - 2) = -4$$

Thus

$$\lim_{x \to -2} \frac{x^2 - 4}{x + 2} = -4$$

even though $(x^2 - 4)/(x + 2)$ is undefined at $x = -2$.

It is instructive to compare the graphs of the functions

$$y = \frac{x^2 - 4}{x + 2} \qquad \text{and} \qquad y = x - 2$$

According to Figure 2.11, both functions represent the same straight line, but $y = (x^2 - 4)/(x + 2)$ has a hole at $(-2, -4)$, while $y = x - 2$ does not.

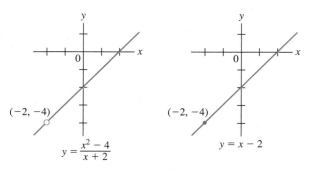

Figure 2.11

Confirm this behavior with a graphing utility by graphing the function

$$f(x) = \frac{x^2 - 4}{x + 2}$$

■

Example 3   Evaluate

$$\lim_{x \to 3} \frac{9 - x^2}{x - 3}$$

Solution. Since direct substitution yields 0/0, we proceed as in Example 2:

$$\lim_{x \to 3} \frac{9 - x^2}{x - 3} = \lim_{x \to 3} \frac{(3 - x)(3 + x)}{x - 3} \qquad \textbf{factoring}$$

$$= \lim_{x \to 3} \frac{(3 - x)(3 + x)}{-1(3 - x)} \qquad \textit{x} - 3 = -1(3 - x)$$

$$= \lim_{x \to 3} \frac{(3 + x)}{-1} \qquad \textbf{reducing the fraction}$$

$$= \frac{3 + 3}{-1} = -6$$

Note again that

$$\frac{(3 - x)(3 + x)}{-1(3 - x)} = \frac{3 + x}{-1}$$

because x only approaches 3 and is never equal to 3, thereby avoiding division by zero.

The graph of

$$y = \frac{9 - x^2}{x - 3}$$

is shown in Figure 2.12.

■

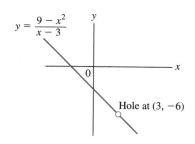

$$y = \frac{9 - x^2}{x - 3}$$

Hole at (3, −6)

Figure 2.12

Continuing our general discussion of limits, we use the notation

$$\lim_{x \to a+} f(x) = L$$

to indicate that x approaches a from the right—that is, through values larger than a. Similarly,

$$\lim_{x \to a-} f(x) = L$$

One-sided limits

indicates that x approaches a from the left. Together these are referred to as *one-sided limits*.

Example 4
$$\lim_{x \to 2+} \sqrt{x - 2} = 0$$

The restriction $x \to 2+$ is necessary to avoid imaginary values.   ∎

Example 5   Evaluate

$$\lim_{x \to \infty} \frac{3x^2 + x + 1}{2x^2 - x + 2}$$

Solution. Here the technique is identical to that for finding horizontal asymptotes: we divide numerator and denominator by x^2. Hence

$$\lim_{x \to \infty} \frac{3x^2 + x + 1}{2x^2 - x + 2} = \lim_{x \to \infty} \frac{3 + \dfrac{1}{x} + \dfrac{1}{x^2}}{2 - \dfrac{1}{x} + \dfrac{2}{x^2}} = \frac{3}{2}$$ ∎

For completeness we will state without proof the following theorems on limits: if $\lim_{x \to a} f(x) = L$ and $\lim_{x \to a} g(x) = M$, then:

A. $\lim_{x \to a} [f(x) \pm g(x)] = L \pm M$

B. $\lim_{x \to a} f(x)g(x) = LM$

C. $\lim_{x \to a} \dfrac{f(x)}{g(x)} = \dfrac{L}{M}$ $(M \neq 0)$

D. $\lim_{x \to a} kf(x) = kL$

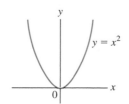

Figure 2.13

(These theorems will be referred to from time to time in later sections.)

Another important concept is continuity. Informally, a function is said to be *continuous* on an interval if its graph has no breaks or gaps. For example, the function $y = x^2$ is certainly continuous since the graph is just a parabola (Figure 2.13). On the other hand,

$$f(x) = \begin{cases} 0 & x \leq 0 \\ 1 & x > 0 \end{cases}$$

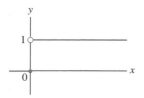

Figure 2.14

has a break at $x = 0$, since the function values suddenly jump from 0 to 1 (Figure 2.14). The break is called a *discontinuity*.

The function $y = x/(x - 1)$ is clearly discontinuous at $x = 1$, since the line $x = 1$ is a vertical asymptote (Figure 2.15).

Making use of the limit concept, continuity may be defined as follows.

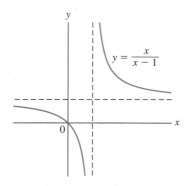

Figure 2.15

Definition of Continuity

A function f is continuous at $x = a$ if f is defined at a and

$$\lim_{x \to a} f(x) = f(a)$$

If a function is continuous at all points in an interval, it is said to be continuous in the interval.

Example 6 Show that $f(x) = x^2 + x$ is continuous at $x = 1$.

Solution. Since

$$\lim_{x \to 1} (x^2 + x) = 2 = f(1)$$

$f(x)$ is continuous at $x = 1$ by definition (Figure 2.16).

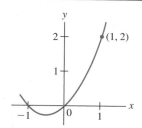

Figure 2.16

Example 7 Show that the function

$$f(x) = \begin{cases} 2 & x \geq 1 \\ 1 & x < 1 \end{cases}$$

is discontinuous at $x = 1$.

Solution. The graph of the function is shown in Figure 2.17.
Note that $f(1) = 2$—since $f(x) = 2$ for $x \geq 1$. From the graph, we have

$$\lim_{x \to 1^-} f(x) = 1$$

In other words,

$$\lim_{x \to 1^-} f(x) = 1, \text{ while } f(1) = 2$$

This violates the condition

$$\lim_{x \to a} f(x) = f(a)$$

so that the function is not continuous at $x = 1$.

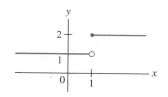

Figure 2.17

■ Exercises / Section 2.2

In Exercises 1–6, find the limits by using a calculator.

1. $\lim\limits_{x \to 0} \dfrac{\tan x}{x}$

2. $\lim\limits_{x \to 0} \dfrac{1 - \cos x}{x}$

3. $\lim\limits_{x \to 0+} x^x$

4. $\lim\limits_{x \to 0} (1 + x)^{1/x}$

5. $\lim\limits_{x \to \pi/2} (\sec x - \tan x)$

6. $\lim\limits_{x \to 1-} 2^{1/(x-1)}$

In Exercises 7–10, find the limits indicated. Identify the two functions that agree on all but one point and graph the functions. (See Example 2.)

7. $\lim\limits_{x \to 0} \dfrac{x^2 + 4x}{x}$

8. $\lim\limits_{x \to 1} \dfrac{x^2 - 1}{x - 1}$

9. $\lim\limits_{x \to 4} \dfrac{x^2 - 3x - 4}{x - 4}$

10. $\lim\limits_{x \to -3} \dfrac{x^2 + x - 6}{x + 3}$

In Exercises 11–52, find the limits indicated.

11. $\lim\limits_{x \to 0} x^2$

12. $\lim\limits_{x \to 0} (x^3 + 3)$

13. $\lim\limits_{x \to 1} (x^4 - 3x + 2)$

14. $\lim\limits_{x \to 2} \dfrac{x^2 - 1}{x + 2}$

15. $\lim\limits_{x \to 2} \dfrac{x^2 - 4}{x - 2}$

16. $\lim\limits_{x \to 4} \dfrac{x^2 - 16}{x - 4}$

17. $\lim\limits_{x \to 0} \dfrac{x^2 - x}{x}$

18. $\lim\limits_{x \to 0} \dfrac{2x^2 + x}{x}$

19. $\lim_{x \to -4} \dfrac{x^2 - 16}{x + 4}$

20. $\lim_{x \to 8} \dfrac{x^2 - 64}{x - 8}$

21. $\lim_{x \to 2} \dfrac{x^2 - 4x + 4}{x - 2}$

22. $\lim_{x \to 8} \dfrac{x^2 - 64}{x + 8}$

23. $\lim_{x \to 2} \dfrac{x^2 + x - 6}{x - 2}$

24. $\lim_{x \to -2} \dfrac{x^2 + x - 5}{x - 2}$

25. $\lim_{x \to 4} \dfrac{x^2 + x - 8}{x + 4}$

26. $\lim_{x \to -4} \dfrac{x^2 + 2x - 8}{x + 4}$

27. $\lim_{x \to 1} \dfrac{x^3 - 2x^2 + x}{x - 1}$

28. $\lim_{x \to 1/2} \dfrac{4x^2 - 4x + 1}{2x - 1}$

29. $\lim_{r \to 3} \dfrac{r^2 - 9}{3 - r}$

30. $\lim_{s \to -5} \dfrac{s^2 - 25}{s + 5}$

31. $\lim_{x \to 5} \dfrac{25 - x^2}{5 - x}$

32. $\lim_{x \to 2} \dfrac{x^2 - 6}{x + 2}$

33. $\lim_{x \to 1} \dfrac{4x^2 - 2}{x}$

34. $\lim_{x \to 3} \dfrac{x^2 - x - 6}{x - 3}$

35. $\lim_{x \to 4} \dfrac{x^2 + 6x + 4}{x + 4}$

36. $\lim_{x \to \infty} \dfrac{2x^2 - 3}{4x^2 + 5x - 7}$

37. $\lim_{x \to \infty} \dfrac{4x^3 - 2x + 1}{5x^3 + 3x^2 - x}$

38. $\lim_{x \to \infty} \dfrac{2x^2 - 3x + 1}{3x^2 - 4}$

39. $\lim_{x \to \infty} \dfrac{3x^2 + 2x}{4x^2 - 3}$

40. $\lim_{x \to \infty} \dfrac{5x - 7}{3x^2 + 5x - 3}$

41. $\lim_{y \to \infty} \dfrac{1 - 4y}{y^2 + 1}$

42. $\lim_{z \to \infty} \dfrac{2z^2 - 3z + 1}{4z^2 - 2}$

43. $\lim_{x \to \infty} \dfrac{3x^2 + 4x}{9x^2 - 7x + 6}$

44. $\lim_{x \to \infty} \dfrac{1 - 3x - 10x^2}{2 - 5x^2}$

45. $\lim_{x \to \infty} \dfrac{1 - 12x^2}{2 + 6x - 6x^2}$

46. $\lim_{x \to \infty} \dfrac{\sqrt{x + 1}}{x + 1}$

47. $\lim_{x \to \infty} \dfrac{\sqrt{x^2 + 6}}{x + 1}$

48. $\lim_{x \to \infty} (\sqrt{x^2 + 4} - x)$ $\left(\textit{Hint: multiply by } \dfrac{\sqrt{x^2 + 4} + x}{\sqrt{x^2 + 4} + x} \right)$

49. $\lim_{x \to \infty} (x - \sqrt{x^2 - 1})$

50. $\lim_{x \to 3+} \sqrt{x - 3}$

51. a. $\lim_{x \to 0+} 3^{1/x}$ **b.** $\lim_{x \to 0-} 3^{1/x}$

52. a. $\lim_{x \to 0+} \dfrac{1}{1 + 2^{-1/x}}$ **b.** $\lim_{x \to 0-} \dfrac{1}{1 + 2^{-1/x}}$

In Exercises 53–56, show that the given functions are not continuous at the given point.

53. $f(x) = \begin{cases} -1 & x \le 2 \\ 1 & x > 2 \end{cases}$ at $x = 2$

(See Figure 2.18.)

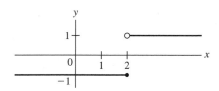

Figure 2.18

54. $f(x) = \begin{cases} x & x < 1 \\ 2 & x \ge 1 \end{cases}$ at $x = 1$

(See Figure 2.19.)

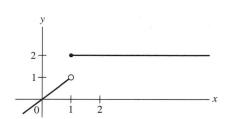

Figure 2.19

55. $f(x) = \dfrac{x^2 - 1}{x + 1}$ at $x = -1$

56. $f(x) = \begin{cases} \dfrac{x^2 - 1}{x + 1} & x \ne -1 \\ -3 & x = -1 \end{cases}$ at $x = -1$

2.3 The Derivative

In this section we are going to introduce the concept of the slope of a line tangent to a curve. This concept leads to the definition of a derivative.

The notion of *derivative* originated with Newton's attempt to study the velocity of a falling object and other quantities whose values change continuously. To appreciate the difficulty, suppose a boxlike container with a square base 2 ft by 2 ft is filled with water from a faucet. If water pours in at the rate of 4 ft³/min, then during the course of one minute the volume will increase by 4 ft³. If x stands for the increase in the liquid level during one minute, then we get from $2 \cdot 2 \cdot x = 4$ that $x = 1$ ft. So the liquid level rises at the rate of 1 ft/min. But what if the container has the shape of a cone? Then the water level will rise at an ever-changing rate, a rate that cannot be determined algebraically. (We will consider this case in the next chapter.) The problem of the falling body leads to similar difficulties; both were beyond the scope of the mathematical methods known before Newton's time.

Newton recognized that the problem of the falling body, involving so-called instantaneous rates of change, was equivalent to finding the slope of a line tangent to a curve. For fairly simple functions, slopes of tangent lines had already been studied by Fermat, while the acceleration due to gravity had been determined experimentally by Galileo.

To facilitate our discussion we first turn our attention to the geometric problem of the tangent to a curve. We will briefly return to the idea of rate of change at the end of the next section.

First, recall that the slope of a line

$$m = \frac{y_2 - y_1}{x_2 - x_1}$$

through the points (x_1, y_1) and (x_2, y_2) is constant. This is not true of the slope of nonlinear curves. For example, in Figure 2.20 the curve is steeper at point P than at point Q. Yet the two ideas can be connected. Note that the line T, which is drawn tangent to the curve at P, has the same direction as the curve at that point. So the problem of finding the slope of a curve is reduced to finding the slope of the line tangent to the curve.

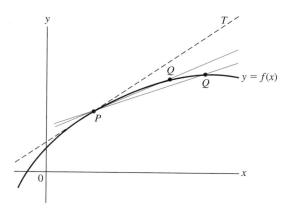

Figure 2.20

We can identify the tangent line to a curve at a given point in the following manner. Consider the function $y = f(x)$ in Figure 2.20. Let P and Q be two distinct points on the graph and let PQ be the line determined by the chord PQ. Suppose P is held fixed while Q is free to vary. As Q moves toward P, PQ moves toward the tangent line T. (See Figure 2.20). If P and Q coincide, we no longer have two points to determine the chord. In other words: Q only approaches P; it never coincides with P. Thus T is the *limiting position* of line PQ as Q approaches P. In symbols:

$$\lim_{Q \to P} \text{line } PQ = T$$

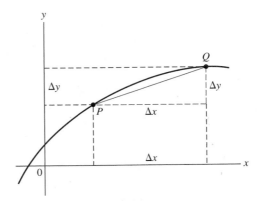

Figure 2.21

The next step is to convert this limit to a computational formula. To be able to do so, we need a simple notational device: if x is a number to which we add a number Δx (read "delta x"), we obtain $x + \Delta x$ and say that x has been given an **increment** Δx. (One could simply say that Δx is the change in x.) If $y = f(x)$, a change Δx in x induces a corresponding change Δy in y. (See Figure 2.21.)

Increment

Let P be the fixed point (x_0, y_0). Then, making use of the delta notation, Q can be represented by $(x_0 + \Delta x, y_0 + \Delta y)$ (Figure 2.22). As a result,

$$y_0 = f(x_0) \qquad \text{and} \qquad y_0 + \Delta y = f(x_0 + \Delta x)$$

(See Figure 2.22.) Furthermore, the slope of line PQ can now be written

$$\frac{\Delta y}{\Delta x}$$

From

$$y_0 + \Delta y = f(x_0 + \Delta x)$$

we get

$$\Delta y = f(x_0 + \Delta x) - f(x_0) \text{ (See Figure 2.22.)}$$

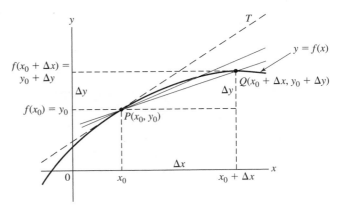

Figure 2.22

Hence

$$\text{Slope of line } PQ = \frac{\Delta y}{\Delta x} = \frac{f(x_0 + \Delta x) - f(x_0)}{\Delta x}$$

The descriptive statement $Q \to P$ (Q approaches P) can now be replaced by the numerical statement $\Delta x \to 0$ (Δx approaches zero), since

$$(x_0 + \Delta x, \ y_0 + \Delta y) \to (x_0, y_0) \text{ as } \Delta x \to 0$$

Consequently,

$$\text{Slope of } T = \lim_{\Delta x \to 0} \frac{\Delta y}{\Delta x} = \lim_{\Delta x \to 0} \frac{f(x_0 + \Delta x) - f(x_0)}{\Delta x}$$

Denoting this limit by dy/dx or by $f'(x_0)$, we can now write

$$\frac{dy}{dx} = \lim_{\Delta x \to 0} \frac{\Delta y}{\Delta x}$$

or

$$f'(x_0) = \lim_{\Delta x \to 0} \frac{\Delta y}{\Delta x} = \lim_{\Delta x \to 0} \frac{f(x_0 + \Delta x) - f(x_0)}{\Delta x}$$

Derivative

This limit, denoted by $f'(x_0)$ (read "f prime of x_0"), is called the **derivative** of $f(x)$ at x_0. If $y = f(x)$, the derivative is also denoted by y'.

If the limit exists at x_0, then $f(x)$ is said to be *differentiable* at x_0. If $f'(x)$ exists at every point in (a, b), then $f(x)$ is *differentiable* in (a, b). The process of finding

Differentiation

$f'(x)$ is called **differentiation.**

2.4   The Derivative by the Four-Step Process

In the previous section we defined the derivative of $y = f(x)$ at $x = x_0$. By omitting the subscript zero, we obtain a general expression for the derivative that is valid for all x.

Definition of the Derivative

$$f'(x) = \lim_{\Delta x \to 0} \frac{\Delta y}{\Delta x} = \lim_{\Delta x \to 0} \frac{f(x + \Delta x) - f(x)}{\Delta x} \tag{2.1}$$

To use this definition to find the derivative of a given function $y = f(x)$, we first replace x by $x + \Delta x$. For example, if $f(x) = 2x$, then $f(x + \Delta x) = 2(x + \Delta x)$. Now we subtract $f(x)$ from this expression to get $2(x + \Delta x) - 2x = 2x + 2\Delta x - 2x = 2\Delta x$. Dividing by Δx, we get $2\Delta x/\Delta x = 2$. Finally, $\lim_{\Delta x \to 0} 2 = 2$, so that $f'(x) = 2$. (Observe that the slope of the line $y = 2x$ is indeed equal to 2.)

This procedure, which requires four steps, can be carried out systematically by a process called the *delta process* or **four-step process.**

The Four-Step Process

Step 1. Replace x by $x + \Delta x$ and y by $y + \Delta y$ in the function $y = f(x)$:

$$y + \Delta y = f(x + \Delta x)$$

Step 2. Subtract $y = f(x)$ from both sides:

$$\Delta y = f(x + \Delta x) - f(x)$$

Step 3. Divide both sides of the resulting expression by Δx:

$$\frac{\Delta y}{\Delta x} = \frac{f(x + \Delta x) - f(x)}{\Delta x}$$

Step 4. Obtain $f'(x)$ by evaluating

$$\lim_{\Delta x \to 0} \frac{\Delta y}{\Delta x}$$

Particular attention should be paid to Step 3. Note that we cannot simply evaluate the limit by substituting zero for Δx since we will always get $0/0$. So the expression has to be simplified (or at least rewritten in some way) to permit the cancelation of Δx in the denominator. This simplification can be done at various points in the process, as the following examples show.

Example 1   Differentiate the function $y = 1 - x^2$.

Solution. To obtain an expression for $f(x + \Delta x)$, we leave a blank space for x:

$$f(\ \ ) = 1 - (\ \ )^2$$

Now we substitute $x + \Delta x$:

Step 1. $y + \Delta y = f(x + \Delta x) = 1 - (x + \Delta x)^2$ $f(x + \Delta x)$

$$y + \Delta y = 1 - (x + \Delta x)^2$$

Step 2. $\Delta y = 1 - (x + \Delta x)^2 - (1 - x^2)$ $\Delta y = f(x + \Delta x) - f(x)$

Step 3. $\dfrac{\Delta y}{\Delta x} = \dfrac{1 - (x + \Delta x)^2 - (1 - x^2)}{\Delta x}$ dividing by Δx

Before going to Step 4, we simplify the expression and cancel Δx:

$$\frac{\Delta y}{\Delta x} = \frac{1 - [x^2 + 2x\,\Delta x + (\Delta x)^2] - 1 + x^2}{\Delta x} \qquad (a+b)^2 = a^2 + 2ab + b^2$$

$$= \frac{-2x\,\Delta x - (\Delta x)^2}{\Delta x} = \frac{\Delta x(-2x - \Delta x)}{\Delta x} \qquad \text{factoring } \Delta x$$

$$= -2x - \Delta x \qquad \text{reducing}$$

Step 4. $f'(x) = \lim_{\Delta x \to 0} \dfrac{\Delta y}{\Delta x} = \lim_{\Delta x \to 0} (-2x - \Delta x) = -2x$

Since $f'(x) = -2x$, we have found a general expression for the derivative, which is itself a function that can be evaluated for any x. For example, $f'(1) = -2$, which is the slope of the tangent line to the curve $y = 1 - x^2$ at the point $(1, 0)$ (Figure 2.23).
    To get the equation of the line, we use the point-slope form, Equation (1.7): $y - y_1 = m(x - x_1)$. Since $m = -2$ and the point of tangency is $(1, 0)$, we get

$$y - 1 = -2(x - 0)$$

or $y = -2x + 1$

This is the dashed line in Figure 2.23. ∎

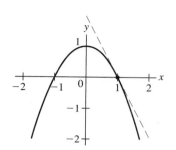

Figure 2.23

Example 2   Find the derivative of $y = 2x^3 - x$.

Solution.

Step 1. $y + \Delta y = f(x + \Delta x) = 2(x + \Delta x)^3 - (x + \Delta x)$
Step 2. $\Delta y = 2(x + \Delta x)^3 - (x + \Delta x) - (2x^3 - x)$

This time it is better to simplify even before going on to Step 3. Expanding the right side by multiplying, we get

$$2[x^3 + 3x^2\,\Delta x + 3x(\Delta x)^2 + (\Delta x)^3] - (x + \Delta x) - (2x^3 - x)$$
$$= 2x^3 + 6x^2\,\Delta x + 6x(\Delta x)^2 + 2(\Delta x)^3 - x - \Delta x - 2x^3 + x$$
$$= 6x^2\,\Delta x + 6x(\Delta x)^2 + 2(\Delta x)^3 - \Delta x$$

Note that all the remaining terms contain at least one Δx, which can be factored. Thus

Step 3. $\dfrac{\Delta y}{\Delta x} = \dfrac{\Delta x[6x^2 + 6x\Delta x + 2(\Delta x)^2 - 1]}{\Delta x}$

$$= 6x^2 + 6x\,\Delta x + 2(\Delta x)^2 - 1$$

Step 4. $f'(x) = \lim_{\Delta x \to 0} \dfrac{\Delta y}{\Delta x} = \lim_{\Delta x \to 0} [6x^2 + 6x\,\Delta x + 2(\Delta x)^2 - 1] = 6x^2 - 1$ ∎

Example 3   Find the derivative of $y = 1/(x - 1)$ at $x = 2, x = -1$.

Solution. First we obtain a general expression for y':

Step 1. $y + \Delta y = f(x + \Delta x) = \dfrac{1}{x + \Delta x - 1}$

Step 2. $\Delta y = \dfrac{1}{x + \Delta x - 1} - \dfrac{1}{x - 1}$

Again it is better to simplify before going on to Step 3. In a case like this we need to combine the two fractions:

$$\Delta y = \frac{1}{x + \Delta x - 1}\frac{x-1}{x-1} - \frac{1}{x-1}\frac{x+\Delta x-1}{x+\Delta x-1}$$

$$= \frac{(x-1)-(x+\Delta x-1)}{(x+\Delta x-1)(x-1)} = \frac{-\Delta x}{(x+\Delta x-1)(x-1)}$$

Step 3. $\dfrac{\Delta y}{\Delta x} = \dfrac{-\Delta x}{(x+\Delta x-1)(x-1)} \cdot \dfrac{1}{\Delta x} = \dfrac{-1}{(x+\Delta x-1)(x-1)}$

Step 4. $f'(x) = \lim\limits_{\Delta x \to 0} \dfrac{\Delta y}{\Delta x} = \lim\limits_{\Delta x \to 0} \dfrac{-1}{(x+\Delta x-1)(x-1)} = \dfrac{-1}{(x-1)^2}$

Since

$$f'(x) = \frac{-1}{(x-1)^2}$$

the slope of the tangent line at $x = 2$ is $f'(2) = -1/(2-1)^2 = -1$. At $x = -1$, $f'(-1) = -1/4$. (See Figure 2.24.)

For the latter, the point of tangency is $(-1, -1/2)$. Since $m = -1/4$, we get from the point-slope form, $y + 1/2 = -1/4(x+1)$ or $y = -(1/4)x - 3/4$. As before, we can now graph the curve and the tangent line with a graphing utility, as shown in Figure 2.25. (See also Exercises 25–28.)

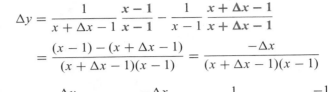

Figure 2.24

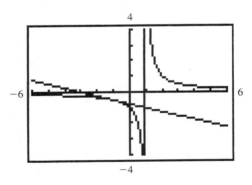

Figure 2.25

Example 4   Differentiate the function

$$f(x) = \sqrt{x+2}$$

Solution.

Step 1. $y + \Delta y = f(x + \Delta x) = \sqrt{x + \Delta x + 2}$

Step 2. $\Delta y = \sqrt{x + \Delta x + 2} - \sqrt{x + 2}$

Step 3. $\dfrac{\Delta y}{\Delta x} = \dfrac{\sqrt{x + \Delta x + 2} - \sqrt{x + 2}}{\Delta x}$

This expression cannot really be simplified in the usual sense of algebra. However, to eliminate Δx we can resort to the simple trick of rationalizing the numerator:

$$\frac{\Delta y}{\Delta x} = \frac{\sqrt{x + \Delta x + 2} - \sqrt{x + 2}}{\Delta x} \cdot \frac{\sqrt{x + \Delta x + 2} + \sqrt{x + 2}}{\sqrt{x + \Delta x + 2} + \sqrt{x + 2}}$$

$$= \frac{(\sqrt{x + \Delta x + 2})^2 - (\sqrt{x + 2})^2}{\Delta x(\sqrt{x + \Delta x + 2} + \sqrt{x + 2})}$$

$$= \frac{(x + \Delta x + 2) - (x + 2)}{\Delta x(\sqrt{x + \Delta x + 2} + \sqrt{x + 2})} = \frac{1}{\sqrt{x + \Delta x + 2} + \sqrt{x + 2}}$$

Step 4. $f'(x) = \lim\limits_{\Delta x \to 0} \dfrac{1}{\sqrt{x + \Delta x + 2} + \sqrt{x + 2}} = \dfrac{1}{2\sqrt{x + 2}}$ ∎

Notations for the Derivative

In addition to $f'(x)$ and y', some commonly used notations for the derivative are

$$\frac{dy}{dx} \qquad \frac{df(x)}{dx} \qquad \frac{d}{dx}f(x) \qquad D_x f(x)$$

Newton also employed $\dot{y}$, which is sometimes seen in books on mechanics to denote dy/dt. As noted earlier, dy/dx was introduced by Leibniz. This form suggests that the derivative may be viewed as a quotient; in a limited sense this is indeed the case, as we will see in the section on differentials in Chapter 3.

The Derivative as a Rate of Change

Returning now to Newton's attempt to measure a rate of change, suppose a particle moves according to the equation $s = 2t + 1$, where s is the distance covered as a function of time t. The graph of the function is a straight line (Figure 2.26) with a slope of 2. Consider two points on the graph, say $(0, 1)$ and $(1, 3)$. Calculating the slope again, we get

$$m = \frac{y_2 - y_1}{x_2 - x_1} = \frac{3 - 1}{1 - 0} = \frac{2}{1} = 2$$

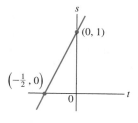

Figure 2.26

by the familiar formula. Now, since the graph has a physical interpretation, what about the slope itself? Measuring s in meters and t in seconds, the preceding calculation tells us that

$$m = \frac{2 \text{ m}}{1 \text{ s}}$$

That is, s changes by 2 meters whenever t changes by 1 second, or *the rate of change of s with respect to t is 2 m/s*. This rate of change is the velocity!

This interpretation, although of some interest, does not tell us what we really need to know about velocity. Suppose, for example, that you are traveling along a road at 30 mi/h and suddenly speed up to 50 mi/h. In that case you will see the

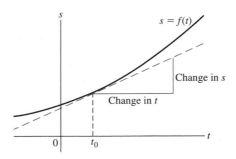

Figure 2.27

speedometer needle move continuously from 30 to 50. In other words, we do not actually have to travel for a whole hour to say that we are doing 40 mi/h; we could be traveling at that speed for only an *instant*. The graph of a function $s = f(t)$ describing a nonconstant velocity is a smooth curve rather than a straight line. To find the velocity at some instant $t = t_0$, we need the rate of change *only* at that instant. Intuitively, then, this rate of change will be given by the slope of the line tangent to the curve at the point with abscissa $t = t_0$ (Figure 2.27).

Suppose a particle moves along a line according to the function $s = t^2$, where s is the distance (feet) and t the time (seconds). It is easily checked that $ds/dt = 2t$, which is the velocity of the particle as a function of time. At the instant $t = 2$ s, $v = ds/dt = 4$ ft/s; at $t = 5$ s, $v = ds/dt = 10$ ft/s.

As a final comment, the *difference quotient*

$$\frac{\Delta y}{\Delta x} = \frac{f(x + \Delta x) - f(x)}{\Delta x}$$

is called the *average rate of change of y with respect to x*, while $f'(x)$ is called the *instantaneous rate of change of y with respect to x*. As a consequence, the general definition of the drivative is the **instantaneous rate of change of one variable with respect to another.** (Rates of change will be studied in greater detail in Section 2.6.)

■ Exercises / Section 2.4

In Exercises 1–20, use the four-step process to differentiate each function.

1. $y = 2x + 1$

2. $y = 1 - 5x$

3. $y = 2 - 3x$

4. $y = -x^2$

5. $y = x^2 + 1$

6. $y = 2x^2 + 2$

7. $y = 2x^2 - x$

8. $y = 1 - 4x^3$

9. $y = x^3 - 3x^2$

10. $y = \dfrac{1}{x}$

11. $y = \dfrac{1}{x + 1}$

12. $y = \dfrac{1}{2x - 4}$

13. $y = \dfrac{1}{x^2}$

14. $y = \dfrac{1}{x^2 - 2}$

15. $y = \dfrac{1}{1 - x^2}$

16. $y = \dfrac{1}{x - x^2}$

17. $y = \sqrt{x}$ (See Example 4.)

18. $y = \sqrt{x - 2}$

19. $y = \sqrt{1 - x}$

20. $y = \sqrt{2 - 3x}$

In Exercises 21–24, evaluate $f'(x)$ at the given points and interpret each geometrically.

21. $f(x) = x^2 + 2$; $f'(-1), f'(2), f'(3)$

22. $f(x) = \dfrac{1}{x^2 + 1}$; $f'(0), f'(1)$

23. $f(x) = \dfrac{1}{\sqrt{x}}$; $f'(4), f'(9)$

24. $f(x) = \dfrac{1}{\sqrt{x + 1}}$; $f'(0), f'(7)$

In Exercises 25–28, find the equation of the tangent line at the given point. Use a graphing utility to graph both the curve and the tangent line. (See Example 1.)

25. $y = x^2 - 1$; $(2, 3)$

26. $y = 3 - x^2$; $(1, 2)$

27. $y = \dfrac{2}{x}$; $(2, 1)$

28. $y = 1 - \dfrac{1}{x}$; $\left(2, \dfrac{1}{2}\right)$

2.5 Derivatives of Polynomials

So far we have found derivatives by the four-step process. This process is lengthy even for fairly simple functions. In this section we will develop some rules for differentiating polynomials directly. (However, we will need the four-step process later to obtain additional shortcut methods.)

The first special function to be considered is $y = c$, where c is a constant. Since the graph of $y = c$ is a horizontal line, which is its own tangent, we would expect y' to equal zero. Indeed,

Step 1. $y + \Delta y = f(x + \Delta x) = c$ Step 2. $\Delta y = c - c = 0$

Step 3. $\dfrac{\Delta y}{\Delta x} = 0$ Step 4. $\lim\limits_{\Delta x \to 0} \dfrac{\Delta y}{\Delta x} = 0$

Constant Rule

$$\frac{dc}{dx} = 0 \qquad \text{where } c \text{ is a constant} \tag{2.2}$$

The function $y = x$ is a straight line having a slope of 1, so that $y' = 1$. Although the function is a special case of Equation (2.4), we will list it separately:

$$\frac{dx}{dx} = 1 \tag{2.3}$$

For the function $y = x^n$, where n is a positive integer, we need the binomial theorem to expand the expression in Step 2.

Step 1. $y + \Delta y = f(x + \Delta x) = (x + \Delta x)^n$

Step 2. $\Delta y = (x + \Delta x)^n - x^n$

$$= x^n + nx^{n-1}\Delta x + \frac{n(n-1)}{2!}x^{n-2}(\Delta x)^2 + \cdots + (\Delta x)^n - x^n$$

$$= nx^{n-1}\Delta x + \frac{n(n-1)}{2!}x^{n-2}(\Delta x)^2 + \cdots + (\Delta x)^n$$

Step 3. $\dfrac{\Delta y}{\Delta x} = nx^{n-1} + \dfrac{n(n-1)}{2!}x^{n-2}\Delta x + \cdots + (\Delta x)^{n-1}$

Step 4. $\displaystyle\lim_{\Delta x \to 0} \dfrac{\Delta y}{\Delta x} = nx^{n-1}$

since all terms except the first contain at least one factor Δx.

Thus we have the following formula, called the *power rule:*

Derivative of x^n, $n > 0$

$$\frac{dx^n}{dx} = nx^{n-1} \tag{2.4}$$

Since the binomial theorem can be extended to rational powers, Formula (2.4) is actually valid for rational n. This case will be checked later by different methods.

The next formula, called the *sum rule,* is also needed to differentiate polynomials, but it is really of more general interest.

Suppose $u = f(x)$ and $v = g(x)$ are two differentiable functions. We wish to find the derivative of the sum $f(x) + g(x)$. By definition

$$
\begin{aligned}
\frac{d}{dx}[f(x) + g(x)] &= \lim_{\Delta x \to 0} \frac{f(x + \Delta x) + g(x + \Delta x) - [f(x) + g(x)]}{\Delta x} \\
&= \lim_{\Delta x \to 0} \left[\frac{f(x + \Delta x) - f(x)}{\Delta x} + \frac{g(x + \Delta x) - g(x)}{\Delta x} \right]
\end{aligned}
$$

Since the limit of a sum is equal to the sum of the limits (Theorem A on limits), we get

$$\frac{d}{dx}[f(x) + g(x)] = \frac{d}{dx}f(x) + \frac{d}{dx}g(x)$$

In words: *the derivative of a sum is equal to the sum of the derivatives.*

This derivation is quite straightforward, but to make our task easier later on, suppose we return to the delta notation. Since $u = f(x)$ and $v = g(x)$,

$$\Delta u = f(x + \Delta x) - f(x)$$

$$\Delta v = g(x + \Delta x) - g(x)$$

and

$$y = u + v$$

Then the steps above can be written in the following compact form:

Step 1. $y + \Delta y = u + \Delta u + v + \Delta v$

Step 2. $\Delta y = u + \Delta u + v + \Delta v - u - v = \Delta u + \Delta v$

Step 3. $\dfrac{\Delta y}{\Delta x} = \dfrac{\Delta u}{\Delta x} + \dfrac{\Delta v}{\Delta x}$

Step 4. $\dfrac{dy}{dx} = \displaystyle\lim_{\Delta x \to 0} \dfrac{\Delta y}{\Delta x} = \lim_{\Delta x \to 0} \dfrac{\Delta u}{\Delta x} + \lim_{\Delta x \to 0} \dfrac{\Delta v}{\Delta x} = \dfrac{du}{dx} + \dfrac{dv}{dx}$

or $(d/dx)(u + v) = du/dx + dv/dx$, which will be called the *sum rule.*

Sum Rule

$$\frac{d}{dx}(u + v) = \frac{du}{dx} + \frac{dv}{dx}$$

(2.5)

Our final formula in this section is easily obtained (Exercise 27).

Constant Multiplier Rule

$$\frac{d}{dx}(cu) = c\frac{du}{dx}$$

(2.6)

We are now ready to differentiate polynomials.

Example 1   Differentiate

  a. $y = x^3$ **b.** $y = 3x^4$

Solution.

  a. Since $y = x^3$,

$$\frac{dy}{dx} = 3x^{3-1} = 3x^2 \qquad \textbf{by Formula (2.4)}$$

  b. By Formulas (2.4) and (2.6),

$$\frac{dy}{dx} = 3 \cdot 4x^{4-1} = 12x^3 \qquad \blacksquare$$

Example 2   Differentiate

$$y = 5x^4 - 3x^3 + x^2 - x + 10$$

Solution. Since the derivative of a sum is equal to the sum of the derivatives, we can differentiate each term separately and add the results:

$$y' = 5(4x^3) - 3(3x^2) + 2x - 1 + 0 = 20x^3 - 9x^2 + 2x - 1 \qquad \blacksquare$$

A line is said to be *normal* to the curve $y = f(x)$ at (x_1, y_1) if it is perpendicular to the tangent line through (x_1, y_1). Since two perpendicular lines have slopes that are negative reciprocals, the slope of the normal line is $-1/f'(x_1)$, as long as $f'(x_1) \neq 0$. (See Exercises 23–26.)

■ Exercises / Section 2.5

Differentiate each of the following functions:

1. $y = x + 1$

2. $y = 3x + 2$

3. $y = x^2 - 2$

4. $y = x^3 - 2x$

5. $y = x^2 + x$

6. $y = 4x^2 - x + 2$

7. $y = 3x^2 + 4x$

8. $y = 5x^2 - 3x + 6$

9. $y = 5x^3 - 7x^2 + 2$

10. $y = 10x^5 + 6x^4 - 12x^2$

11. $y = 7x^3 - x^2 - x + 2$

12. $y = 5x^7 - 8x^6 + x^2 - 3$

13. $y = \frac{1}{3}x^3 + \frac{1}{2}x^2 + x$

14. $y = \frac{1}{4}x^4 + \frac{1}{3}x^2 - \sqrt{2}$

15. $y = \frac{x^2}{2} - \frac{x^3}{3}$

16. $y = \frac{x^4}{4} - \frac{x^2}{7}$

17. $y = 20x^{10} - 24x^6 + 2x^3 - \sqrt{3}$

18. $y = \sqrt{5}x^4 - \sqrt{2}x^3 + \pi$

19. $y = \frac{1}{5}t^7 - \frac{1}{\sqrt{2}}t^5 + \frac{1}{3}$

20. $y = \sqrt{5}t^{10} - \sqrt{7}t^7 - 5\sqrt{2}$

21. $y = \frac{1}{6}R^6 + \frac{1}{5}R^4 - \sqrt{3}$

22. $y = 3V^{12} + \sqrt{5}V^5 - 2V^3 - \sqrt{3}$

In Exercises 23–26, find the equations of the tangent and normal lines to each curve at the given point. Use a graphing utility to graph the curve and the lines.

23. $y = 4 - x^2; (2, 0)$

24. $y = \frac{1}{2}x^2 + \frac{3}{2}; (1, 2)$

25. $y = \frac{1}{3}x^3 + x; \left(1, \frac{4}{3}\right)$

26. $y = 2x^2 - x; (1, 1)$

27. Derive Formula (2.6).

2.6   Instantaneous Rates of Change

We observed in Section 2.4 that **the derivative is the instantaneous rate of change of one variable with respect to another.** So for the function $y = f(x)$, $f'(x)$ is the instantaneous rate of change of y with respect to x. For example, if y denotes distance and x time, then y' is the velocity. In this section we will consider several other cases where the instantaneous rate of change has an actual physical meaning.

For our first application we will return to the derivative as a *velocity* and examine *the motion of a particle along a line.*

Velocity

$$v = \frac{ds}{dt} \qquad \text{where } s \text{ denotes distance and } t \text{ denotes time}$$

Example 1   Suppose a particle moves along the s-axis in Figure 2.28 according to the equation

$$s = 2t^2 - 4t$$

(Assume s to be in meters and t in seconds.) Describe the motion.

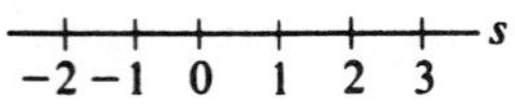

Figure 2.28

Solution. Since the velocity v is given by

$$v = \frac{ds}{dt}$$

(from Section 2.4), it follows that

$$v = 4t - 4$$

To see how the particle moves, let us construct the following "time table."

t(s)	s(m)	v(m/s)
−1	6	−8
0	0	−4
1	−2	0
2	0	4
3	6	8

To locate the particle, we can either look at the table (as if it were a train schedule) or calculate the position from the function s. The third column gives the velocity. If we interpret $t = 0$ as the present instant, then $t = -1$ means a second ago and $t = 2$ means two seconds from now. We see that the particle moves to the left, turns around, then moves to the right. As we will check below, a negative velocity indicates that it moves to the left. To see precisely where the particle turns around, we need to find the value of t for which $v = 0$; thus

$$v = 4t - 4 = 0$$

or $t = 1$. From the preceding table we see that this value of t corresponds to $s = -2$. The motion is indicated in Figure 2.29. (The path is drawn above the s-axis for easier visualization.)

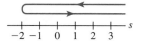

Figure 2.29

Another important concept in the study of motion is acceleration. We know from experience that if we step on the accelerator of a car, the car speeds up—that is, the velocity increases. This statement suggests defining the *acceleration* as the rate of change of velocity, or $a = dv/dt$.

Acceleration

$$a = \frac{dv}{dt} \qquad \text{where } v \text{ denotes velocity and } t \text{ denotes time}$$

In our example,

$$a = \frac{dv}{dt} = 4 \text{ m/s}^2$$

so that the velocity increases by 4 m/s every second. If this seems to conflict with the observation that the particle is slowing down near $t = 1$, recall that v is *negative* for $t < 1$, so v actually *increases* from −8 to −4 between $t = -1$ and $t = 0$. Now, the absolute value of the velocity is called the **speed**. So while the velocity increases from −8 to −4, the speed decreases from 8 to 4.   ∎

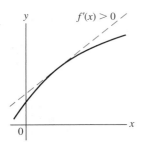

Figure 2.30

Returning now to the sign of the velocity of the particle described in Example 1, observe that whenever the derivative of $y = f(x)$ is positive on an interval, the slope of the tangent line is positive, and the graph rises as we move to the right. In other words, if x increases then y increases, and f is said to be an **increasing** function (Figure 2.30). Similarly, f is **decreasing** on an interval if f' is negative in the interval. So if v is positive, ds/dt is positive and s increases. Obviously, if s gets larger, then the particle must be moving to the right. Analogous statements apply to negative v.

Rates of change may arise in practically any situation. Consider the following examples.

Example 2   Suppose A is the area of a circle and r the radius; then $A = \pi r^2$. If the circle is allowed to expand, then the instantaneous rate of change of the area with respect to the radius is

$$\frac{dA}{dr} = 2\pi r \qquad \text{since } \pi \text{ is a numerical coefficient}$$

(Remember that the derivative is the instantaneous rate of change of one variable with respect to another.)

If $r = 2$ cm, then $dA/dr = 4\pi$ cm^2 per cm; or if $r = 4$ cm, $dA/dr = 8\pi$ cm^2 per cm, and so on.   ∎

Example 3   The voltage V (in volts) of a certain thermocouple is given by

$$V = 3.90T + 0.000480T^3$$

where T is measured in degrees Celsius. Find the instantaneous rate of change of V with respect to T at the instant when $T = 50.0°$C.

Solution. The instantaneous rate of change of V with respect to T is the derivative dV/dT:

$$\frac{dV}{dT} = 3.90(1) + 0.000480(3T^2)$$

At the instant when $T = 50.0°$C, we have

$$\frac{dV}{dT} = 3.90(1) + 0.000480(3)(50.0)^2 = 7.50\frac{\text{V}}{°\text{C}}$$ ∎

Rates of change also occur in the study of electrical circuits. The current, for example, is the instantaneous rate of change of the charge. Suppose we let i be the current (amperes), q the charge (coulombs), v the voltage (volts), L the inductance (henrys), and C the capacitance (farads). Then the following list includes some of these relationships:

1. **Current** in a circuit: $i = dq/dt$.
2. **Voltage** across an inductor: $v = L\,di/dt$.
3. **Charge** on a capacitor: $q = Cv$, so that $i = C\,dv/dt$, the current to a capacitor.

> **SI (International System of Units) Notations**
> A for ampere; C for coulomb; V for volt; F for farad; H for henry; and Ω for ohm. Other abbreviations to be used later are J for joule and N for newton. (For a list of SI units, see Table 1 in Appendix A.)

Example 4 For a short time interval, the current through a 0.040-H inductor is given by $i = 1.60t^2$. Find the voltage across the inductor at $t = 0.50$ s.

Solution. From $v = L\, di/dt$, we have

$$v = 0.040 \frac{d}{dt}(1.60t^2) = 0.040(1.60)(2t)$$

When $t = 0.50$ s, $v = 0.040(1.60)(2 \cdot 0.50) = 0.064$ V ∎

■ Exercises / Section 2.6

In Exercises 1–10, s is the position of a particle along the s-axis. Find an expression for the velocity and acceleration and determine when $v = 0$. (See Example 1.)

1. $s = 2t^2$
2. $s = 1 - 3t^2$
3. $s = t^2 - 2t + 1$
4. $s = 2t^2 - 8$
5. $s = 2t - t^2$
6. $s = 4t^2 + 4t - 3$
7. $s = 12t - 2t^2$
8. $s = t^3 - \frac{3}{2}t^2$
9. $s = 3t^3 + 2t^2 + 2$
10. $s = 8t^3 + 2t^2 - 3t$

11. Suppose that $s = 50t^2$ is the distance in meters a rocket has traveled from the launching pad in t seconds. Find its velocity after 10 s.

12. An object is dropped from a building 100 ft tall. The distance s in feet from the top t seconds later is given by $s = 16t^2$. What is its velocity when it hits the ground?

13. A light signal is shot upward on the moon with an initial velocity of 50 m/s. Its height y as a function of time is given by $y = 50t - 0.83t^2$. Find its velocity after 10 s.

14. A population of rats moves into a new industrial complex at time $t = 0$. At time t (in years) the population is $P(t) = 500(2 + 0.5t + 0.02t^2)$. Find the instantaneous rate of change of P at $t = 3$ years.

15. Find the instantaneous rate of change of the area of a square with respect to the length of the side at the instant when the side is 2 cm in length. (See Example 2.)

16. Find the instantaneous rate of change of the volume of a sphere with respect to the radius at the instant when the radius is 1.50 cm. (Recall that $V = (4/3)\pi r^3$.)

17. The resistance R (in ohms) in a certain wire as a function of temperature T (in degrees Celsius) is $R = 20.0 + 0.520T + 0.00973T^2$. Find the instantaneous rate of change of R with respect to T at the instant when $T = 125°C$.

18. Water is poured into a cylindrical tank of radius 3.0 ft. Find the instantaneous rate of change of the volume with respect to the depth.

19. The power P (in watts) delivered to a 15-Ω resistor is $P = 15i^2$. Find the instantaneous rate of change of P with respect to i when $i = 2.1$ A.

20. The energy radiated by a certain blackbody is given by $E = 0.0042T^4 - 16$, where E is measured in joules and T in degrees Celsius. Find the instantaneous rate of change of E with respect to T when $T = 8.0°C$.

21. Within a limited range, the tensile strength S (in pounds) of a piece of material as a function of temperature is $S = 520 - 0.000085T^2$. Find the instantaneous rate of change of S with respect to T at the instant when $T = 145°F$.

22. The rate of change of the cost $C(x)$ of producing a commodity with respect to the number x of units produced is called the *marginal cost*. If the cost of producing a certain commodity is $C(x) = x^2 - 4x + 5$, find an expression for the marginal cost.

23. For a rotating body, let θ, given in radians, be the amount of rotation from the positive x-axis. Then $\omega = d\theta/dt$ is called the *angular velocity* and $\alpha = d\omega/dt$ the *angular acceleration*. If $\theta = 3t^2$, find ω and α when $t = 1$ s.

24. A projectile shot directly upward with a velocity of 96 ft/s moves according to the equation $s = 96t - 16t^2$. Determine the velocity after 2.5 s.

25. Find the instantaneous current $i = dq/dt$ if $q = 10.0t^2 + 2.0t$ at $t = 0.01$ s.

26. Find the voltage across an inductor of 5.00×10^{-3} H if the current is given by $i = 3.00t$ A.

27. Find the current to a capacitor of 1.0×10^{-2} F after 1 s if the voltage is given by $v = 10t^2$ (for a short time interval).

28. Power is sometimes defined as the time rate of doing work. If W denotes the work done and P the power, then $P = dW/dt$. Suppose the work done in moving an object is $W = 10t^3 + 3t^4$ J. Find the power in joules per second at $t = 2$ s.

29. Refer to Exercise 28. If the work done (in joules) in moving an object is $W = 5t^4 + 2t$, find the power in joules per second when $t = 1$ s.

30. Consider a fluid moving through a cylindrical tube of radius a and length L (Figure 2.31). If r is the distance from the center, then the velocity v is given by the *law of laminar flow,*

$$v = \frac{p_0 - p_1}{4\eta L}(a^2 - r^2)$$

where η is the coefficient of viscosity and p_0 and p_1 are the respective pressures at the ends of the tube. Find an expression for the *velocity gradient,* which is the instantaneous rate of change of the velocity with respect to r.

31. In thermodynamics the isothermal compressibility β is defined to be

$$\beta = -\frac{1}{V}\frac{dV}{dP}$$

For a sample of a certain gas the volume as a function of pressure was found to be $V = 6.4/P^{1/2}$. Find an expression for β.

32. Suppose a tank holds 5000 gal of water, which can be drained from the bottom of the tank in 1 h. Then *Torricelli's Law* gives the volume V remaining after t minutes:

$$V = 5000\left(1 - \frac{t}{60}\right)^2, \qquad 0 \le t \le 60$$

Find the instantaneous rate of change of V with respect to t when $t = 30$ min.

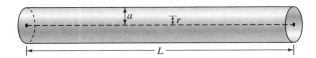

Figure 2.31

2.7   Differentiation Formulas

The Product and Quotient Rules

We learned in Section 2.5 that the derivative of a sum of two functions is the sum of their derivatives. In this section we will obtain the formulas for differentiating a product or quotient of two functions.

Suppose u and v are two differentiable functions of x. Then $y = uv$ can be differentiated and the derivative expressed in terms of derivatives of u and v. As in Section 2.4 we will use the delta notation:

Step 1. $y + \Delta y = (u + \Delta u)(v + \Delta v)$

(See Figure 2.32.)

Step 2. $\Delta y = (u + \Delta u)(v + \Delta v) - uv$

$\qquad = uv + u\Delta v + v\Delta u + \Delta u\Delta v - uv = u\Delta v + v\Delta u + \Delta u\Delta v$

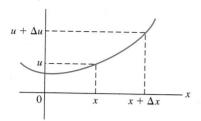

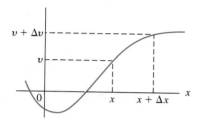

u and v are functions of x

Figure 2.32

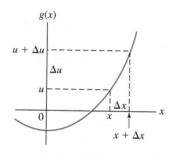

As $\Delta x \to 0$, $\Delta u \to 0$

Figure 2.33

Step 3. $\dfrac{\Delta y}{\Delta x} = u\dfrac{\Delta v}{\Delta x} + v\dfrac{\Delta u}{\Delta x} + \Delta u\dfrac{\Delta v}{\Delta x}$

Step 4. $\dfrac{dy}{dx} = \lim\limits_{\Delta x \to 0}\dfrac{\Delta y}{\Delta x} = u\lim\limits_{\Delta x \to 0}\dfrac{\Delta v}{\Delta x} + v\lim\limits_{\Delta x \to 0}\dfrac{\Delta u}{\Delta x} + 0$

$\qquad\quad = u\dfrac{dv}{dx} + v\dfrac{du}{dx}$

by Theorem A on limits.

To see why the last term in Step 3 vanishes, recall that if $u = g(x)$, then

$$\Delta u = g(x + \Delta x) - g(x)$$

Hence

$$\lim\limits_{\Delta x \to 0}\Delta u = \lim\limits_{\Delta x \to 0}[g(x + \Delta x) - g(x)]$$

$$= \lim\limits_{\Delta x \to 0}\left[\frac{g(x + \Delta x) - g(x)}{\Delta x}\right] \cdot \Delta x$$

$$= \lim\limits_{\Delta x \to 0}\frac{g(x + \Delta x) - g(x)}{\Delta x} \cdot \lim\limits_{\Delta x \to 0}\Delta x = g'(x) \cdot 0 = 0$$

because the limit of a product is equal to the product of the limits (Theorem B on limits). So $\lim_{\Delta x \to 0}\Delta u = 0$. (See also Figure 2.33.) It follows that

$$\lim\limits_{\Delta x \to 0}\Delta u\frac{\Delta v}{\Delta x} = 0 \cdot \frac{dv}{dx} = 0$$

(This argument actually proves that a differentiable function is continuous.)

We have shown that if u and v are differentiable, then $d(uv)/dx = u(dv/dx) + v(du/dx)$, which is called the *product rule*.

Product Rule

$$\frac{d(uv)}{dx} = u\frac{dv}{dx} + v\frac{du}{dx} \tag{2.7}$$

Using the product rule, we can obtain the *quotient rule*.

Quotient Rule

$$\frac{d}{dx}\left(\frac{u}{v}\right) = \frac{v\dfrac{du}{dx} - u\dfrac{dv}{dx}}{v^2} \tag{2.8}$$

To derive this rule, we let $w = u/v$. Then $u = vw$ and, by the product rule (2.7),

$$\frac{du}{dx} = v\frac{dw}{dx} + w\frac{dv}{dx}$$

$$v\frac{dw}{dx} = \frac{du}{dx} - w\frac{dv}{dx}$$

$$\frac{dw}{dx} = \frac{\dfrac{du}{dx} - w\dfrac{dv}{dx}}{v}$$

$$\frac{d}{dx}\left(\frac{u}{v}\right) = \frac{\dfrac{du}{dx} - \dfrac{u}{v}\dfrac{dv}{dx}}{v} \qquad \textbf{letting } \textbf{\textit{w}} = \textbf{\textit{u/v}}$$

$$\frac{d}{dx}\left(\frac{u}{v}\right) = \frac{v\dfrac{du}{dx} - u\dfrac{dv}{dx}}{v^2} \qquad \textbf{multiplying numerator and denominator by } \textbf{\textit{v}}$$

Example 1 Find the derivative of

$$y = (x^3 + 2x^2 - 3x)(x^2 - 4x)$$

Solution. We are certainly not forced to use the product rule here since we could multiply out the terms and obtain a polynomial of fifth degree, but the product rule makes the task easier.

Let $u = x^3 + 2x^2 - 3x$ and $v = x^2 - 4x$; then

$$\frac{dy}{dx} = u\frac{dv}{dx} + v\frac{du}{dx}$$

Since we already know how to differentiate polynomials, we get directly

$$\frac{dy}{dx} = (x^3 + 2x^2 - 3x)(2x - 4) + (x^2 - 4x)(3x^2 + 4x - 3)$$ ∎

Example 2 Differentiate

$$y = \frac{x^3 - 3x}{x + 1}$$

Solution. Since the function has the form of a quotient, we apply the quotient rule:

$$\frac{d}{dx}\left(\frac{u}{v}\right) = \frac{v\dfrac{du}{dx} - u\dfrac{dv}{dx}}{v^2} \qquad \textbf{\textit{u}} = \textbf{\textit{x}}^{\textbf{3}} - \textbf{3\textit{x}}, \textbf{\textit{v}} = \textbf{\textit{x}} + \textbf{1}$$

It follows that

$$\frac{dy}{dx} = \frac{(x+1)\dfrac{d}{dx}(x^3 - 3x) - (x^3 - 3x)\dfrac{d}{dx}(x+1)}{(x+1)^2}$$

$$= \frac{(x+1)(3x^2 - 3) - (x^3 - 3x)\cdot 1}{(x+1)^2} = \frac{2x^3 + 3x^2 - 3}{(x+1)^2}$$

∎

The Generalized Power Rule

Next we will obtain the formula for differentiating the power of a function, thereby extending Formula (2.4), the derivative of $y = x^n$.

Let u be a differentiable function of x, and consider the function $y = u^n$, where n is a positive integer.

Step 1. $y + \Delta y = (u + \Delta u)^n$

Step 2. $\Delta y = (u + \Delta u)^n - u^n$

$$= u^n + nu^{n-1}\Delta u + \frac{n(n-1)}{2!}u^{n-2}(\Delta u)^2 + \cdots + (\Delta u)^n - u^n$$

Step 3. $\dfrac{\Delta y}{\Delta x} = nu^{n-1}\dfrac{\Delta u}{\Delta x} + \dfrac{n(n-1)}{2!}u^{n-2}\dfrac{\Delta u}{\Delta x}\Delta u + \cdots + \dfrac{\Delta u}{\Delta x}(\Delta u)^{n-1}$

As noted earlier, $\lim_{\Delta x \to 0}\Delta u = 0$; thus

Step 4. $\dfrac{dy}{dx} = \lim_{\Delta x \to 0}\dfrac{\Delta y}{\Delta x} = nu^{n-1}\dfrac{du}{dx}$

If $u = x$, we get $dx^n/dx = nx^{n-1}(dx/dx) = nx^{n-1}$, which is Formula (2.4).

If n is a negative integer, we let $n = -m(m > 0)$ and apply the quotient rule. Then we get

$$y = u^n = u^{-m} = \frac{1}{u^m}$$

and

$$\frac{dy}{dx} = \frac{u^m\dfrac{d}{dx}(1) - \dfrac{du^m}{dx}}{(u^m)^2} = \frac{-mu^{m-1}\dfrac{du}{dx}}{u^{2m}}$$

$$= -mu^{m-1-2m}\frac{du}{dx} = -mu^{-m-1}\frac{du}{dx}$$

Substituting n for $-m$, we have

$$\frac{dy}{dx} = nu^{n-1}\frac{du}{dx}$$

The case where n is a rational number will be left as an exercise in the next section. (Recall that a number is rational if it has the form p/q, where p and q are integers.) The formula is known as the *generalized power rule*.

> **Generalized Power Rule**
>
> $$\frac{du^n}{dx} = nu^{n-1}\frac{du}{dx} \qquad \text{(for any rational } n)$$ (2.9)

Example 3   Differentiate

$$f(x) = \sqrt{x^2 + 1}$$

Solution. Since $f(x) = (x^2 + 1)^{1/2}$, the generalized power rule applies. Thus

$$f'(x) = \frac{1}{2}(x^2 + 1)^{-1/2}\frac{d}{dx}(x^2 + 1) \qquad u = x^2 + 1, n = \frac{1}{2}$$

$$= \frac{1}{2}(x^2 + 1)^{-1/2}(2x) = \frac{x}{\sqrt{x^2 + 1}}$$

Don't forget to multiply by the derivative of $x^2 + 1$. (Even though it is customary to simplify algebraic expressions, the last step is not part of the differentiation procedure.) ∎

The generalized power rule turns out to be a special case of the *chain rule,* which involves composite functions (Section 2.1). If f and g are differentiable functions, and $y = f(g(x))$, then $y' = f'(g(x)) \cdot g'(x)$. To see why, let us first obtain a more convenient form of the chain rule by letting $u = g(x)$. Then

$$f'(g(x)) = f'(u) = \frac{dy}{du} \qquad \text{and} \qquad g'(x) = \frac{du}{dx}$$

So

$$\frac{dy}{dx} = \frac{dy}{du}\frac{du}{dx}$$

> **Chain Rule**
>
> $$\frac{dy}{dx} = \frac{dy}{du}\frac{du}{dx}$$ (2.10)

We obtain the chain rule as follows:

$$\frac{dy}{dx} = \lim_{\Delta x \to 0}\frac{\Delta y}{\Delta x} = \lim_{\Delta x \to 0}\frac{\Delta y}{\Delta u}\frac{\Delta u}{\Delta x} = \lim_{\Delta x \to 0}\frac{\Delta y}{\Delta u} \cdot \lim_{\Delta x \to 0}\frac{\Delta u}{\Delta x}$$

Since $\Delta u \to 0$ as $\Delta x \to 0$ whenever $u = g(x)$ is continuous,

$$\lim_{\Delta x \to 0}\frac{\Delta y}{\Delta u} = \lim_{\Delta u \to 0}\frac{\Delta y}{\Delta u} = \frac{dy}{du}$$

and Formula (2.10) follows. The chain rule will be needed in Chapter 3. (We have assumed that $\Delta u \neq 0$ if $\Delta x \neq 0$. However, it can be shown that the chain rule holds for this case also.)

Example 4   Differentiate $y = u^n$ by the chain rule.

Solution.

$$\frac{dy}{dx} = \frac{du^n}{du}\frac{du}{dx} = nu^{n-1}\frac{du}{dx}$$

which is the generalized power rule.   ∎

The remaining examples illustrate the various rules in this section.

Example 5   If $y = x\sqrt{x^2 - 2x}$, find y'.

Solution. We may treat the function as a product and use the product rule with $u = x$ and $v = (x^2 - 2x)^{1/2}$, so that

$$y' = x\frac{d}{dx}(x^2 - 2x)^{1/2} + (x^2 - 2x)^{1/2}\frac{dx}{dx} \qquad \textbf{product rule}$$

$$= x \cdot \frac{1}{2}(x^2 - 2x)^{-1/2}(2x - 2) + (x^2 - 2x)^{1/2} \qquad \textbf{generalized power rule}$$

$$= x(x^2 - 2x)^{-1/2}(x - 1) + (x^2 - 2x)^{1/2} \qquad \tfrac{1}{2}\textbf{ (2x − 2) = x − 1}$$

$$= \frac{x(x - 1)}{(x^2 - 2x)^{1/2}} + \frac{(x^2 - 2x)^{1/2}}{1}$$

Adding these fractions, we get

$$y' = \frac{x(x - 1)}{(x^2 - 2x)^{1/2}} + \frac{(x^2 - 2x)^{1/2}}{1}\frac{(x^2 - 2x)^{1/2}}{(x^2 - 2x)^{1/2}}$$

$$= \frac{x(x - 1)}{(x^2 - 2x)^{1/2}} + \frac{(x^2 - 2x)^{1}}{(x^2 - 2x)^{1/2}} \qquad \textbf{a}^{1/2} \cdot \textbf{a}^{1/2} = \textbf{a}^{1}$$

$$= \frac{x^2 - x + x^2 - 2x}{(x^2 - 2x)^{1/2}} = \frac{2x^2 - 3x}{\sqrt{x^2 - 2x}} \qquad\qquad ∎$$

If you have access to a symbolic differentiation utility, you may want to use it to compare answers. Please note that the forms may vary, especially with functions containing radicals.

Example 6   Find the derivative of

$$y = \frac{x}{\sqrt{x^2 + 1}}$$

Solution. We write

$$y = \frac{x}{(x^2 + 1)^{1/2}}$$

and apply the quotient rule:

$$\frac{dy}{dx} = \frac{(x^2 + 1)^{1/2}\dfrac{dx}{dx} - x\dfrac{d}{dx}(x^2 + 1)^{1/2}}{[(x^2 + 1)^{1/2}]^2}$$

Then, by the generalized power rule,

$$\frac{dy}{dx} = \frac{(x^2+1)^{1/2} - x \cdot \frac{1}{2}(x^2+1)^{-1/2}(2x)}{x^2+1} = \frac{(x^2+1)^{1/2} - \frac{x^2}{(x^2+1)^{1/2}}}{x^2+1}$$

This expression can be simplified by multiplying numerator and denominator by $(x^2+1)^{1/2}$; note that $(x^2+1)^{1/2}(x^2+1)^{1/2} = x^2+1$. Thus

$$\frac{dy}{dx} = \frac{(x^2+1)^{1/2}(x^2+1)^{1/2} - \frac{x^2}{(x^2+1)^{1/2}}(x^2+1)^{1/2}}{(x^2+1)(x^2+1)^{1/2}}$$

$$= \frac{(x^2+1) - x^2}{(x^2+1)(x^2+1)^{1/2}} = \frac{1}{(x^2+1)^{3/2}} \qquad\blacksquare$$

Suggestion: To differentiate a function with a constant numerator

$$y = \frac{k}{g(x)} \qquad (k \text{ a constant})$$

write

$$y = k[g(x)]^{-1}$$

and use the generalized power rule (instead of the quotient rule) as shown in Example 7.

Example 7   Differentiate

$$y = \frac{1}{\sqrt[3]{x^3+x}}$$

Solution. We could use the quotient rule, as in the previous example, but it is much simpler to write the function in the form

$$y = (x^3+x)^{-1/3}$$

and apply the generalized power rule. We obtain at once

$$\frac{dy}{dx} = -\frac{1}{3}(x^3+x)^{-4/3}(3x^2+1) = -\frac{3x^2+1}{3(x^3+x)^{4/3}} \qquad \begin{matrix} u = x^3+x \\ n = -\frac{1}{3} \end{matrix} \qquad \blacksquare$$

■ Exercises / Section 2.7

GROUP A   Differentiate the given functions (Exercises 1–40).

1. $y = 4x^4 - 4x^2 + 8$

2. $y = 4x^{-4} - 4x^{-2} + 8$

3. $y = \dfrac{1}{x}$ (or $y = x^{-1}$)

4. $y = \dfrac{1}{\sqrt[3]{x}}$ (or $y = x^{-1/3}$)

5. $y = x^5 - 3x^{-3} + 2x^{-2}$

6. $y = x\sqrt{x}$ (or $y = x^{3/2}$)

7. $y = x^2\sqrt{x} + \dfrac{1}{\sqrt{x}}$ (or $y = x^{5/2} + x^{-1/2}$)

8. $y = 3x^{-2} - 2x^{-3} + x + \sqrt{2}$

9. $y = \dfrac{2}{x^4} - \dfrac{4}{\sqrt{x}} + \dfrac{1}{\sqrt{2}}$

10. $y = \dfrac{2}{x^2} - \dfrac{1}{x^3} + \dfrac{1}{2^3}$

11. $y = (2x^2 - 3)^4$

12. $y = (x^2 - 3x + 2)^3$

13. $y = (4 - 3x^2)^4$

14. $y = (4 + x^4)^5$

15. $y = (x^{10} + 1)^{10}$

16. $y = \left(1 - \dfrac{1}{3}x^3\right)^4$

17. $y = \dfrac{2}{\sqrt{x^3 - 3x}}$ (See Example 7.)

18. $y = \dfrac{1}{\sqrt{1 - x^2}}$

19. $v = \sqrt[3]{t^3 - 3}$

20. $z = \sqrt[4]{v + 2}$

21. $y = x^3(x + 1)^2$

22. $y = x(x - 1)^4$

23. $y = 2x^4(x + 2)^2$

24. $y = 3x^3(x + 3)^4$

25. $y = x^2(x^2 - 5)^2$

26. $y = 3x^3(x - 1)^3$

27. $y = 4x(x + 5)^4$

28. $y = x(x^3 + 2)^3$

29. $y = 2x^3(x - 2)^3$

30. $y = 5x^2(x - 1)^4$

31. $y = \dfrac{x}{x - 1}$

32. $y = \dfrac{4 + x^2}{4 - x^2}$

33. $P = \dfrac{t - 2}{t^2 + 4}$

34. $V = \dfrac{s^2}{s^2 - 4}$

35. $y = x^2\sqrt{x + 1}$

36. $y = x^2\sqrt{x^2 - 1}$

37. $y = x\sqrt{x^2 + 2}$

38. $y = 4x\sqrt{x + 4}$

39. $R = \dfrac{s^2 - 3}{s - 2}$

40. $V = \dfrac{q^2 + q}{q^2 - 4}$

41. Use a symbolic differentiation utility to check the answers to Exercises 29–35.

GROUP B   Differentiate the given functions (Exercises 1–28).

1. $y = \sqrt{1 - x}$

2. $y = \sqrt{x^2 + 2}$

3. $y = \sqrt[3]{1 - x^2}$

4. $y = \sqrt[3]{x^3 - 4}$

5. $y = \dfrac{4}{\sqrt[4]{x - x^2}}$

6. $y = -\dfrac{2}{3\sqrt[4]{x^3 - 6}}$

7. $n = \dfrac{m^3 + 2m}{m^2 - 8}$

8. $Q = \dfrac{2r^2 - r}{r^3 - 4}$

9. $y = x\sqrt{x^2 - 1}$

10. $y = x\sqrt{x^3 + 2}$

11. $y = 2x\sqrt{1 - x}$

12. $y = x\sqrt{x + 1}$

13. $T = \theta^3\sqrt{\theta + 7}$

14. $L = R^2\sqrt{2 - R}$

15. $y = \dfrac{\sqrt{x}}{x - 4}$

16. $y = \dfrac{\sqrt{x}}{1 - x}$

17. $y = \dfrac{x^2}{\sqrt{x + 1}}$

18. $y = \dfrac{\sqrt{x + 1}}{x^2}$

19. $y = \dfrac{\sqrt{x^2 - 1}}{x^2}$

20. $y = \dfrac{x\sqrt{x - 1}}{x + 2}$

21. $y = \dfrac{x^2\sqrt{x}}{x^2 + 3}$

22. $y = \dfrac{\sqrt{x}}{1 + \sqrt{x}}$

23. $y = (x - 1)\sqrt{x - 2}$

24. $y = (x - 2)\sqrt{x^2 + 1}$

25. $y = \dfrac{x\sqrt{x - 1}}{2x + 3}$

26. $y = \dfrac{x^2\sqrt{x + 3}}{x - 1}$

27. $y = \dfrac{x^2\sqrt{x - 5}}{\sqrt{x + 3}}$

28. $y = \dfrac{x\sqrt[3]{x^2 + 1}}{x - 8}$

29. Find the slope of the line tangent to the curve $y = (x^3 - 1)(x^2 + 3x + 2)$ at the point $(1, 0)$.

30. Find the point at which the slope of the line tangent to $y = 1/\sqrt{3x^2 + 3}$ is equal to zero.

31. If R ohms resistance is in series with X ohms reactance, then the impedance $Z = \sqrt{R^2 + X^2}$ ohms. If $R = 4\,\Omega$, find the instantaneous rate of change of Z with respect to X.

32. Find the current $i = dq/dt$ in a circuit at $t = 8$ s if the charge $q = 3.3t^{4/3}$ C.

33. Starting at $t = 0$, the charge on a certain capacitor is given by $q(t) = t/(t^2 + 4)$. At what instant is the current to the capacitor equal to zero?

2.8   Implicit Differentiation

So far all our functions have been expressed in the **explicit** form $y = f(x)$. In many relationships y is not expressed explicitly as a function of x, but only in **implicit** form. In this section we will develop a technique for differentiating a function in implicit form, leading to the **implicit derivative.**

To see what all these terms mean, consider the circle $x^2 + y^2 = 25$. On the interval $(-5, 5)$ we get two y values for every x value, so the relation is not a function. If we solve the equation for y in terms of x, we get

$$y = \pm\sqrt{25 - x^2}$$

which can be written as two equations,

$$y = +\sqrt{25 - x^2} \quad \text{and} \quad y = -\sqrt{25 - x^2}$$

Each separate equation is now a function; in fact, $y = +\sqrt{25 - x^2}$ is the upper semi-circle and $y = -\sqrt{25 - x^2}$ is the lower semicircle in Figure 2.34.

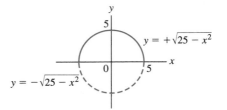

Figure 2.34

If we want to find the slope of the tangent at $(4, -3)$, for example, we take the lower branch $y = -\sqrt{25 - x^2}$ and find dy/dx. The result is

$$\frac{dy}{dx} = -\frac{1}{2}(25 - x^2)^{-1/2}(-2x) = \frac{x}{\sqrt{25 - x^2}}$$

and at $x = 4$, we find that $dy/dx = 4/3$.

The conclusion is that even though the equation $x^2 + y^2 = 25$ is not a function, it actually contains two functions, and we say that y is an **implicit function** of x. With this understanding, it becomes totally unnecessary to solve for y in terms of x—we just use the generalized power rule. For example, while $dx^2/dx = 2x$, as usual, $dy^2/dx = 2y(dy/dx)$, since y *is a function of* x. Many students find this step troublesome, but all we need to do is keep in mind that the u in Formula (2.9) has been replaced by y:

Generalized Power Rule

$$\frac{dy^n}{dx} = ny^{n-1}\frac{dy}{dx} \qquad\qquad (2.11)$$

Returning to the equation $x^2 + y^2 = 25$, we now differentiate both sides with respect to x to obtain

$$2x + 2y\frac{dy}{dx} = 0 \qquad \frac{d}{dx}x^2 = 2x \qquad \frac{d}{dx}y^2 = 2y\frac{dy}{dx}$$

whence

$$\frac{dy}{dx} = -\frac{x}{y} \qquad \text{solving for } \frac{dy}{dx}$$

Implicit derivative

which is called the **implicit derivative.** At the point $(4, -3)$,

$$\frac{dy}{dx} = -\frac{4}{-3} = \frac{4}{3}$$

as before. This result should not be surprising, for if we substitute $-\sqrt{25-x^2}$ in the expression for the implicit derivative, then

$$\frac{dy}{dx} = -\frac{x}{y} = -\frac{x}{-\sqrt{25-x^2}} = \frac{x}{\sqrt{25-x^2}}$$

That is, dy/dx collapses to the ordinary derivative.

So far we seem to have been doing the same thing twice, but the real reason for considering implicit derivatives is that some equations are very difficult, or even impossible, to solve for y in terms of x.

Example 1   Find dy/dx for each of the following equations:

   a. $3y^3 = x$ **b.** $x^2y^2 = 1$ **c.** $x^2y = 4$

Solution.

   a. By the power rule we get

$$3 \cdot 3y^2 \frac{dy}{dx} = 1 \qquad \text{since } \frac{d}{dx}y^3 = 3y^2\frac{dy}{dx}$$

so that

$$\frac{dy}{dx} = \frac{1}{9y^2} \qquad \text{solving for } \frac{dy}{dx}$$

   b. Using the product and power rules, we have

$$x^2\frac{d(y^2)}{dx} + y^2\frac{d(x^2)}{dx} = \frac{d}{dx}(1)$$

or

$$x^2 \cdot 2y\frac{dy}{dx} + y^2 \cdot 2x = 0 \qquad \text{and} \qquad 2x^2y\frac{dy}{dx} + 2xy^2 = 0$$

It follows that

$$2x^2y\frac{dy}{dx} = -2xy^2$$

and

$$\frac{dy}{dx} = -\frac{2xy^2}{2x^2y} = -\frac{y}{x}$$

   c. Using the product rule to find the implicit derivative of $x^2y = 4$, we get

$$x^2\frac{d}{dx}(y) + y\frac{d}{dx}(x^2) = \frac{d}{dx}(4)$$

and

$$x^2\frac{dy}{dx} + 2xy = 0 \qquad \frac{d}{dx}(y) = \frac{dy}{dx}$$

whence

$$\frac{dy}{dx} = -\frac{2xy}{x^2} = -\frac{2y}{x}$$

∎

Example 2   Find dy/dx implicitly:

$$3x^2 + 4x^3y^4 + 2y = 4$$

Solution. Differentiating each term, we get

$$\frac{d}{dx}(3x^2) + \frac{d}{dx}(4x^3y^4) + \frac{d}{dx}(2y) = \frac{d}{dx}(4)$$

Now use the product rule on the second term:

$$6x + 4x^3\frac{d}{dx}(y^4) + y^4\frac{d}{dx}(4x^3) + 2\frac{d}{dx}(y) = 0$$

$$6x + 4x^3 \cdot 4y^3\frac{dy}{dx} + y^4(12x^2) + 2\frac{dy}{dx} = 0$$

Now collect all terms containing dy/dx on the left side:

$$16x^3y^3\frac{dy}{dx} + 2\frac{dy}{dx} = -6x - 12x^2y^4$$

$$(16x^3y^3 + 2)\frac{dy}{dx} = -6x - 12x^2y^4 \qquad \textbf{factoring } \frac{dy}{dx}$$

$$\frac{dy}{dx} = \frac{-6x - 12x^2y^4}{16x^3y^3 + 2} \qquad \textbf{dividing by } (16x^3y^3 + 2)$$

$$\frac{dy}{dx} = -\frac{3x + 6x^2y^4}{8x^3y^3 + 1}$$

∎

Example 3   Find the slope of the line tangent to the graph of the equation

$$x^2 - 3xy + y^2 + 4x - 2y = 1$$

at the point $(1, 4)$.

Solution. Differentiating both sides, we obtain

$$2x - 3\left(x\frac{dy}{dx} + y\frac{dx}{dx}\right) + 2y\frac{dy}{dx} + 4 - 2\frac{dy}{dx} = 0$$

or

$$2x - 3x\frac{dy}{dx} - 3y + 2y\frac{dy}{dx} + 4 - 2\frac{dy}{dx} = 0$$

To solve this equation for dy/dx, we keep all the terms containing dy/dx on the left side and transpose the rest:

$$-3x\frac{dy}{dx} + 2y\frac{dy}{dx} - 2\frac{dy}{dx} = -2x + 3y - 4$$

or

$$(-3x + 2y - 2)\,\frac{dy}{dx} = -2x + 3y - 4$$

and

$$\frac{dy}{dx} = \frac{-2x + 3y - 4}{-3x + 2y - 2} = \frac{2x - 3y + 4}{3x - 2y + 2}$$

Finally, at the point $(1, 4)$,

$$\frac{dy}{dx} = 2$$ ∎

Example 4   Find the equations of the tangent and normal lines to the parabola $y^2 = 3x$ at the point $(3, -3)$.

Solution. We differentiate implicitly:

$$2y\frac{dy}{dx} = 3 \qquad \frac{dy}{dx} = \frac{3}{2y}$$

For the slope of the tangent line, we then get

$$\left.\frac{dy}{dx}\right|_{(3,-3)} = \frac{3}{2(-3)} = \frac{3}{-6} = -\frac{1}{2}$$

and for the slope of the normal line, we get 2, since $-1/2$ and 2 are negative reciprocals. The equations of the tangent and normal lines are

$$y + 3 = -\frac{1}{2}(x - 3) \qquad \text{and} \qquad y + 3 = 2(x - 3)$$

respectively.

Let's use a graphing utility to graph the curve and both lines. (See Figure 2.35.)

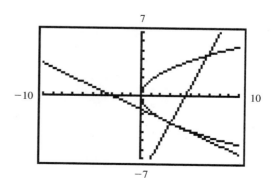

Figure 2.35   ∎

■ Exercises / Section 2.8

In Exercises 1–26, find dy/dx implicitly.

1. $2x + 3y = 3$ **2.** $4x - 5y = 3$

3. $x^2 - y^2 = 2$ **4.** $3x^2 + 5y^2 = x$

5. $2x^2 - 3y^2 = 1$ **6.** $y^2 - x^2 - 2x = 0$

7. $2y^3 + x^2 + 1 = 0$ **8.** $7y^4 - 3x^2 = 3$

9. $4y^4 - 3x^3 + 2 = 0$ **10.** $3y^3 - 7x^3 + 2x = 0$

11. $x - 5x^2 - 6y^3 = 0$ **12.** $x^2 + y^2 = r^2$

13. $\dfrac{x^2}{a^2} + \dfrac{y^2}{b^2} = 1$ **14.** $y^2 = 4px$

15. $xy = 3$ **16.** $xy + x = 4$

17. $x^2 y = 7$ **18.** $x^2 y = x + 1$

19. $x^2 + x^2 y^2 + x = 0$ **20.** $2x^3 + xy^2 + x = 4$

21. $x^3 - 4x^2 y^2 + y^2 = 1$ **22.** $2x^2 + 3x^2 y = y^3$

23. $5x^2 y^3 - y^4 = 2x^3$ **24.** $2y^4 - x^2 y^3 + x^6 = 3$

25. $x^4 y^4 - 3y^2 + 5x = 6$

26. $3x^4 + 3xy^3 + y - 3x + 6 = 0$

In Exercises 27–33, find the equations of the tangent and normal lines to each curve at the given point. Use a graphing utility to graph the curve and the lines.

27. $x^2 + 4y^2 = 5;\ (1, -1)$ **28.** $y^2 = -12x;\ (-3, 6)$

29. $y^2 = -4x;\ (-1, -2)$ **30.** $3x^2 + 2y^2 = 14;\ (2, -1)$

31. $x^2 - 2y^2 = 2;\ (-2, 1)$ **32.** $y^2 - x^2 = 5;\ (2, -3)$

33. $2x^2 + y^2 = 17;\ (-2, -3)$

34. Derive the power rule (Equation 2.9) for rational n. (Let $y = u^{p/q}$, where p and q are integers, so that $y^q = u^p$. Now differentiate implicitly and solve for dy/dx.)

2.9 Higher Derivatives

We saw in the earlier sections that the derivative of a function is also a function, as the notation $f'(x)$ suggests. It is sometimes necessary to differentiate a derivative. For example, the acceleration is defined to be the derivative of the velocity, which is itself a derivative, and so acceleration is called the **second derivative.** The process can be continued indefinitely, as long as the resulting functions are differentiable. The derivative of the second derivative is called the **third derivative,** and so forth. Collectively, these are known as **higher derivatives.**

Notations

If $y = f(x)$, then

$$\frac{d}{dx}\left(\frac{dy}{dx}\right) = \frac{d^2 y}{dx^2}$$

and

$$\frac{d}{dx}\left(\frac{d^2 y}{dx^2}\right) = \frac{d^3 y}{dx^3}$$

nth derivative

In general, $d^n y/dx^n$ is the notation for the **nth derivative.**
Alternately, we use the notations

$$f'(x),\ f''(x),\ f'''(x),\ f^{(4)}(x),\ \ldots,\ f^{(n)}(x)$$

or

$$y',\ y'',\ y''',\ y^{(4)},\ \ldots,\ y^{(n)}$$

For example, if $y = 3x^3 - 2x + 5$, then

$$y' = 9x^2 - 2 \qquad y'' = 18x \qquad y''' = 18 \qquad y^{(4)} = 0$$

Also, if $f(x) = 3/x$, then

$$f'(x) = -\frac{3}{x^2} \qquad f''(x) = \frac{6}{x^3}$$

The main advantage of this notation is that we can write $f'(1) = -3$, $f''(1) = 6$, and so forth.

It is possible to find higher derivatives implicitly. Here it is best to use the prime notation y', y'', etc. As an illustration, since yy' is a product, the product rule yields

$$\frac{d}{dx}(yy') = yy'' + y'y' = yy'' + (y')^2$$

Example 1   Find y'' for the implicit function $xy - y^2 = 2$.

Solution.   By the product and power rules

$$xy' + y - 2yy' = 0 \qquad\qquad \frac{d}{dx}(xy) = xy' + y \cdot 1$$
$$y'(x - 2y) = -y$$
$$y' = \frac{y}{2y - x} \qquad\qquad \frac{d}{dx}y^2 = 2y\frac{dy}{dx} = 2yy'$$

Differentiating again,

$$y'' = \frac{(2y - x)y' - y(2y' - 1)}{(2y - x)^2} \qquad\qquad \textbf{quotient rule}$$

Substituting for y', we get

$$y'' = \frac{(2y - x)\dfrac{y}{2y - x} - 2y\dfrac{y}{2y - x} + y}{(2y - x)^2}$$

$$= \frac{y - \dfrac{2y^2}{2y - x} + y}{(2y - x)^2} \cdot \frac{2y - x}{2y - x}$$

$$= \frac{y(2y - x) - 2y^2 + y(2y - x)}{(2y - x)^3}$$

$$= \frac{2y^2 - xy - 2y^2 + 2y^2 - xy}{(2y - x)^3}$$

$$= \frac{2y^2 - 2xy}{(2y - x)^3} = -\frac{2(xy - y^2)}{(2y - x)^3}$$

$$= -\frac{2(2)}{(2y - x)^3} = -\frac{4}{(2y - x)^3} \qquad\qquad \textbf{given equation: } xy - y^2 = 2 \qquad \blacksquare$$

■ Exercises / Section 2.9

In Exercises 1–9, find the indicated higher derivatives of the given functions.

1. $y = 5x^4 + 5x^3 - 3x + 1$; find y''.

2. $f(x) = \dfrac{1}{\sqrt{x}}$; find $f''(x)$.

3. $y = \sqrt{x - 1}$; find y''.

4. $f(x) = (x^3 - 2x)^2$; find $f''(x)$.

5. $y = x^6 - 2x^5 - x^4$; find $\dfrac{d^3y}{dx^3}$.

6. $f(x) = \dfrac{x}{x+1}$; find $f^{(4)}(x)$.

7. $f(x) = \sqrt{5 + x}$; find $f'''(x)$.

8. $y = \sqrt{x^2 - 1}$; find $\dfrac{d^2y}{dx^2}$.

9. $y = \dfrac{3 + 2x}{3 - 2x}$; find $\dfrac{d^2y}{dx^2}$.

10. a. Find $y^{(4)}$ if $y = x^4$.
   b. Show that $d^9x^9/dx^9 = 9!$, where
   $9! = 9 \cdot 8 \cdot 7 \cdot 6 \cdot 5 \cdot 4 \cdot 3 \cdot 2 \cdot 1$.

In Exercises 11–18, find y'' implicitly.

11. $x^2 + y^2 = 4$
12. $x^2 + 2y^2 = 5$
13. $x^2 - y^2 = 4$
14. $2x^2 + 3y^2 = 9$
15. $y^2 = 4px$
16. $y^2 = 2x + 1$
17. $\sqrt{x} + \sqrt{y} = 1$
18. $y^2 - xy = 6$

■ Review Exercises / Chapter 2*

1. If $f(x) = x^2 - 1$, find $f(0), f(1), f(\sqrt{2})$.

2. If $f(x) = \sqrt[3]{x}$ and $g(x) = x^2 + 2$, find $f(g(x))$ and $g(f(x))$.

3. If $f(x) = \begin{cases} 0 & 0 \le x < 1 \\ 2 & x > 1 \end{cases}$, find $f(0), f\left(\dfrac{1}{2}\right), f\left(\dfrac{5}{2}\right), f(1)$.

4. State the domain and range of the function in Exercise 3.

5. Find the domain and range of the following functions:
   a. $y = \sqrt{x - 1}$ **b.** $y = \sqrt[3]{x - 1}$

In Exercises 6–18, find the limits indicated.

6. $\lim\limits_{x \to 9} \dfrac{x^2 - 81}{x - 9}$

7. $\lim\limits_{x \to 4} \dfrac{16 - x^2}{4 - x}$

8. $\lim\limits_{x \to 3} \dfrac{15 - 2x - x^2}{3 - x}$

9. $\lim\limits_{x \to 0} \dfrac{x^3 - x^2 + 3x}{x}$

10. $\lim\limits_{x \to 0} \dfrac{x^2 + 2x}{x}$

11. a. $\lim\limits_{x \to 2}(1 - x^2)$ **b.** $\lim\limits_{x \to 1} \dfrac{x^2 - 5x + 4}{x - 1}$

12. $\lim\limits_{x \to 3} \dfrac{2x^2 - 5x - 3}{x - 3}$

13. $\lim\limits_{x \to 1} \dfrac{\sqrt{x} - 1}{x - 1}$

14. $\lim\limits_{x \to 0} \dfrac{x^2 - 2}{x - 1}$

15. $\lim\limits_{x \to \infty} \dfrac{2x^2 - 3x + 2}{x^2 - 10x + 1}$

16. $\lim\limits_{x \to \infty} \dfrac{x - 1}{x^2 + x + 2}$

17. $\lim\limits_{x \to 4+} \sqrt{x - 4}$

18. If $f(x) = \begin{cases} 1 & x \le 1 \\ 2 & x > 1 \end{cases}$, find: **a.** $\lim\limits_{x \to 1+} f(x)$ **b.** $\lim\limits_{x \to 1-} f(x)$

In Exercises 19–22, use the four-step process to differentiate the given functions.

19. $y = x - 3x^2$
20. a. $y = x^3$ **b.** $y = \dfrac{2}{x}$

21. a. $y = \dfrac{1}{4 - x}$ **b.** $y = \sqrt{x}$

22. $y = \sqrt{3 - x}$

In Exercises 23–29, differentiate.

23. $y = (x^3 - 2)^4$
24. $y = \dfrac{1}{x^4 + 3}$

25. $y = \dfrac{x - 4}{x + 1}$
26. $y = \dfrac{1}{\sqrt{7 - x^5}}$

27. $y = \dfrac{x^2}{\sqrt{4 - x^2}}$
28. $y = (x^2 + 1)^2(x - 3)$

29. $y = x\sqrt{4 - x^2}$

In Exercises 30–32, find the implicit derivatives.

30. $y^2 + x^2 + 3x = 1$
31. $x^2y + xy^2 + y^3 = 1$
32. $2x^2y^2 - 4xy = x$
33. Find y'': $y^3 - xy = 4$.

34. Find the slope of the line tangent to the hyperbola $3x^2 - y^2 = 23$ at $(3, -2)$.

*All the answers to the review exercises are given in the answer section (Appendix B).

35. Find the slope of the line tangent to the curve $8x^2 + 4xy + 5y^2 + 28x - 2y + 20 = 0$ at $(-1, 6/5)$.

36. Find $f'''(x)$ if $f(x) = \sqrt{x + 3}$.

37. **a.** Find the value of x for which the derivative of $f(x) = x/\sqrt{x - 1}$ is equal to zero.

   b. If $f(x) = 2x^3 - 6x^2 + 4$, find x such that $f''(x) = 0$.

38. Find the value of x for which the slope of the line tangent to the curve $y = x/\sqrt{4 + x^2}$ is $1/2$.

39. The resistance of a certain wire is given by $R = k/r^2$, where k is a constant and r the radius of the wire. Find the expression for the rate of change of R with respect to r.

40. Two curves are perpendicular at their point of intersection if their tangent lines are perpendicular at that point. Show that $y = (1/3)x^{-3}$ and $y = (1/3)x^3$ are perpendicular at $(1, 1/3)$.

41. Recall that the voltage V (volts) across a resistor in a circuit is given by $V = IR$. If $R = 0.010t^2 \ \Omega$ and $I = 4.12 + 0.020t$ A, find the rate of change of the voltage with respect to time when $t = 2.5$ s.

42. The specific heat of a certain gas as a function of temperature is $c = 12.5 + 0.65T + 0.000019T^2$. Find an expression for the instantaneous rate of change of c with respect to T.

43. It costs the Ace Widget Company

$$C(y) = 50 + \frac{1}{2}y - \frac{1}{500}y^2$$

dollars to produce y widgets. What is the marginal cost of producing 50 widgets? (See Exercise 22, Section 2.6.)

44. An object is moving vertically according to the equation $s = 100t - t^2$, where t is the time in seconds and s the distance in feet above the ground. Find the value of t for which the velocity is zero.

45. Two unlike charges of 1 electrostatic unit each exert an attractive force of $F = 1/r^2$ dynes on each other at a distance of r cm. What is the instantaneous rate of change of the force with respect to distance at a distance of 5.02 cm?

3

Applications of the Derivative

In this chapter we will further explore the physical meaning of the derivative as an instantaneous rate of change. Various applications to graphing and optimization are also considered.

The First-Derivative Test

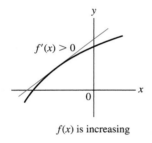

$f'(x) > 0$

$f(x)$ is increasing

Figure 3.1

In this section we will develop a technique for using the derivative as an aid to curve sketching. Mainly, we can use the derivative to determine where a graph reaches its highest and lowest points, called the *maximum* and *minimum points* respectively. To this end, we need the definition of an increasing and decreasing function.

> **Definition of Increasing and Decreasing Functions**
>
> A function f is **increasing** on an interval if, for any two numbers x_1 and x_2 in the interval,
>
> $$x_1 < x_2 \text{ implies that } f(x_1) < f(x_2)$$
>
> A function f is **decreasing** on the interval if
>
> $$x_1 < x_2 \text{ implies that } f(x_1) > f(x_2)$$

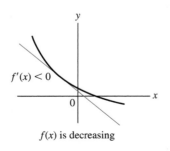

$f'(x) < 0$

$f(x)$ is decreasing

Figure 3.2

Recall from Section 2.6 that a function $f(x)$ is increasing on an interval if $f'(x)$ is positive in the interval (Figure 3.1) and $f(x)$ is decreasing if $f'(x)$ is negative (Figure 3.2).

It is possible for a function to be increasing on one interval and decreasing on another. In that case the graph must rise to a peak and then fall again. Consider the following example.

Example 1   Determine the intervals on which the function $y = 1 - x^2$ is increasing and decreasing and find the highest point on the curve.

Solution. Since $f(x) = 1 - x^2$, $f'(x) = -2x$. To determine where $f(x)$ is increasing and decreasing, we need to determine where $f'(x) = -2x$ is positive and negative. Note that

$$-2x > 0 \qquad \text{if } x < 0$$

and

$$-2x < 0 \qquad \text{if } x > 0$$

Hence f is increasing on $(-\infty, 0)$ and decreasing on $(0, \infty)$. (By our definition, f is actually increasing on $(-\infty, 0]$ and decreasing on $[0, \infty)$.) At $x = 0, -2x = 0$ so that the graph has a horizontal tangent line at $x = 0$. Since $f(0) = 1 - 0^2 = 1$, the point is $(0, 1)$. This point has a special property. As we noted above:

1.   To the left of the point f is increasing.
2.   To the right of the point f is decreasing.

Therefore:

3. f reaches a peak at the point $(0, 1)$.

This peak is called a **maximum point** or simply a **maximum**, and $f(0) = 1$ is called a **maximum value.** This graph is shown in Figure 3.3.   ∎

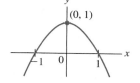

Figure 3.3

Just as a graph may have a highest point (Example 1), it may have a lowest point or *minimum.*

> **Definition of Relative Maximum and Minimum**
>
> A point is called a **relative (or local) maximum** if it has a larger y-value than any point near it.
>   A point is called a **relative (or local) minimum** if it has a smaller y-value than any point near it.

Example 2   Determine the intervals on which the function $y = x^3 - 3x + 2$ is increasing and decreasing. From this information determine the maximum and minimum points.

Solution. To determine where the function is increasing and decreasing, we find the intervals on which $f'(x) > 0$ and $f'(x) < 0$. To this end, we set $f'(x)$ equal to zero and solve for x. From

$$f'(x) = 3x^2 - 3 = 3(x - 1)(x + 1) = 0$$

we get $x = 1$ or -1. Now observe that

$$f'(x) > 0 \text{ on } (-\infty, -1) \qquad (f \text{ is increasing})$$
$$f'(x) < 0 \text{ on } (-1, 1) \qquad (f \text{ is decreasing})$$

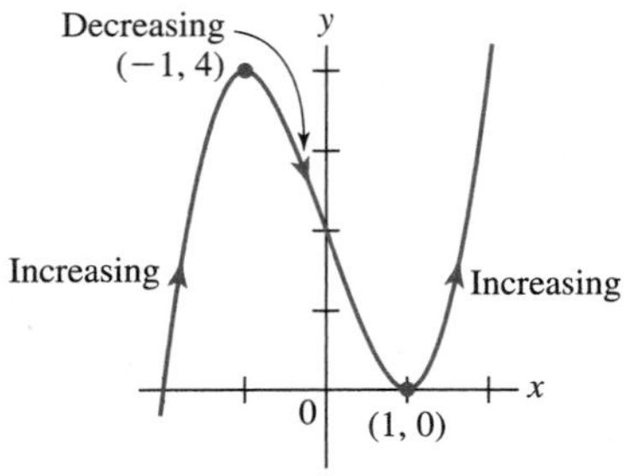

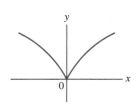

Figure 3.4

and

$$f'(x) > 0 \text{ on } (1, \infty) \qquad (f \text{ is increasing})$$

Moreover, when $x = -1$, $y = (-1)^3 - 3(-1) + 2 = 4$, and when $x = 1$, $y = (1)^3 - 3(1) + 2 = 0$. Hence the graph has horizontal tangent lines at $(-1, 4)$ and $(1, 0)$. Now observe that:

1. To the left of $(-1, 4)$ f is increasing ($f' > 0$).
2. To the right of $(-1, 4)$ f is decreasing ($f' < 0$).

Therefore:

3. The point $(-1, 4)$ is a relative maximum (Figure 3.4).

For the point $(1, 0)$ we have the following:

1. To the left of $(1, 0)$ f is decreasing ($f' < 0$).
2. To the right of $(1, 0)$ f is increasing ($f' > 0$).

Therefore:

3. The point $(1, 0)$ is a relative minimum.

The graph is shown in Figure 3.4. ∎

We can see from Example 2 that a relative maximum is not necessarily the highest point on the curve—it may be a maximum only in the vicinity of the point. On the other hand, the point $(0, 1)$ in Example 1 is higher than any other point on the curve. Such a point is called an **absolute maximum.** The term *global maximum* is also used. (Similar comments apply to minimum points.)

For convenience we will use the term *maximum* for an absolute or a relative maximum and the term *minimum* for an absolute or a relative minimum. Collectively, maximum and minimum values will be referred to as **extreme values.**

Before summarizing these results we need to consider one more case. Suppose $f(x) = x^{2/3}$; then

Extreme values

$$f'(x) = \frac{2}{3}x^{-1/3} = \frac{2}{3x^{1/3}}$$

Note that for $x = 0$ the derivative is infinitely large (undefined), which means that the graph has a vertical tangent line at $(0, 0)$. To the left of the origin $f'(x) < 0$ and to the right $f'(x) > 0$. Consequently, the origin must be a minimum (Figure 3.5).

It is clear from these examples that to find the extreme values of a function f, we need to examine those values c for which $f'(c) = 0$ or for which $f'(c)$ does not exist. Such values are called **critical values** and the corresponding points $(c, f(c))$ on the graph are called **critical points.**

Figure 3.5

> **Definition of Critical Value**
>
> A **critical value** is a number c in the domain of f for which $f'(c) = 0$ or for which $f'(c)$ does not exist. The points $(c, f(c))$ are called **critical points.**

The foregoing procedure for testing the critical points will be referred to as the **first-derivative test.**

First-Derivative Test

If c is a critical value, test the derivative with two values of x, one slightly less and the other slightly more than c. If, as x increases, the sign of the derivative changes from $+$ to $-$, then $f(c)$ is a maximum value and $(c, f(c))$ is a maximum point. If the sign changes from $-$ to $+$, then $f(c)$ is a minimum value. If the sign does not change, then $(c, f(c))$ is neither a minimum nor a maximum.

That a critical point need not lead to an extreme value can be seen from the following example.

Example 3

Test the function $f(x) = x^3$ for extreme values and sketch the graph.

Solution. Since $f'(x) = 3x^2 > 0$ for $x \neq 0$, the function is increasing everywhere. As a consequence, even though $x = 0$ is a critical value, $f(0)$ is neither a minimum nor a maximum value. In sketching the graph we must keep in mind that even though the point $(0, 0)$ is neither a minimum nor a maximum, the line $y = 0$ is a tangent line. (See Figure 3.6.) Note that the tangent line passes through the graph. ∎

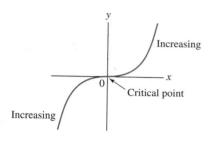

Figure 3.6

These examples show the importance of extreme values in curve sketching: if you know the behavior of the function at the critical points, you have a good idea of the shape of the graph. The next example illustrates a convenient method for testing the critical points by means of a chart.

Example 4

Find the extreme values of the function $y = (-2/3)x^3 + x^2 + 4x - 5$ and sketch the graph.

Solution. To locate the critical points, we find the derivative of the function and set it equal to zero:

$$\frac{dy}{dx} = -2x^2 + 2x + 4 = 0$$

or

$$\frac{dy}{dx} = -2(x^2 - x - 2) = -2(x - 2)(x + 1) = 0$$

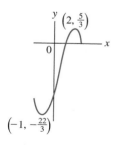

Figure 3.7

Solving for x, we get $x = 2$ and $x = -1$. Substituting in the given equation, we find that $y = 5/3$ when $x = 2$ and $y = -22/3$ when $x = -1$. Hence $(2, 5/3)$ and $(-1, -22/3)$ are the critical points. (See Figure 3.7.)

A simple way to determine the signs on the derivative is the following:

$$\frac{dy}{dx} = -2(x - 2)(x + 1) = 0$$

when x is either 2 or -1. Moreover, these values are the *only* values for which $dy/dx = 0$. Consequently, dy/dx is different from zero at all other points. This observation allows us to substitute arbitrary test values for x to determine the signs. For example, if $x = 3$, then $dy/dx = -8$, which is <0, so that $dy/dx < 0$ for *all* $x > 2$. (If this were not so, then there would have to exist another $x > 2$ for which $dy/dx = 0$, but 2 and -1 are the only critical values.) Similarly, if $x = 1$, then $dy/dx = 4$, which is >0. So $dy/dx > 0$ for all x in the interval $(-1, 2)$. Finally, if $x = -3$, then $dy/dx = -20$, which is <0, and $dy/dx < 0$ on $(-\infty, -1)$.

Using 3, 1, and -3 as test values, we can carry out this process systematically by means of the following chart:

	Test values	$x - 2$	$x + 1$	$f'(x) = -2(x - 2)(x + 1)$	
$x > 2$	3	$+$	$+$	$-$	**f decreasing**
$-1 < x < 2$	1	$-$	$+$	$+$	**f increasing**
$x < -1$	-3	$-$	$-$	$-$	**f decreasing**

According to the chart:

1. To the left of $x = -1$, f is decreasing ($f' < 0$).
2. To the right of $x = -1$, f is increasing ($f' > 0$).

Therefore:

3. The point $(-1, -22/3)$ is a minimum.

Moreover:

1. To the left of $x = 2$, f is increasing ($f' > 0$).
2. To the right of $x = 2$, f is decreasing ($f' < 0$).

Therefore:

3. The point $(2, 5/3)$ is a maximum.

The graph is shown in Figure 3.7. ■

■ Exercises / Section 3.1

In Exercises 1–4, find the intervals on which the given function is increasing and decreasing.

1. $y = -x^2 - 2x$

2. $y = (x - 2)^2$

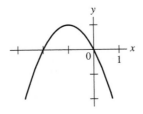

Figure 3.8

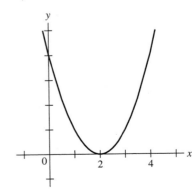

Figure 3.9

3. $y = \frac{1}{4}x^3 - 3x + 2$

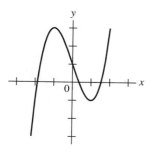

Figure 3.10

4. $y = x^4 - 2x^2 - 1$

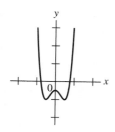

Figure 3.11

Find the extreme values of each of the following functions and sketch the curves.

5. $y = x^2 - 2x + 1$

6. $y = x^2 - 4x + 3$

7. $y = 8 - 2x - x^2$

8. $y = -x^2 + 5x - 6$

9. $y = 2x^3 + 3x^2 - 12x + 6$

10. $y = 2x^3 + 7x^2 + 4x + 1$

11. $y = -x^3 + 6x^2 - 9x - 5$

12. $y = x^3 - 3x + 7$

13. $y = x^4 - 2x^2 - 2$

14. $y = (x + 1)^3$

15. $y = x^4 + \frac{4}{3}x^3$

16. $y = 3x^4 - 4x^3 + 1$

17. $y = 4 - 4x^3 - 3x^4$

18. $y = \sqrt{x}$

19. $y = \sqrt[3]{x}$

20. $y = 4\sqrt{x} - x$

21. A projectile shot directly upward with a velocity of 80 m/s moves according to the equation $s = 80t - 5t^2$. Determine the maximum altitude.

22. If a resistor of 4 Ω is linked parallel with a variable resistor of R Ω, then the resistance R_T of the combination is given by

$$R_T = \frac{4R}{R + 4}$$

Show that R_T starts at zero and increases steadily as R increases.

23. The power in a circuit with variable resistance R is given by

$$P = \frac{16R}{(R + 2)^2} \text{ watts}$$

Find the setting of the variable resistor that allows it to take maximum power.

24. The output P of a certain battery is given by $P = 20I - 10I^2$, where I is the current in amperes. Find the current for which the output is a maximum.

3.2 The Second-Derivative Test

We observed in the last section that critical points give us information that greatly facilitates curve sketching. In this section we are going to develop a procedure that will give us even more information about the graph by using the second derivative.

Suppose a function f is increasing on some interval. Then $f'(x) > 0$ on that interval. The same statement can be made about f', since f' is itself a function: if f' is increasing on an interval, then $df'(x)/dx = f''(x) > 0$. What does this tell us about the graph? Consider the curve in Figure 3.12. The curve is drawn in such a way that the slope is steadily increasing, which is possible only if the curve remains above the tangent line at each point. In the language of calculus we say that the curve in

Concave up
Concave down

Figure 3.12 is **concave up.** Similarly, if f' is decreasing, then $f''(x) < 0$ and the graph is **concave down** (Figure 3.13).

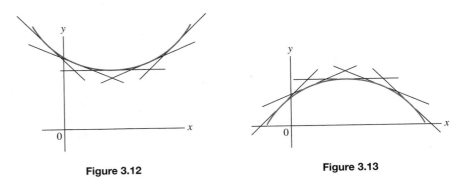

Figure 3.12   Figure 3.13

Concavity and Inflection Points

For all x in an interval $[a, b]$, the graph of a function f is:

1. Concave up if $f''(x) > 0$
2. Concave down if $f''(x) < 0$

A point on the graph at which the concavity changes is called an **inflection point.**

Example 1

Suppose we return to the function $f(x) = x^3 - 3x + 2$ whose graph appears in Figure 3.4. From $f''(x) = 6x$ we see that $f''(x) < 0$ if $x < 0$, and $f''(x) > 0$ if $x > 0$. So the graph is concave down to the left of the point $(0, 2)$ and concave up to the right. Since the concavity changes at $(0, 2)$, this point is an inflection point. ∎

To find the point of inflection, we normally set $f''(x) = 0$, solve for x, and determine the sign of $f''(x)$ to the left and right of the point. (A point $(x, f(x))$ for which $f''(x)$ does not exist may also be an inflection point, as we will see in Exercise 37; we will concentrate mainly on the former case.) Once the values of x for which $f''(x) = 0$ are known, it is a simple matter to determine the sign of $f''(x)$ elsewhere. In Example 1, $f''(x) = 6x = 0$ when $x = 0$. Hence $f''(x) \neq 0$ at all other points. Consequently, since $f''(1) = 6 > 0$, $f''(x)$ must be positive for all $x > 0$ and since $f''(-1) = -6 < 0$, $f''(x)$ must be negative for all $x < 0$.

Caution: If $(c, f(c))$ is an inflection point, then $f''(c) = 0$ or $f''(c)$ does not exist. The converse is not necessarily true: if $f''(c) = 0$ for some $x = c$, then $(c, f(c))$ is not necessarily an inflection point, as we can see from the function $f(x) = x^4$. (See the graph in Figure 3.14.) Here $f'(x) = 4x^3$ and $f''(x) = 12x^2$; $f''(0) = 0$, but the point $(0, 0)$ is not an inflection point.

A final key observation from our discussion of concavity concerns the determination of extreme values. Suppose a curve is known to be concave up on some interval. If for some point c inside the interval we have a horizontal tangent line at the corresponding point $(c, f(c))$, then this point must be a minimum. For example, if

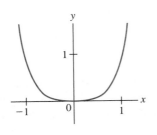

Figure 3.14

$f(x) = x^2$, then $f'(x) = 2x$, so that $(0, 0)$ is a critical point. Since $f''(x) = 2 > 0$, the graph is concave up everywhere and $(0, 0)$ must be a minimum. This criterion is known as the **second-derivative test.**

Second-Derivative Test

Suppose $f'(c) = 0$.

1.   If $f''(c) > 0$, then $f(c)$ is a minimum value.
2.   If $f''(c) < 0$, then $f(c)$ is a maximum value.
3.   If $f''(c) = 0$, the test fails.

Caution: Regarding part (3), if the test fails, we mean just that—the second derivative gives us no information about the critical point. For $f(x) = x^4$ we already saw that $f''(0) = 0$; yet f attains a minimum at the critical point $(0, 0)$. In Example 3 of Section 3.1 the function $f(x) = x^3$ has critical value $x = 0$ and again $f''(0) = 0$. This time, however, $f(0)$ is not an extreme value. Occasionally, then, we need to fall back on the first-derivative test. (At times it may also be inconvenient to compute f''. In that case, refer back to the first-derivative test.)

Before considering further examples, let us summarize the procedure for sketching curves.

Suggested Procedure for Curve Sketching

1.   Find all critical values and critical points.
2.   Test the critical values:
     a. Use the second-derivative test.
     b. If the second-derivative test fails, use the first-derivative test.
3.   Use the second derivative to determine the intervals for which the graph is concave up and concave down.
4.   Determine the points of inflection from Step 3.
5.   Other: Find any easily determined intercepts and asymptotes; test for symmetry.
6.   Plot the critical points, inflection points, and (if available) the intercepts. Draw any existing asymptotes. Sketch the curve, using a few additional points.

Example 2   Discuss the function $y = x^3 - 3x$ for minima and maxima, concavity, and inflection points, and sketch the function.

Solution.   Derivatives: $f'(x) = 3x^2 - 3$; $f''(x) = 6x$.

Step 1.   Critical points: $3x^2 - 3 = 0$ whenever $x = \pm 1$; the points are $(1, -2)$ and $(-1, 2)$, determined from $y = x^3 - 3x$.
Step 2.   Test of critical points:

$f''(1) = 6 > 0$; $(1, -2)$ is a minimum point.

$f''(-1) = -6 < 0$; $(-1, 2)$ is a maximum point.

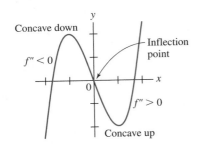

Concave down

$f'' < 0$

Inflection point

$f'' > 0$

Concave up

Figure 3.15

Step 3. Concavity: we need to determine where $f'' > 0$ and where $f'' < 0$. To this end we first find those values of x for which $f''(x) = 0$:

$$f''(x) = 6x = 0 \text{ when } x = 0$$

Since $x = 0$ is the only value for which $f''(x) = 0$, it follows that $f''(x) \neq 0$ at all other points. Since $f''(1) = 6$, $f''(x) > 0$ whenever $x > 0$. Hence the graph is concave up on $[0, \infty)$. Similarly, since $f''(-1) = -6$, the graph is concave down on $(-\infty, 0]$. (See Figure 3.15.)

Using $x = 1$ and $x = -1$ as test values, we summarize these observations in the following chart:

	Test values	$f''(x) = 6x$	
$x > 0$	1	+	concave up
$x < 0$	-1	−	concave down

Step 4. Inflection point: because of the change in concavity, there is an inflection point at $x = 0$. If $x = 0$, then $y = 0$; so the inflection point is $(0, 0)$.

Step 5. Other: the intercepts are $(0, 0)$ and $(\pm\sqrt{3}, 0)$; the graph is symmetric with respect to the origin.

Step 6. Plotting all the points found and perhaps a few additional ones, we obtain the graph in Figure 3.15. ∎

Example 3   Discuss and sketch the graph of $y = x^4 + (4/3)x^3$.

Solution. Derivatives: $f'(x) = 4x^3 + 4x^2$; $f''(x) = 12x^2 + 8x$.

Step 1. Critical points:

$$4x^3 + 4x^2 = 0 \quad \text{or} \quad 4x^2(x + 1) = 0$$

Solving for x, we find that $x = 0$ and $x = -1$. Hence $(-1, -1/3)$ and $(0, 0)$ are the critical points.

Step 2. Test of critical points:

$$f''(-1) = 4; \ f(-1) \text{ is a minimum value.}$$

$$f''(0) = 0; \text{ the test fails.}$$

Using the first-derivative test, we conclude that since $f'(x) = 4x^2(x + 1) > 0$ on $(-1, \infty)$, $x = 0$ does not lead to an extreme value.

Step 3. Concavity:

$$f''(x) = 12x^2 + 8x = 0 \quad \text{or} \quad 4x(3x + 2) = 0$$

whenever $x = 0$ and $x = -2/3$. These are the only values for which $f''(x) = 0$, so that $f''(x) \neq 0$ at all other points. Since $f''(-1) > 0$, $f''(x) > 0$ on $(-\infty, -2/3)$.

Similarly, from $f''(-1/2) < 0$, we conclude that $f''(x) < 0$ on $(-2/3, 0)$. Finally, since $f''(1) = 20$, $f''(x) > 0$ on $(0, \infty)$.

These observations are summarized in the following chart:

	Test values	$4x$	$3x + 2$	$f''(x) = 4x(3x + 2)$	
$x > 0$	1	+	+	+	concave up
$-\dfrac{2}{3} < x < 0$	$-\dfrac{1}{2}$	−	+	−	concave down
$x < -\dfrac{2}{3}$	-1	−	−	+	concave up

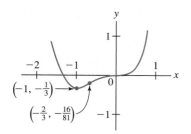

Figure 3.16

Step 4. Inflection points: since the concavity changes, the points $(0, 0)$ and $(-2/3, -16/81)$ are inflection points, confirming that $(0, 0)$ is not a minimum or maximum.

Step 5.   Other: the intercepts are $(0, 0)$ and $(-4/3, 0)$.

Step 6.   The graph is shown in Figure 3.16. ∎

Example 4   Discuss and sketch the graph of

$$y = \frac{x}{\sqrt{x - 1}}$$

Solution. Because of the radical, it is wise to start with the general discussion. Since $x - 1$ has to be strictly positive to avoid imaginary numbers and division by zero, the domain of the function is the interval $(1, \infty)$. Setting the denominator equal to zero, we find that $x = 1$ is a vertical asymptote. Finally, $y = 0$ only if $x = 0$, but $x = 0$ is not in the domain. Consequently, the graph has no intercepts.

The derivatives can be found by the quotient rule and will be left as an exercise:

$$f'(x) = \frac{x - 2}{2(x - 1)^{3/2}} \quad \text{and} \quad f''(x) = -\frac{x - 4}{4(x - 1)^{5/2}}$$

Step 1.   The only critical value is $x = 2$.

Step 2.   Since $f''(2) > 0$, the point $(2, 2)$ is a minimum.

Step 3.   Concavity:

$$f''(x) = 0 \quad \text{when } x = 4$$

It is easy to check that $f''(x) > 0$ on $(1, 4)$ so that the graph is concave up on the interval $(1, 4]$. On the interval $[4, \infty)$ the graph is concave down. Using $x = 5$ and $x = 3$ as test values, we summarize these observations in the following chart:

	Test values	$x - 4$	$x - 1$	$f''(x) = -\dfrac{x - 4}{4(x - 1)^{5/2}}$	
$x > 4$	5	+	+	−	concave down
$x < 4$	3	−	+	+	concave up

Step 4. Inflection point: since the concavity changes, the point $(4, 4/\sqrt{3})$ is an inflection point.

Steps 5 and 6. The graph is shown in Figure 3.17.

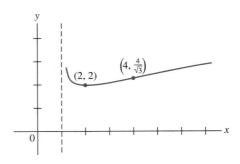

Figure 3.17   ■

Example 5   Discuss and sketch the graph of $y = (x - 2)^{2/3}$.

Solution. Derivatives:

$$f'(x) = \frac{2}{3(x-2)^{1/3}} \qquad f''(x) = -\frac{2}{9(x-2)^{4/3}}$$

Step 1. Critical points: Since $f'(2)$ is undefined, $x = 2$ is a critical value and $(2, 0)$ is a critical point; the line $x = 2$ is a vertical tangent.

Step 2. Test of critical points: Since f' is not differentiable at $x = 2$ (that is, $f''(2)$ does not exist), the second-derivative test cannot be employed. It is easy to see, however, that $f'(x) < 0$ for $x < 2$ and $f'(x) > 0$ for $x > 2$. Hence $(2, 0)$ is a minimum by the first-derivative test.

Steps 3 and 4. Concavity and points of inflection: Since $f''(2)$ does not exist, the point $(2, 0)$ is a possible point of inflection. At no point is $f''(x) = 0$. Substituting two values, such as $x = 1$ and $x = 3$, in the expression for $f''(x)$, we find that $f''(x) < 0$ for all $x \neq 2$. Hence $(2, 0)$ is not an inflection point. (The graph is concave down on $(-\infty, 2]$ and $[2, \infty)$.)

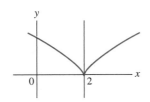

Figure 3.18

Steps 5 and 6. The graph is shown in Figure 3.18.   ■

Example 6   Discuss and sketch the graph of

$$y = x^2 + \frac{16}{x^2}$$

Solution. Derivatives. From $f(x) = x^2 + 16x^{-2}$ we get

$$f'(x) = 2x - 32x^{-3} \qquad \text{and} \qquad f''(x) = 2 + 96x^{-4}$$

or

$$f'(x) = 2x - \frac{32}{x^3} \quad \text{and} \quad f''(x) = 2 + \frac{96}{x^4}$$

Step 1. Critical points: Setting $2x - (32/x^3) = 0$, we get

$$2x - \frac{32}{x^3} = 0$$

$$2x^4 - 32 = 0 \quad \textbf{multiplying by } x^3$$

$$x^4 - 16 = 0 \quad \textbf{dividing by 2}$$

$$x = \pm 2$$

The critical points are $(\pm 2, 8)$.

Step 2. Test of critical points: See Step 3.

Steps 3 and 4. Concavity: Since $f''(x) = 2 + (96/x^4) > 0$, $x \neq 0$, the graph is concave up everywhere. Consequently, there cannot be any inflection points. Moreover, both critical points are minima.

Step 5. Other: from

$$y = x^2 + \frac{16}{x^2} = \frac{x^4 + 16}{x^2}$$

we see that the graph has no intercepts and that the y-axis is a vertical asymptote. (See Figure 3.19.)

Step 6. As $x \to \infty$, $16/x^2 \to 0$. So the graph of $y = x^2 + (16/x^2)$ gets closer to the parabola $y = x^2$. The graph of $y = x^2$ may be called an *asymptotic curve*, shown as the dashed curve in Figure 3.19. ∎

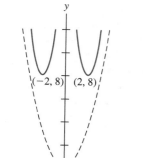

Figure 3.19

■ Exercises / Section 3.2

Find the intervals on which the graph is concave up and concave down.

1. $y = 2x - x^2$

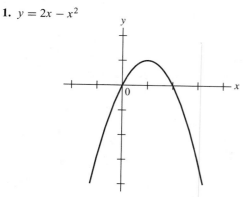

Figure 3.20

2. $y = 2(x - 2)^2$

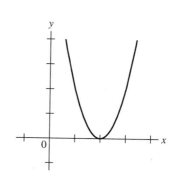

Figure 3.21

3. $y = x^3 - 12x + 6$

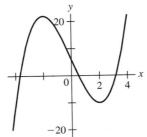

Figure 3.22

4. $y = -\frac{1}{3}x^3 + x^2$

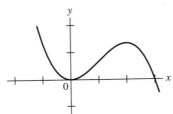

Figure 3.23

5. $y = x^4 - 4x^3 + 8x - 5$

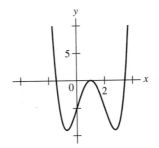

Figure 3.24

6. $y = 13 + 16x - 8x^3 - 2x^4$

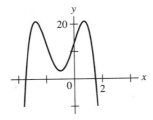

Figure 3.25

7. $y = x^3 - 9x^2 + 27x - 27$

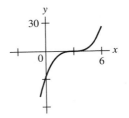

Figure 3.26

8. $y = \dfrac{x}{x + 2}$

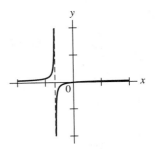

Figure 3.27

Find the minima and maxima, the intervals on which the graph is concave up and concave down, and the inflection points of each of the following functions; sketch the curves.

9. $y = 2x^2 - 4x$ **10.** $y = x^2 - 2x + 1$

11. $y = 6x - 6x^2$ **12.** $y = 2x - x^2$

13. $y = -4 - 3x - \frac{1}{2}x^2$ **14.** $y = 3x - 3x^2$

15. $y = 2x^3 - 6x + 1$ **16.** $y = x^3 - 3x$

17. $y = x^3 - 6x^2 + 9x - 3$ **18.** $y = 2x^3 + 3x^2 - 12x + 2$

19. $y = x^3 - 3x^2 - 9x + 11$ **20.** $y = 1 + 9x - 3x^2 - x^3$

21. $y = 2 + 3x - x^3$ **22.** $y = x^3 - 4x^2 + 4x$

23. $y = 3x^4 - 4x^3 + 2$ **24.** $y = \dfrac{1}{x^3}$

25. $y = x^4 + x^3$ **26.** $y = (x - 2)^3$

27. $y = (x - 3)^4$ **28.** $y = (x^2 - 9)^2$

29. $y = \dfrac{x}{x - 3}$ **30.** $y = x + \dfrac{1}{x}$

31. $y = x^2 + \dfrac{8}{x}$ **32.** $y = x^3 + \dfrac{1}{x^2}$

33. $y = \dfrac{x+1}{x-2}$

34. $y = x^{4/3}$

35. $y = x^{3/4}$

36. $y = x^2(4 - x^2)$

37. $y = (x - 3)^{1/3}$

38. $y = x - 2\sqrt{x}$

39. $y = \dfrac{2x}{(x+1)^2}$

40. $y = \dfrac{x^2}{(x^2+1)^2}$

41. $y = \dfrac{6x}{x^2+3}$

42. $y = \dfrac{2x}{x^2+1}$

3.3 Exploring with Graphing Utilities (Optional)

In this section we will see how graphing utilities can be used as an effective tool for exploring and analyzing a graph using the following plan: (1) obtain a complete graph using a graphing utility; (2) visually determine the approximate location of all critical points and inflection points; (3) use the first and second derivatives to determine all the candidates for critical and inflection points; and (4) use the graph to test these points. Study the next example.

Example 1

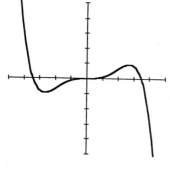

[−1, 1] by [−1, 1]

Figure 3.28

Draw the graph of the function

$$y = 3.1x^3 - 6.5x^5$$

Discuss the critical points and determine the inflection points.

Solution. Using the viewing rectangle [−1, 1] by [−1, 1], we get the graph in Figure 3.28.

Judging from the figure, we see there are three critical points (where the tangent lines are horizontal), with a minimum point to the left of the origin and a maximum point to the right:

$$y' = 9.3x^2 - 32.5x^4 = 0$$
$$x^2(9.3 - 32.5x^2) = 0$$

$$x^2 = 0 \qquad\qquad 9.3 - 32.5x^2 = 0$$

$$x = 0 \qquad\qquad x^2 = \frac{9.3}{32.5}$$

$$x = \pm 0.53$$

The points are (0, 0), (−0.53, −0.19), and (0.53, 0.19). Since the graph is already known, no further tests are necessary.

Returning to the graph, the origin looks like an inflection point (in addition to being a critical point.) There appear to be two additional inflection points on opposite sides of the origin:

$$y'' = 18.6x - 130x^3 = 0$$
$$x(18.6 - 130x^2) = 0$$

$$x = 0 \qquad\qquad 18.6 - 130x^2 = 0$$

$$x^2 = \frac{18.6}{130}$$

$$x = \pm 0.38$$

So the inflection points are $(0, 0)$, $(-0.38, -0.12)$, and $(0.38, 0.12)$. As before, since the graph is already known, no further tests are necessary.   ∎

Observe the surprising behavior of the graph in the next example.

Example 2   Graph the function

$$y = 1.2x^4 + 2.7x^3 + 6.0x^2 + 1.0$$

and discuss the critical points.

Solution. Based on our earlier experience, we might expect three critical points. However, the graph (Figure 3.29) suggests the existence of a single minimum point:

$$y' = 4.8x^3 + 8.1x^2 + 12x = 0$$
$$x(4.8x^2 + 8.1x + 12) = 0$$
$$x = 0 \qquad 4.8x^2 + 8.1x + 12 = 0$$

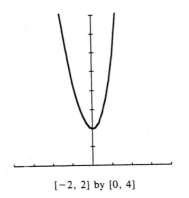

[−2, 2] by [0, 4]

Figure 3.29

By the quadratic formula,

$$x = \frac{-8.1 \pm \sqrt{8.1^2 - 4(4.8)(12)}}{2(4.8)}$$

$$x = -0.84 \pm 1.34j \qquad \textbf{complex roots}$$

Since $x = 0$ is the only real root, the point $(0, 1)$ is the only critical point.   ∎

Example 3   Draw the graph of

$$y = x^6 - 2x^4 - 2$$

Discuss the critical points and determine the inflection points.

Solution. Using the viewing rectangle $[-4, 4]$ by $[-4, 4]$, we get the graph in Figure 3.30.

Judging from the graph, we have one maximum point, two minimum points, and two inflection points:

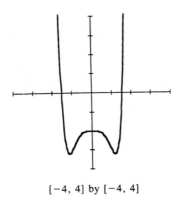

[−4, 4] by [−4, 4]

Figure 3.30

$$y' = 6x^5 - 8x^3 = 0 \qquad\qquad y'' = 30x^4 - 24x^2 = 0$$
$$x^3(6x^2 - 8) = 0 \qquad\qquad x^2(30x^2 - 24) = 0$$
$$x = 0 \qquad x^2 = \frac{4}{3} \qquad\qquad x = 0 \qquad x^2 = \frac{4}{5}$$
$$x = \pm\frac{2\sqrt{3}}{3} \qquad\qquad\qquad x = \pm\frac{2\sqrt{5}}{5}$$

So the critical points are at $x = 0$ and at $x = \pm 2\sqrt{3}/3$. What is particularly noteworthy here is that there are indeed two inflection points at $x = \pm 2\sqrt{5}/5$, but not at $x = 0$. In other words, **while $f''(0) = 0$, the point $(0, -2)$ is not an inflection point.** ∎

■ Exercises / Section 3.3

In Exercises 1–14, draw each graph. Discuss the critical points and determine the inflection points.

1. $y = x^5 + x^3 + 1$

2. $y = x^4 + x^2 - 2$

3. $y = 4x^2 - \dfrac{4}{5}x^5 + 2$

4. $y = 2.0x^5 - 10.0x^2$

5. $y = 3.0x^2 + 1.2x^5 - 1.0$

6. $y = 1.0x^6 + 2.5x^3 + 0.50$

7. $y = 3.7x^6 + 2.4x^4 + 1.5$

8. $y = 0.30x^6 - 1.0x^4 + 1.0$

9. $y = 1.5x^4 - 0.50x^6 + 0.20$

10. $y = 1.6x^5 + 2.0x^4 - 2.5$

11. $y = \dfrac{3}{5}x^5 - 2x^3$

12. $y = 2.1x^5 - 4.5x^3 + 1.0$

13. $y = \dfrac{1}{7}x^7 - \dfrac{1}{5}x^5$

14. $y = 2x^5 - x^7$

In Exercises 15–24, draw each graph. Find any vertical asymptotes and discuss the critical points.

15. $y = 1.2x + \dfrac{4.0}{\sqrt{x}}$

16. $y = 4.00\sqrt{x} - 1.50x$

17. $y = \dfrac{x}{x^2 - 1}$

18. $y = \dfrac{x^3}{x^2 + 3}$

19. $y = \dfrac{x^3}{x^2 - 3}$

20. $y = \dfrac{x^2}{2 - x^2}$

21. $y = 0.50x^3 + \dfrac{2.34}{x}$

22. $y = 2.0x^3 + \dfrac{4.0}{x^3}$

23. $y = 2x - \dfrac{1}{x}$

24. $y = x^2 - \dfrac{2}{x}$

3.4 Applications of Minima and Maxima

Have you ever wondered why bubbles are round? The reason is that a bubble encloses a certain volume of air and the surface tension contracts the surface of the bubble to its smallest possible area, which is a spherical surface. Many other situations in nature offer examples of minimum and maximum values. For example, the second law of thermodynamics may be stated in the following form: there is a tendency in nature for all systems to proceed toward a state of maximum molecular disorder. This accounts for the observations that well-formed crystals dissolve in a solvent, organisms decay after death, and rocks weather.

In our study of applied minima and maxima we are naturally confined to problems that can be analyzed with the techniques of Sections 3.1 and 3.2. If the function is known, these techniques can be applied directly, as illustrated in the first two examples.

Example 1 The formula for the output P of a battery is given by

$$P = VI - RI^2$$

where V is the voltage, I the current, and R the resistance. Find the current for which the output is a maximum if $V = 12$ V and $R = 5.0$ Ω.

Solution. After substituting the given values, we get

$$P = 12I - 5.0I^2$$

We now maximize P by the method of Section 3.2; that is, we find the derivative with respect to I and set it equal to zero;

$$\frac{dP}{dI} = 12 - 10I = 0$$

so that $I = 1.2$ A is the critical value. Since

$$\frac{d^2 P}{dI^2} = -10 < 0$$

it follows that $I = 1.2$ does correspond to the maximum output. ∎

Example 2

A turbine for generating power is rotated by means of a high-speed jet of water striking circularly mounted blades. The speed of the jet is normally fixed, but the speed (rate of rotation) of the turbine can be adjusted by changing the blade angle. If J is the speed of the jet and T the speed of the turbine, then the power is given by

$$P = kJT(J - T)$$

where k is a constant. What speed of the turbine will yield maximum power?

Solution. The equation can be written

$$P = kJ^2 T - kJT^2$$

Since J is fixed, both J and k are treated as constants. The critical value is found from

$$\frac{dP}{dT} = kJ^2 \frac{d}{dT}(T) - kJ \frac{d}{dT}(T^2) \qquad \textbf{k, J constants}$$

or

$$\frac{dP}{dT} = kJ^2 - 2kJT = 0$$

Solving for T, we obtain

$$2kJT = kJ^2 \qquad \text{and} \qquad T = \frac{kJ^2}{2kJ} = \frac{J}{2}$$

Since $d^2 P/dT^2 = -2kJ < 0$, we see that the power is maximal when the speed of the turbine is numerically equal to one-half the speed of the jet. ∎

In many problems the expression to be minimized or maximized is not known in advance and has to be obtained from the given information. Consider the following example.

Example 3

A rectangle has an area of 100 m². What should the dimensions be so that the perimeter will be as small as possible?

Solution. Let x be the length and y the width of the rectangle. Then the perimeter is given by

$$P = 2x + 2y$$

Before proceeding, we need to eliminate one of the variables: since the area is $xy = 100$, we have $y = 100/x$, so that

$$P = 2x + 2y$$

$$= 2x + 2\left(\frac{100}{x}\right)$$

$$= 2x + \frac{200}{x} = 2x + 200x^{-1}$$

P is now a function of x alone, and we can find the minimum by the usual method. Thus

$$\frac{dP}{dx} = 2 - 200x^{-2} \qquad \text{and} \qquad \frac{d^2P}{dx^2} = 400x^{-3}$$

Setting $dP/dx = 0$, we obtain

$$2 - 200x^{-2} = 0, \qquad 2x^2 - 200 = 0 \qquad \textbf{multiplying by } x^2$$

solving, $x = 10$. (The negative root has no meaning in this problem because the length of a rectangle cannot be a negative number.) Since

$$\left.\frac{d^2P}{dx^2}\right|_{x=10} = \frac{2}{5} > 0$$

we see that $x = 10$ leads to a minimum. Since $x = 10$, $y = 100/10 = 10$. So the desired dimensions are 10 m × 10 m. ∎

Example 3 is quite typical and gives us a plan of attack for all such problems:

To Solve Minimum-Maximum Problems

1. Write an expression for the quantity F to be minimized or maximized, using appropriate variables. (Drawing a figure may help.)
2. If the expression for F contains two variables, eliminate one of them by using the information in the problem.
3. Minimize or maximize F.

Example 4

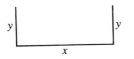

Figure 3.31

The manager of a shop has an order for making a gutter from a long sheet of metal 16 cm wide by bending up equal widths along the edges into vertical positions. The order states that the gutter is to have the largest possible carrying capacity. What should the dimensions be?

Solution. In this problem a figure is indeed helpful (Figure 3.31). Making use of the auxiliary variables, the quantity to be maximized is the cross-sectional area of the gutter,

$$A = xy$$

where x and y are the variables shown in the figure. The cross-section must be as large as possible. To eliminate one of the variables, we need a relation between them. From the given information,

$$x + 2y = 16 \quad \text{or} \quad x = 16 - 2y$$

Substituting in the equation $A = xy$, we obtain

$$A = yx = y(16 - 2y) = 16y - 2y^2$$

which is a function of y alone. Thus

$$\frac{dA}{dy} = 16 - 4y = 0$$

so that $y = 4$ is the critical value. Also, $x = 16 - 2(4) = 8$. Since $d^2A/dy^2 = -4$, A has a maximum value at $y = 4$. Consequently, the desired dimensions are 4 cm by 8 cm.

That the function $A = 16y - 2y^2$ attains a maximum value can also be confirmed with a graphing utility (Figure 3.32). ■

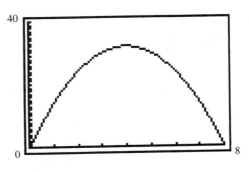

Figure 3.32

Example 5

A wholesaler finds that he can make a profit of $5 for each crate of peaches for orders of 100 crates or fewer. Since he gives a discount for large orders, he finds that he makes 2¢ less profit for each crate above 100. (For example, if an order is for 105 crates, he makes $4.90 per crate.) What size order will yield the maximum profit?

Solution. Let x be the number of crates *above* 100. The profit for every crate is $5 - 0.02x$. Since he sells a total of $100 + x$ crates, his profit is

$$P = (100 + x)(5 - 0.02x) = 500 + 3x - 0.02x^2$$

Thus

$$\frac{dP}{dx} = 3 - 0.04x = 0$$

and $x = 75$. Since $d^2P/dx^2 = -0.04$, P is a maximum when $x = 75$. Hence an order of $75 + 100 = 175$ crates will yield the maximum profit.   ■

Example 6

A ray of light from point A reflected to point B from a plane mirror will follow a path requiring the least time (Figure 3.33). Show that the angle of incidence α is equal to the angle of reflection β.

Solution. Let D and E be the points on the mirror nearest A and B, respectively, and let C be the point where the ray strikes. (See Figure 3.33.) Denote the distance from D to E by L. So if x is the distance from D to C, then $L - x$ is the distance from C and E. We now let c be the velocity of light and use the fact that distance equals rate times time or time equals distance over rate. In particular, because AC is a distance and c a rate, AC/c is the time required for light to traverse the distance AC. (Similar comments apply to BC.) So if T is the total time, we get

$$T = \text{time of trip along } AC + \text{time of trip along } CB = AC/c + CB/c.$$

But $AC = \sqrt{x^2 + a^2}$ and $CB = \sqrt{(L - x)^2 + b^2}$. So

$$T = \frac{\sqrt{x^2 + a^2}}{c} + \frac{\sqrt{(L - x)^2 + b^2}}{c}$$

is the quantity we wish to make a minimum. We now write

$$T = \frac{1}{c}\{(x^2 + a^2)^{1/2} + [(L - x)^2 + b^2]^{1/2}\}$$

Then

$$\frac{dT}{dx} = \frac{1}{c}\left\{\frac{1}{2}(x^2 + a^2)^{-1/2} \cdot 2x + \frac{1}{2}[(L - x)^2 + b^2]^{-1/2} \cdot 2(L - x)(-1)\right\} = 0$$

or

$$\frac{dT}{dx} = \frac{1}{c}\left(\frac{x}{\sqrt{x^2 + a^2}} - \frac{L - x}{\sqrt{(L - x)^2 + b^2}}\right) = 0$$

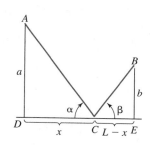

Figure 3.33

Normally we would now solve for x, but in this problem that turns out to be entirely unnecessary: referring to Figure 3.33, we see that the last expression implies that

$$\frac{x}{\sqrt{x^2 + a^2}} = \frac{L - x}{\sqrt{(L - x)^2 + b^2}} \qquad \text{or} \qquad \frac{DC}{AC} = \frac{CE}{CB}$$

That is, $\cos \alpha = \cos \beta$, which implies that $\alpha = \beta$. ∎

■ Exercises / Section 3.4

1. The power P (in watts) delivered to an electrical element as a function of the current i is given by $P = 4.50 + 12.8i - 3.20i^2$. Determine the maximum power.

2. For a short time interval, the current i (in amperes) in a circuit containing an inductor is given by

$$i = 30.0t^2 - 115.0t^3 \qquad (t > 0)$$

   Determine the maximum current.

3. A beam 40 in. long is clamped at $x = 0$ and $x = 40$ and carries a uniform load. The deflection (in inches) as a function of the distance from the left end is given by $d(x) = 1.5 \times 10^{-6}x^2(40 - x)^2$. Find the distance from the left end at which the deflection is a maximum.

4. The efficiency E of a screw is given by

$$E = \frac{T(1 - 0.35T)}{T + 0.35}$$

   where T is the tangent of the pitch angle of the screw and 0.35 is the coefficient of friction. For what value of $T > 0$ is the efficiency the greatest?

5. According to Kelvin's law, the power lost in a transmission line is inversely proportional to the cross-sectional area (that is, the larger the cross-sectional area, the smaller the heat loss). The cost of the wire, however, is directly proportional to its area. In symbols, the total cost is

$$C = k_1 A + \frac{k_2}{A}$$

   where k_1 and k_2 are positive constants. Show that C is a minimum when $A = \sqrt{k_2/k_1}$. (Since $C = k_1 A + k_2 A^{-1}$, $dC/dA = k_1 + k_2(-1)A^{-2}$.)

6. The drag on an airplane traveling at velocity v is

$$D = av^2 + \frac{b}{v^2}$$

where a and b are positive constants. At what speed does the airplane experience the least drag?

7. The total charge in an electrical circuit as a function of time is given by $q = t/(t^2 + 1)$ coulombs. Find the maximum charge q.

8. By Kirchhoff's voltage law the sum of the voltages across the components of a circuit is equal to the applied voltage in the circuit. Consider the circuit with variable resistor R in Figure 3.34.

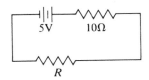

Figure 3.34

   a. Find a relationship between I and R in the circuit, using $E = IR$ and Kirchhoff's voltage law.
   b. The power to a resistor is given by $I^2 R$ watts. Write P as a function of R.
   c. Find the setting of the variable resistor R so that it takes maximum power.

9. For a certain manufacturer, the cost C of producing x machine parts is $C(x) = (2.0 \times 10^{-6})x^3 - 0.0015x^2 + 2.5x + 500$ (in dollars). How many units should be produced to minimize the marginal cost? (The marginal cost is the instantaneous rate of change of the cost.)

10. Water reaches its maximum density above its freezing point. The volume V (in cubic centimeters) of 1 kg of water at temperature T between $0°$ C and $30°$ C can be approximated by the formula $V = 999.87 - 0.06426T + 0.0085043T^2 - 0.0000679T^3$. Find the temperature at which water has its maximum density.

11. Find the two positive numbers whose sum is 60 and whose product is a maximum.

12. A rectangle has a perimeter of 8 cm. What should the dimensions be so that its area is a maximum?

13. A rectangular area adjacent to a river is to be enclosed by a fence on the other three sides. If the area enclosed is to be 200 m^2 and if no fencing is needed along the river, what dimensions require the least amount of fencing? (See Figure 3.35.)

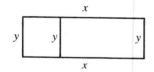

Figure 3.35

14. A rectangular field is to be enclosed by a fence and separated into two parts by a fence parallel to one of the sides. If 600 m of fencing is available, what should the dimensions be so that the area is a maximum? (See Figure 3.36.)

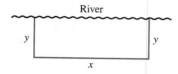

Figure 3.36

15. Suppose the area in Exercise 14 is divided by two fences parallel to one of the sides. What dimensions will maximize the area?

16. A rectangular area adjacent to a wall is to be enclosed on the other three sides and separated into two parts by a fence perpendicular to the wall. If no fencing is needed along the wall and if the total enclosed area is to be 1200 ft^2, what should the dimensions be so that the total amount of fencing used is a minimum?

17. Suppose the area in Exercise 16 is separated into three parts by two fences perpendicular to the wall. If the total area is 1600 ft^2, what should the dimensions be so that the total amount of fencing used is a minimum?

18. A property owner wants to build a rectangular enclosure around some land that is next to the lot of a neighbor who is willing to pay for half the fence that actually divides the two lots. If the area is A, what should the dimensions of the enclosure be so that the cost to the *owner* is a minimum?

19. A box with an open top is to be made from a square piece of cardboard by cutting equal squares from the corners and turning up the sides. If the piece of cardboard measures 12 cm on the side, find the size of the squares that must be cut out to yield the maximum volume for the box. (See Figure 3.37.)

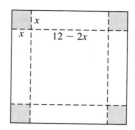

Figure 3.37

20. A box with an open top is to be made from a rectangular piece of tin by cutting equal squares from the corners and turning up the sides. The piece of tin measures 1 m × 2 m. Find the size of the squares that yields a maximum capacity for the box.

21. The *strength* of a beam with rectangular cross-section is directly proportional to the product of the width and the square of the depth (thickness from top to bottom of the beam). Find the shape of the strongest beam that can be cut from a cylindrical log of diameter $d = 3$ ft. (See Figure 3.38.)

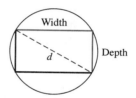

Figure 3.38

22. A closed box is to be a rectangular solid with a square base. If the enclosed volume is 32 in.^{3}, determine the dimensions for which the surface area is minimum. (See Figure 3.39.)

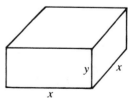

Figure 3.39

23. Repeat Exercise 22 for a box open at the top.

24. A wire of length 50 cm is to be cut into two pieces. One of the pieces is to be bent into the form of a circle and the other into the form of a square. How should the wire be cut so that the sum of the enclosed areas is a minimum?

25. An arched window is in the shape of a rectangle surmounted by a semicircle. If the perimeter is 5 m, what should the radius of the semicircular part be if the window is to admit as much light as possible? (See Figure 3.40.)

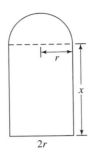

Figure 3.40

26. Show that a tin can having a fixed volume V will require the least amount of material if the height equals the diameter of the base. (*Hint:* Minimize $A = 2\pi rh + 2\pi r^2$ after eliminating h. The critical value is $r^3 = V/(2\pi)$; use this value to calculate $h/r = hr^2/r^3$.)

27. Repeat Exercise 26 for a cylinder with open top.

28. A closed box is to be a rectangular solid with a square base and a volume of 12 ft³. Find the most economical dimensions if the top of the box is twice as expensive as the sides and bottom.

29. Find the point on the curve $y = x^2/4$ nearest the point $(1, 2)$. (*Hint:* if d is the distance from $(1, 2)$ to a point on the curve, minimize d^2 to avoid radicals.)

30. Find the point in the first quadrant on the curve $xy = 2$ nearest the origin.

31. Find the largest possible rectangle in the first quadrant such that its sides lie along the axes and it has one vertex on the curve $y = 9 - x^2$. (See Figure 3.41.)

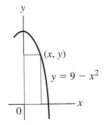

Figure 3.41

32. Find the largest possible rectangle that can be inscribed in a circle of radius 10.

33. A bus company will take 30 passengers on an excursion trip for $400 per passenger. If more than 30 passengers (up to 50) sign up, the company will reduce the price by $10 for every

person above 30. (For example, if 32 passengers sign up, the ticket price is $380 per passenger for all 32 passengers.) What number will maximize the intake?

34. An oil field with 30 wells produces 3000 barrels per well per day. For each new well drilled the daily production decreases by 25 barrels per day for each well. (For example, if 32 wells are drilled, the daily production is 2950 barrels.) Determine the number of new wells that must be drilled to maximize the daily production.

35. A rectangular box with an open top is to have a volume of 486 in.³ and its base is to be exactly three times as long as it is wide. Find the dimensions for which the surface area is a minimum.

36. Repeat Exercise 35 with a box closed at the top, assuming now that the volume is 972 in.³

37. The postal service requires that for a rectangular package the combined length and perimeter of a cross-section cannot exceed 108 in. Find the dimensions of the package of maximum volume, assuming a square cross-section. (See Figure 3.42.)

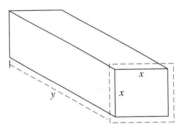

Figure 3.42

38. A small dog shelter in the shape of a rectangular solid is to have a square base and an open front. If the volume is to be 36 ft³, find the dimensions for which the amount of material needed is a minimum. (See Figure 3.43.)

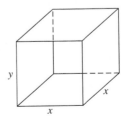

Figure 3.43

39. A rectangular poster is to contain a rectangular picture in the center with an area of 72 in.² The margins at the top and bottom are to be 2 in. and the left and right margins 1 in. (See Figure 3.44.) Determine the dimensions of the poster requiring the least amount of material.

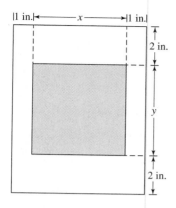

Figure 3.44

40. The manager of a store finds that if she charges $20 for an item, she can sell an average of 120 per week. For each $1 increase in price, the average number of sales per week drops by 4 units. What price should she charge for maximum revenue?

41. A person in a boat 6 km from the nearest point on the shore wants to reach a point P on the shore 10 km from that nearest point. He can walk 5 km/h but row only at the rate of 4 km/h.

Determine the place where the boat must land if he wants to reach point P in the least possible time. (See Example 6 and Figure 3.45.)

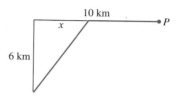

Figure 3.45

42. A girl on a horse 4 km from a river wants to ride to the stable, which is 2 km from the river. The respective points on the bank nearest the horse and stable are 6 km apart. Where should she stop to water the horse in order to reach the stable in the least time? (See Example 6.)

43. What is the altitude of a cylinder of maximum volume that can be inscribed in a right circular cone?

44. Find the altitude of a cone of maximum volume that can be inscribed in a sphere of radius r.

3.5 Related Rates

In this section we are going to continue our study of rates of change with respect to time, but instead of confining ourselves to motion in a straight line, we will consider rates of change in a general setting. As an example, suppose the radius r of a circle is allowed to expand at some known rate, say 2 cm/min. Since $A = \pi r^2$, it ought to be possible to find the rate of change of A in terms of cm²/min. To see how this can be done, we must emphasize that in this and all other problems in this section *every variable quantity is a function of time,* so that all derivatives are taken with respect to t. In our example, the equation $A = \pi r^2$ can be written

$$A(t) = \pi [r(t)]^2$$

Differentiating with respect to t, we get

$$\frac{d}{dt} A(t) = \pi \cdot 2[r(t)] \frac{d}{dt} r(t) \qquad \textbf{power rule}$$

If we remember that A and r are *functions of time,* we can get a simpler expression from the original formula $A = \pi r^2$:

$$A = \pi r^2 \qquad \text{given formula}$$

$$\frac{dA}{dt} = \pi \cdot 2r \frac{dr}{dt} \qquad \text{power rule}$$

$$\frac{dA}{dt} = 2\pi r \frac{dr}{dt}$$

Substituting the known rate of change $dr/dt = 2$, we get

$$\frac{dA}{dt} = 4\pi r$$

So if $r = 1$ cm,

$$\frac{dA}{dt} = 4\pi \frac{\text{cm}^2}{\text{min}}$$

Since the rates of change are related, this type of problem is referred to as a problem in **related rates.**

> In a problem in **related rates,** one or more rates are given, and another, related rate has to be found.

Example 1   An experimenter has determined that the relationship between the tensile strength (in pounds) of a piece of material and the temperature is

$$S = 620 - 0.08\sqrt{T}$$

If the temperature is increasing at the rate of $0.2°\text{F}/\text{min}$, how fast is the tensile strength changing when $T = 100°\text{F}$?

Solution. We are given that $dT/dt = 0.2$ and are asked to find dS/dt when $T = 100$. To this end we differentiate both sides of the equation with respect to t. We have

$$S = 620 - 0.08T^{1/2}$$

and by the power rule, which is a special case of the chain rule, Formula (2.10),

$$\frac{dS}{dt} = \frac{dS}{dT} \frac{dT}{dt}$$

or

$$\frac{dS}{dt} = (-0.08)\frac{1}{2}T^{-1/2}\frac{dT}{dt} \qquad \frac{d}{dt}[T(t)]^{1/2} = \frac{1}{2}[T(t)]^{-1/2}\frac{dT(t)}{dt}$$

Simplifying, we have

$$\frac{dS}{dt} = \frac{-0.04}{\sqrt{T}}\frac{dT}{dt}$$

Substituting the given rate of change, $dT/dt = 0.2$, it follows that

$$\frac{dS}{dt} = \frac{-0.04}{\sqrt{T}}(0.2) = \frac{(-0.04)(0.2)}{\sqrt{T}}$$

Finally, letting $T = 100$,

$$\frac{dS}{dt} = \frac{(-0.04)(0.2)}{\sqrt{100}} = -0.0008\frac{\text{lb}}{\text{min}}$$

(The negative sign indicates that the tensile strength is decreasing.) ∎

Before forming a general plan, consider another example.

Example 2 The relationship between the resistance in a wire and the temperature is

$$R = a + bT + cT^2 + \cdots$$

For temperatures that are not too great, the terms in T^3 and higher powers may be neglected. Suppose for a certain wire that

$$R = 10.123 + 1.320T + 0.00400T^2$$

If T increases at the rate of 2.00°C/min, how fast is R increasing when $T = 20.0$°C?

Solution. As before, we differentiate with respect to t:

$$\frac{dR}{dt} = 1.320\frac{dT}{dt} + (0.00400)(2T)\frac{dT}{dt} \qquad \frac{dR}{dt} = \frac{dR}{dT}\frac{dT}{dt}$$

Substituting $dT/dt = 2.00$ and $T = 20.0$, we find that

$$\frac{dR}{dt} = 1.320(2.00) + (0.00400)(2)(20.0)(2.00) = 2.96\ \Omega/\text{min}$$

to three significant digits. ∎

> **General Strategy for Solving Problems in Related Rates**
> Whenever possible:
> 1. Draw a diagram.
> 2. Label all quantities that vary with time by letters.
> 3. Label all numerical quantities—that is, those quantities that remain fixed throughout the problem.
> 4. Obtain a relationship (equation) between the variables involved.
> 5. List all given rates of change; state the desired rate of change, as well as the instant at which it is to be found.
> 6. Differentiate with respect to t.
> 7. Substitute the known quantities and solve for the unknown rate of change.

Example 3   Boyle's law states that for an ideal gas the pressure is inversely proportional to the volume at a constant temperature; that is,

$$P = \frac{k}{V}$$

where P is the pressure, V the volume, and k a constant. If a certain gas occupies 3.0 m³ when the pressure is 25 Pa (N/m²) and the volume is increasing at the rate of 0.10 m³/min, how fast is the pressure decreasing at the instant when $V = 4.0$ m³?

Solution. In this problem a diagram is not needed. The constant k in the formula can be calculated from the given information: substituting $V = 3.0$ and $P = 25$, we get

$$25 = \frac{k}{3.0} \quad \text{and} \quad k = 75$$

Thus

$$P = 75V^{-1}$$

We are given that $dV/dt = 0.10$ and must find dP/dt when $V = 4.0$. Differentiating, we get

$$\frac{dP}{dt} = 75(-V^{-2})\frac{dV}{dt} = -\frac{75}{V^2}\frac{dV}{dt} \qquad \frac{dP}{dt} = \frac{dP}{dV}\frac{dV}{dt}$$

Since $dV/dt = 0.10$ and $V = 4.0$, we get

$$\frac{dP}{dt} = -\frac{75}{(4.0)^2}(0.10) = -0.47$$

We conclude that the pressure is decreasing at the rate of 0.47 Pa/min at the instant when $V = 4.0$ m³. ∎

Example 4   A metal cube contracts when it is cooled. If the edge of the cube decreases at the rate of 1.0 mm/h, how fast is the volume decreasing at the instant when the edge is 50 mm long?

Solution. In the diagram (Figure 3.46) we label the edge x and the volume V, since both quantities change continuously. (The length 50 mm does not appear in the diagram since x is not a constant.) We are given that $dx/dt = -1.0$ (since x is decreasing) and are asked to find dV/dt when $x = 50$. From

$$V = x^3$$

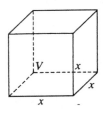

Figure 3.46

we get

$$\frac{dV}{dt} = 3x^2\frac{dx}{dt} \qquad \frac{dV}{dt} = \frac{dV}{dx}\frac{dx}{dt}$$

Since $dx/dt = -1.0$, we obtain

$$\frac{dV}{dt} = 3x^2(-1.0) = -3.0x^2$$

Finally, letting $x = 50$, we get

$$\frac{dV}{dt} = (-3.0)(2500) = -7500 \text{ mm}^3/\text{h}$$

■

Example 5   A woman is walking at the rate of 4.0 ft/s toward a tower 89.6 ft high. Determine the rate of change of the distance to the top of the tower at the instant when she is 50.0 ft from the base.

Solution. In the diagram (Figure 3.47) we label the distance to the base x and the distance to the top z. These are the variable quantities. The height of the tower does not change and is therefore labeled 89.6 ft. In the language of the diagram, we are given that

$$\frac{dx}{dt} = -4.0 \qquad \text{(since } x \text{ is decreasing)}$$

and are asked to find dz/dt when $x = 50.0$. The distance 50.0 ft does not appear in the diagram since x is not a constant.

By the Pythagorean theorem,

$$x^2 + (89.6)^2 = z^2$$

Differentiating, we get

$$2x\frac{dx}{dt} + 0 = 2z\frac{dz}{dt}$$

or

$$x\frac{dx}{dt} = z\frac{dz}{dt}$$

Substituting the given rate of change $(dx/dt = -4.0)$, we find that

$$x(-4.0) = z\frac{dz}{dt}$$

or

$$\frac{dz}{dt} = -4.0\frac{x}{z}$$

This is the general expression for dz/dt. To obtain the rate of change at the instant in question, we substitute the values for x and z. From the diagram (Figure 3.48), we see that

$$z^2 = (89.6)^2 + (50.0)^2 \qquad \text{or} \qquad z = 103$$

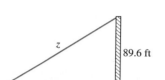

Figure 3.47

z 89.6 ft

50.0 ft

Figure 3.48

Hence $z = 103$ ft at the instant when $x = 50.0$ ft. Substituting $x = 50.0$ and $z = 103$, we get

$$\frac{dz}{dt} = -4.0 \left(\frac{50.0}{103} \right) = -1.9 \text{ ft/s}$$

(The negative sign shows that z is decreasing.)

Caution: A common error in this kind of problem is premature use of the quantity $x = 50.0$. In most problems on related rates, the desired rate of change is not constant. In our example, we first found a general expression for dz/dt in terms of the variable quantities x and z. From the general expression we obtained the rate of change at the instant in question. As a result, the substitution for x and z was not made until after the differentiation was completed. ∎

Example 6   A chemical is poured into a tank in the shape of a cone with vertex down at the rate of $1/5$ m³/min. If the radius of the cone is 3 m and its height 5 m, how fast is the liquid level rising at the instant when the depth of the liquid is 1 m in the center?

Solution. We first draw the diagram in Figure 3.49 and label the variable quantities $V, r,$ and h. The dimensions of the cone do not change and are labeled with constants. The distance $h = 1$ does not appear in the diagram since h is not constant. In the language of the diagram, we are given that $dV/dt = 1/5$ and wish to find dh/dt when $h = 1$.

It is desirable to express the volume of the liquid as a function of h. To this end we make use of similar triangles. Note that

$$\frac{r}{h} = \frac{3}{5}$$

so that $r = (3/5)h$. It follows that

$$V = \frac{1}{3}\pi r^2 h = \frac{1}{3}\pi \left(\frac{3}{5}h \right)^2 h \qquad \text{since } r = \frac{3}{5}h$$

and

$$V = \frac{3\pi}{25}h^3$$

Differentiating, we get

$$\frac{dV}{dt} = \frac{3\pi}{25}(3h^2)\frac{dh}{dt}$$

We now substitute the given rate of change:

$$\frac{1}{5} = \frac{3\pi}{25}(3h^2)\frac{dh}{dt} \qquad \text{or} \qquad \frac{dh}{dt} = \frac{1}{5}\left(\frac{25}{3\pi} \right)\frac{1}{3h^2} = \frac{5}{9\pi h^2}$$

Finally, when $h = 1$,

$$\frac{dh}{dt} = \frac{5}{9\pi}\frac{\text{m}}{\text{min}}$$

∎

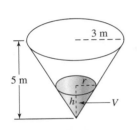

Figure 3.49

■ Exercises/Section 3.5

1. $y = x^2$. Given: $dx/dt = 1$; find: dy/dt when $x = 2$.

2. $y = x^3 + 2$. Given: $dx/dt = 2$; find: dy/dt when $x = 1$.

3. $x^2 + y^2 = 25$. Given: $dy/dt = 3$; find: dx/dt when $y = 3 \, (x > 0)$.

4. $w^2 = x^2 + y^2$. Given: $dx/dt = dy/dt = 2$; find: dw/dt when $x = 5$ and $y = 12 (w > 0)$.

5. Suppose that $E = 100$ V in a DC circuit and that a rheostat is used to increase R at the rate of $2 \, \Omega/s$. Find the resulting rate of change of $I (I = E/R)$ when $R = 10 \, \Omega$.

6. The impedance Z in a series circuit is given by $Z^2 = R^2 + X^2$, where X is the reactance. If $X = 10.0 \, \Omega$ and R increases at the rate of $2.0 \, \Omega/s$, find the rate at which Z is changing when $R = 5.0 \, \Omega$.

7. The power, in watts, in a circuit varies according to $P = Ri^2$. If $R = 100 \, \Omega$ and i varies at the rate of 0.20 A/s, find the rate of change of the power when $i = 2.0$ A.

8. If two variable resistances R_1 and R_2 are linked in parallel, then the effective resistance R of the combination is such that $1/R = 1/R_1 + 1/R_2$. If R_1 increases at the rate of $0.33 \, \Omega/s$ and R_2 at $0.25 \, \Omega/s$, what is the rate of change of R at the instant that $R_1 = 2.00 \, \Omega$ and $R_2 = 3.00 \, \Omega$?

9. Suppose a ladder 5 m long is leaning against a wall. If the bottom of the ladder is being pulled away at the rate of 2 m/min, how fast is the top of the ladder slipping down at the instant that the foot of the ladder is 4 m from the wall? (See Figure 3.50 and Example 5.)

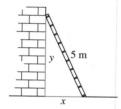

Figure 3.50

10. A balloon is rising from the ground at the rate of 1.5 m/s from a point 30 m from where an observer is standing. How fast is the balloon receding from the observer when it is 40 m above the ground? (See Figure 3.51.)

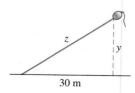

Figure 3.51

11. A balloon is being inflated at the rate of $20.0 \, cm^3/min$. Find the rate at which the radius is increasing when it is 10.0 cm. (Assume the balloon is a sphere.)

12. A balloon is being inflated in such a way that its radius is increasing at the rate of 1.00 cm/s. How fast is the volume increasing when the radius is 5.00 cm? (Assume the balloon is a sphere.)

13. The adiabatic law (explaining changes in a system that can occur with no change in temperature) for the expansion of a diatomic gas is $PV^{1.4} = k$, where P is the pressure, V the volume of the container, and k a constant. At a certain instant the volume is $2.0 \, m^3$ and decreasing at the rate of $1.0 \, m^3/min$, and the pressure is 76 Pa (N/m^2). Find the rate at which the pressure is changing at this instant.

14. The natural frequency f in an LC circuit is given by

$$f = \frac{1}{2\pi} \sqrt{\frac{1}{LC}}$$

If $L = 1.00 \times 10^{-2}$ H and C increases at the rate of 1.00×10^{-6} F/s, find the rate of change of f when $C = 1.00 \times 10^{-2}$ F.

15. A tractor is moving at the rate of 11.8 ft/s away from a building 90.8 ft high (Figure 3.52). How fast is the distance to the top of the building increasing when the tractor is 151 ft away from the base of the building?

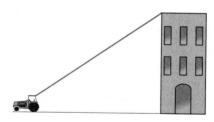

Figure 3.52

16. A circular metal plate is being heated and expands so that its radius increases at the rate of 0.25 mm/min. How fast is the area increasing when the radius is 10 cm?

17. An airplane is cruising at 350 km/h at an altitude of 4000 m. If the plane passes directly over an observer on the ground, how fast is the distance from the plane to the observer changing when it is 3000 m away from the point directly above the observer? (See Figure 3.53.)

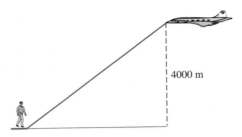

Figure 3.53

18. A student is walking at the rate of 5.0 ft/s toward the Margaret Loock Residence Hall, which is 123 ft high. How fast is the distance to the top of the building changing when he is 60.0 ft from the base of the building?

19. Two ships are leaving port at the same time. The first ship is sailing due east at 20 km/h and the other due north at 15 km/h. How fast are the ships moving away from each other 2.0 h later?

20. A ladder 4 m long is leaning against a wall. If the top of the ladder is slipping down at the rate of 2 m/s, how fast is the bottom moving away from the wall when it is 3 m from the wall?

21. A point moves along the parabola $y^2 = x$ in such a way that the abscissa is decreasing at the rate of 2 units/min. Find the rate of change of the ordinate at the point (16, 4).

22. A point moves along the hyperbola $x^2 - y^2 = 4$. The ordinate is increasing uniformly at 3 units/s. Find the rate of change of the abscissa at $(-3, \sqrt{5})$.

23. A boat is being pulled toward a wharf by a rope attached to the boat's deck from a point 8.0 m above the deck. If the rope is being pulled in at the rate of 2.0 m/min, how fast is the boat approaching the wharf when it is 12.0 m away? (See Figure 3.54.)

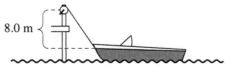

Figure 3.54

24. Mineral waste is being poured onto a pile forming a cone whose radius and height are equal at all times. If the mineral waste is poured at the rate of 50 cm³/s, what is the rate of change of the radius when the pile is 3.0 m high?

25. Wheat is poured on the ground at the rate of 12 ft³/s. The pile forms a cone whose altitude is always three-fourths of the radius. Find the rate at which the altitude is increasing when the altitude is 6.0 ft.

26. At noon, ship A is 129 km due north of ship B. Ship A is moving due west at 18 km/h and ship B due north at 25 km/h. Find the rate of change of the distance between them at 3 P.M.

27. A baseball player is coming in from third base at 24 ft/s. What is the rate of change of his distance from second base when he is 25 ft from home plate? A baseball diamond is a 90-ft square.

28. A piston is moving inside a sealed cylindrical compartment having a diameter of 20.0 cm. If gas is pumped into the cylinder at the rate of 80 cm³/min, determine the rate at which the piston is moving. (See Figure 3.55.)

Figure 3.55

29. A man 170 cm tall walks away from a 3.0-m-high street light at the rate of 2.0 m/s. Find the rate at which his shadow is growing. (See Figure 3.56.)

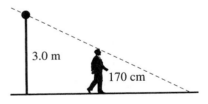

Figure 3.56

30. A storage tank is in the shape of a cone with vertex down. Suppose the altitude of the cone is 10.0 m and the radius of the base is 5.0 m. If water is poured into the tank at the rate of 3.0 m³/min, find the rate at which the water is rising when the water is 4.0 m deep in the center. (See Example 6.)

31. A conical tank with vertex down has a radius of 6.0 ft and a height of 8.0 ft. Water is poured into the tank at the rate of 9.0 ft³/min. Find the rate at which the level is increasing at the instant when the water is 2.0 ft deep in the center.

32. Oil is poured into a conical tank with vertex down at the rate of 2.0 m³/min. The radius of the tank is 2.5 m and the height 5 m. Find the rate at which the level is rising at the instant when the oil is 1/2 m deep in the center.

33. A trough 10.0 m long has a triangular cross-section 5.0 m across the top and 4.0 m deep. If water is poured in at the rate of 2.0 m³/min, how fast is the water level rising when the water is 3.0 m deep?

34. If two ships leave a point at the same time and travel in mutually perpendicular directions at respective velocities a and b, show that they move apart at the constant rate of $\sqrt{a^2 + b^2}$.

35. A swimming pool is 50 ft long and 15 ft wide. Its depth varies from 1 ft to 13 ft, as shown in Figure 3.57. If the pool is filled at the rate of 120 ft³/min, find the rate at which the water level is rising at the deep end at the instant when the depth is 8 ft.

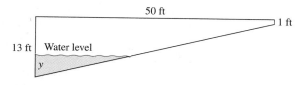

Figure 3.57

3.6   Differentials

In this section we are going to return to the Leibniz notation $dy/dx = f'(x)$ to see in what sense dy/dx may be viewed as a quotient. The question leads quite naturally to the definition of a differential. While the concepts studied in calculus could be developed without the differentials, their use has become so common, particularly in physical science, that we need to familiarize ourselves with the basic ideas.

Formally solving $dy/dx = f'(x)$ for dy, we have $dy = f'(x)\,dx$, called the **differential** of f.

By itself the formula $dy = f'(x)\,dx$ has no obvious meaning. But suppose we choose to interpret dx as an increment, usually denoted by Δx. Then it becomes possible to interpret dy geometrically and to establish a connection between Δy and dy. Consider the graph of $y = f(x)$ in Figure 3.58.

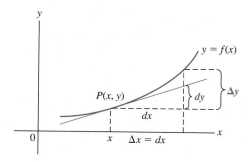

Figure 3.58

The slope of the tangent line at $P(x, y)$ is given by $f'(x)$. Since dx is a finite quantity, we can see from the figure that $f'(x)$ is equal to the vertical distance, denoted by dy, divided by dx, or $dy = f'(x)\,dx$. To repeat, dy is a vertical distance. Furthermore, $dy \approx \Delta y$ if dx is relatively small. The distinction between dy/dx and dy is now seen: dy/dx is the instantaneous *rate* of change of y with respect to x, while dy is the *amount* of change (approximately equal to Δy).

> **Definition of Differential**
>
> $$dy = f'(x)\,dx \qquad \text{(where } dx = \Delta x\text{)} \tag{3.1}$$

Example 1 If $y = x^2$, compute Δy and dy if $x = 2$ and $dx = \Delta x = 0.01$.

Solution. Since $x = 2$, $y = 2^2 = 4$; now recall that

$$\Delta y = f(x + \Delta x) - f(x)$$

Hence

$$\Delta y = (x + \Delta x)^2 - x^2 = (2 + 0.01)^2 - 2^2 \qquad \text{or } \Delta y = \text{new } y - \text{old } y$$
$$= (2.01)^2 - 2^2 = 4.0401 - 4 = \mathbf{0.0401}$$

In other words, when x increases from 2 to 2.01, the value of y increases from 4 to 4.0401, an increase of 0.0401 unit. Now,

$$dy = f'(x)\,dx = 2x\,dx = 2(2)(0.01) = \mathbf{0.04}$$

Thus dy is indeed a good *approximation* for Δy and is easier to calculate than Δy. ∎

We now turn to the usual applications of the differential.

Example 2 Resistance is known to vary with temperature. Suppose that for a certain resistor, $R = 3.5 + 0.002T^2\ \Omega$. If T is measured to be $100°C$ with a possible error of $\pm1°C$, what is the approximate maximum possible error in R?

Solution. Here we simply interpret the error in T as an increment dT. The resulting error in R is ΔR, or approximately dR. Since

$$\frac{dR}{dT} = 0.004T$$

we get for the differential

$$dR = 0.004T\,dT$$

We substitute the given values $T = 100$ and $dT = \pm1$:

$$\Delta R \approx dR = (0.004)(100)(\pm1) = \pm0.4\ \Omega \qquad ∎$$

Relative error

In many cases the actual error is not as important as the size of the error relative to the quantity being measured. To determine the seriousness of the error, we find the quantity dy/y, called the *approximate relative error,* often expressed as a percentage.

Example 3   In Example 2 the actual error was found to be approximately $0.4\,\Omega$. At $T = 100$, $R = 3.5 + 0.002(100)^2 = 23.5$. So the approximate relative error is

$$\frac{dR}{R} = \frac{0.4}{23.5} = 0.017 = 1.7\%$$

■

Example 4   A protective coat of thickness 0.5 mm is applied evenly to the surface of a metal sphere of radius 20.00 cm. Find the approximate number of cubic centimeters of coating used.

Solution. Since $V = \dfrac{4}{3}\pi r^3$,

$$dV = 4\pi r^2\, dr$$

We let $dr = 0.05$ cm and $r = 20.00$ cm. Then

$$dV = 4\pi(20.00)^2(0.05) = 251 \text{ cm}^3$$

which is the approximate amount of coating used.

■

■ Exercises / Section 3.6

In Exercises 1–4, find the differential of the given functions.

1. $y = x^3 - x$

2. $y = x^2 - \dfrac{1}{x}$

3. $y = \dfrac{x}{x-1}$

4. $y = \dfrac{1}{\sqrt{x^2+1}}$

In Exercises 5 and 6, find Δy and dy for the given values of x and Δx.

5. $y = x^2 - x$, $x = 2$, $dx = 0.1$

6. $y = 1/x^2$, $x = 3$, $dx = 0.2$

7. Find the approximate error and percentage error in the area of a square 6.00 cm on a side if an error of 0.02 cm is made in measuring the edge.

8. The side of a square is measured to be 5.00 inches with an error of 0.04 in. Find the approximate error and percentage error in the area.

9. The edge of a cube is measured to be 5.00 inches with a possible error of ±0.01 in. Find the approximate error and percentage error in the volume.

10. The edge of a cube is measured to be 6.00 inches with a possible error of ±0.03 in. Find the approximate error and percentage error in the surface area.

11. Find the approximate percentage error in the volume and in the surface of a sphere of radius 10.00 cm, if an error of ±1 mm is made in measuring the radius.

12. The radius of a sphere is measured to be 3.40 m with an error of ±0.05 m. Find the approximate error and percentage error in the volume.

13. The period of a simple pendulum is given by

$$T = 2\pi\sqrt{\frac{L}{10}}$$

where L is in meters. If $L = 2.0$ m, with an error in measurement of ±0.1 m, what is the approximate error and percentage error in T?

14. A DC circuit has a constant voltage source of 20 V. Find the approximate change in the current if the resistance changes from 8.0 to 8.1 Ω.

15. The power P (in watts) delivered to a resistor is $P = 10.0i^2$. If i changes from 2.1 A to 2.2 A, determine the approximate change in P.

16. The resistance in a certain wire is $R = 60.0 + 0.020T^2$. If T changes from $50.4°$F to $50.7°$F, find the approximate change in R.

17. Find the approximate formula for the area of a circular ring of radius r and width dr.

18. Let A be the area of a square of side s. Draw a figure showing the square, dA, and ΔA.

■ Review Exercises / Chapter 3

1. Find the equations of the tangent and normal lines to the curve $y = \sqrt{x - 2}$ at the point $(3, 1)$.

2. Find the equations of the tangent and normal lines to the hyperbola $x^2 - 3y^2 + 23 = 0$ at $(2, -3)$.

In Exercises 3–10, find the minima and maxima, the intervals on which the graph is concave up and concave down, and the inflection points. Sketch the curve in each case.

3. $y = x^2 - 4x + 3$ 4. $y = x^3 - 6x^2 + 9x$

5. $y = -x^3 + 12x + 2$ 6. $y = 3x^4 - 8x^3 + 9$

7. $y = 3x^4 - 4x^3 + 1$ 8. $y = \dfrac{x^2}{x^2 - 1}$

9. $y = x^2 - \dfrac{1}{x}$ 10. $y = \dfrac{x^2 - 1}{x^3}$

11. The velocity of air through a bronchial tube under pressure is given by $v = kr^2(a - r)$, where a is the radius of the tube when no pressure is applied and k is a constant. Find the radius of the tube for which the velocity is a maximum. (Experiments by means of x-ray photographs have shown that the velocity is indeed maximized during a cough.)

12. A right circular cylinder is inscribed in a sphere of radius a. Show that its maximum volume is $1/\sqrt{3}$ times the volume of the sphere.

13. The *stiffness* of a beam with rectangular cross-section is directly proportional to the product of the width and the cube of the depth. Show that the ratio of depth to width of the stiffest beam that can be cut from a circular log is $\sqrt{3}$.

14. A farmer has 240 ft of fence to build four adjacent rectangular pig pens. The pens are to be constructed as follows: a large rectangular enclosure is subdivided into four parts by three dividing fences parallel to one of the sides. What should the overall dimensions be in order to maximize the total area?

15. Recall that if $C(x)$ is the cost of producing x units of a certain commodity, then $C'(x)$ is the *marginal cost*. Similarly, if $R(x)$ is the revenue derived from the sale of x units, then $R'(x)$ is the *marginal revenue*. The profit is $P(x) = R(x) - C(x)$. Show that for a company to realize maximum profit, the marginal revenue must equal the marginal cost.

16. Assume that the rate at which a rumor spreads through a company is directly proportional to the product of the number of people who have heard the rumor and the number of those who have not. Show that the rumor spreads most rapidly when half the people have heard it.

17. A nuclear plant is located on the edge of a straight river 20 m wide. An electric cable is to be run to a factory on the opposite edge, 50 m downstream. If it costs $30 per meter to run

the cable on land and $45 per meter across the river, how should the cable be laid to minimize the cost?

18. A messenger is in a motorboat 24 mi from a straight shore. He wants to reach a point P that is 150 mi along the shore from a point on the shore nearest the boat. His boat can travel at the rate of 11 mi/h and, upon reaching the shore, he will travel by car at 55 mi/h. Where should he land to reach P in minimum time?

19. A trough is to be made from a long rectangular sheet of metal 16 in. wide by bending the sheet lengthwise to form a V. Find the depth of the trough that gives the largest possible carrying capacity. (See Figure 3.59.)

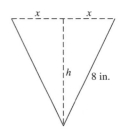

Figure 3.59

20. A rectangular box with an open top has a surface area of 726 cm^2. The base is twice as long as it is wide. Find the dimensions for which the volume is a maximum.

21. The resistance (in ohms) of a resistor is $R = 25 + 0.020T^2$. If the temperature is increasing at the rate of $1.6°$C/min, how fast is the resistance changing at the instant when $T = 50.0°$C?

22. Oil is leaking from an offshore well forming a circular slick. The area of the slick is increasing at the rate of 3 mi^2/day. Determine the rate at which the radius is increasing at the instant when the radius is 5 mi.

23. The electromotive force E in volts that produces a current of I amperes in a wire of diameter d inches is given by $E = 0.1138I/d^2$. If $d = 0.060$ inch and E is increasing at the rate of 3.00 V/s, find the time rate of change of I.

24. The thin-lens equation $1/s_1 + 1/s_2 = 1/f$ relates the distances s_1 and s_2 from an object and its image to the lens. The constant f is called the *focal length*. An object is moving away from a lens with focal length 30.0 cm at the rate of 15.0 cm/min. How fast is the image moving when the object is 90.0 cm from the lens?

25. Water is entering a conical tank 10 m deep and 20 m across the top at 9 m^3/min. How fast is the water rising when 3 m deep?

26. The ends of a trough are equilateral triangles with vertex down. The trough is 9 m long. When the water in the trough is 2 m deep, its depth increases at the rate of 1/2 m/min. At what rate is water being poured in at that instant?

27. The volume V and the pressure P in the cylinder of a diesel engine are related by the equation $PV^{1.4} = k$, where k is a constant. At a certain instant during the compression stroke the cylinder contains 40 in.^{3} of gas vapor under a pressure of 300 lb/in.^{2} and the volume is decreasing at the rate of 80 in.^{3}/s. At what rate is the pressure increasing at that instant?

28. If $y = x - \sqrt{x}$, compute Δy and dy if $x = 4$ and $dx = \Delta x = 0.02$.

29. The edge of a cube is measured to be 10.00 cm with a possible error of ± 0.2 mm. Find the approximate error and percentage error in the volume.

30. A spherical balloon has a radius of 2.0 ft with a possible error of ± 0.1 ft. What is the approximate percentage error in the volume?

31. The radius of a circle is measured to be 5.00 in. with an error of ± 0.02 in. Find the approximate error and percentage error in the area.

4

The Integral

Antiderivatives

Antiderivative

So far we have been concerned primarily with the following problem: given a function, find its derivative. We will now consider the inverse problem: given the derivative of a function, find the function—or, in symbols, given f, find F such that $F' = f$. F is called the **antiderivative** of f. To see how antiderivatives may arise, recall that if v is the velocity of a particle, then the acceleration is $a = dv/dt$. Now g, the acceleration due to gravity, is 32 ft/s^2 or 10 m/s^2. (A more precise value is 9.8 m/s^2.) It follows that the velocity in meters per second is the antiderivative $v = 10t$, since $dv/dt = 10 \, \text{m/s}^2$.

Antiderivatives are not unique. If $a = 10 \, \text{m/s}^2$, then $v = 10t$, as already noted. However, $v = 10t + 1$ and $v = 10t + 3$ are also antiderivatives. In fact, $v = 10t + C$, for an arbitrary constant C, is an antiderivative. (The physical significance of the arbitrary constant will be discussed in Section 4.8.)

Even though antidifferentiation is, in general, much more complicated than differentiation, simple functions can be handled routinely. For example, an antiderivative $F(x)$ of $f(x) = x^2$ has to be a cubic, but since $dx^3/dx = 3x^2$, we must have $F(x) = (1/3)x^3 + C$. Thus $(d/dx)F(x) = x^2$, as desired. In general, then, if

$$f(x) = x^n \qquad n \text{ rational}$$

then

$$F(x) = \frac{x^{n+1}}{n+1} + C \qquad n \neq -1 \tag{4.1}$$

where C is an arbitrary constant. It is a simple exercise to show that $F'(x) = f(x)$. It follows that for any constant k, if

$$f(x) = kx^n \qquad n \text{ rational}$$

then

$$F(x) = \frac{kx^{n+1}}{n+1} + C \qquad n \neq -1 \tag{4.2}$$

Finally, if $f(x) = k$, then

$$F(x) = kx + C \tag{4.3}$$

Example 1   If $f(x) = 2x^2 + x^3$, find an antiderivative $F(x)$.

Solution. By Formulas (4.1) and (4.2) we get

$$F(x) = 2 \cdot \frac{x^3}{3} + \frac{x^4}{4} + C = \frac{2x^3}{3} + \frac{x^4}{4} + C$$

As a check, observe that $F'(x) = f(x)$ by the sum rule.   ■

Example 2   Find $F(x)$ if $f(x) = 3x^4 + x + 2$.

Solution. In this example, we also need Formula (4.3) which gives $2x$, the anti-derivative of $k = 2$:

$$F(x) = \frac{3x^5}{5} + \frac{x^2}{2} + 2x + C$$ ■

Example 3   Find $F(x)$ if $f(x) = \sqrt{x} - (2/x^2) - 6$.

Solution. We write

$$f(x) = x^{1/2} - 2x^{-2} - 6$$

and obtain

$$F(x) = \frac{x^{3/2}}{\frac{3}{2}} - \frac{2x^{-1}}{-1} - 6x + C = \frac{2}{3}x^{3/2} + \frac{2}{x} - 6x + C$$ ■

Example 4   Find $F(x)$, given that $f(x) = x^2\sqrt{x} - (1/\sqrt[3]{x})$.

Solution. Again the function has to be rewritten:

$$f(x) = x^2 x^{1/2} - \frac{1}{x^{1/3}} = x^{5/2} - x^{-1/3}$$

Hence

$$F(x) = \frac{x^{7/2}}{\frac{7}{2}} - \frac{x^{2/3}}{\frac{2}{3}} + C = \frac{2}{7}x^{7/2} - \frac{3}{2}x^{2/3} + C$$ ■

■ Exercises / Section 4.1

Find the antiderivatives of the following functions.

1. $f(x) = 3$

2. $f(x) = 2x$

3. $f(x) = 1 - 3x^2$

4. $f(x) = x^2 + x + 4$

5. $f(x) = 2x^3 - 3x^2 + x$

6. $f(x) = 2x^4 - 6x^2 + x + 5$

7. $f(x) = x^3 - 3x^2$

8. $f(x) = x - 7x^4$

9. $f(x) = x^5 - 6x^4 + 2x^3 + 3$

10. $f(x) = x^{-3} + 2x^{-2} + 4$ **11.** $f(x) = \dfrac{1}{x^2} - 2$ **14.** $f(x) = x - \dfrac{1}{2x^2} + 7$ **15.** $f(x) = \dfrac{2}{3x^2} + \dfrac{5}{4x^3} + \sqrt{x}$

12. $f(x) = \dfrac{1}{x\sqrt{x}}$ **13.** $f(x) = \dfrac{3}{x^2} + \dfrac{2}{\sqrt[3]{x}}$ **16.** $f(x) = 2x^{-5/2} + 7x^{-9/2}$

If additional practice in finding antiderivatives is desired at this point, moving ahead to Exercises 1–52 in Section 4.5 is entirely feasible. The notation for the indefinite integral (Section 4.4) is needed, however.

4.2   The Area Problem

In this section we are going to study a limit procedure for determining the area under a curve. This procedure will lead to the definition of the definite integral.

Consider a region bounded by a function, the x-axis, and the vertical lines $x = a$ and $x = b$ (Figure 4.1). Such a region can be approximated by rectangles. The larger the number of rectangles, the better the approximation.

To get an expression for the sum of the areas of the rectangles, we need to introduce the appropriate notation. The number of subdivisions of the interval $[a, b]$ will be denoted by n. Let $x'_0 = a$ and $x'_n = b$, while $x'_1, x'_2, \ldots, x'_{n-1}$ are distinct points between a and b; this way we obtain n subdivisions of the interval. (See Figure 4.2.) Note that the subintervals need not be of equal length. The length of the ith subinterval will be denoted by $\Delta x_i = x'_i - x'_{i-1}$.

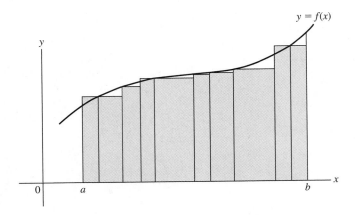

Figure 4.1

Figure 4.2

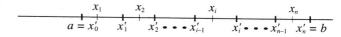

Figure 4.3

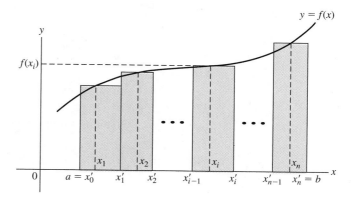

Figure 4.4

To construct the desired rectangles we select an arbitrary point x_i in each subdivision (Figure 4.3). Each x_i can then be used to erect the altitude of the ith rectangle in the region (Figure 4.4). The altitude of each rectangle is the dashed line of length $f(x_i)$. Consequently, the area of the first rectangle is $f(x_1)\Delta x_1$, that of the second, $f(x_2)\Delta x_2$, and so on. The sum of all the areas is

$$f(x_1)\Delta x_1 + f(x_2)\Delta x_2 + \cdots + f(x_n)\Delta x_n \qquad (4.4)$$

This sum can be written more compactly by means of the *sigma notation*. The sum

$$a_1 + a_2 + \cdots + a_n$$

of n terms can be denoted by

Sigma notation

$$\sum_{i=1}^{n} a_i$$

which means "the summation of all terms a_i, where i assumes all integral values from 1 to n inclusive." For example:

$$\sum_{i=1}^{n} i^2 = 1^2 + 2^2 + 3^2 + \cdots + n^2$$

$$\sum_{i=0}^{3} 2^i = 2^0 + 2^1 + 2^2 + 2^3$$

$$\sum_{i=2}^{5} \frac{1}{i} = \frac{1}{2} + \frac{1}{3} + \frac{1}{4} + \frac{1}{5}$$

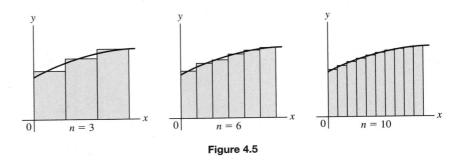

Figure 4.5

Similarly, the Sum (4.4) can be written

$$\sum_{i=1}^{n} f(x_i)\Delta x_i \qquad (4.5)$$

Our intuition will now lead us to the definition of the area of a region. We have deliberately avoided assigning a specific value to n, so that Formula (4.5) remains valid for any number of subdivisions of $[a, b]$. If n gets large, the sum of the areas of the rectangles, $\sum_{i=1}^{n} f(x_i)\Delta x_i$, will approximate the area of the region more and more closely. (See Figure 4.5.) If n increases without limit, and if the length of each subinterval approaches zero, then the sum approaches the area of the region in the limit. This can be stated simply as

$$\text{Area under the curve} = \lim_{n\to\infty} \sum_{i=1}^{n} f(x_i)\Delta x_i$$

Note: It will be understood from now on that any time we use the notation

$$\lim_{n\to\infty} \sum_{i=1}^{n}$$

all the subintervals will shrink to zero.

Since the region under the curve is bounded by the lines $x = a$ and $x = b$, we will employ the following notation for the area:

$$\text{Area} = \int_a^b f(x)\,dx = \lim_{n\to\infty} \sum_{i=1}^{n} f(x_i)\Delta x_i \qquad (4.6)$$

Definite integral
Integrand
Limits of integration
Integral sign

The middle expression for the area is called the **definite integral of f from a to b** and $f(x)$ is called the **integrand;** the numbers a and b are called the **limits of integration.** The symbol $\int$ is actually an old-fashioned S (for sum) and is called an **integral sign.** The symbolism is attributed to Leibniz. Although the simplified notation $\int_a^b f$ would express the definite integral perfectly satisfactorily, it is customary to include letters, such as x, in the expression of the integrand. Since the letters themselves are arbitrary,

$$\int_a^b f(x)\,dx = \int_a^b f(y)\,dy = \int_a^b f(w)\,dw$$

The variables in the integral are sometimes referred to as *dummy variables*.

To clarify the definition, we are going to compute the area of a region by dividing it into subintervals of equal length. The following formulas will be needed:

A. $\displaystyle\sum_{i=1}^n i = \frac{n(n+1)}{2}$

B. $\displaystyle\sum_{i=1}^n i^2 = \frac{n(n+1)(2n+1)}{6}$

C. $\displaystyle\sum_{i=1}^n i^3 = \left[\frac{n(n+1)}{2}\right]^2$

To illustrate Formula A, note that

$$1 + 2 + 3 + \cdots + 15 = \frac{15(15+1)}{2} = 120$$

as can be readily checked.

Example 1   Evaluate $\int_0^3 x^2\,dx$ by use of Definition (4.6).

Solution. Subdivide the interval $[0, 3]$ into n subintervals of equal length. Then $\Delta x_i = 3/n$ for all i. We choose the right-hand endpoint for x_i in each subinterval, although the left-hand endpoint could be chosen instead (Figure 4.6). Then the values of x_i are found to be

$$x_1 = 1 \cdot \frac{3}{n},\, x_2 = 2 \cdot \frac{3}{n},\, \ldots,\, x_i = i \cdot \frac{3}{n},\, \ldots,\, x_n = n \cdot \frac{3}{n} = 3$$

The altitudes of the individual rectangles are computed from the function $f(x) = x^2$. Thus

$$f(x_1) = 1^2 \cdot \frac{9}{n^2} \qquad \text{since } f(x_1) = x_1{}^2 = \left(1 \cdot \frac{3}{n}\right)^2$$

$$f(x_2) = 2^2 \cdot \frac{9}{n^2} \qquad \text{since } f(x_2) = x_2{}^2 = \left(2 \cdot \frac{3}{n}\right)^2$$

$$f(x_i) = i^2 \cdot \frac{9}{n^2}$$

$$f(x_n) = n^2 \cdot \frac{9}{n^2}$$

Since $\Delta x_i = 3/n$, the sum of the areas of the rectangles can now be written

$$\sum_{i=1}^n f(x_i)\Delta x_i = \sum_{i=1}^n i^2 \frac{9}{n^2}\frac{3}{n}$$

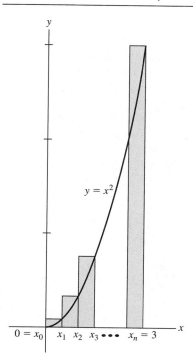

y

$y = x^2$

$0 = x_0$ x_1 x_2 x_3 $\bullet\bullet\bullet$ $x_n = 3$ x

Figure 4.6

This expression reduces to

$$\sum_{i=1}^{n} f(x_i)\Delta x_i = \frac{27}{n^3}\sum_{i=1}^{n} i^2 = \frac{27}{n^3}\frac{n(n+1)(2n+1)}{6}$$

by Formula B. Consequently,

$$\int_0^3 x^2\,dx = \lim_{n\to\infty}\frac{27n(n+1)(2n+1)}{6n^3}$$

$$= \lim_{n\to\infty}\frac{27(2n^3+3n^2+n)}{6n^3}$$

$$= \lim_{n\to\infty}\frac{27\left(2+\frac{3}{n}+\frac{1}{n^2}\right)}{6} = 9 \qquad \textbf{dividing by } n^3$$

We conclude that the area of the region is 9 square units. ∎

■ Exercises / Section 4.2 (Optional)

Use the method of Example 1 to evaluate the following definite integrals.

1. $\int_0^1 x\,dx$ 2. $\int_0^3 x\,dx$ 3. $\int_0^1 x^2\,dx$ 4. $\int_0^1 x^3\,dx$ 5. $\int_0^2 3x^2\,dx$ 6. $\int_0^2 (1+2x)\,dx$

4.3 The Fundamental Theorem of Calculus

In Section 4.1 we encountered antiderivatives. In Section 4.2 we considered the definition of an area under a curve. In this section we combine the concepts of antiderivative and area, which leads us to the fundamental theorem of calculus.

Let B be a fixed number. Consider the area under the curve in Figure 4.7. This area is bounded on the left by the line $x = B$ and on the right by a vertical line through x, where x is allowed to vary. As long as B is fixed, every value of x determines a unique value for the area. Consequently, the area is a function of the variable x, denoted by $A(x)$. Let x take on a small increment Δx, so that $A(x)$ changes by an amount ΔA. These increments are pictured in Figure 4.8.

Now ΔA is the area $PQUT$ under the curve. This area, which is not necessarily rectangular, is wedged between the rectangles $RQUT$ and $PSUT$ in such a way that area $PSUT \le \Delta A \le$ area $RQUT$.

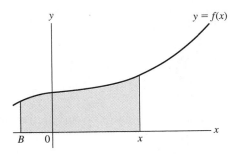

Figure 4.7

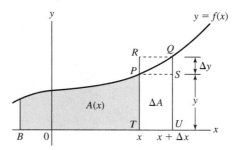

Figure 4.8

If we think about the increment Δx that generated the area $PQUT$, we realize that area $PSUT = TP\ \Delta x$ and area $RQUT = UQ\ \Delta x$. It follows that

$$TP\ \Delta x \leq \Delta A \leq UQ\ \Delta x$$

or

$$TP \leq \frac{\Delta A}{\Delta x} \leq UQ \qquad \text{dividing by } \Delta x$$

Noting that TP is the value of y at x and that $UQ = y + \Delta y$, the inequalities can be written

$$y \leq \frac{\Delta A}{\Delta x} \leq y + \Delta y$$

Finally, if $\Delta x \to 0$, then $\Delta y \to 0$, so that $y + \Delta y \to y$. It follows that $\Delta A / \Delta x \to y$ as $\Delta x \to 0$, or

$$y = \lim_{\Delta x \to 0} \frac{\Delta A}{\Delta x}$$

But this is the very definition of derivative of the function $A(x)$, that is,

$$y = \lim_{\Delta x \to 0} \frac{\Delta A}{\Delta x} = \frac{dA}{dx} \qquad \text{Definition (2.1)}$$

We now reach the surprising conclusion that $dA/dx = f(x)$, which makes $A(x)$ an antiderivative of $f(x)$. Moreover, from our definition of area, we have

$$\int_B^x f(x)\,dx = A(x)$$

Using $F(x)$ to denote an antiderivative, we can rewrite the last equation:

$$\int_B^x f(x)\,dx = F(x) + C$$

(The nonuniqueness of the area function can be explained geometrically by the fact that we could have chosen a different value for B.)

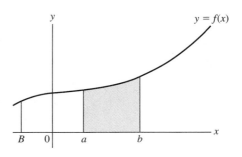

Figure 4.9

To find the area under the curve between two specific lines $x = a$ and $x = b$ (Figure 4.9), we subtract the smaller from the larger area. The larger area, bounded by B and b, is $F(b) + C$ and the smaller area is $F(a) + C$, so that the area bounded by $x = a$ and $x = b$ is given by

$$[F(b) + C] - [F(a) + C] = F(b) - F(a)$$

Since this area is denoted by the definite integral $\int_a^b f(x)\,dx$, we conclude that

$$\int_a^b f(x)\,dx = F(b) - F(a)$$

This result is known as the **fundamental theorem of calculus** and was discovered independently by Newton and Leibniz.

Fundamental Theorem of Calculus

If f is continuous on the interval $[a, b]$, then

$$\int_a^b f(x)\,dx = F(x)\Big|_a^b = F(b) - F(a) \tag{4.7}$$

where F is a function such that $F' = f$ on $[a, b]$—that is, F is an antiderivative of f.

The fundamental theorem enables us to find areas under a curve by antidifferentiation, thereby establishing a profound connection between areas and antidifferentiation.

According to Formula (4.7) we evaluate a definite integral by finding an antiderivative $F(x)$ and evaluating $F(x)$ for $x = b$ and $x = a$. The difference between the two is the value of the integral. Note that the arbitrary constant C does not have to be included since it cancels when the values are subtracted. From now on the process of finding an antiderivative will be called **integration.** (To *integrate* means to find an antiderivative.)

Integration
Integrate

Notation

In Formula (4.7) the expression $F(b) - F(a)$ is denoted by

$$F(x)\Big|_a^b$$

The purpose of this notation is to keep track of the limits of integration.

Example 1

Let us find the area evaluated in Example 1 of the last section by integration. By Formula (4.7)

$$\int_0^3 x^2 \, dx = \frac{x^3}{3}\Big|_0^3 = \frac{3^3}{3} - \frac{0^3}{3} = 9$$

The new procedure is somewhat shorter. ∎

Example 2

Find the area under the curve $y = 1/x^2$ between the lines $x = 1$ and $x = 3$. (See Figure 4.10.)

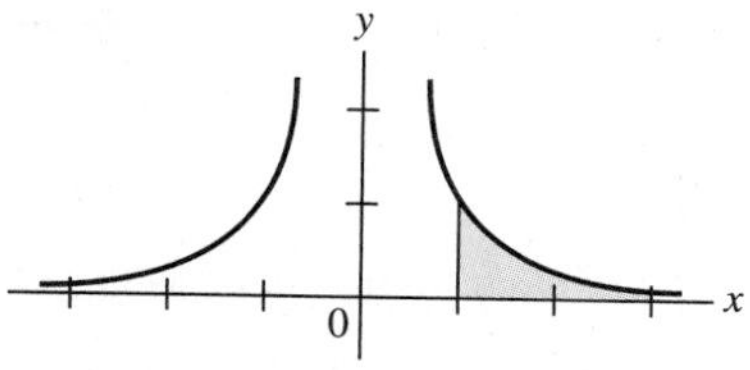

Solution. Since $f(x) = 1/x^2$, we have

$$\int_1^3 \frac{1}{x^2} \, dx = \int_1^3 x^{-2} \, dx = \frac{x^{-1}}{-1}\Big|_1^3$$

$$= -\frac{1}{x}\Big|_1^3 = \left(-\frac{1}{3}\right) - (-1) = \frac{2}{3} \text{ square unit}$$ ∎

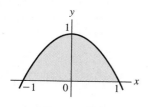

Figure 4.10

Example 3

Find the area of the region bounded by $y = 1 - x^2$ and the x-axis.

Solution. Since $1 - x^2 = 0$ when $x = \pm 1$, the x-intercepts are $(-1, 0)$ and $(1, 0)$, so that the region extends from $x = -1$ to $x = 1$ (Figure 4.11). Thus

$$\int_{-1}^1 (1 - x^2) \, dx = x - \frac{x^3}{3}\Big|_{-1}^1$$

$$= \left(1 - \frac{1}{3}\right) - \left(-1 + \frac{1}{3}\right) = \frac{4}{3} \text{ square units}$$ ∎

Figure 4.11

■ Exercises / Section 4.3

1. Evaluate the definite integrals in the exercises in Section 4.2 by integration.

In Exercises 2–10, find each area bounded by the indicated curves and x-axis.

2. $y = x$, $x = 1$, $x = 4$

3. $y = \frac{1}{2}x$, $x = 0$, $x = 2$

4. $y = x^2$, $x = 1$, $x = 3$

6. $y = x - x^2$

8. $y = \frac{2}{x^2}$, $x = 1$, $x = 2$

10. $y = 1 - x^4$

5. $y = x^3 + 1$, $x = 0$, $x = 1$

7. $y = 4 - x^2$

9. $y = \frac{1}{x^3}$, $x = 1$, $x = 3$

4.4   The Integral: Notation and General Definition

The purpose of this section is twofold: to discuss the notation used for the integral and to generalize the definition of the definite integral.

Since the fundamental theorem enables us to find areas by antidifferentiation, the antiderivative is denoted by $\int f(x) \, dx$, called the **indefinite integral.** Note especially that the integration symbol uses the differential notation. So if $F(x)$ is the antiderivative of $f(x)$, then

Indefinite integral

$$dF(x) = d \int f(x) \, dx = f(x) \, dx$$

The differential notation, due to Leibniz, is particularly useful in applications of integration. To see why, recall that we obtained an approximation for the area under a curve by using the sum

$$\sum_{i=1}^{n} f(x_i)\Delta x_i$$

Riemann sum

in Section 4.2. A sum of this form is called a **Riemann sum** in honor of the German mathematician G. F. B. Riemann (1826–1866). This concept enables us to generalize the definition of the definite integral.

Definition of Definite Integral

A definite integral is the limit of a Riemann sum:

$$\int_a^b g(x)\,dx = \lim_{n \to \infty} \sum_{i=1}^{n} g(x_i)\Delta x_i$$

Once this definition has been formulated, *we no longer care how a particular Riemann sum is obtained*—only the form matters. In fact, the Riemann sum enables us to set up different types of integrals in a way that has a definite intuitive appeal. Consider the area under the graph of $y = f(x)$ and draw a **typical element** (Figure 4.12) of height $f(x)$ and thickness dx. The area of the typical element is $f(x)\,dx$, suggesting a shortcut to the Riemann sum procedure.

Typical element

Shortcut to Riemann Sum

We can think of $\int_a^b$ as an operation summing up the little areas $f(x)\,dx$ to get

$$\int_a^b f(x)\,dx$$

This shortcut, which may be called a *sloppy Riemann sum,* is used extensively in Chapter 5, where the definition of the definite integral is carried far beyond the study of areas. Put another way, every definite integral is the limit of a Riemann sum, and its value can have many different interpretations, both geometric and physical.

The sloppy Riemann sum is sometimes referred to as the sum of an infinite number of infinitesimally thin rectangles. (See Figure 4.12.) This appears to be the way

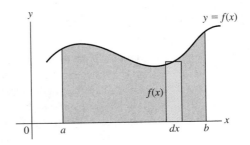

Figure 4.12

Leibniz interpreted his own creation. Since the procedure using Riemann sums is always available to us, the reference to infinitesimals can be avoided. On the other hand, the shortcut method is independent of this interpretation of dx. As a heuristic device, thinking of dx as infinitely small cannot really be faulted.

4.5   Basic Integration Formulas

In this section we are going to develop some basic techniques for integrating algebraic functions. Using the notation for the definite integral, Formulas (4.1) and (4.3) in Section 4.1 can be written

$$\int x^n dx = \frac{x^{n+1}}{n+1} + C \qquad (n \neq -1) \tag{4.8}$$

and

$$\int k\, dx = kx + C \tag{4.9}$$

Formula (4.2) can be expressed in the form

$$\int kx^n dx = k \int x^n dx = \frac{kx^{n+1}}{n+1} + C \qquad n \neq -1$$

and, more generally,

$$\int kf(x)\, dx = k \int f(x)\, dx \tag{4.10}$$

since $d \int kf(x)\, dx = k[d \int f(x)\, dx] = kf(x)\, dx$.

Formula (4.10) is a special case of the following property:

$$\int [c_1 f(x) + c_2 g(x)]\, dx = c_1 \int f(x)\, dx + c_2 \int g(x)\, dx$$

Because of this property, the integral is said to be *linear*. (We will see the reason for this name in Chapter 12.)

Most of Chapter 7 is devoted to developing different integration techniques. Being confined to algebraic functions at this point, we will consider only one more special case, *the generalized power rule in reverse*. Recall that for

$$f(x) = (x^2 + 1)^2$$

the generalized power rule yields

$$f'(x) = 2(x^2 + 1)\frac{d}{dx}(x^2 + 1) = 2(x^2 + 1)(2x)$$

or, in differential form,

$$df(x) = 2(x^2 + 1)(2x\,dx)$$

From the differential form it follows that

$$\int 2(x^2 + 1)(2x\,dx) = (x^2 + 1)^2 + C$$

Now consider the integral

$$\int \frac{1}{2}(x^2 + 1)^{-1/2} 2x\,dx$$

Keeping the previous example in mind, we first observe that $d(x^2 + 1) = 2x\,dx$. It follows that

$$\int \frac{1}{2}(x^2 + 1)^{-1/2} 2x\,dx = (x^2 + 1)^{1/2} + C$$

The general form of this integral, which is readily verified by the generalized power rule, will be called the *general power formula*.

General Power Formula

$$\int u^n\,du = \frac{u^{n+1}}{n+1} + C \qquad (n \neq -1) \tag{4.11}$$

Formula (4.11) suggests a systematic procedure for performing an integration of this type.

Example 1 Use Formula (4.11) to compute the integral $\int (1/2)(x^2 + 1)^{-1/2}\, 2x\,dx$.

Solution. We let $u = x^2 + 1$ and find that

$$du = 2x\,dx$$

These expressions may be substituted in the integral, which becomes

$$\int \frac{1}{2} u^{-1/2}\,du = \frac{1}{2}\frac{u^{1/2}}{\frac{1}{2}} + C = u^{1/2} + C$$

by Formula (4.11). Hence (substituting back)

$$\int \frac{1}{2}(x^2 + 1)^{-1/2}\, 2x\,dx = (x^2 + 1)^{1/2} + C \qquad \blacksquare$$

Example 2 Integrate $\int x^2\sqrt{x^3 + 1}\,dx$.

Solution. The integral can be written in the form

$$\int (x^3 + 1)^{1/2}x^2\,dx$$

As before, let $u = x^3 + 1$, whence $du = 3x^2\,dx$.

Direct substitution in the integral is now impossible since $x^2\,dx$ does not match du. Yet all that is required here is a simple trick: by Formula (4.10)

$$\int (x^3+1)^{1/2}x^2\,dx = \int (x^3+1)^{1/2}\,\frac{1}{3}\cdot 3x^2\,dx$$

$$= \frac{1}{3}\int (x^3+1)^{1/2}\,3x^2\,dx \qquad \int kf(x)\,dx = k\int f(x)\,dx$$

Substitution now gives

$$\frac{1}{3}\int u^{1/2}\,du = \frac{1}{3}\frac{u^{3/2}}{\frac{3}{2}}+C = \frac{2}{9}(x^3+1)^{3/2}+C \qquad\blacksquare$$

Example 3   Integrate

$$\int \frac{x\,dx}{\sqrt[3]{1-x^2}}$$

Solution.   The integral can be written in the form

$$\int (1-x^2)^{-1/3}x\,dx$$

Let $u = 1-x^2$, so that $du = -2x\,dx$. We insert the number -2 and place $-1/2$ in front of the integral:

$$-\frac{1}{2}\int (1-x^2)^{-1/3}(-2x\,dx) = -\frac{1}{2}\int u^{-1/3}\,du = -\frac{1}{2}\frac{u^{2/3}}{\frac{2}{3}}+C$$

$$= -\frac{3}{4}(1-x^2)^{2/3}+C \qquad\blacksquare$$

Caution: Since we are dealing with a very special type of integral, its form has to match the form in Formula (4.11), except for a multiplicative constant. Consider, for example, the integral

$$\int \frac{dx}{\sqrt{1-x^2}}$$

If we let $u = 1-x^2$, then $du = -2x\,dx$. Since $-2x$ is not a constant, our previous trick no longer works. In other words, the integral cannot be worked out with our present techniques.

Example 4   Integrate $\int (x^2-1)^2\,dx$.

Solution.   If $u = x^2-1$, then $du = 2x\,dx$. As noted above, since $2x$ is not a constant, this integral is not of the form

$$\int u^n\,du$$

Consequently, we must proceed by multiplying out the expression in the integrand and integrating term by term:

$$\int (x^2 - 1)^2 dx = \int (x^4 - 2x^2 + 1)\, dx = \frac{1}{5}x^5 - \frac{2}{3}x^3 + x + C \qquad \blacksquare$$

A common error is ignoring the differential du and writing

$$\int (2x^2 + 1)^4 x\, dx \qquad \text{as} \qquad \frac{(2x^2 + 1)^5}{5} + C$$

Since $u = 2x^2 + 1$ and $du = 4x\, dx$, the correct procedure is to insert 4 and place $1/4$ in front of the integral. Thus

$$\int (2x^2 + 1)^4 x\, dx = \frac{1}{4}\int (2x^2 + 1)^4 (4x)\, dx = \frac{1}{4}\frac{(2x^2 + 1)^5}{5} + C$$

The next example illustrates the evaluation of a definite integral.

Example 5   Evaluate

$$\int_{-\sqrt{6}}^{-1} \frac{x\, dx}{\sqrt{10 - x^2}}$$

Solution. The variable x is changed to the variable u by the substitution

$$u = 10 - x^2$$
$$du = -2x\, dx$$

The equation $u = 10 - x^2$ is also used to change the limits of integration:

Lower limit: If $x = -\sqrt{6}$, then $u = 10 - x^2 = 10 - (-\sqrt{6})^2 = 4.$
Upper limit: If $x = -1$, then $u = 10 - x^2 = 10 - (-1)^2 = 9.$

It follows that

$$\int_{-\sqrt{6}}^{-1} \frac{x\, dx}{\sqrt{10 - x^2}} = -\frac{1}{2}\int_{-\sqrt{6}}^{-1} \frac{-2x\, dx}{\sqrt{10 - x^2}} = -\frac{1}{2}\int_{4}^{9} \frac{du}{\sqrt{u}}$$

$$= -\frac{1}{2}\int_{4}^{9} u^{-1/2}\, du = -u^{1/2}\Big|_{4}^{9}$$

$$= -(9)^{1/2} - [-(4)^{1/2}] = -\sqrt{9} + \sqrt{4}$$

$$= -3 + 2 = -1 \qquad \blacksquare$$

■ Exercises / Section 4.5

Perform the following integrations.

1. $\int \sqrt{x}\, dx$

2. $\int (x\sqrt{x} - x)\, dx$ (Recall that $x\sqrt{x} = x^{3/2}$.)

3. $\int \left(\dfrac{1}{x^3} - \dfrac{3}{x^2} \right) dx$

4. $\int (x^{-2/3} + 3x^{1/2})\, dx$

5. $\int (2\sqrt{x} - 3x^2 + 1)\, dx$

6. $\int \left(\dfrac{2}{\sqrt{x}} - 3x\sqrt{x} + 2 \right) dx$

7. $\int \left(\dfrac{1}{x^4} + \dfrac{1}{\sqrt{x}} - 4 \right) dx$

8. $\int (x^2 + 1)^4 (2x)\, dx$

9. $\int (2x^2 - 3)^3 (4x)\, dx$

10. $\int (2x^2 - 3)^3 x\, dx$

11. $\int (2 - x^2)^4 x\, dx$

12. $\int (4 - x^3)^2 x^2\, dx$

13. a. $\int (1 - x)\, dx$ b. $\int (1 - x)^4\, dx$

14. $\int (x^4 + 1)^3 x^3\, dx$

15. $\int \dfrac{x\, dx}{(x^2 - 1)^2}$

16. $\int \sqrt{1 + x}\, dx$

17. $\int (2x^2 + x)^3 (4x + 1)\, dx$

18. $\int (x^3 - 3x)^5 (x^2 - 1)\, dx$

19. $\int \dfrac{dt}{\sqrt{1 - t}}$

20. $\int \sqrt{1 - 2t}\, dt$

21. $\int \dfrac{x\, dx}{\sqrt{1 - x^2}}$

22. $\int t\sqrt[3]{t^2 + 1}\, dt$

23. $\int (2x^3 - 1)\sqrt[5]{x^4 - 2x}\, dx$

24. $\int \dfrac{2x - 1}{\sqrt{x^2 - x}}\, dx$

25. $\int (x^2 + 1)^2\, dx$ (See Example 4.)

26. $\int x(x^2 + 1)^2\, dx$

27. $\int (1 - x^2)^2 x\, dx$

28. $\int (1 - x^2)^2\, dx$

29. $\int (1 + \sqrt{x})^2\, dx$

30. $\int \dfrac{1 + \sqrt{x}}{\sqrt{x}}\, dx$

31. $\int (1 - 5s)^{4/3}\, ds$

32. $\int \left(2\sqrt{z} - \dfrac{1}{2}z^{-1/2} + 2z \right) dz$

33. $\int \left(x^{-1/4} + \dfrac{1}{x\sqrt{x}} - x \right) dx$

34. $\int \dfrac{1 - 3v}{3\sqrt{v}}\, dv$

35. $\int (x^3 + 1)^2 (3x)\, dx$

36. $\int 7x\sqrt[3]{x^2 - 10}\, dx$

37. $\int (x^3 + 1)^3 (5x^2)\, dx$

38. $\int (3 - 2x^3)^2 (-6x)\, dx$

39. $\int (4x^3 - 1)^2 (12x)\, dx$

40. $\int (4x^3 - 1)^2 (10x^2)\, dx$

41. $\int (1 + x^3)^4 (3x^2)\, dx$

42. $\int (1 - 2x^2)^3 (-4x)\, dx$

43. $\int (1 + x^3)^4 (x^2)\, dx$

44. $\int (1 - 2x^2)^3 (x)\, dx$

45. $\int (2x^3 + 1)^2 (6x)\, dx$

46. $\int (2x^3 + 1)^2 (6x^2)\, dx$

47. $\int (x^4 + 2)^2 (4x^3)\, dx$

48. $\int (x^4 + 2)^2 (4x)\, dx$

49. $\int (x^3 + 1)^2 x\, dx$

50. $\int (x^3 + 1)^2 x^2\, dx$

51. $\int (3 - t^4)^2 t^3\, dt$

52. $\int (3 - s^4)^2 s^2\, ds$

53. $\int_0^1 (1 - x)\, dx$

54. $\int_0^2 (2x - 1)\, dx$

55. $\int_1^8 \sqrt[3]{x}\, dx$

56. $\int_1^4 2\sqrt{x}\, dx$

57. $\int_0^1 \sqrt{1 - x}\, dx$

58. $\int_1^6 \sqrt{x + 3}\, dx$

59. $\int_0^1 (x^2 - 1)^2\, dx$

60. $\int_0^2 \dfrac{x^2 + 2}{2}\, dx$

61. $\int_2^7 \dfrac{dx}{\sqrt{x + 2}}$

62. $\int_{-2}^1 \dfrac{dx}{\sqrt{2 - x}}$

63. $\int_{-4}^0 \sqrt{1 - 2x}\, dx$

64. $\int_0^1 x\sqrt{1 - x^2}\, dx$

65. $\int_4^9 \dfrac{1 + \sqrt{r}}{\sqrt{r}}\, dr$

66. $\int_0^4 (2 + \sqrt{z})^2\, dz$

67. $\int_1^2 \theta\sqrt{4 - \theta^2}\, d\theta$

68. $\int_1^3 w\sqrt{w^2 - 1}\, dw$

4.6   Area Between Curves

In this section we are going to continue our study of areas by using the shortcut method discussed in Section 4.4.

Example 1   Find the area of the region bounded by $y = \sqrt{x}$, $x = 4$, and $y = 0$.

Solution.   Since $y = 0$ is the x-axis, we obtain the region in Figure 4.13.

Now draw the **typical element** of height y and thickness dx (Figure 4.13). Note that the area of the typical element is

$$y\,dx = \sqrt{x}\,dx \qquad \text{area of typical element}$$

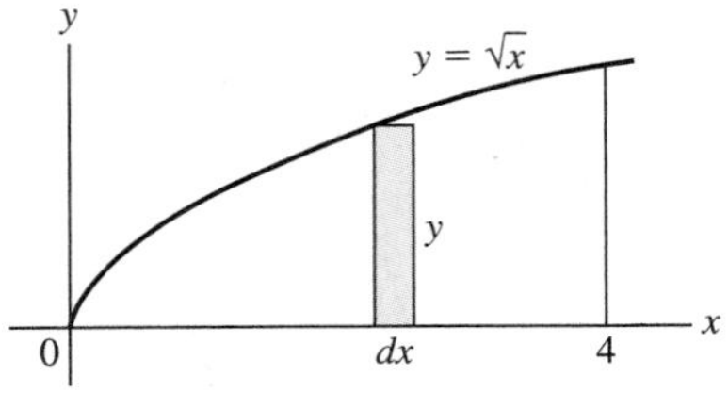

The area we wish to find lies between the lines $x = 0$ and $x = 4$. Summing the little areas from $x = 0$ to $x = 4$ by the operation

$$\int_0^4$$

we get

$$A = \int_0^4 \sqrt{x}\,dx = \int_0^4 x^{1/2}\,dx$$

$$= \frac{2}{3}x^{3/2}\Big|_0^4 = \frac{2}{3}(4^{3/2} - 0^{3/2})$$

$$= \frac{2}{3}(8) = \frac{16}{3} \qquad\blacksquare$$

Typical element

y

$y = \sqrt{x}$

y

0   *dx*   4   *x*

Figure 4.13

Example 1 is a special case of a general problem: finding areas of **regions bounded by two curves,** as shown in Figure 4.14. The typical element of thickness dx now extends between the two curves and has a height given by $f(x) - g(x)$. Consequently, the area is given by the integral

$$\int_a^b [f(x) - g(x)]\,dx$$

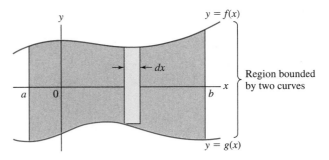

Figure 4.14

Area Between Two Curves

Draw a typical element of thickness dx extending between $f(x)$ and $g(x)$ (Figure 4.14). Sum the small areas $[f(x) - g(x)] dx$ by the operation $\int_a^b$ to obtain

$$\int_a^b [f(x) - g(x)] \, dx$$

When summing typical elements to get an area, we have to make sure that each such element is positive. This is particularly important when a curve, or part of a curve, lies below the x-axis. Consider the next example.

Example 2

Find the area bounded by the curves $y = (-1/2)\,x$, $x = 2$, and the x-axis.

Solution. The region is shown in Figure 4.15.

If we now integrate the function $y = (-1/2)\,x$ from $x = 0$ to $x = 2$, we get

$$\int_0^2 \left(-\frac{1}{2}x\right) dx = -\frac{1}{2} \frac{x^2}{2} \bigg|_0^2 = -1$$

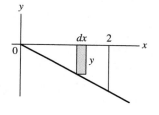

Figure 4.15

The resulting value is negative because each y is negative, yielding a negative sum.

To avoid this problem, we use the principle shown in Figure 4.14. We obtain the height of each element by subtracting $y = (-1/2)x$ from $y = 0$:

$$0 - \left(-\frac{1}{2}x\right) \qquad \textbf{height of element}$$

(See Figure 4.16.)

The area of the typical element is

$$\left[0 - \left(-\frac{1}{2}x\right)\right] dx \qquad \textbf{height} \cdot \textbf{base}$$

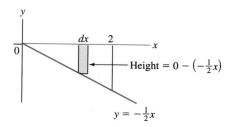

Figure 4.16

Summing from $x = 0$ to $x = 2$, we get

$$A = \int_0^2 \left[0 - \left(-\frac{1}{2}x\right)\right] dx = \int_0^2 \frac{1}{2}x \, dx = 1$$

∎

In the next example the curve crosses the x-axis, so that part of the region lies above the x-axis and part of the region below. To find the area, we break the integral into two parts by using the following additive property: if f is integrable on the three closed intervals determined by a, b, and c, then

$$\int_a^b f(x)\, dx = \int_a^c f(x)\, dx + \int_c^b f(x)\, dx$$

Example 3   Find the area of the region bounded by $y = x^3$, $x = -1$, $x = 2$, and the x-axis.

Solution. The region is shown in Figure 4.17.

Since part of the region lies below the x-axis, we need to find the two areas separately.

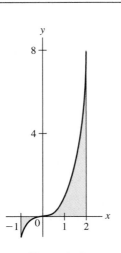

Left region: $\displaystyle\int_{-1}^0 (0 - x^3)\, dx = -\frac{x^4}{4}\bigg|_{-1}^0 = \frac{1}{4}$

Right region: $\displaystyle\int_0^2 x^3\, dx = \frac{x^4}{4}\bigg|_0^2 = 4$

Total area: $\displaystyle \frac{1}{4} + 4 = \frac{17}{4}$ ∎

Figure 4.17

Caution: In Example 3 the direct evaluation of the integral yields

$$\int_{-1}^2 x^3\, dx = \frac{x^4}{4}\bigg|_{-1}^2 = 4 - \frac{1}{4} = \frac{15}{4}$$

Although not equal to the area, the value of the definite integral $\int_{-1}^2 x^3\, dx$ is $15/4$, regardless of the geometric interpretation. In other words, *every definite integral is equal to a number*, and this number can have many different interpretations, just as the value of the derivative at a point can have different physical meanings.

Example 4   Find the area of the region bounded by the parabola $y^2 = 4x$ and the line $y = x$.

Solution.   To determine the limits of integration we have to solve the two equations simultaneously. From $y = x$ we get $y^2 = x^2$. On substituting in the other equation we find that

$$x^2 = 4x \qquad x^2 - 4x = 0 \qquad x(x - 4) = 0$$

whence $x = 0$ and $x = 4$. So the points of intersection are $(0, 0)$ and $(4, 4)$. The graphs are now easily sketched (Figure 4.18). The upper half of the parabola $y^2 = 4x$ is the function $y = \sqrt{4x} = 2\sqrt{x}$, so that the typical element has height $2\sqrt{x} - x$. Summing from $x = 0$ to $x = 4$, we get

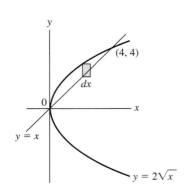

$$\int_0^4 (2\sqrt{x} - x)\, dx = \frac{4}{3}x^{3/2} - \frac{1}{2}x^2\bigg|_0^4 = \frac{8}{3}$$ ∎

Figure 4.18

So far we have been consistently using the letter x for the independent variable and y for the dependent variable, so that, bowing to the usual convention, y is always a function of x. Yet there are times when it is much more convenient to reverse the roles of x and y. For example, in the equation $x = y^3 - 3y + 1$ we might as well regard x as a function of y instead of solving for y in terms of x. In such a case we write $x = f(y)$, keeping the functional notation.

Example 5   The equation $y = 1/(x-1)$ is a function of x. Solving for x, we find that $x = (y+1)/y$, which represents x as a function of y.   ∎

For some regions it is convenient to reverse the roles of x and y, as shown in the next example.

Example 6   Find the area of the region bounded by $x = 8y - 2y^2$ and the y-axis.

Solution. To sketch the parabola $x = 8y - 2y^2$, we need to find the y-intercepts. Letting $x = 0$, we get

$$8y - 2y^2 = 0$$
$$2y(4 - y) = 0$$
$$y = 0, 4$$

The region is shown in Figure 4.19.

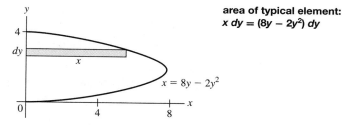

area of typical element:
$x\,dy = (8y - 2y^2)\,dy$

Figure 4.19

The given equation $x = 8y - 2y^2$ represents x as a function of y. So rather than solving the equation for y in terms of x, we interchange the roles of x and y and draw the **typical element in the horizontal position,** as shown in Figure 4.19. Note that the height of the typical element is $x = 8y - 2y^2$ and that the thickness is dy. The area of the typical element is therefore

$$x\,dy = (8y - 2y^2)\,dy \qquad \text{area of typical element}$$

Summing from $y = 0$ to $y = 4$, we get

$$A = \int_0^4 (8y - 2y^2)\,dy$$

$$= 4y^2 - 2\,\frac{y^3}{3}\,\Big|_0^4 = 4\cdot 4^2 - 2\cdot\frac{4^3}{3} - 0$$

$$= 4^3 - 2\cdot\frac{4^3}{3} = 4^3\left(1 - \frac{2}{3}\right) = 64\cdot\frac{1}{3} = \frac{64}{3}$$ ∎

Example 7   Find the area of the region bounded by $x = y^2 - y - 2$ and $y = x + 2$.

$x = y^2 - y - 2$
$x = y - 2$
────────────
$0 = y^2 - 2y$ **(subtracting)**
$y(y - 2) = 0$
$y = 0, 2$

Solution. The equation $x = y^2 - y - 2 = (y - 2)(y + 1)$ represents a parabola with intercepts $(0, 2)$, $(0, -1)$, and $(-2, 0)$. Solving the equations simultaneously we find that $(-2, 0)$ and $(0, 2)$ are the points of intersection. The graphs are shown in Figure 4.20.

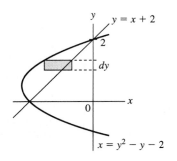

Figure 4.20

Writing $y = x + 2$ as $x = y - 2$, both equations are functions of y. With this point of view we **must draw the typical element in a horizontal position with thickness dy and height** $(y - 2) - (y^2 - y - 2)$. For the limits of integration we refer to the y-axis: the region we are concerned with lies between the lines $y = 0$ and $y = 2$. So the integral becomes

$$\int_0^2 [(y - 2) - (y^2 - y - 2)]\, dy = \int_0^2 (-y^2 + 2y)\, dy$$

$$= -\frac{y^3}{3} + y^2 \Big|_0^2 = \frac{4}{3} \qquad \blacksquare$$

Using a Graphing Utility

Graphing the curves that bound a given region can be facilitated by using a graphing utility.

Example 8   Find the area of the region bounded by $y = (1/2)x^2 - 2$ and $y = x + 2$.

Solution. To see the region, we graph both functions at the same time (Figure 4.21). The equations can now be solved simultaneously, yielding $x = -2$ and $x = 4$. So the area is

$$A = \int_{-2}^4 \left[(x + 2) - \left(\frac{1}{2}x^2 - 2 \right) \right] dx = 18 \qquad \blacksquare$$

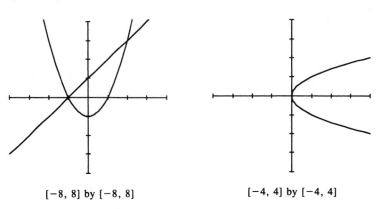

[−8, 8] by [−8, 8]   [−4, 4] by [−4, 4]

Figure 4.21   **Figure 4.22**

To graph an equation that is not a function of x, such as $y^2 = x$, we first solve for y to get $y = \pm\sqrt{x}$ and then graph $y = \sqrt{x}$ and $y = -\sqrt{x}$ at the same time (Figure 4.22).

Example 9   Graph the equation in Example 6, $x = 8y - 2y^2$.

Solution. To graph this equation, we first solve for y in terms of x using the quadratic formula:

$$x = 8y - 2y^2$$

$$2y^2 - 8y + x = 0$$

$$y = \frac{-(-8) \pm \sqrt{(-8)^2 - 4(2)(x)}}{2 \cdot 2}$$

$$y = \frac{8 \pm \sqrt{64 - 8x}}{4} = \frac{1}{4}(8 \pm \sqrt{64 - 8x})$$

Now we graph the equations $y = (1/4)(8 + \sqrt{64 - 8x})$ and $y = (1/4) \times (8 - \sqrt{64 - 8x})$ at the same time to get the graph in Figure 4.23.

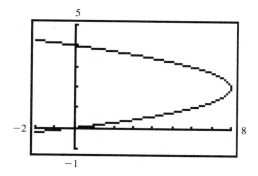

Figure 4.23

■ Exercises / Section 4.6

Find the area of the region bounded by the given curves. Sketch the curves and show a typical horizontal or vertical element of area.

1. $y = 2x, x = 1$, x-axis   **2.** $y = \frac{1}{2}x, x = 2$, x-axis

3. $y = 2x, y = 2$, y-axis   **4.** $y = \frac{1}{2}x, y = 2$, y-axis

5. $y = x, x = -1, x = 1, y = 0$

6. $y = 2x, x = -1, x = 1, y = 0$

7. $y = 3x, x = 1, x = 2, y = 0$

8. $y = x^2, x = 1, y = 0$ **9.** $y = -x, x = 1, y = 0$

10. $y = x^2, x = -1, y = 0$ **11.** $y = x^3, x = -1, y = 0$

12. $y = -x, y = 1, x = 0$

13. $y = x^2 + 1, x = 1, x = 3$, x-axis

14. $y = x^3, x = -1, x = 1$, x-axis

15. $y = x^3, x = -1, x = 2$, x-axis

16. $y = (x - 1)^3, x = 0, x = 3$, x-axis

17. $y = x(x - 2)^2, x = -1, x = 2$, x-axis

18. $y = x(x - 1)(x + 1), x = -1, x = 1$, x-axis

19. $y = \sqrt[3]{x}, x = 0, x = 8$, x-axis

20. $y = \frac{1}{x^2}, x = 1, x = 2$, x-axis

21. $x = y^2, y = 1$, y-axis

22. $y = x, y = x^2$ **23.** $y = x^2, y = x^3$

24. $y^2 = x, y^3 = x$ **25.** $y = x^2 - 4$, x-axis

26. $y^2 = x + y$, y-axis   **27.** $y = x^2 - 1, y = 3$

28. $y = \sqrt[3]{x}, y = -x, x = 0, x = 8$

29. $y^2 = x - 1, y = x - 3$ **30.** $x = \sqrt{y}, x = -y, y = 1$

31. $x^2 + 4y = 0, x^2 - 4y - 8 = 0$

32. $x = y^2 - 4y + 2, 2x - y + 5 = 0$

33. Use integration to obtain the formula for the area of
    a. a square of side s
    b. a rectangle of length a and width b

34. Use integration to obtain the formula for the area of a right triangle with base b and height h.

In the remaining exercises use a graphing utility to graph the region bounded by the given curves. Use the integration capability of the utility to evaluate the integrals.

35. $y = \dfrac{1}{2}x^2, y = x + 4$ **36.** $y = x^2 + 1, y = 2x + 1$

37. $y = 2 - x^2, y = -x$

38. $y = 3 - x^2, y = x + 1$

39. $y = x^2 + 2x, y = 2x + 4$

40. $y = x^2 - 4x, y = -x^2 + 2x$

41. $y = x^2 - 4x + 2, y = 2 + 4x - x^2$

42. $y = x^4 - 2x^2 + 1, y = 2x^2 + 1$

43. $y = x^4 - 4x^2 + 4, y = x^2$

44. $y = x^3 - 3x^2 + 3x + 1, y = x^2 + 1$

4.7   Improper Integrals

So far we have assumed that the limits of integration of the integral $\int_a^b f(x)\,dx$ are finite and that $f(x)$ is bounded. In this section we will extend the definition to include infinite limits or infinite discontinuities (vertical asymptotes).

    Consider the unbounded region determined by $y = 1/x^2$, $x = 1$, and the x-axis to the right of 1 in Figure 4.24. To find the area of the region we could use our intuition and write the integral formally as

$$\int_1^{\infty} \frac{1}{x^2}\,dx$$

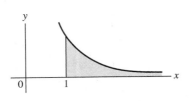

Figure 4.24

The problem is that given the definition of the integral as a limit of a sum, the interval $[a, b]$ has to be finite. To get around this problem we define

$$\int_1^{\infty} \frac{dx}{x^2} = \lim_{b \to \infty} \int_1^b \frac{1}{x^2}\,dx$$

Improper integral

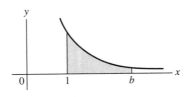

Figure 4.25

which is called an **improper integral**. Geometrically, we are finding the area under the graph bounded by the lines $x = 1$ and $x = b$ (Figure 4.25) and then letting $b \to \infty$. If the limit exists, this limit is defined to be the area. The integration yields.

$$\int_1^{\infty} \frac{dx}{x^2} = \lim_{b \to \infty} \int_1^b x^{-2}\,dx = \lim_{b \to \infty} \left.\frac{x^{-1}}{-1}\right|_1^b = \lim_{b \to \infty} \left(-\frac{1}{b} + 1\right) = 1$$

If the limit exists, the integral is said to be *convergent;* otherwise it is *divergent.*

Example 1   Evaluate the improper integral

$$\int_0^{\infty} \frac{dx}{(x + 2)^{3/2}}$$

Solution.

$$\int_0^\infty \frac{dx}{(x+2)^{3/2}} = \lim_{b\to\infty} \int_0^b \frac{dx}{(x+2)^{3/2}}$$

$$= \lim_{b\to\infty} \frac{(x+2)^{-1/2}}{-\frac{1}{2}}\bigg|_0^b = \lim_{b\to\infty} \frac{-2}{\sqrt{x+2}}\bigg|_0^b$$

$$= \lim_{b\to\infty} \left[\frac{-2}{\sqrt{b+2}} + \frac{2}{\sqrt{2}} \right] = 0 + \frac{2}{\sqrt{2}} = \frac{2}{\sqrt{2}}\frac{\sqrt{2}}{\sqrt{2}} = \sqrt{2} \qquad \blacksquare$$

If one of the limits is $-\infty$, the procedure is similar, as shown in the next example.

Example 2   Evaluate the improper integral

$$\int_{-\infty}^1 \frac{dx}{(3-x)^{5/3}}$$

Solution.

$$\int_{-\infty}^1 \frac{dx}{(3-x)^{5/3}} = \lim_{b\to-\infty} \int_b^1 \frac{dx}{(3-x)^{5/3}} = \lim_{b\to-\infty} \int_b^1 (3-x)^{-5/3}\, dx$$

$$= \lim_{b\to-\infty} \left(-\int_b^1 (3-x)^{-5/3}(-dx) \right) \qquad \left[\begin{array}{l} \boldsymbol{u = 3 - x} \\ \boldsymbol{du = -dx} \end{array}\right]$$

$$= \lim_{b\to-\infty} \left(-\frac{(3-x)^{-2/3}}{-\frac{2}{3}} \right)\bigg|_b^1 = \lim_{b\to-\infty} \left(\frac{3}{2}(2)^{-2/3} - \frac{3}{2}(3-b)^{-2/3} \right)$$

$$= \lim_{b\to-\infty} \left(\frac{3}{2\cdot 2^{2/3}} - \frac{3}{2(3-b)^{2/3}} \right) = \frac{3}{2\cdot 2^{2/3}} + 0 \approx 0.945 \qquad \blacksquare$$

An integral is also improper if the integrand has an infinite discontinuity (vertical asymptote) somewhere in the interval $[a, b]$. Consider the next example.

Example 3   Find the area "bounded" by $y = 1/\sqrt{x}$, $x = 1$, and the coordinate axes. (See Figure 4.26.)

Solution.   We can avoid the vertical asymptote $x = 0$ by integrating from $x = \epsilon$ to $x = 1$ (as shown in Figure 4.26) and then letting $\epsilon \to 0$:

$$\int_0^1 \frac{1}{\sqrt{x}}\, dx = \lim_{\epsilon\to 0} \int_\epsilon^1 \frac{1}{\sqrt{x}}\, dx = \lim_{\epsilon\to 0} \int_\epsilon^1 x^{-1/2}\, dx$$

$$= \lim_{\epsilon\to 0} 2x^{1/2}\bigg|_\epsilon^1 = \lim_{\epsilon\to 0} (2 - 2\epsilon^{1/2}) = 2$$

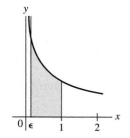

Figure 4.26

Since the limit exists, the area is defined to be 2. $\qquad \blacksquare$

In Example 3 the asymptote is located at the left end of the interval. In the next example the asymptote is inside the interval.

Example 4   Evaluate the integral

$$\int_{-1}^{2} \frac{dx}{(x-1)^2}$$

Solution.   We avoid the vertical asymptote at $x = 1$ by integrating from $x = -1$ to $x = 1 - \epsilon$ and from $x = 1 + \eta$ to 2 and then letting ϵ and η go to 0:

$$\int_{-1}^{2} \frac{dx}{(x-1)^2} = \lim_{\epsilon \to 0} \int_{-1}^{1-\epsilon} \frac{dx}{(x-1)^2} + \lim_{\eta \to 0} \int_{1+\eta}^{2} \frac{dx}{(x-1)^2}$$

$$= \lim_{\epsilon \to 0} \left(-\frac{1}{x-1} \right)\Big|_{-1}^{1-\epsilon} + \lim_{\eta \to 0} \left(-\frac{1}{x-1} \right)\Big|_{1+\eta}^{2}$$

$$= \lim_{\epsilon \to 0} \left(-\frac{1}{1-\epsilon-1} + \frac{1}{-2} \right) + \lim_{\eta \to 0} \left(-1 + \frac{1}{1+\eta-1} \right)$$

$$= \lim_{\epsilon \to 0} \left(\frac{1}{\epsilon} - \frac{1}{2} \right) + \lim_{\eta \to 0} \left(-1 + \frac{1}{\eta} \right)$$

Observe that neither of the limits exists. If either limit fails to exist, the integral is divergent.

Remark: If the integral

$$\int_{-1}^{2} \frac{dx}{(x-1)^2}$$

were treated as an ordinary integral, we would get

$$-\frac{1}{x-1}\Big|_{-1}^{2} = -\frac{1}{2-1} + \frac{1}{-1-1} = -\frac{3}{2}$$

This result is absurd since the function $f(x) = 1/(x-1)^2$ is never negative. This example shows that an improper integral should never be treated as an ordinary integral.   ■

■ Exercises / Section 4.7

In Exercises 1–8, evaluate the given improper integrals.

1. $\displaystyle\int_{1}^{\infty} \frac{2}{x^4}\, dx$

2. $\displaystyle\int_{-\infty}^{-1} \frac{2}{x^4}\, dx$

3. $\displaystyle\int_{-\infty}^{0} \frac{4}{(x-3)^2}\, dx$

4. $\displaystyle\int_{5}^{\infty} \frac{6}{(x+4)^3}\, dx$

5. $\displaystyle\int_{0}^{\infty} \frac{x}{(x^2+4)^2}\, dx$

6. $\displaystyle\int_{2}^{\infty} \frac{x}{(1-x^2)^2}\, dx$

7. $\displaystyle\int_{-\infty}^{0} \frac{dz}{(2z-3)^3}$

8. $\displaystyle\int_{-\infty}^{0} \frac{9\, dt}{(t-3)^4}$

In Exercises 9–24, find the area of each given region.

9. $y = 2/x^2$, $x = 3$, the x-axis to the right of $x = 3$

10. $y = 1/x^{3/2}$, $x = 4$, the x-axis to the right of $x = 4$

11. $y = 1/(x-1)^2$, $x = 2$, the x-axis to the right of $x = 2$

12. $y = 1/(x + 2)^2$, the coordinate axes, area in first quadrant

13. $x^3y = 2$, $x = 1$, the x-axis to the right of $x = 1$

14. $y = x^{-4/3}$, $x = 8$, the x-axis to the right of $x = 8$

15. $y = 1/(x + 1)^{3/2}$, the coordinate axes, area in first quadrant

16. $y = 2/(2x - 1)^2$, coordinate axes, area in second quadrant

17. $y = 1/(2x - 3)^3$, coordinate axes, area in third quadrant

18. $y = 3/x^{4/3}$, $x = -8$, x-axis, to the left of $x = -8$

19. $y = 3x^2/(x^3 + 1)^2$, $x = -2$, x-axis, to the left of $x = -2$

20. $y = x^{-3/4}$, $x = 0$, $x = 16$, x-axis. *Hint:* Avoid the vertical asymptote by evaluating

$$\lim_{\epsilon \to 0} \int_{\epsilon}^{16} x^{-3/4}\, dx \qquad \epsilon > 0 \qquad \text{(See Example 3.)}$$

21. $y = 1/\sqrt{x - 1}$, $x = 1$, $x = 5$, x-axis

22. $y = 1/\sqrt[3]{x - 2}$, $x = 2$, $x = 3$, x-axis

23. $y = 2/\sqrt{4 - x}$, $x = 4$, coordinate axes

24. $y = 3/(3 - x)^2$, $x = 3$, coordinate axes

25. Show that the integral $\int_1^{\infty} (1/\sqrt{x})\, dx$ does not exist.

Evaluate the following improper integrals:

26. $\displaystyle\int_1^5 \frac{x}{(x^2 - 4)^2}\, dx$

27. $\displaystyle\int_{-1}^2 \frac{dx}{\sqrt[3]{x}}$

28. $\displaystyle\int_0^3 \frac{2}{\sqrt[3]{x - 1}}\, dx$

29. $\displaystyle\int_0^4 \frac{x}{(9 - x^2)^2}\, dx$

30. $\displaystyle\int_1^3 \frac{x^2\, dx}{(x^3 - 8)^3}$

31. $\displaystyle\int_1^4 \frac{3\, dx}{(x - 3)^{2/5}}$

32. $\displaystyle\int_1^5 \frac{x}{(x^2 - 4)^2}\, dx$

33. $\displaystyle\int_{-3}^{-1} \frac{1}{(x + 2)^4}\, dx$

4.8   The Constant of Integration

The purpose of this section is to discuss some common applications of the indefinite integral.

Example 1   Find the family of curves such that each member has a derivative equal to $2x$.

Solution.   For each member of the desired family,

$$\frac{dy}{dx} = 2x$$

By integration,

$$y = x^2 + C$$

Since $y - C = x^2$ has the form $(y - k) = (x - 0)^2$, we need to recall from Section 1.11 that the family is generated by translating the parabola $y = x^2$ by a distance C units in the vertical direction. (Each value of C corresponds to a member of the family.) The graphs are shown in Figure 4.27.

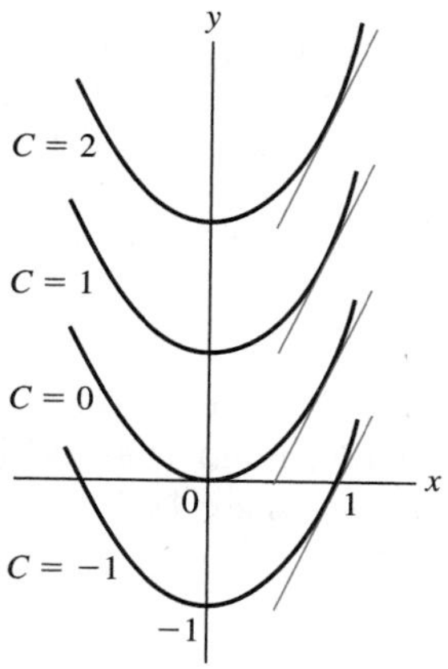

Now if x has a specific value, say $x = 1$, then

$$\frac{dy}{dx} = 2x\Big|_{x=1} = 2$$

for each member of the family. Consequently, for $x = 1$ all the tangents have the same slope and must be parallel. (See Figure 4.27.)   ∎

Figure 4.27

Example 2   Find the curve in Example 1 passing through the point $(2, 5)$.

Solution.   After substituting $x = 2$ and $y = 5$ in the equation $y = x^2 + C$, we get $5 = 4 + C$. It follows that $C = 1$, so that $y = x^2 + 1$ is the desired curve.   ∎

Another important application of indefinite integrals is the study of velocity and acceleration. Recall from Section 2.6 that

$$v = \frac{ds}{dt} \quad \text{and} \quad a = \frac{dv}{dt}$$

Since integration is the inverse of differentiation, we may also write

$$s = \int v\, dt \quad \text{and} \quad v = \int a\, dt$$

For objects acting under the influence of gravity, we are going to adopt the following conventions. The acceleration due to gravity is $g = 10\,\text{m/s}^2$ (rather than the more accurate value of $9.8\,\text{m/s}^2$). All distances are measured from the ground with upward direction positive. Finally, $t = 0$ corresponds to the instant when the motion begins. As a result of these assumptions:

Basic assumptions

1. $g = -10\,\text{m/s}^2$.
2. $s = 0$ on the ground.
3. $t = 0$ at the instant when the motion begins.
4. A distance from the ground to a point above the ground is positive.
5. An object moving in the upward direction has a positive velocity.
6. An object moving in the downward direction has a negative velocity.

Remark: We assume in this section that the air resistance is negligible. Motion that takes air resistance into account will be studied in Section 11.4.

Example 3   An object is hurled upward from the ground with a velocity of 25 m/s.

a. How long does the object stay in the air?
b. How high does it rise?

Solution. As already noted, the upward direction is positive, so that

$$g = -10\,\text{m/s}^2$$

According to the given conditions, when $t = 0$ the velocity is $+25$ m/s. (The velocity is positive because the object is moving in the upward direction.) The velocity at $t = 0$ is called the **initial velocity** and is denoted by v_0. (See Figure 4.28.)

Initial velocity

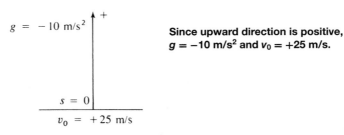

$g = -10\,\text{m/s}^2$

$s = 0$

$v_0 = +25$ m/s

Since upward direction is positive, $g = -10\,\text{m/s}^2$ and $v_0 = +25$ m/s.

Figure 4.28

From the formula

$$v = \int a\,dt$$

we get

$$v = \int (-10)\,dt \qquad \text{or} \qquad v = -10t + C$$

where C is an arbitrary constant. To evaluate C, we let $t = 0$ and $v = +25$:

$$v = -10t + C$$
$$+25 = -10 \cdot 0 + C \qquad \text{or} \qquad C = +25$$

It follows that

$$v = -10t + 25$$

From the formula

$$s = \int v\,dt$$

we now get

$$s = \int (-10t + 25)\,dt \qquad \text{or} \qquad s = -5t^2 + 25t + k$$

where k is an arbitrary constant. Since all distances are measured from the ground, we have $s = 0$ when $t = 0$. (See Figure 4.28.) It follows that

$$s = -5t^2 + 25t + k$$
$$0 = -5(0)^2 + 25(0) + k \qquad \text{or} \qquad k = 0$$

so that

$$s = -5t^2 + 25t$$

a. To determine how long the object stays in the air, we let $s = 0$ in the last equation and solve for t:

$$0 = -5t^2 + 25t$$
$$-5t(t - 5) = 0$$
$$t = 0, 5$$

So $s = 0$ when $t = 0$ and $t = 5$. Since $t = 0$ when the motion begins, it takes 5 s for the object to return to the ground.

b. To determine how high the object rises, we need to find the value of t for which $v = 0$ (because at the highest point the object momentarily stops):

$$v = -10t + 25$$
$$0 = -10t + 25$$
$$t = 2.5$$

Thus the object takes **2.5** s to reach the highest point. From

$$s = -5t^2 + 25t$$

we get

$$s = -5(2.5)^2 + 25(2.5) = 31.25$$

The object therefore reaches a height of 31.25 m.

Example 4   An object is hurled downward from the top of a building 20 m high with an initial velocity of 5 m/s. Find the time it takes to reach the ground.

Solution. Consider the diagram in Figure 4.29.

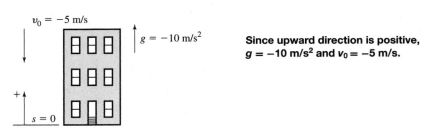

$v_0 = -5$ m/s

$g = -10$ m/s²

Since upward direction is positive, $g = -10$ m/s² and $v_0 = -5$ m/s.

$s = 0$

Figure 4.29

Since the upward direction is positive, $g = -10$ m/s². When $t = 0$, $s = 20$ m (since the object starts at a point 20 m above the ground). Since the object is hurled downward (in the negative direction) $v_0 = -5$ m/s. (See Figure 4.29.)

We now have

$$v = \int (-10)\, dt = -10t + C$$

Substituting $t = 0$ and $v = -5$:

$$v = -10t + C$$
$$-5 = -10 \cdot 0 + C \qquad \text{or} \qquad C = -5$$

Hence

$$v = -10t - 5$$

Also,

$$s = \int (-10t - 5)\, dt \qquad \text{and} \qquad s = -5t^2 - 5t + k$$

Since $s = +20$ when $t = 0$,

$$+20 = -5(0)^2 - 5(0) + k \qquad \text{or} \qquad k = +20$$

Thus

$$s = -5t^2 - 5t + 20$$

To determine the time required to reach the ground, we let $s = 0$ in the last equation and solve for t:

$$0 = -5t^2 - 5t + 20$$
$$t^2 + t - 4 = 0 \qquad \textbf{dividing by } -5$$

and by the quadratic formula,

$$t = \frac{-1 \pm \sqrt{17}}{2} \qquad \textbf{See inside front cover.}$$

Hence $t \approx 1.6$ s. (We ignore the negative root since the motion begins at $t = 0$.)   ∎

Example 5 If the object in Example 4 is hurled upward at 5 m/s, with what velocity will it strike the ground?

Solution. As before, $g = -10$ m/s^2. This time, however, $v_0 = +5$ m/s, since the object is moving in the positive direction. We get

$$v = -10t + 5 \quad \text{and} \quad s = -5t^2 + 5t + 20$$

To determine the velocity with which the object strikes the ground, we must first determine the time required to reach the ground. Letting $s = 0$, we get

$$0 = -5t^2 + 5t + 20$$
$$t^2 - t - 4 = 0$$
$$t = \frac{1 + \sqrt{17}}{2} \approx 2.6 \text{ s}$$

It follows that the velocity is

$$v = -10t + 5$$
$$= -10(2.6) + 5 = -21$$

Thus the object attains a velocity of 21 m/s. (The negative sign indicates that the object is moving in the downward direction.) ∎

We recall from Section 2.6 that the current in a circuit is given by $i = dq/dt$. In differential form $dq = i\,dt$, so that $q = \int i\,dt$. Also, since the voltage v across a capacitor is q/C,

$$v = \frac{1}{C}\int i\,dt$$

Example 6 The voltage across a 2.0×10^{-3}-F capacitor is 20 V initially. If the current to the capacitor is $i = 0.3\sqrt{t}$ A, what is the voltage after 1 s?

Solution. From

$$v = \frac{1}{C}\int i\,dt$$

we have

$$v = \frac{1}{2.0 \times 10^{-3}}\int 0.3 t^{1/2}\,dt = \frac{0.3}{2.0 \times 10^{-3}}\left(\frac{2}{3}t^{3/2}\right) + k$$

Letting $v = 20$ and $t = 0$, we find that $k = 20$. At $t = 1$,

$$v = \frac{0.3\left(\frac{2}{3}\right)}{2.0 \times 10^{-3}} + 20 \approx 120 \text{ V}$$ ∎

■ Exercises / Section 4.8

In Exercises 1–6, find the function satisfying the given conditions.

1. $dy/dx = 3x$, passing through $(0, 1)$

2. $dy/dx = 3x^2$, passing through $(1, 2)$

3. $dy/dx = 6x^2 + 1$, passing through $(-1, 1)$

4. $dy/dx = 3x^3 - 1$, passing through $(-2, 3)$

5. $dy/dx = 3x^2 + 2$, passing through $(1, 0)$

6. $dy/dx = x^3 - 4$, passing through $(2, 3)$

7. A stone is tossed up in the air at 15 m/s. How long does it stay in the air?

8. An object is tossed up in the air at 12.5 m/s. How long does it stay in the air?

9. A ball is thrown upward with a velocity of 30 m/s. How high does it rise?

10. A ball is thrown upward with a velocity of 20 m/s. How high does it rise?

11. An object is dropped from a height of 125 m. How long will it take to hit the ground?

12. An object is dropped from a height of 245 m. How long will it take to hit the ground?

13. An object is hurled downward with an initial velocity of 10 m/s from a height of 50 m. How long does the object take to hit the ground and with what speed does it strike?

14. A stone is hurled downward with an initial velocity of 12 m/s from a height of 80 m. How long does the stone take to hit the ground and with what speed does it strike?

15. A woman standing on a cliff 50 m high tosses a rock upward with an initial velocity of 15 m/s. How long does the rock remain in the air and with what speed does it hit the ground?

16. A man standing on a cliff 60 m high hurls a stone upward at the rate of 20 m/s. How long does the stone remain in the air and with what speed does it strike the ground?

17. An object is hurled up in the air from a height of 40 m with an initial velocity of 10 m/s. How long does the object remain in the air and with what speed does it strike the ground?

18. An object is hurled up in the air from a height of 30 m with an initial velocity of 20 m/s. How long does the object remain in the air and with what speed does it strike the ground?

19. A stone is hurled from the bottom of a pit 20 m deep with an initial velocity of 25 m/s. How long does the stone take to reach the top of the pit?

20. A rock is dropped from a height of 50 m into a hole 15 m deep. How long does the rock take to hit the bottom of the hole and with what speed does it strike?

21. A car is moving along a road at 28 m/s. Suddenly the driver sees a child on the road about 60 m ahead of him. He slams on the brakes and decelerates at the rate of 7 m/s². Will he hit the child?

22. A window washer accidentally falls off his scaffold 110 m above the ground. Two seconds later Superman arrives on the scaffold and dives after the man with an initial velocity of 45 m/s. Will he catch the man before he hits the ground?

23. A student in front of the Roy W. Johnson Residence Hall wants to toss his keys to his roommate, who is leaning out the dorm window 15 m above the student's hand. With what initial velocity must the keys be tossed?

24. Determine the initial velocity of an object that reaches a height of 35 m.

25. An object has an acceleration of $t/(t^2 + 1)^2$. Find an expression for v if $v_0 = 10$.

26. Derive the equations of motion for a particle moving in a straight line with constant acceleration. Use a for acceleration and v_0 and s_0 for the initial velocity and position, respectively.

27. For a short time interval the current in a circuit is given by $i = \sqrt{t + 1.0}$. How many coulombs of charge pass a point in the first 3.0 s if $q = 0$ when $t = 0$?

28. The angular velocity of an object is $\omega = (0.10t^{1/3} + 1.0)$ rad/s. Find the angular displacement θ at $t = 8.0$ s if $\theta = 0$ when $t = 0$.

29. The current to a capacitor initially charged to 0.030 C is given by $i = 0.010t + 0.10$. Find the charge after 3.0 s.

30. The voltage across a 2.0-H inductor is $2.0 - 0.40t$ V. Find the current in the circuit after 5.0 s if $i = 0$ when $t = 0$. (Recall that the voltage is given by $L\,di/dt$.)

31. The voltage across a 0.00300-F capacitor is 30.0 V initially. If the current to the capacitor is $i = 0.0100t$ A, find the voltage after 4.00 s.

32. Find the capacitance of a capacitor with 150 V across it initially and 300 V one second later. The current as a function of time is given by $i = 0.030t^{1/3}$ A.

33. The rate of change of the resistance with respect to the temperature of a certain resistor is given by

$$\frac{dR}{dT} = 0.00060T^2 + 0.0080T + 0.14$$

Given that $R = 50\ \Omega$ when $T = 0°C$, find R when $T = 20°C$.

34. The density of a certain rod of length 4 m is given by

$$\rho(x) = 20 + \frac{3}{8}\sqrt{x}$$

where x is the distance from one end of the rod; ρ is measured in kilograms per meter. Find the mass of the rod.

35. Recall that power P is defined as the time rate of doing work ($P = dW/dt$) or as the rate at which energy is expanded. If $P = 3\sqrt{t + 1}$, find W when $t = 8$ s, given that $W = 5$ J when $t = 0$ s.

4.9 **Numerical Integration**

In this section we are going to study two numerical methods for evaluating definite integrals that cannot be integrated by present techniques.

Trapezoidal Rule

The first technique, called the **trapezoidal rule,** approximates the area under a curve by means of trapezoids. Consider the area under the curve in Figure 4.30. Subdivide the interval $[a, b]$ into n equal parts of length

$$h = \frac{b - a}{n}$$

such that

$$a = x_0 < x_1 < \cdots < x_n = b$$

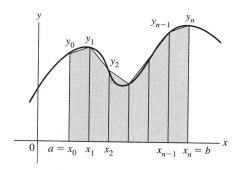

Figure 4.30

Now consider the corresponding y-values $y_0, y_1, \ldots, y_n$ on the curve in Figure 4.30. Connect each pair of adjacent points by a line to form n trapezoids. The area under the curve is approximately equal to the total area of the trapezoids. (Recall that the area of a trapezoid is equal to one-half the product of the altitude and the sum of the bases.) From Figure 4.30,

$$\int_a^b f(x)dx$$

$$\approx \frac{1}{2}(y_0 + y_1)h + \frac{1}{2}(y_1 + y_2)h + \frac{1}{2}(y_2 + y_3)h + \cdots + \frac{1}{2}(y_{n-1} + y_n)h$$

$$= h\left(\frac{1}{2}y_0 + y_1 + y_2 + \cdots + y_{n-1} + \frac{1}{2}y_n\right)$$

Using the more convenient functional notation, the formula can be written as follows:

> **Trapezoidal Rule**
>
> $$\int_a^b f(x)\,dx \approx h\left[\frac{1}{2}f(x_0) + f(x_1) + \cdots + f(x_{n-1}) + \frac{1}{2}f(x_n)\right] \quad (4.12)$$
>
> where $h = (b-a)/n$.

Example 1

Use the trapezoidal rule with $n = 4$ to find the approximate value of

$$\int_0^1 x^2\,dx$$

Solution. For $n = 4$ we have

$$h = \frac{b-a}{4} = \frac{1}{4}$$

Hence

$$x_0 = 0,\ x_1 = \frac{1}{4},\ x_2 = \frac{1}{2},\ x_3 = \frac{3}{4},\ \text{and}\ x_4 = 1.$$

So by Formula (4.12)

$$\int_0^1 x^2\,dx \approx \frac{1}{4}\left[\frac{1}{2}(0)^2 + \left(\frac{1}{4}\right)^2 + \left(\frac{1}{2}\right)^2 + \left(\frac{3}{4}\right)^2 + \frac{1}{2}(1)^2\right] = 0.344$$

compared to the exact value of $1/3$ obtained by direct integration. ∎

Example 2

Use the trapezoidal rule with $n = 8$ to approximate the following integral to three decimal places:

$$\int_{-2}^4 \frac{2}{\sqrt{x^4 + 5}}\,dx$$

Solution. Since $n = 8$, $h = (b-a)/8 = [4 - (-2)]/8 = 0.75$. So $x_0 = -2$, $x_1 = -2 + 0.75 = -1.25$, $x_2 = -1.25 + 0.75 = -0.5$, and so on.

The function values are listed next:

$$f(x_0) = \frac{2}{\sqrt{(-2)^4 + 5}} = 0.43644$$

$$f(x_1) = \frac{2}{\sqrt{(-1.25)^4 + 5}} = 0.73317$$

$$f(x_2) = \frac{2}{\sqrt{(-0.5)^4 + 5}} = 0.88889$$

$$f(x_3) = \frac{2}{\sqrt{(0.25)^4 + 5}} = 0.89408$$

$$f(x_4) = \frac{2}{\sqrt{(1.0)^4 + 5}} = 0.81650$$

$$f(x_5) = \frac{2}{\sqrt{(1.75)^4 + 5}} = 0.52743$$

$$f(x_6) = \frac{2}{\sqrt{(2.5)^4 + 5}} = 0.30130$$

$$f(x_7) = \frac{2}{\sqrt{(3.25)^4 + 5}} = 0.18524$$

$$f(x_8) = \frac{2}{\sqrt{4^4 + 5}} = 0.12380$$

So by Formula (4.12)

$$\int_{-2}^{4} \frac{2}{\sqrt{x^4 + 5}}\, dx \approx 0.75\left[\frac{1}{2}(0.43644) + 0.73317 + \cdots + \frac{1}{2}(0.12380)\right]$$

$$= 3.470 \qquad\qquad \blacksquare$$

Simpson's Rule

For our next approximation technique consider the graph of $y = f(x)$ in Figure 4.31. The interval $[a, b]$ is divided into n equal subdivisions, where n is an **even** number. Looking at the first two subdivisions, we can obtain an approximation to the curve by passing a parabola through (x_0, y_0), (x_1, y_1), and (x_2, y_2). Then the area under the curve between $x = x_0$ and $x = x_2$ is approximately equal to the area under the parabola. It can be shown that the area under the parabola determined by (x_0, y_0), (x_1, y_1), and (x_2, y_2) is

$$\frac{h}{3}(y_0 + 4y_1 + y_2) \qquad \text{where } h = \frac{b - a}{n}$$

This is the approximate area under the curve from x_0 to x_2.

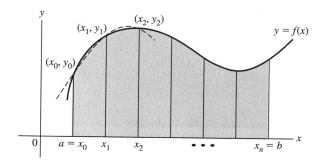

Figure 4.31

Similarly, the area under the parabola determined by (x_2, y_2), (x_3, y_3), and (x_4, y_4) is

$$\frac{h}{3}(y_2 + 4y_3 + y_4)$$

This is the approximate area under the curve from x_2 to x_4. Continuing this process, the total area under the curve from a to b is approximately

$$\frac{h}{3}[(y_0 + 4y_1 + y_2) + (y_2 + 4y_3 + y_4) + (y_4 + 4y_5 + y_6)$$
$$+ \cdots + (y_{n-2} + 4y_{n-1} + y_n)]$$
$$= \frac{h}{3}(y_0 + 4y_1 + 2y_2 + 4y_3 + 2y_4 + 4y_5 + 2y_6$$
$$+ \cdots + 2y_{n-2} + 4y_{n-1} + y_n)$$

This approximation procedure is known as **Simpson's rule** and leads to a better approximation than the trapezoidal rule for the same n (although in the trapezoidal rule n can be even or odd).

Simpson's Rule

$$\int_a^b f(x)\,dx \approx \frac{h}{3}[f(x_0) + 4f(x_1) + 2f(x_2) + 4f(x_3)$$
$$+ \cdots + 2f(x_{n-2}) + 4f(x_{n-1}) + f(x_n)] \qquad (4.13)$$

where $h = (b - a)/n$ and n is an **even** integer.

Example 3   Use Simpson's rule with $n = 4$ to approximate the value of the integral in Example 1.

Solution. Since the x values and h are the same, from Formula (4.13) we get

$$\int_0^1 x^2\,dx \approx \frac{\frac{1}{4}}{3}\left[0^2 + 4\left(\frac{1}{4}\right)^2 + 2\left(\frac{1}{2}\right)^2 + 4\left(\frac{3}{4}\right)^2 + 1^2\right] = 0.333$$

which is better than the approximation obtained using the trapezoidal rule.   ∎

Example 4   Find the approximate value of the following definite integral by Simpson's rule with $n = 6$:

$$\int_{-1}^2 \frac{dx}{x^2 + 2}$$

Solution. For $n = 6$ we have $h = (b - a)/6 = 3/6 = 1/2$. Hence $x_0 = -1$, $x_1 = -1/2$, $x_2 = 0$, $x_3 = 1/2$, $x_4 = 1$, $x_5 = 3/2$, and $x_6 = 2$. So by (4.13)

$$\int_{-1}^2 \frac{dx}{x^2 + 2} \approx \frac{\frac{1}{2}}{3}\left(\frac{1}{(-1)^2 + 2} + 4 \cdot \frac{1}{\left(-\frac{1}{2}\right)^2 + 2} + 2 \cdot \frac{1}{0^2 + 2} + 4 \cdot \frac{1}{\left(\frac{1}{2}\right)^2 + 2}\right.$$
$$\left. + 2 \cdot \frac{1}{1^2 + 2} + 4 \cdot \frac{1}{\left(\frac{3}{2}\right)^2 + 2} + \frac{1}{2^2 + 2}\right) = 1.111 \qquad ∎$$

These numerical techniques can also be used to integrate a function known only at a discrete set of points, as in the case of experimental data. (See Exercises 17–20.)

Exercises / Section 4.9

In Exercises 1–6, find the approximate value of each of the given integrals with the specified value of n by **(a)** the trapezoidal rule; **(b)** Simpson's rule. Use a calculator and round off to three decimal places.

1. $\int_1^3 x^2\, dx$, $n = 6$ (Check by direct integration.)

2. $\int_0^4 x^3\, dx$, $n = 6$ (Check by direct integration.)

3. $\int_0^1 \dfrac{dx}{1 + x^2}$, $n = 4$

4. $\int_1^4 \dfrac{dx}{x}$, $n = 4$

5. $\int_0^2 \sqrt{1 + x}\, dx$, $n = 4$

6. $\int_2^6 \dfrac{dx}{1 + x}$, $n = 8$

7. Use the trapezoidal rule with $n = 5$ to approximate the value of $\int_0^3 \sqrt{x}\, dx$. Check by direct integration.

In Exercises 8–12, find the approximate values of the given integrals by using the trapezoidal rule with the specified value of n. Round off the answers to three decimal places.

8. $\int_0^3 \sqrt{x^2 + 2}\, dx$, $n = 10$

9. $\int_{-1}^2 \dfrac{dx}{x^3 + 2}$, $n = 12$

10. $\int_{-2}^2 \dfrac{x^2}{x^2 + 4}\, dx$, $n = 10$

11. $\int_0^6 \dfrac{\sqrt{x} + 3}{x + 7}\, dx$, $n = 8$

In Exercises 12–16, find the approximate value of the given integrals by using Simpson's rule with the specified value of n. (Use a calculator and round off to three decimal places.)

12. $\int_{-1}^1 \sqrt{x + 1}\, dx$, $n = 6$

13. $\int_1^4 \sqrt{1 + x^2}\, dx$, $n = 6$

14. $\int_1^3 \sqrt{x^2 - 1}\, dx$, $n = 4$

15. $\int_0^2 \sqrt{1 + x^4}\, dx$, $n = 6$

16. $\int_1^4 \dfrac{\sqrt{x}\, dx}{1 + x}$, $n = 6$

In Exercises 17 and 18, find the approximate area under the curve defined by each of the following sets of experimental data by using Simpson's rule.

17.

x:	3	4	5	6	7	8	9	10	11
y:	1.3	1.9	3.2	3.8	4.7	6.8	10.2	15.6	20.3

18.

x:	0.3	0.6	0.9	1.2	1.5	1.8	2.1
y:	7.3	8.2	9.8	12.3	10.2	8.7	6.5

19. The current i (in amperes) to a capacitor as a function of time t (in seconds) was determined experimentally. Use the trapezoidal rule to approximate the value of the integral determined by these measurements. (The value of the integral is the charge accumulated in the interval 1.0 s to 2.4 s.)

t:	1.0	1.2	1.4	1.6	1.8	2.0	2.2	2.4
i:	0.10	0.40	0.46	0.57	0.64	0.54	0.43	0.31

20. The force F (in pounds) required to stretch a spring x ft was determined experimentally. Use the trapezoidal rule to approximate the value of the integral determined by these measurements. (The value of the integral is the energy in foot-pounds required to stretch the spring from a length of 2.0 ft to 2.7 ft.)

x:	2.0	2.1	2.2	2.3	2.4	2.5	2.6	2.7
F:	4.9	5.3	5.3	5.7	6.1	6.2	6.4	6.8

Review Exercises / Chapter 4

1. Evaluate the integral $\int_0^3 3x^2\, dx$ as a limit of a sum. Check the result by integration.

In Exercises 2–13, perform the integrations.

2. $\int_1^4 2x\sqrt{x}\, dx$

3. $\int (3\sqrt{x} - x^{-4} + 1)\, dx$

4. $\int (5x^2 + 4)^3 (10x)\, dx$

5. $\int (1 - x^2)^5 x\, dx$

6. $\int \dfrac{dx}{\sqrt{x - 4}}$

7. $\int \dfrac{3x\, dx}{\sqrt{x^2 - 2}}$

8. $\int (x^3 + 1)^2 (4x^2)\, dx$

9. $\int (x^3 + 1)^2 (3x)\, dx$

10. $\int \dfrac{(\sqrt{x} - 1)^2}{\sqrt{x}}\, dx$

11. $\int (\sqrt{x} - 1)^2\, dx$

12. $\int x^3 \sqrt{3 - x^4} \, dx$

13. $\int (x - 2)\sqrt{x^2 - 4x} \, dx$

In Exercises 14–20, find the area of the region bounded by the given curves.

14. $y = x^2, y = -x^2, x = 2$

15. $y = \sqrt{x}, y = -x, x = 4$

16. $4x + y^2 = 0, y^2 - 4x - 8 = 0$

17. $x = 4 - y^2, y\text{-axis}$

18. $y = x(x - 1)(x - 3), x\text{-axis}$

19. $x = 4 - y, x = y^2 - 4y + 4$

20. $y = x^2 + 4, x + y = 6$

In Exercises 21–24, evaluate the given improper integrals.

21. $\int_{-\infty}^{0} \frac{2x}{(x^2 + 4)^2} \, dx$

22. $\int_{2}^{6} \frac{dx}{\sqrt{x - 2}}$

23. $\int_{-1}^{1} \frac{3 \, dx}{4x^2}$

24. $\int_{1}^{5} \frac{t \, dt}{(t^2 - 4)^2}$

25. $y = 1/x^3, x = 2, x\text{-axis, to the right of } x = 2$

26. $y = 1/\sqrt{x}, x = 1, x\text{-axis, to the right of } x = 1$

27. $y = 1/\sqrt{x}, x = 1, \text{coordinate axes}$

28. $x = 1/(y - 3)^{3/2}, y = 4, y\text{-axis, above the line } y = 4$

29. If the current in a circuit is given by $i = 3.08t^{1/2}$, how many coulombs of charge will pass a point in the first 1.75 s if $q = 0$ when $t = 0$?

30. The voltage across a 4.00-microfarad (μF) capacitor is 10.0 V initially. If the current to the capacitor is $i = 0.310 \sqrt{t} - 1.23t$, what is the voltage after 10.0 milliseconds? (1 μF $= 10^{-6}$ F; 1 millisecond $= 10^{-3}$ s.)

31. A woman standing on the roof of a building 121 ft high tosses a stone upward with a velocity of 29 ft/s. How long does it take for the stone to reach the ground? ($g = 32$ ft/s^2)

32. A man hurls an object downward with an initial velocity of 20 m/s from a height of 63 m. How long does the object take to reach the ground and with what speed does it strike? ($g = 10$ m/s^2)

33. A minivan is moving at an unknown velocity. When the driver applied the brakes, the minivan skidded 180 ft before it came to a complete stop. Suppose that during the skid the deceleration was a constant 20 ft/s^2. Was the driver exceeding the posted speed limit of 60 mi/h (88 ft/s) when the brakes were first applied?

34. A flywheel has an angular velocity given by $\omega = 10 - 2t + 3t^2$, where ω is measured in radians per second. Find the angular displacement θ at $t = 2$ s if $\theta = 0$ when $t = 0$. (Recall that $\omega = d\theta/dt$.)

35. Find the approximate value of the integral $\int_{-1}^{2} dx/(x + 3)$ by Simpson's rule with $n = 6$.

Applications of the Integral

The purpose of this chapter is to discuss the basic applications of the integral, which are many and varied. Problems dealing with applications of additional integration techniques are included in the exercises in Chapter 7.

5.1 Means and Root Mean Squares

Everyone is familiar with the idea of arithmetic **average** or **mean.** We hear about batting averages, average per capita income, average salaries, and even average students. Less obvious is the meaning of the average value of a continuous function. Yet we use this idea all the time: if it takes you half an hour to drive a distance of 25 mi, then you must be averaging 50 mi/h.

To obtain the average value of a continuous function $f(x)$ on $[a, b]$, we divide the interval $[a, b]$ into n *equal* subintervals of length Δx_i (Figure 5.1). Next we take the average of the n values $f(x_1), f(x_2), \ldots, f(x_n)$ by adding the values and dividing by n:

$$\frac{\sum_{i=1}^{n} f(x_i)}{n}$$

Multiplying numerator and denominator by Δx_i, we get

$$\frac{\sum_{i=1}^{n} f(x_i)\Delta x_i}{n\Delta x_i} = \frac{\sum_{i=1}^{n} f(x_i)\Delta x_i}{b - a}$$

since

$$n\Delta x_i = n\frac{b - a}{n} = b - a$$

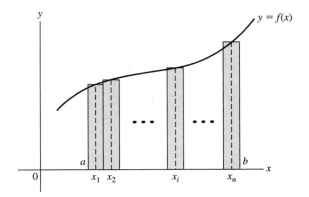

Figure 5.1

The mean value of $f(x)$, denoted by f_{av}, is then defined to be the limit of this average as $n \to \infty$. Since the numerator is a Riemann sum, we get the following formula:

Mean Value of $f(x)$ on $[a, b]$

$$f_{av} = \frac{\lim\limits_{n \to \infty} \sum\limits_{i=1}^{n} f(x_i) \Delta x_i}{b - a} = \frac{\int_a^b f(x)\,dx}{b - a} \qquad (5.1)$$

Graphically, f_{av} is the ordinate of a point on a curve such that the area of the rectangle of height f_{av} and base $b - a$ is equal to the area beneath the curve on $[a, b]$ (Figure 5.2).

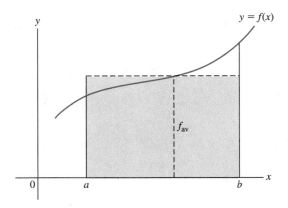

Figure 5.2

Example 1   Find the mean value of the function $y = \sqrt[3]{x}$ on $[0, 8]$.

Solution. By Formula (5.1)

$$f_{av} = \frac{\displaystyle\int_0^8 \sqrt[3]{x}\,dx}{8 - 0} = \frac{1}{8}\int_0^8 x^{1/3}\,dx = \frac{3}{2}$$

■

Example 2   If the current in a circuit is given by $i = \sqrt[3]{t}$ A, then the mean current from $t = 0.0$ s to $t = 8.0$ s is 1.5 A by Example 1.

■

If the current in a circuit is alternating, then it changes directions periodically. Hence the mean current of any complete cycle is always zero. A more useful measure of the "average" in this case is the **root mean square** (rms), defined as follows:

> **Root Mean Square**
>
> If f is continuous on $[a, b]$, then
>
> $$f_{\text{rms}} = \left\{ \frac{1}{b - a}\int_a^b [f(x)]^2\,dx \right\}^{1/2} \qquad (5.2)$$

Effective current

If i is the current, then i_{rms}, called the **effective current,** is the value of the direct current that generates the same amount of heat. Moreover, if the resistance is constant, then the mean power produced by the current is

$$P = i_{\text{rms}}^2 \cdot R \text{ (in watts)}$$

Example 3   If a current $i = t - t^2$ A flows through a resistor of 30 Ω from $t = 0$ s to $t = 2$ s, find i_{rms} and the power generated.

Solution. From the definition of root mean square we get

$$i_{\text{rms}} = \left[\frac{1}{2 - 0}\int_0^2 (t - t^2)^2\,dt \right]^{1/2}$$

$$= \left[\frac{1}{2}\int_0^2 (t^2 - 2t^3 + t^4)\,dt \right]^{1/2}$$

$$= \left[\frac{1}{2}\left(\frac{1}{3}t^3 - \frac{1}{2}t^4 + \frac{1}{5}t^5 \right)\bigg|_0^2 \right]^{1/2}$$

$$= \left[\frac{1}{2}\left(\frac{8}{3} - 8 + \frac{32}{5} \right) \right]^{1/2} = \left(\frac{8}{15} \right)^{1/2} \text{ A}$$

Hence

$$P = i_{\text{rms}}^2 \cdot R = \left[\left(\frac{8}{15} \right)^{1/2} \right]^2 \cdot 30 = \frac{8}{15} \cdot 30 = 16 \text{ W}$$

■

■ Exercises / Section 5.1

In Exercises 1–4, find the mean value of each function on the given interval.

1. $y = \sqrt{x}; [1, 16]$

2. $y = 1 - x; [1, 4]$

3. $y = x\sqrt{x^2 + 1}; [-2, 2]$

4. $y = \dfrac{1}{\sqrt{x+2}}; [-1, 3]$

In Exercises 5–8, find the root mean square in each case.

5. $y = \dfrac{1}{x}; [1, 2]$

6. $y = x^{2/3}; [-1, 1]$

7. $y = \sqrt{x}\,(x^2 + 1); [0, 1]$

8. $y = \sqrt[4]{x}; [0, 4]$

9. An object is dropped from a height of 180 m. Find its mean velocity. $(g = 10 \text{ m/s}^2)$

10. A rocket travels a distance $s = 3t^{5/2}$ meters from a launching pad during the first 10 s. Find its mean velocity during that time.

11. Find the mean current from $t = 0.0$ s to $t = 4.0$ s if $i = 1.0t + 1.0\sqrt{t}$ A.

12. Find the root mean square current for the current in Exercise 11.

13. A current $i = 1.0 - 1.0t^2$ A is flowing through a 5.0-Ω resistor. Find the mean power from $t = 0.0$ s to $t = 3.0$ s.

14. The voltage across a resistor of 10 Ω is given by $v = 2\sqrt{t} - t$. Find the mean power generated during the time interval $[0, 8]$.

5.2 Volumes of Revolution: Disk and Washer Methods

We noted in Chapter 4 that finding areas is only one of many applications of the definite integral. In this section we will learn how to find volumes of certain solids, called *solids of revolution*. Examples of such solids are bottles, axles, and funnels.

Let f be a function continuous on $[a, b]$ and consider the region under the curve in Figure 5.3. If the entire region is rotated about the x-axis, we obtain a **solid of revolution.** If the region is approximated by rectangles, then the approximate volume is obtained when each rectangle is rotated about the base. Employing the shortcut method (sloppy Riemann sum), we draw a typical element and rotate this element about the base, forming a cylindrical disk. The volume of this cylinder is

$$\pi(\text{radius})^2 \cdot \text{height} = \pi[f(x)]^2\, dx$$

Summing from $x = a$ to $x = b$, the volume is given by

$$V = \pi \int_a^b [f(x)]^2\, dx = \pi \int_a^b y^2\, dx \tag{5.3}$$

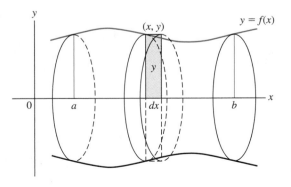

Figure 5.3

Although Formula (5.3) can be employed directly to find the volume of a solid of revolution, it is better to draw a typical element and obtain the integral from the figure. The reason is that in this chapter we will be dealing with many different integrals. It is poor practice to rely entirely on memorizing set formulas, and, above all, *a given formula may not be applicable to every situation encountered.*

The Disk Method

To find the volume of a solid of revolution by the **disk method,** draw a typical element, construct a disk, and obtain the desired integral from this disk.

Volume of disk: $\pi(\text{radius})^2 \cdot (\text{thickness})$

Example 1

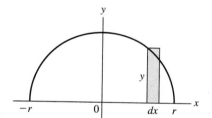

Figure 5.4

Derive the formula for the volume of a sphere of radius r.

Solution. A sphere is a solid of revolution that can be conveniently generated by the semicircle $y = \sqrt{r^2 - x^2}$ as in Figure 5.4. Taking advantage of the symmetry, we may rotate the quarter-circle in the first quadrant and double the result. Following the plan given above, we draw a typical rectangle in Figure 5.4 and obtain the volume of each cylindrical disk:

$$\pi(\text{radius})^2 \cdot (\text{thickness}) = \pi(\sqrt{r^2 - x^2})^2 \, dx$$

Summing from $x = 0$ to $x = r$ by the operation $\int_0^r$ and multiplying by 2, we get

$$V = 2 \int_0^r \pi(\sqrt{r^2 - x^2})^2 \, dx$$

$$= 2\pi \int_0^r (r^2 - x^2) \, dx \qquad \textbf{Note that } r \textbf{ is a constant.}$$

$$= 2\pi \left(r^2 x - \frac{1}{3} x^3 \right) \Big|_0^r = 2\pi \left(r^3 - \frac{1}{3} r^3 \right) = \frac{4}{3} \pi r^3 \qquad \blacksquare$$

If the region to be rotated is bounded by two curves (Figure 5.5), then the typical rectangle extends between the curves. Upon rotation the typical element generates a *washer* (Figure 5.6) with outer radius $f(x)$ and inner radius $g(x)$. Hence the volume of the washer is

$$\pi[f(x)]^2 \, dx - \pi[g(x)]^2 \, dx = \pi\{[f(x)]^2 - [g(x)]^2\} \, dx$$

Volume of a Washer

$$\pi\{[f(x)]^2 - [g(x)]^2\} \, dx \qquad\qquad\qquad (5.4)$$

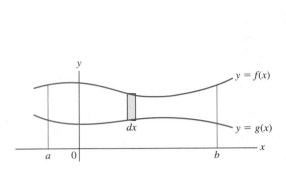

Figure 5.5

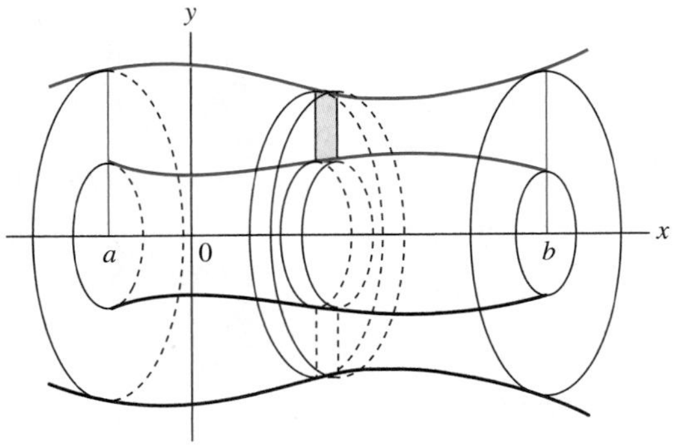

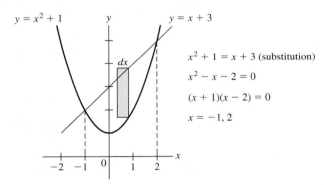

Figure 5.6

Example 2   The line $y = x + 3$ and the parabola $y = x^2 + 1$ form a bounded region. Find the volume of the solid of revolution formed by revolving the region about the x-axis (Figure 5.7).

$y = x^2 + 1$ y $y = x + 3$

$x^2 + 1 = x + 3$ (substitution)

$x^2 - x - 2 = 0$

$(x + 1)(x - 2) = 0$

$x = -1, 2$

Figure 5.7

Solution. Solving the equations simultaneously, we find that $x = -1$ and $x = 2$ are the limits of integration. The volume of the typical washer is

$$\pi[(x+3)^2 - (x^2+1)^2]\,dx$$

Summing from -1 to 2, we have

$$V = \int_{-1}^{2} \pi[(x+3)^2 - (x^2+1)^2]\,dx$$

$$= \pi \int_{-1}^{2} (-x^4 - x^2 + 6x + 8)\,dx$$

$$= \pi \left(-\frac{1}{5}x^5 - \frac{1}{3}x^3 + 3x^2 + 8x \right)\Big|_{-1}^{2}$$

$$= \pi \left[\left(-\frac{32}{5} - \frac{8}{3} + 12 + 16 \right) - \left(\frac{1}{5} + \frac{1}{3} + 3 - 8 \right) \right]$$

$$= \frac{117\pi}{5} \approx 73.5$$ ∎

In the next example the given region is rotated about the y-axis.

Example 3

Find the volume of the solid obtained by rotating the region bounded by $x = \sqrt{y}$, $y = 4$, and the y-axis about the y-axis.

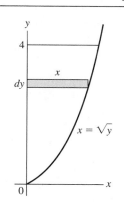

Figure 5.8

Solution. The region is shown in Figure 5.8.

Since the region is to be rotated about the y-axis, we need to interchange the roles of x and y. To this end, we *draw the typical element in the horizontal position with height x and thickness dy*, as shown in Figure 5.8. We now get

$$\text{Volume of disk} = \pi(\text{radius})^2 \cdot (\text{thickness})$$
$$= \pi x^2 dy = \pi(\sqrt{y})^2\, dy \qquad \boldsymbol{x = \sqrt{y}}$$

Summing from $y = 0$ to $y = 4$ by the operation $\int_0^4$, we get

$$V = \int_0^4 \pi(\sqrt{y})^2 dy = \pi \int_0^4 y\, dy = \pi \frac{y^2}{2}\Big|_0^4 = 8\pi \approx 25.1 \qquad \blacksquare$$

Example 4

Revolve the region bounded by $y = x^{2/3}$, $x = 8$, and the x-axis about the y-axis (Figure 5.9).

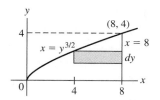

Figure 5.9

Solution. Since the region is to be revolved about the y-axis, we need to draw the typical rectangle horizontally. Solving the equation $y = x^{2/3}$ for x, we get $x = y^{3/2}$. The volume of the typical washer is given by

$$\pi(8^2 - x^2)\, dy = \pi[8^2 - (y^{3/2})^2]\, dy \qquad \textbf{washer}$$

(See Figure 5.9.)

Since $y = 4$ when $x = 8$, we sum from $y = 0$ to $y = 4$:

$$V = \int_0^4 \pi[8^2 - (y^{3/2})^2]\, dy$$
$$= \pi \int_0^4 (64 - y^3)\, dy = \pi \left(64y - \frac{1}{4}y^4\right)\Big|_0^4$$
$$= \pi \left(64 \cdot 4 - \frac{1}{4} \cdot 4^4\right) = \pi(4^4 - 4^3) = 4^3\pi(4 - 1) = 192\pi \qquad \blacksquare$$

■ Exercises / Section 5.2

In Exercises 1–11, a region R is bounded by the given curves. Find the volume of the solid of revolution obtained by rotating the region R about the x-axis.

1. $y = 2x$, $x = 1$, $x = 4$, x-axis

2. $y = x^3$, $x = 0$, $x = 1$, x-axis

3. $y = x^3$, $y = x$, the part of the region in the first quadrant

4. $y = x^4$, $x = -2$, $x = 2$, x-axis

5. $y = x^{3/2}$, $x = 0$, $x = 2$, x-axis

6. $y^2 = x$, $x = a$, the part of the region in the first quadrant. (The solid is called a *paraboloid*.)

7. $y = x^2 + 1$, $x = 2$, coordinate axes

8. $y = x^{3/2}$, $y = x^2$

9. $y = \sqrt{x^2 + 1}$, $x = 1$, $x = 3$, x-axis

10. $y = \dfrac{1}{x}$, $y = x$, $x = 3$

11. $y = \dfrac{1}{x + 2}$, $x = -1$, $x = 1$, x-axis

In Exercises 12–25, find the volume of the solid of revolution obtained by rotating R, defined by the bounds given in each exercise, about the y-axis.

12. $y = x^2$, $y = 4$ (first quadrant)

13. $y = \frac{1}{2}x^2$, $y = 2$ (first quadrant)

14. $x = \sqrt{y+1}$, coordinate axes

15. $x = \frac{1}{2}y$, $y = 4$, y-axis

16. $x = \sqrt{y}$, $y = 6$, y-axis

17. $x = \frac{1}{2}y$, $x = 2$, x-axis

18. $x = \sqrt{y}$, $x = 2$, x-axis

19. $y = \frac{1}{2}x$, $y = 2$, y-axis

20. $y = \frac{1}{3}x$, $y = 2$, y-axis

21. $y = \frac{1}{2}x$, $x = 4$, x-axis

22. $y = \frac{1}{3}x$, $x = 6$, x-axis

23. $y = x$, $x + y = 2$, x-axis

24. $y^2 = 4x$, $4x - 3y - 4 = 0$

25. $x = y^2$, $x = y + 2$

26. Derive the formula for the volume of a cone of radius r and height h.

27. The headlight on a car has a parabolic mirror with a depth of 12.0 cm and a diameter of 20.0 cm. Determine the volume.

28. A whispering gallery has a flat floor and a ceiling with elliptical cross-sections. The room is 50 ft long and 15 ft high in the center. Find the volume of the room.

29. A hemispherical tank has a radius of 12.0 ft. The water is 3.0 ft deep in the center. Find the volume of the water.

30. A wheel for a certain machine is made by cutting out the center portion of a sphere of radius 9.74 inches. (See Figure 5.10.) Find the volume of the wheel.

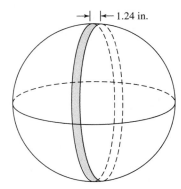

Figure 5.10

In Exercises 31–34, the region R is bounded by the given curves and rotated about the x-axis. Find the volume of the solid in each case, noting that *the resulting integrals are improper.*

31. $y = 1/x$, $x = 1$, x-axis to the right of $x = 1$

32. $x^2 y = 2$, $x = 2$, x-axis to the right of $x = 2$

33. $y = 1/x^{3/4}$, $x = 4$, x-axis to the right of $x = 4$ (Show that the area of the region does not exist.)

34. $y = 1/(x + 1)$, coordinate axes (first quadrant)

5.3   **Volumes of Revolution: Shell Method**

In this section we are going to find the volume of a solid of revolution by a different method: **the shell method.** Consider the region in Figure 5.11 between a and b. Employing the shortcut method again, we draw a typical element of height $f(x)$ and thickness dx. If the typical element is revolved about the y-axis, we obtain a **cylindrical shell.** To estimate its volume, we cut the shell vertically and lay it out flat to form a box (Figure 5.12). Note that the radius of the shell is x and that the resulting box is extremely thin; its length is approximately equal to the circumference $2\pi x$ of the shell. So the volume of the box is given by

$$\text{Length} \times \text{height} \times \text{width} = 2\pi x \, f(x)\, dx = 2\pi xy \, dx \tag{5.5}$$

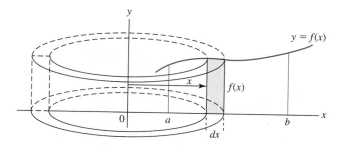

Figure 5.11

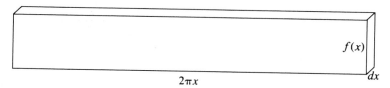

Figure 5.12

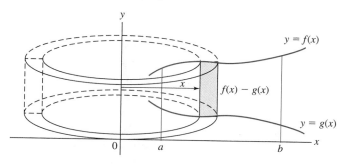

Figure 5.13

If the region is bounded by two functions f and g (Figure 5.13), then the volume of the typical shell becomes

$$2\pi x[f(x) - g(x)]\,dx \tag{5.6}$$

Summing from $x = a$ to $x = b$, we get

$$V = 2\pi \int_a^b x[f(x) - g(x)]\,dx \tag{5.7}$$

As in the case of the disk method, *it is best to remember how to construct the shell rather than to memorize the Formula (5.7).*

The Shell Method

To find the volume of a solid of revolution by the **shell method,** draw a typical element, construct a shell, and obtain the desired integral from this shell.

Volume of a shell: $2\pi(\text{radius}) \cdot (\text{height}) \cdot (\text{thickness})$

Example 1   Find the volume of the solid of revolution in Example 4 of Section 5.2 by the shell method.

Solution. We draw the graph of the given function over again (Figure 5.14). This time, however, the typical rectangle has to be drawn vertically to generate shells.

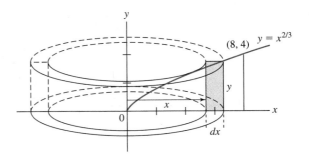

Figure 5.14

Since $y = x^{2/3}$, the volume of the typical shell is

$$2\pi(\text{radius}) \cdot (\text{height}) \cdot (\text{thickness}) = 2\pi xy \, dx = 2\pi x \cdot x^{2/3} \, dx$$

Now sum from $x = 0$ to $x = 8$:

$$V = \int_0^8 2\pi x \cdot x^{2/3} dx$$

$$= 2\pi \int_0^8 x^{5/3} \, dx = 192\pi$$

In this particular case the method of shells turns out to be somewhat simpler. It is to be expected, though, that in other problems the disk method will be more convenient than the shell method. ∎

Example 2   Use the method of shells to find the volume of the solid obtained by revolving the region bounded by $y = (1/2)x$, $y = 2$, and the y-axis about the x-axis.

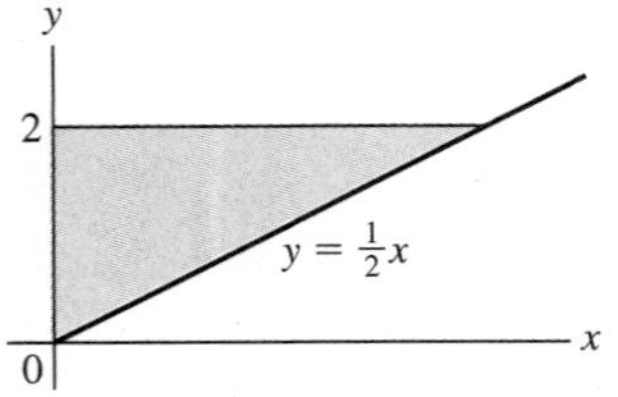

Solution. The region is shown in Figure 5.15.

The region is to be rotated about the x-axis. To generate shells, we must draw the typical element in the horizontal position and interchange the roles of x and y. (See Figure 5.16.)

By Figure 5.16, the volume of the typical shell is

$$2\pi(\text{radius}) \cdot (\text{height}) \cdot (\text{thickness}) = 2\pi \cdot y \cdot x \, dy = 2\pi y(2y) \, dy \qquad x = 2y$$

Summing from $y = 0$ to $y = 2$, we get

$$V = \int_0^2 2\pi y(2y) \, dy = 4\pi \int_0^2 y^2 \, dy$$

$$= 4\pi \left. \frac{y^3}{3} \right|_0^2 = \frac{32\pi}{3}$$

∎

Figure 5.15

$y = \frac{1}{2}x$

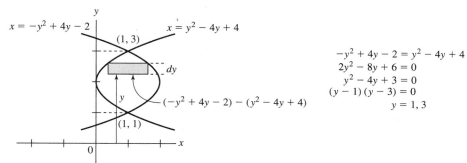

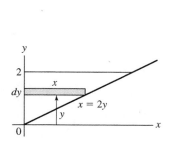

Figure 5.16   **Figure 5.17**

Example 3
Using the method of cylindrical shells, find the volume of the solid of revolution obtained by revolving the region bounded by

$$x = -y^2 + 4y - 2 \qquad \text{and} \qquad x = y^2 - 4y + 4$$

about the x-axis.

Solution. Solving the equations simultaneously, we find that the curves intersect at $(1, 1)$ and $(1, 3)$. The region is pictured in Figure 5.17.

Since the region is to be revolved about the x-axis, we need to draw the typical element in a horizontal position. Note that the thickness of the typical element is dy and the circumference of the typical shell $2\pi y$. The height of the shell is

$$(-y^2 + 4y - 2) - (y^2 - 4y + 4)$$

so that the volume of the typical shell becomes

$$2\pi(\text{radius}) \cdot (\text{height}) \cdot (\text{thickness})$$
$$= 2\pi y[(-y^2 + 4y - 2) - (y^2 - 4y + 4)] \, dy$$
$$= 2\pi y(-2y^2 + 8y - 6) \, dy$$

Summing from $y = 1$ to $y = 3$, we get

$$V = \int_1^3 2\pi y(-2y^2 + 8y - 6) \, dy = \frac{32\pi}{3}$$ ■

It is possible to revolve a region about a line other than a coordinate axis. Returning to basic principles is especially important here, since Formula (5.7) fails us.

Example 4
Find the volume of the solid obtained by revolving the region bounded by $y = 2x$, $x = 1$, and the x-axis about the line $x = -1$.

Solution. The region is shown in Figure 5.18.

The height of the typical element is y and the thickness is dx. Since we are rotating about the line $x = -1$, the radius of the typical shell is the distance from the typical element to the line $x = -1$; that is,

$$\text{Radius} = x - (-1) = x + 1$$

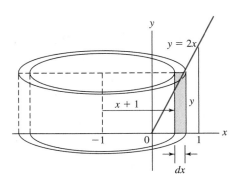

Figure 5.18

The volume of the typical shell now becomes

$$2\pi(\text{radius}) \cdot (\text{height}) \cdot (\text{thickness}) = 2\pi(x+1)y\,dx$$
$$= 2\pi(x+1)(2x)\,dx$$

Summing from $x = 0$ to $x = 1$, we get

$$V = \int_0^1 2\pi(x+1)(2x)\,dx$$
$$= 4\pi \int_0^1 (x^2 + x)\,dx = \frac{10\pi}{3}$$

■

Example 5   Find the volume of the solid generated by revolving the region bounded by $y = 8 + 4x - x^2$, $y = 2x$, and the y-axis (first quadrant) about the line $x = 4$ (Figure 5.19).

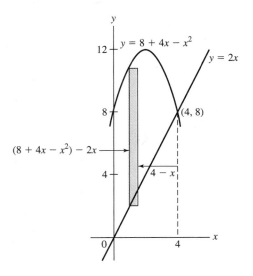

Figure 5.19

Solution. Since the region is to be revolved about the line $x = 4$, the radius of the typical shell is $4 - x$. The height of each shell is

$$(8 + 4x - x^2) - 2x$$

so that the volume of the shell becomes

$$2\pi(\text{radius}) \cdot (\text{height}) \cdot (\text{thickness})$$
$$= 2\pi(4 - x)[(8 + 4x - x^2) - 2x]\,dx$$
$$= 2\pi(4 - x)(8 + 2x - x^2)\,dx$$

Summing from $x = 0$ to $x = 4$, we obtain

$$V = 2\pi \int_0^4 (4 - x)(8 + 2x - x^2)\,dx = 128\pi$$

Summary of Disk and Shell Methods

In finding a volume of revolution by the **disk method,** place the typical element **perpendicular** to the axis of rotation.

In finding a volume of revolution by the **shell method,** place the typical element **parallel** to the axis of rotation.

In both methods, if the width of the element is dx, then the variable of integration is x. If the width of the element is dy, then the variable of integration is y.

■ Exercises / Section 5.3

In Exercises 1–12, use the method of shells to find the volume of the solid obtained by revolving the region bounded by the given curves about the y-axis.

1. $y = x, x = 3$, x-axis

2. $y = 4x^2, x = 1, x = 4$, x-axis

3. $y = x^2, y = x$

4. $y = 2\sqrt{x}, x = 4$, x-axis

5. $y = x^2, y^2 = 8x$

6. $y = \sqrt{x^2 - 1}, x = 1, x = 4$, x-axis

7. $y = x + 2, y = x^2$ (first quadrant)

8. $y = 2x, y = 3 - x$, y-axis

9. $y = x^2 - 2x$, x-axis

10. $y = -\dfrac{1}{2}x, x = 2$, x-axis

11. $y = x^2 - 4x + 3, x = 0, x = 2$, x-axis

12. $y = -x^2 + 3x - 2, x = 0, x = 2$, x-axis

In Exercises 13–23, rotate the region bounded by the given curves about the line indicated. Obtain the volume of the solid by the method of shells.

13. $y = x^2, x = 0, x = 1$, x-axis; about the line $x = 1$

14. $y = x^2, y = x$; about the line $x = -1$

15. $y = x^2, y = \sqrt{x}$; about the x-axis

16. $x = \sqrt{4 - y^2}$, y-axis; about the y-axis

17. $y = x - x^2$, x-axis; about the line $x = 2$

18. $x^2 - y^2 = 1, x = \sqrt{5}$ (first quadrant); about the y-axis

19. $x = 2y - y^2$, y-axis; about the x-axis

20. $y = x^2, y = 4$; about the x-axis

21. $y = 4x^2, x = 4$, x-axis; about the line $x = -1$

22. $y^2 = 8x, x = 2$; about the line $x = 4$

23. $y = 2x - 2x^2$, x-axis; about the line $x = -2$

24. Derive the formula for the volume of a sphere by the shell method. *Hint:* Rotate the first-quadrant region bounded by $y = \sqrt{r^2 - x^2}$ and the coordinate axes about the y-axis.

25. Derive the formula for the volume of a cone by the shell method.

26. Use the shell method to find the volume generated by revolving the region bounded by

$$y = x^{3/2}, x = 0, x = 1, \text{ x-axis}$$

a. about the x-axis
b. about the y-axis

27. Find the volume generated by rotating the region bounded by

$$y = 2\sqrt{x}, x = 4, \text{ x-axis}$$

about the x-axis
a. by the disk method
b. by the shell method

28. Repeat Exercise 27 for the region bounded by $y = \sqrt{x-1}$, $x = 10$, and the x-axis.

29. Consider the first-quadrant region bounded by $y^2 = x^3$, $x = 4$, and the x-axis. Find the volume generated by rotating this region about:
a. the x-axis   **b.** the y-axis
c. the line $x = 4$ **d.** the line $y = 8$

30. Find the volume of the solid obtained by revolving the region bounded by $y = \sqrt{x}$, $x = 4$, and the x-axis about:
a. the x-axis   **b.** the y-axis
c. the line $x = 5$ **d.** the line $y = 3$

31. Find the volume of the solid obtained by revolving the region bounded by $x = \sqrt{y}$, $y = 1$, and the y-axis about:
a. the line $y = -1$ **b.** the line $x = 2$

32. Find the volume of the solid in Exercise 1 by the disk or washer method.

33. Find the volume of the solid in Exercise 3 by the disk or washer method.

In Exercises 34–36, find the volume of the solid obtained by revolving the region bounded by the given curves about the indicated axis
a. by the disk method   **b.** by the shell method

34. $y = x$, $y = 2$, y-axis; about the x-axis

35. $y = 2x$, $x = 4$, x-axis; about the x-axis

36. $y = x^2$, $x = 2$, x-axis; about the y-axis

Use the integration capability of your graphing utility to find the approximate volume of the solid generated by revolving the region bounded by the given curves about the y-axis for Exercises 37–40.

37. $y = \sqrt{1 + \sqrt{x}}$, $x = 0$, $x = 2$, x-axis

38. $y = \sqrt{8 - x^3}$, $x = 0$, x-axis

39. $y = \sqrt[3]{4 - x}$, $x = 0$, x-axis

40. $y = \dfrac{1}{x^3 + 1}$, $x = 0$, $x = 1$, x-axis

41. The tank in Figure 5.20 has a parabolic cross-section. Find its volume by the shell method.

42. The bead shown in Figure 5.21 is a sphere of diameter 1.0 cm with a cylindrical hole drilled through the center. The hole has a diameter of 1.0 mm. Find the volume of the bead by the shell method.

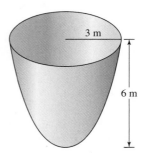

Figure 5.20

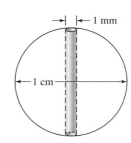

Figure 5.21

5.4 Centroids

In this section we will use integration to find centroids of plates and solids of revolution.

    Suppose a man wants to play seesaw with his three small children. The children climb up on one side of the fulcrum, while he sits on the other side. Where should he place himself to obtain an exact balance? Let w_1, w_2, and w_3 be the weights of the children and d_1, d_2, and d_3 their respective distances from the fulcrum O (Figure 5.22). If the man's weight is denoted by w_4, then he should position himself a distance d_4 from O such that

$$w_4 d_4 = w_1 d_1 + w_2 d_2 + w_3 d_3$$

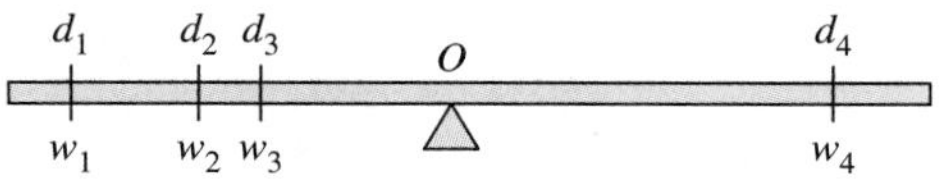

Figure 5.22

Each distance d_i is the length of the **moment arm,** and each product $w_i d_i$ is the **moment** of w_i with respect to O. Note that moments are additive, so that the moment of a system is equal to the sum of the individual moments. Furthermore, we may pretend that all the weights on the left are concentrated at one point whose distance $\bar{d}$ from O is such that

$$(w_1 + w_2 + w_3)\,\bar{d} = w_1 d_1 + w_2 d_2 + w_3 d_3$$

or

$$\bar{d} = \frac{w_1 d_1 + w_2 d_2 + w_3 d_3}{w_1 + w_2 + w_3}$$

In this formula, $\bar{d}$ is called the **center of mass** or **centroid** of the system on the left and is the point where all the weights appear to be concentrated.

For n weights,

$$\bar{d} = \frac{w_1 d_1 + w_2 d_2 + \cdots + w_n d_n}{w_1 + w_2 + \cdots + w_n} = \frac{\displaystyle\sum_{i=1}^{n} w_i d_i}{\displaystyle\sum_{i=1}^{n} w_i} \tag{5.8}$$

If we let

$$M = \sum_{i=1}^{n} w_i d_i$$

the moment of a system with respect to O, and

$$W = \sum_{i=1}^{n} w_i$$

then we may also write

$$\bar{d} = \frac{M}{W} \tag{5.9}$$

Furthermore, $M = \bar{d}W$.

Definition of a Moment

The moment of a system with respect to O is equal to the total weight of the system multiplied by the distance from O to the center of mass.

Moment (margin)

Centroid (margin)

These ideas can be extended to systems in which mass is distributed uniformly throughout a plate. The center of mass of a plate is the point at which the entire plate will balance if supported, for instance, by a sharp nail. If the plate consists of a combination of rectangles, finding the center of mass is simple: for each rectangle the center of mass is just the geometric center, provided, of course, that the mass per unit area is constant throughout the plate. We will illustrate the procedure by an example.

Example 1

Find the center of mass of the plate of uniform density ρ pictured in Figure 5.23.

Solution. First we divide the plate into rectangles numbered I, II, and III as in Figure 5.23. The coordinate locations of the respective geometric centers are $(-1, 3/2)$, $(2, 3)$, and $(1, -1/2)$. Letting ρ be the density (mass/unit area), the weight per unit area is ρg, while the weight of the plate is proportional to the area. Now, the respective areas are 10, 8, and 2, so the corresponding weights must be $10\rho g$, $8\rho g$, and $2\rho g$. The center of mass has two coordinates $(\bar{x}, \bar{y})$, which have to be found separately. Noting that the x-coordinates of the geometric centers are $-1, 2$, and 1, respectively, by Equation (5.8) we have

$$\bar{x} = \frac{[-1(10) + 2(8) + 1(2)]\rho g}{(10 + 8 + 2)\rho g} = \frac{2}{5}$$

Similarly, since the respective y-coordinates of the geometric centers are $3/2, 3$, and $-1/2$, we get

$$\bar{y} = \frac{\left[\frac{3}{2}(10) + 3(8) + \left(-\frac{1}{2}\right)(2)\right]\rho g}{20\rho g} = \frac{19}{10} \qquad \blacksquare$$

Note that the center of mass of the plate in Example 1 lies outside the plate. Also, since the weight ρg cancels, the weight plays no part in locating the center of mass. As a consequence, we assume in this section that the weight per unit area has value 1:

Weight per unit area = 1 unit

Finally, the choice of origin is completely arbitrary; although the values of $\bar{x}$ and $\bar{y}$ depend on the origin, the location of the center of mass will always be the same relative to the plate.

Centroid of an Arbitrary Region

To locate the center of mass of an arbitrary region, it is best to start with Riemann sums rather than the shortcut method. Let the region be bounded by $y = f(x)$, the lines $x = a$ and $x = b$, and the x-axis (Figure 5.24). We divide the interval $[a, b]$ into n subintervals, each of length Δx_i, and let x_i be the center of each subinterval. To calculate the moment M_y about the y-axis, we find the moments of the individual rectangles and add. (Recall that moments are additive.) Since the center of mass of each rectangle is the geometric center $(x_i, (1/2)f(x_i))$, the moment of the ith rectangle with respect to the y-axis is $x_i f(x_i)\Delta x_i$, since x_i is the length of the moment arm and

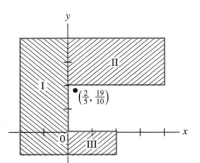

Figure 5.23

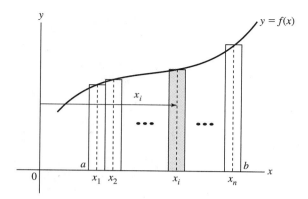

Figure 5.24

$f(x_i)\Delta x_i$ is the area. (Remember that the weight per unit area $= 1$ unit, so that the weight is numerically equal to the area.) Hence

$$M_y \approx \sum_{i=1}^{n} x_i f(x_i)\Delta x_i$$

This expression has the form of a Riemann sum, so that by the definition of the definite integral

$$M_y = \lim_{n \to \infty} \sum_{i=1}^{n} x_i f(x_i)\Delta x_i = \int_a^b x f(x)\, dx \tag{5.10}$$

Finally, since the area A may be assumed equal to the weight,

$$\bar{x} = \frac{M_y}{A} = \frac{\displaystyle\int_a^b x f(x)\, dx}{\displaystyle\int_a^b f(x)\, dx} = \frac{\displaystyle\int_a^b xy\, dx}{\displaystyle\int_a^b y\, dx} \tag{5.11}$$

Let us now see how this result may be obtained by the shortcut method. Draw the typical element in Figure 5.25. The length of the moment arm is now x and the moment of the typical element with respect to the y-axis is $xy\, dx$. Since moments are additive, we sum the moments over $[a, b]$ to find that

$$M_y = \int_a^b xy\, dx \tag{5.12}$$

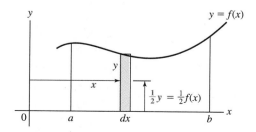

Figure 5.25

To get M_x, the moment with respect to the x-axis, we note that the centroid of the typical rectangle is $(x, (1/2)y)$. Hence the moment with respect to the x-axis is

$$\textbf{Moment arm} \cdot \textbf{area} = \frac{1}{2}y \cdot y\, dx = \frac{1}{2}y^2\, dx$$

so that the sum over $[a, b]$ is

$$M_x = \frac{1}{2}\int_a^b y^2\, dx \tag{5.13}$$

Also

$$\bar{y} = \frac{M_x}{A} = \frac{\dfrac{1}{2}\displaystyle\int_a^b y^2\, dx}{\displaystyle\int_a^b y\, dx} \tag{5.14}$$

For many regions we can find $\bar{y}$ by drawing the typical element in the horizontal position, as shown in Figure 5.26. Since the moment arm is y and the area of each typical element is $x\, dy$, we get

$$\textbf{Moment arm} \cdot \textbf{area} = y \cdot x\, dy$$

Summing from $y = c$ to $y = d$, we have

$$M_x = \int_c^d yx\, dy \tag{5.15}$$

Thus,

$$\bar{y} = \frac{M_x}{A} = \frac{\displaystyle\int_c^d yx\, dy}{A} \tag{5.16}$$

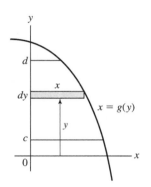

Figure 5.26

Example 2   Find the coordinates of the centroid of the region bounded by $y = 4 - 2x$ and the coordinate axes (Figure 5.27).

Solution. To find $\bar{x}$, we determine the moment of a typical element about the y-axis:

$$x \cdot y\, dx = x\,(4 - 2x)\, dx \qquad \textbf{Moment arm} \cdot \textbf{area}$$

Summing from $x = 0$ to $x = 2$, we get

$$M_y = \int_0^2 x(4 - 2x)\, dx$$

It follows that

$$\bar{x} = \frac{M_y}{A} = \frac{\displaystyle\int_0^2 x(4 - 2x)\, dx}{\displaystyle\int_0^2 (4 - 2x)\, dx} = \frac{\frac{8}{3}}{4} = \frac{8}{3} \cdot \frac{1}{4} = \frac{2}{3}$$

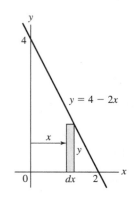

Figure 5.27

To find $\bar{y}$, we determine the moment of a typical element about the x-axis:

$$\frac{1}{2}y \cdot y \, dx = \frac{1}{2}(4 - 2x)(4 - 2x) \, dx \qquad \textbf{Moment arm} \cdot \textbf{area}$$

(See Figure 5.28.)

Summing from $x = 0$ to $x = 2$, we get

$$M_x = \int_0^2 \frac{1}{2}(4 - 2x)(4 - 2x) \, dx$$

It follows that

$$\bar{y} = \frac{M_x}{A} = \frac{\displaystyle\int_0^2 \frac{1}{2}(4 - 2x)(4 - 2x) \, dx}{4}$$

$$= \frac{\frac{16}{3}}{4} = \frac{16}{3} \cdot \frac{1}{4} = \frac{4}{3}$$

We can also find $\bar{y}$ by placing the typical element in the horizontal position (Figure 5.29).

The moment of a typical element about the x-axis can be expressed as

$$y \cdot x \, dy = y \cdot \frac{1}{2}(4 - y) \, dy \qquad \textbf{Moment arm} \cdot \textbf{area}$$

Since the limits of integration along the y-axis are $y = 0$ and $y = 4$, we have

$$\bar{y} = \frac{M_x}{A} = \frac{\displaystyle\int_0^4 y \cdot \frac{1}{2}(4 - y) \, dy}{4} = \frac{4}{3} \qquad \blacksquare$$

The next example illustrates the method for finding the centroid of a region bounded by two curves.

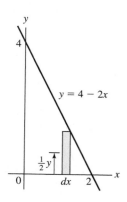

Figure 5.28

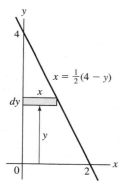

Figure 5.29

Example 3   Find the centroid of the region bounded by $y = 4x^2$ and $y = 8x$.

Solution. To find $\bar{x}$, we use Figure 5.30:

$$\bar{x} = \frac{M_y}{A} = \frac{\displaystyle\int_0^2 x(8x - 4x^2) \, dx}{\displaystyle\int_0^2 (8x - 4x^2) \, dx} = \frac{\frac{16}{3}}{\frac{16}{3}} = 1$$

To find $\bar{y}$, we use Figure 5.31:

$$\bar{y} = \frac{M_x}{A} = \frac{\displaystyle\int_0^{16} y\left(\frac{1}{2}\sqrt{y} - \frac{1}{8}y\right) \, dy}{A} = \frac{\frac{512}{15}}{\frac{16}{3}} = \frac{32}{5}$$

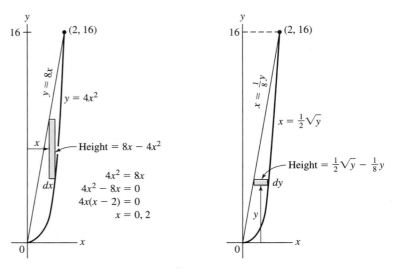

Figure 5.30

Figure 5.31   ■

The region in the next example is also bounded by two curves, but interchanging the roles of x and y to find $\bar{y}$ is not convenient.

Example 4   Find the centroid of the first-quadrant region bounded by $x^2 = 2y$ and $y = x + 4$ (Figure 5.32).

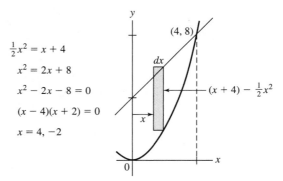

Figure 5.32

Solution. Solving the given equations simultaneously shows that $(4, 8)$ is the point of intersection in the first quadrant. The typical element of thickness dx has a length of $(x + 4) - (1/2)x^2$. As before, the length of the moment arm is x. We add the moments over the interval $[0, 4]$ to obtain

$$M_y = \int_0^4 x \left[(x + 4) - \frac{1}{2}x^2 \right] dx$$

$$= \int_0^4 x \left(-\frac{1}{2}x^2 + x + 4 \right) dx = \int_0^4 \left(-\frac{1}{2}x^3 + x^2 + 4x \right) dx = \frac{64}{3}$$

Computing M_x requires a little more care: for a region bounded below by the x-axis, the centroid of the typical element is $(x, (1/2)y)$. Since the region in this problem is bounded by two functions, the y-coordinate of the centroid of the typical rectangle is the arithmetic average of the upper and lower extremities of the rectangle; that is,

$$\frac{(x + 4) + \frac{1}{2}x^2}{2}$$

Consequently, the moment of the typical rectangle with respect to the x-axis is

$$\frac{1}{2}\left[(x + 4) + \frac{1}{2}x^2\right]\left[(x + 4) - \frac{1}{2}x^2\right] dx = \frac{1}{2}\left[(x + 4)^2 - \left(\frac{1}{2}x^2\right)^2\right] dx$$

$$= -\frac{1}{8}x^4 + \frac{1}{2}x^2 + 4x + 8$$

Summing, we have

$$M_x = \int_0^4 \left(-\frac{1}{8}x^4 + \frac{1}{2}x^2 + 4x + 8\right) dx = \frac{736}{15}$$

The area A is easily found to be

$$A = \int_0^4 \left(x + 4 - \frac{1}{2}x^2\right) dx = \frac{40}{3}$$

Hence

$$\bar{x} = \frac{M_y}{A} = \frac{64}{3} \cdot \frac{3}{40} = \frac{8}{5}$$

and

$$\bar{y} = \frac{M_x}{A} = \frac{736}{15} \cdot \frac{3}{40} = \frac{92}{25} \qquad \blacksquare$$

Centroid of a Solid of Revolution

The method for finding the centroid of a plate can be extended to find the centroid of a solid of revolution. Suppose that the region in Figure 5.33 is rotated about the x-axis. Then the volume of the typical disk is $\pi y^2\, dx$, again assumed to be equal to

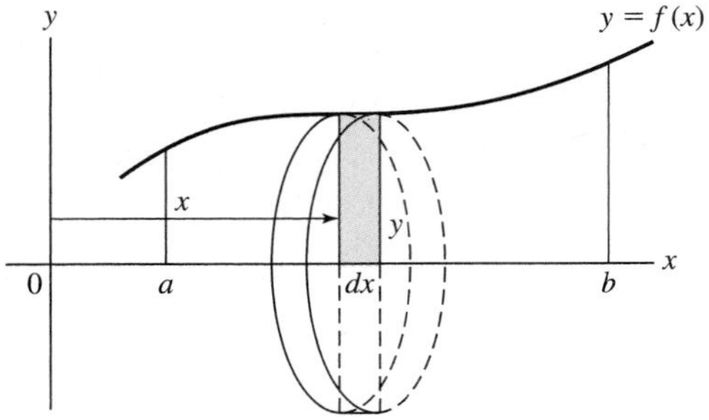

Figure 5.33

the weight. Since the length of the moment arm is x, the moment of the typical disk with respect to the y-axis is $x \cdot \pi y^2 \, dx = \pi x y^2 \, dx$. Summing, we get

$$\bar{x} = \frac{\pi \displaystyle\int_a^b x y^2 \, dx}{\pi \displaystyle\int_a^b y^2 \, dx} \tag{5.17}$$

By symmetry, it follows that $\bar{y} = 0$.

Example 5 Find the centroid of the solid of revolution obtained by rotating the first-quadrant region bounded by $y = 4 - x^2$ and the coordinate axes about the x-axis (Figure 5.34).

Solution. The volume of the typical cylinder is

$$\pi(\text{radius})^2 \cdot (\text{thickness}) = \pi(4 - x^2)^2 \, dx$$

and the moment M_y about the y-axis* is

$$x \cdot \pi(4 - x^2)^2 \, dx = \pi x(4 - x^2)^2 \, dx$$

We now sum from $x = 0$ to $x = 2$ to get

$$M_y = \pi \int_0^2 x(4 - x^2)^2 \, dx = \frac{32\pi}{3}$$

Since

$$V = \pi \int_0^2 (4 - x^2)^2 \, dx = \frac{256\pi}{15}$$

we have

$$\bar{x} = \frac{M_y}{V} = \frac{32\pi}{3} \cdot \frac{15}{256\pi} = \frac{5}{8}$$

By symmetry, $\bar{y} = 0$. ∎

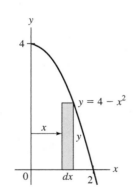

Figure 5.34

*Strictly speaking, this should be the moment about a plane perpendicular to the plane of the paper and containing the y-axis. For convenience, we will retain the notation M_y.

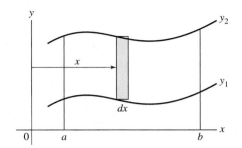

Figure 5.35

If the region is bounded by two functions y_1 and y_2 (Figure 5.35), then the washer method yields

$$\pi x \left(y_2^2 - y_1^2 \right) dx \tag{5.18}$$

for the moment of the typical washer with respect to the y-axis.

■ Exercises / Section 5.4

In Exercises 1–4 (Figures 5.36–5.39), find the coordinates of the centroid of each given plate. (See Example 1.)

1.

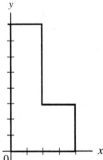

Figure 5.36

2.

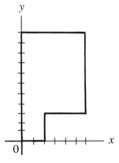

Figure 5.37

3.

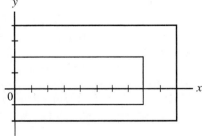

Figure 5.38

4.

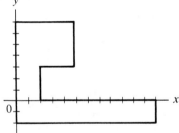

Figure 5.39

In Exercises 5–26, find the centroid of each region bounded by the given curves.

5. $x + y = 1$, coordinate axes

6. $x + 2y = 2$, coordinate axes

7. $2x + y = 2$, coordinate axes

8. $2x + 3y = 12$, coordinate axes

9. $y = x, x = 1$, x-axis

10. $y = 2x, x = 4$, x-axis

11. $y = x, y = 2$, y-axis

12. $y = \dfrac{1}{2}x, y = 1$, y-axis

13. $y = 4 - x^2$, x-axis

14. $y = -\dfrac{1}{2}x, x = 2$, x-axis

15. $y = x^2 - 2x$, x-axis

16. $y = x - x^3$, x-axis (first quadrant)

17. $y = \dfrac{1}{\sqrt[3]{x}}$, $x = 1$, $x = 8$, x-axis

18. $y^2 = x$, $x = 1$

19. $x = y - y^2$, y-axis

20. $x = 1 - y^2$, y-axis

21. $y = \sqrt{a^2 - x^2}$, coordinate axes (first quadrant). Note that $A = (1/4)\pi a^2$ (sector of circle).

22. $y = x^2$, $y^2 = x$

23. $y = 2x$, $y = x^2$

24. $y = x^{3/2}$, $y = x$

25. $y^2 = 4x$, $y = 2x$

26. $y = 6x - x^2$, $y = x$

27. Find the centroid of a semicircle of radius r.

28. Show that the centroid of the region bounded by $\sqrt{x} + \sqrt{y} = \sqrt{a}$ and the coordinate axes is $(a/5, a/5)$.

29. Find the centroid of a right triangle with sides a and b.

In Exercises 30–38, find the centroid of the solid obtained by rotating the region about the axis indicated.

30. $y = \sqrt{x}$, $x = 1$, $x = 4$, x-axis; about the x-axis

31. **a.** $y = x$, $x = 1$, x-axis; about the x-axis
   b. $y = x^2$, $y = 2$, y-axis (first quadrant); about the y-axis

32. $y = \dfrac{1}{2}x$, $x = 4$, x-axis; about the x-axis

33. $y = 2x^2$, $x = 1$, x-axis; about the x-axis

34. $y = 2x$, $y = 2$, y-axis; about the x-axis

35. $y = x^2$, $y = 9$, y-axis (first quadrant); about the x-axis

36. $y = x^3$, $x = 0$, $x = 2$, x-axis; about the x-axis

37. $y = \sqrt{r^2 - x^2}$ and the coordinate axes (first quadrant); about the x-axis (centroid of a hemisphere)

38. $x + y = 4$, $y = 2$, coordinate axes; about the y-axis

39. Find the centroid of the solid of revolution obtained by rotating the region bounded by $y = x$ and $y = x^2$
   a. about the x-axis
   b. about the y-axis

40. Find the centroid of the solid obtained by rotating the region in Exercise 36 about the y-axis.

41. Find the centroid of a right circular cone.

42. Find the centroid of the solid obtained by rotating the region bounded by $x = 4/y^2$, $y = 1$, $y = 2$, and the y-axis about the y-axis.

43. Find the centroid of the solid obtained by revolving the region bounded by $y^2 = 4px$, $y = b$, and the y-axis about y-axis.

44. Show that the centroid of a paraboloid of revolution obtained by rotation of the first-quadrant region bounded by $y^2 = 4px$ and $x = h$ about the x-axis is $((2/3)h, 0)$.

45. A company has a logo that can be described as a plate bounded by two parabolas. (See Figure 5.40.) The plate is to be fastened to the ceiling by a single screw. Where should the screw be placed?

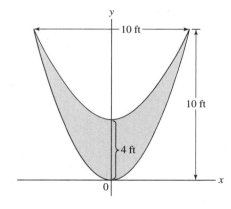

Figure 5.40

5.5   Moments of Inertia

In this section we are going to define *moment of inertia* and then use integration to find moments of inertia of plates and solids of revolution.

   If a particle of mass m is moving in a straight line with velocity v, then its kinetic energy is given by $K = (1/2)mv^2$. Now suppose that the particle is moving along a circle of radius r with a constant angular velocity $\omega = d\theta/dt$. From Figure 5.41, $s = r\theta$, so that $ds/dt = r(d\theta/dt) = r\omega$, the speed of the particle along the circle. Hence

$$K = \frac{1}{2}m\left(\frac{ds}{dt}\right)^2 = \frac{1}{2}m(r\omega)^2 = \frac{1}{2}(mr^2)\omega^2$$

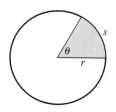

Figure 5.41

Moment of inertia

If we let $I = mr^2$, then

$$K = \frac{1}{2}I\omega^2$$

which has the same form as the formula $K = (1/2)mv^2$. This observation suggests the definition

$$I = mr^2 \tag{5.19}$$

which is called the **moment of inertia** of the particle about the axis of rotation.

The moment of inertia is the rotational analog of the mass, as the preceding example suggests. In another analogy, a particle rotating about a fixed axis has *angular momentum* $L = I\omega$, just as a particle of mass m and velocity v moving in a straight line has momentum mv. Also, the analog of Newton's second law, $F = ma$, is $T = I\alpha$, where α is the *rotational acceleration* and T is called the *torque*.

If n particles of mass $m_1, m_2, \ldots, m_n$ revolve about an axis through O perpendicular to the plane of the paper at respective distances $d_1, d_2, \ldots, d_n$ from O (Figure 5.42), then the moment of inertia is defined to be

$$I = \sum_{i=1}^{n} m_i d_i^2$$

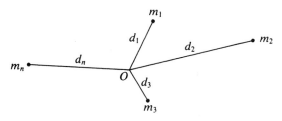

Figure 5.42

We can find a number R such that

$$(m_1 + m_2 + \cdots + m_n)R^2 = m_1 d_1^2 + m_2 d_2^2 + \cdots + m_n d_n^2$$

That is,

$$R = \left[\frac{\displaystyle\sum_{i=1}^{n} m_i d_i^2}{\displaystyle\sum_{i=1}^{n} m_i} \right]^{1/2} = \left(\frac{I}{M}\right)^{1/2} \tag{5.20}$$

Radius of gyration

where M is the total mass of the system. R is called the **radius of gyration.** Since the moment of inertia of the system remains the same if all the masses are a distance R from O, the radius of gyration is the rotational analog of the center of mass.

Example 1

Refer back to Figure 5.42. Suppose $m_1 = 4$ kg, $m_2 = 10$ kg, $d_1 = 5$ m, and $d_2 = 8$ m. If the system rotates about the axis through O with an angular velocity of 4 rad/s, find the kinetic energy of the system.

Solution. The moment of inertia is found to be

$$I = 4 \cdot 5^2 + 10 \cdot 8^2 = 740 \text{ kg} \cdot \text{m}^2$$

Hence

$$K = \frac{1}{2}I\omega^2 = \frac{1}{2}(740)(4)^2 = 5920 \text{ J} \qquad \blacksquare$$

Moment of Inertia of a Region

If the object is a plate of uniform density ρ, then the moment of inertia can be found by a method very similar to the one used for finding moments. Consider the region in Figure 5.43 and let I_y be the moment of inertia about the y-axis. The distance from the y-axis is x, and the mass of the typical rectangle is $\rho A = \rho[f(x) - g(x)]\,dx$. Thus by Formula (5.19) the moment of inertia of the typical element is $x^2 \cdot$ mass, or

$$x^2 \cdot \rho[f(x) - g(x)]\,dx = \rho x^2[f(x) - g(x)]\,dx$$

Finally, since moments of inertia are additive, we get

$$I_y = \rho \int_a^b x^2[f(x) - g(x)]\,dx \qquad (5.21)$$

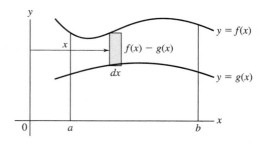

Figure 5.43

Example 2

Find I_y of the region bounded by $y = x^3$, $x = 1$, and the x-axis (Figure 5.44).

Solution. Since the height of the rectangle is y, we see that the moment of inertia of the typical rectangle about the y-axis is

$$x^2 \cdot \text{mass} = x^2 \cdot \rho y \, dx$$

Summing from $x = 0$ to $x = 1$, we get (since $y = x^3$)

$$I_y = \int_0^1 x^2 \cdot \rho x^3 \, dx = \rho \int_0^1 x^2 \cdot x^3 dx = \rho \left(\frac{1}{6}x^6\right)\Big|_0^1 = \frac{\rho}{6} \qquad \blacksquare$$

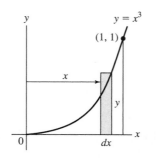

Figure 5.44

Example 3

Find R_x, the radius of gyration about the x-axis, of the region bounded by $y = x^2$ and $y = 3x$ (Figure 5.45).

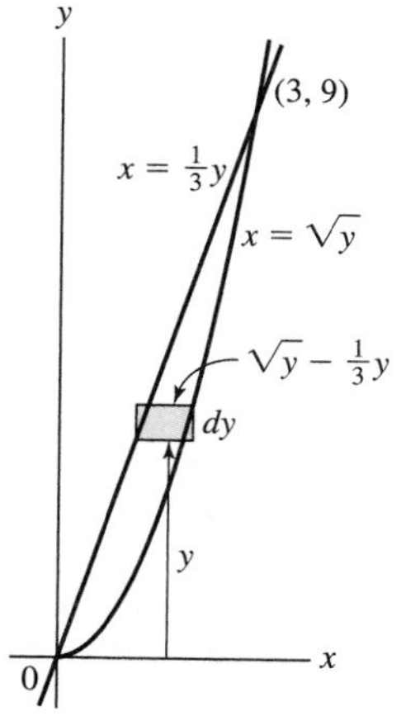

Solution. Since we are seeking the moment of inertia about the x-axis, we are forced to draw the typical element horizontally. Solving the given functions for x, we get $x = \sqrt{y}$ and $x = (1/3)y$. So the mass of the typical rectangle is $\rho(\sqrt{y} - (1/3)y)\,dy$ and its moment of inertia about the x-axis is $y^2 \cdot \rho(\sqrt{y} - (1/3)y)\,dy = \rho y^2(\sqrt{y} - (1/3)y)\,dy$. Summing from $y = 0$ to $y = 9$, we get

$$
I_x = \rho \int_0^9 y^2 \left(\sqrt{y} - \frac{1}{3}y \right) dy = \rho \int_0^9 \left(y^{5/2} - \frac{1}{3}y^3 \right) dy
$$

$$
= \rho \left(\frac{2}{7}y^{7/2} - \frac{1}{12}y^4 \right) \Big|_0^9 = \rho \left[\left(\frac{2}{7} \right) 9^{7/2} - \left(\frac{1}{12} \right) 9^4 \right]
$$

$$
= \rho \left[\left(\frac{2}{7} \right) (3^2)^{7/2} - \left(\frac{1}{12} \right) (3^2)^4 \right] = \rho \left[\left(\frac{2}{7} \right) 3^7 - \left(\frac{1}{12} \right) 3^8 \right]
$$

$$
= \rho(3^7) \left(\frac{2}{7} - \frac{1}{4} \right) = \frac{2187\rho}{28}
$$

The area is found to be

$$
A = \int_0^3 (3x - x^2)\,dx = \frac{9}{2}
$$

so that the mass is $9\rho/2$. Hence

$$
R_x = \left(\frac{2187\rho}{28} \cdot \frac{2}{9\rho} \right)^{1/2} = \frac{9\sqrt{42}}{14} \approx 4.17 \qquad \boxed{R_x = \sqrt{\frac{I_x}{M}}}
$$

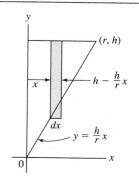

Figure 5.45

Moment of Inertia of a Solid of Revolution

It is possible to find the moment of inertia of a solid of revolution with respect to its axis. For this case the method of shells turns out to be by far the most convenient.

Example 4

Find the moment of inertia of a right circular cone of radius r and height h about its axis.

Solution. The cone can be generated by rotating the region bounded by $y = (h/r)x$, $y = h$, and the y-axis about the y-axis (Figure 5.46). The mass of the typical shell is

$$
\rho \cdot 2\pi (\text{radius}) \cdot (\text{height}) \cdot (\text{thickness}) = \rho \cdot 2\pi x \left(h - \frac{hx}{r} \right) dx
$$

by our earlier work. Since the distance from the y-axis is x, we multiply the last expression by x^2 to get the moment of inertia of the typical element:

$$
x^2 \cdot \rho \cdot 2\pi x \left(h - \frac{hx}{r} \right) dx
$$

Figure 5.46

Summing from $x = 0$ to $x = r$, we have

$$I_y = 2\pi\rho \int_0^r x^2 \cdot x \left(h - \frac{hx}{r} \right) dx$$

$$= 2\pi\rho \int_0^r x^3 \left(h - \frac{hx}{r} \right) dx = \frac{\rho}{10}\pi h r^4$$

We also note that since $V = (1/3)\pi r^2 h$, the mass of the cone is $\rho \cdot (1/3)\pi r^2 h$. So

$$R_y = \left[\frac{\rho \cdot \frac{1}{10}\pi h r^4}{\rho \cdot \frac{1}{3}\pi r^2 h} \right]^{1/2} = \left[\frac{\frac{1}{10}\pi h r^4}{\frac{1}{3}\pi r^2} \right]^{1/2} = \frac{r\sqrt{30}}{10} \qquad \blacksquare$$

For certain standard shapes of uniform density, such as the cone in Example 4, the moment of inertia is often expressed in terms of mass. Since the mass of the cone is $\rho((1/3)\pi r^2 h) = (1/3)\rho\pi h r^2$, we have

$$I_y = \frac{\rho}{10}\pi h r^4 = \frac{3}{10}\left(\frac{1}{3}\rho\pi h r^2 \right) r^2 = \frac{3}{10}m r^2$$

Example 5 If the cone in Example 4 has a mass of 1.0 kg, a radius of 0.20 m, and rotates about its axis at 120 rev/min, find its kinetic energy K and angular momentum L.

Solution. First we need to convert the rotational velocity to radians per second:

$$120 \, \frac{\text{rev}}{\text{min}} \times \frac{2\pi \, \text{rad}}{1 \, \text{rev}} \times \frac{1 \, \text{min}}{60 \, \text{s}} = 4\pi \, \frac{\text{rad}}{\text{s}}$$

We now have

$$K = \frac{1}{2}I\omega^2 = \frac{1}{2}\left(\frac{3}{10}m r^2 \right)\omega^2$$

$$= \frac{1}{2}\left(\frac{3}{10} \cdot 1.0 \cdot 0.20^2 \right)(4\pi)^2 = 0.95 \, \text{J}$$

and

$$L = I\omega = \left(\frac{3}{10}m r^2 \right)\omega$$

$$= \left(\frac{3}{10} \cdot 1.0 \cdot 0.20^2 \right)(4\pi) = 0.15 \, \text{kg} \cdot \text{m}^2/\text{s} \qquad \blacksquare$$

As in the case of linear momentum, the angular momentum $L = I\omega$ is conserved. Figure skaters performing a pirouette on one skate take advantage of this principle: with arms and legs extended, the angular velocity may be fairly small. When the limbs are suddenly pulled in, the moment of inertia decreases, but since $I\omega$ remains constant, the angular velocity ω increases.

■ Exercises / Section 5.5

In Exercises 1–16, R is the region bounded by the given curves.

1. R: $y = x, x = 1$, x-axis; find I_y, R_y

2. R: $y = (1/2)x, y = 1$, y-axis; find I_x, R_x

3. R: $x = \sqrt{y}, y = 1$, y-axis; find R_x

4. R: $x + y = 1$, coordinate axes; find R_y

5. R: $2x + y = 2$, coordinate axes; find I_y, R_y, I_x, R_x

6. R: $y = x^2, x = 0, x = 1$, x-axis; find I_y, R_y, I_x, R_x

7. R: $y = 4 - x^2$, coordinate axes (first quadrant); find I_y, R_y

8. R: $y = 2x, x = 1, x = 2$, x-axis; find R_y

9. R: $y^2 = x, y = (1/2)x$; find I_y, R_y

10. R: $y = 2x^2, y = x^3$; find I_y, R_y, R_x

11. R: $y = 9 - 3x, y = 9 - x^2$; find I_y, R_y

12. R: $y = x^3, y = \sqrt{x}$; find R_y

13. R: $x = y^2 + 2, x = y + 2$; find I_x, R_x

14. R: $y = x^3, y = 4x$ (first quadrant); find I_y, R_y, I_x, R_x

15. R: $y = 2x^2, y = 4x + 6$; find I_y

16. R: $y^2 = x, x = 2$ (first quadrant); find I_y

17. Find the moment of inertia and radius of gyration of a cylinder of radius r and height h about its axis. (*Hint:* Rotate the first-quadrant region bounded by $y = h, x = r$, and the coordinate axes about the y-axis.)

18. Find I_y of the solid generated by rotating the first-quadrant region bounded by $y = ax^2, y = b$, and the y-axis about the y-axis.

19. Find I_x and R_x of the solid generated by revolving the first-quadrant region bounded by $y = x^2$, $y = 4$, and the y-axis about the x-axis.

20. Find R_x of the following solid: the region bounded by $y^2 = x^3$, $x = 4$, and x-axis (first quadrant) rotated about the x-axis.

21. Find the moment of inertia with respect to its axis of the solid obtained by rotating the region bounded by $y = (1/2)x, y = 1$, and the y-axis about the y-axis.

22. Find the radius of gyration with respect to its axis of the solid obtained by revolving the region bounded by $y = x^3, x = 2$, and the x-axis about the y-axis.

23. Find the radius of gyration with respect to its axis of the solid obtained by revolving the region bounded by $y = \sqrt{4 - x}$ and the coordinate axes about the x-axis.

24. Find the radius of gyration with respect to its axis of the solid obtained by revolving the region bounded by $y = x$ and $y = \sqrt{x}$ about the y-axis.

25. Repeat Exercise 24 for the region bounded by $y = x^2$ and $x = y^2$.

26. Suppose a certain *paraboloid* is obtained by rotating the first-quadrant region bounded by $y = x^2$ and $y = 1$ about the y-axis (x and y are measured in meters). Assuming that $\rho = 2$ kg/m³, determine the kinetic energy and angular momentum if the paraboloid rotates at the rate of 20 rev/s.

27. A cylinder has a mass of 2.0 kg, a radius of 0.10 m, and rotates at the rate of 360 rev/min about its axis. Find its kinetic energy and angular momentum. Refer to Exercise 17.

5.6 Work and Fluid Pressure

Work

The problems in this chapter require a lot of work, but in physics the term **work** is used in a different sense: if an object is pushed a distance s along a line by a force acting in the direction of motion, then

$$\text{Work} = \text{force} \times \text{distance}$$

The concept of work is important in technology for determining the energy required to perform certain tasks. If the force is in newtons and s is in meters, then the work is measured in joules. (One joule = 1 newton · meter.) If the force is in pounds and s is in feet, then the work is measured in foot-pounds.

What happens if the force is not constant? In that case we use the type of approximation procedure that has become familiar. Suppose a body is pushed to the right along the x-axis by a force $f(x)$ from a to b. We divide $[a, b]$ into n subintervals and consider the point x_i in the ith subinterval. As usual, the ith subinterval has length Δx_i. Then the force is approximately constant in the ith subinterval and equal to $f(x_i)$. Consequently, the work done in moving the body across the ith subinterval is approximately $f(x_i)\Delta x_i$ and the total work W is given by the following approximation:

$$W \approx \sum_{i=1}^{n} f(x_i)\Delta x_i$$

This expression has the form of a Riemann sum. Hence

$$W = \lim_{n\to\infty} \sum_{i=1}^{n} f(x_i)\Delta x_i = \int_a^b f(x)\,dx \qquad (5.22)$$

These principles can be applied to finding the work done in stretching a **spring.** If an ideal spring is stretched x units beyond its natural length, then according to Hooke's law the spring pulls back with a force $f(x) = kx$. (The law also holds if the spring is compressed.) The proportionality constant k depends on the stiffness of the spring and can be determined experimentally.

Example 1 A spring has a natural length of 5 ft. A force of 4 lb stretches the spring 1/4 ft. Determine how much work is done in stretching the spring:

    a. from its natural length to a length of 7 ft
    b. from a length of 6 ft to 8 ft

Solution. We first determine the constant k. Since the force is 4 lb when $x = 1/4$ ft, from Hooke's law we have

$$F = kx$$

$$4\text{ lb} = k\left(\frac{1}{4}\text{ ft}\right) \qquad \text{or} \qquad k = 16 \text{ lb/ft}$$

Thus $f(x) = kx = 16x$.
    a. Here the spring is stretched from $x = 0$ (no extension) to $x = 2$ (natural length of 5 ft to 7 ft). So by formula (5.22),

$$W = \int_0^2 16x\,dx = 8x^2\Big|_0^2 = 32 \text{ ft-lb}$$

    b. Here the spring is stretched from $x = 1$ to $x = 3$:

$$W = \int_1^3 16x\,dx = 8x^2\Big|_1^3 = 8(9-1) = 64 \text{ ft-lb} \qquad \blacksquare$$

Example 2 A cable is 50 m long and has a density of 3 kg/m. If the cable is hanging from a winch, how much work is done in winding it up?

Solution. In this problem we are going to return to our shortcut method. To visualize the problem we would need to draw a vertical line representing the cable in the

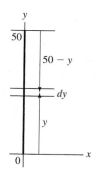

Figure 5.47

coordinate system shown in Figure 5.47. We then subdivide the interval [0, 50]. However, it will be easier if, instead of the actual subdivision, we draw the typical element of length dy shown. It will symbolize one subdivision. Each element will have a mass of 3 dy kg, so that 3 dy kg × 10 m/s^2 = **30** dy newtons. Since the typical element is 50 − y meters from the top, the work done in moving the element from its initial position to the winch at the top is $(50 - y) \cdot$ **30** dy = 30$(50 - y)\,dy$ joules (J). Then, summing from $y = 0$ to $y = 50$, we get

$$W = \int_0^{50} 30(50 - y)\,dy = 37{,}500 \text{ J}$$

■

Example 3   A cylindrical tank full of water is 10.0 m high and has a radius of 6.0 m. Find the work done in pumping the water to a level 5.0 m above the tank.

Solution. Since the motion is vertical, we subdivide the side of the tank (Figure 5.48) and let the typical element have a thickness dy. The tank itself is subdivided into cylindrical slabs, where the typical slab has a volume of $\pi(6)^2\,dy = 36\pi\,dy$ (with final zeros omitted). Taking the mass of water to be 1000 kg/m^3, each cubic meter of water weighs 10,000 N. Hence the weight of each slab is

10,000(36π dy)

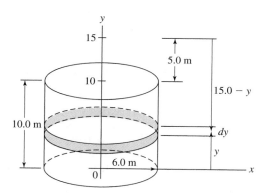

Figure 5.48

Referring to the coordinate system in Figure 5.48, each slab is moved 5 m above the tank, a distance of 15 − y meters. So the work done is

10,000(36 π dy)(15 − y) = 10,000(36π)(15 − y) dy

To get the total work done we sum over all the slabs in the tank from **$y = 0$ to $y = 10$**:

$$W = \int_0^{10} 10{,}000(36\,\pi)(15 - y)\,dy = 360{,}000\,\pi \int_0^{10} (15 - y)\,dy$$
$$= 3.6 \times 10^7\,\pi \approx 1.1 \times 10^8 \text{ J}$$

■

Example 4

A conical tank (vertex down) has a radius of 3.5 m and a height of 7.0 m. If the tank is filled with water, find the work done in pumping the water out over the top.

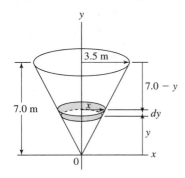

Figure 5.49

Solution. We subdivide the tank into cylindrical slabs of thickness dy (Figure 5.49). Since the slabs vary in size, the main problem is to find the radius x as a function of y. To this end, observe that in the coordinate system in Figure 5.49 the line on the right has slope

$$m = \frac{7.0}{3.5} = 2.0$$

So the equation of the line is $y = 2.0x$.
   We now obtain the following:

1.   Radius of slab: $x = 0.50y$
2.   Volume of slab: $\pi(0.50y)^2 dy$
3.   Weight of slab: $w[\pi(0.50y)^2 dy]$, where $w = 10,000$ N/m^3

Each slab is moved a distance $7.0 - y$. So the work done is

$$w[\pi(0.50y)^2 dy](7.0 - y) \tag{5.23}$$

To get the total work done, we sum from $y = 0$ to $y = 7.0$:

$$W = \int_0^{7.0} w\pi(0.50y)^2(7.0 - y)\, dy = w\pi(0.50)^2 \int_0^{7.0} y^2(7.0 - y)\, dy$$
$$= 1.6 \times 10^6 \text{ J} \qquad\blacksquare$$

Example 5

In Example 4, if water is pumped through a hole in the bottom of the tank, determine the amount of work required to **(a)** fill the tank to a level of 2.0 m; **(b)** fill the entire tank.

Solution. Referring to Figure 5.49, each slab is moved a distance y (rather than $7.0 - y$). So from Equation (5.23), the work done in moving the typical element is

$$w[\pi(0.50y)^2 dy]y$$

In part **(a)**, since the tank is filled to a level of 2.0 m, we get

$$W = \int_0^{2.0} wy\pi(0.50y)^2\, dy = 3.1 \times 10^4 \text{ J}$$

Similarly, for part **(b)**,

$$W = \int_0^{7.0} wy\pi(0.50y)^2\, dy = 4.7 \times 10^6 \text{ J} \qquad\blacksquare$$

Fluid Pressure

A well-known law in hydrostatics states that the *pressure of a fluid in an open container is directly proportional to the depth,* or

Pressure

$$p = wy \tag{5.24}$$

where p is the pressure, y the distance below the surface, and w the weight per unit volume. If ρ is the density of the fluid (mass/unit volume), then by Newton's second law, $F = mg$, we have

$$p = \rho g y$$

which is the force per unit area expressed in N/m² or lb/in². Moreover, the pressure is independent of the shape and size of the container, and at any point in the container the pressure is the same in all directions.

If a flat plate is submerged in a fluid and placed in a horizontal position, then the pressure on the plate is the same at all points. Since p is the force per unit area, the total force on one side of the plate is pA, where A is the area. If the plate is y units below the surface, then the force becomes

Force

$$F = \rho g y\,A \quad \text{or} \quad F = wy\,A \quad \textbf{pressure × area} \tag{5.25}$$

If the plate is in a vertical position, the calculation of the total force is more complicated since the pressure now varies as we move downward. But suppose the plate is subdivided into n strips; then the pressure is approximately constant on each strip if n is large.

Example 6

Suppose that a rectangular plate 4 m × 8 m is placed in water with the long side parallel to and 2 m below the surface. What is the total force against the plate?

Solution. We divide the plate into horizontal strips and let the typical strip have a thickness dy (Figure 5.50). We have seen that water weighs approximately 10,000 N/m³, to be denoted by w. Each strip is located $6 - y$ meters below the surface. So the pressure on each strip is $w(6 - y)$ by (5.24). The area is $8\,dy$, so that the force against each strip is

$$\text{Pressure} \cdot \text{area} = w(6 - y) \cdot 8\,dy \quad \textbf{by (5.25)}$$
$$= 8w(6 - y)\,dy$$

Adding the forces over the plate from $y = 0$ to $y = 4$, we get

$$F = \int_0^4 8w(6 - y)\,dy = 128w \approx 1{,}280{,}000 \text{ N}$$

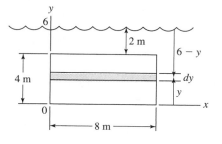

Figure 5.50

In some problems the plate is below the x-axis, as the next example shows.

Example 7

A semicircular gate on a dam has a diameter of 6.0 m and is positioned with the straight portion on top. If the surface of the water is level with the top, find the total force against the gate.

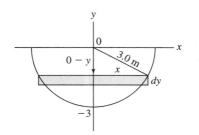

Figure 5.51

Solution. We subdivide the gate into horizontal strips (Figure 5.51). Letting w be the weight per unit volume of water, the force against the typical strip is $w(0 - y)$ times the area. The real problem is to find the area of the strip. Using the letters in the diagram, the area is given by $2x\,dy$. Since the radius of the gate is 3, we get from the Pythagorean theorem $x^2 + y^2 = 9$. The equation of the right semicircle is therefore

$$x = \sqrt{9 - y^2} \qquad \text{so that} \qquad 2x = 2\sqrt{9 - y^2}$$

Since the distance to the top is $0 - y$, the force against the strip is

$$\text{Pressure} \cdot \text{area} = w(0 - y)(2x\,dy)$$
$$= -wy(2x)\,dy = -2wy\sqrt{9 - y^2}\,dy$$

We now sum from $y = -3$ to $y = 0$:

$$F = -2w \int_{-3}^{0} y\sqrt{9 - y^2}\,dy$$

To integrate

$$\int y\sqrt{9 - y^2}\,dy = \int y(9 - y^2)^{1/2}\,dy$$

we let $u = 9 - y^2$ and find that $du = -2\,y\,dy$. Hence

$$\int y\sqrt{9 - y^2}\,dy = -\frac{1}{2} \int (9 - y^2)^{1/2}(-2y)\,dy$$
$$= -\frac{1}{2} \int u^{1/2}\,du = -\frac{1}{2} \cdot \frac{2}{3} u^{3/2} = -\frac{1}{3} u^{3/2}$$

and (since $w = 10,000\,\text{N/m}^3$),

$$F = -2w \left(-\frac{1}{3}\right)(9 - y^2)^{3/2} \Big|_{-3}^{0} = \frac{2w}{3}(9^{3/2} - 0)$$
$$= \frac{2w}{3} \cdot 27 = 18w = 180,000\,\text{N}$$

■ Exercises / Section 5.6

In the following exercises, let 10,000 N/m^3 be designated by w.

1. A force of 6 lb stretches a spring 1/8 ft. How much work is done in stretching the spring 2 ft?

2. A spring has a natural length of 6 ft, and a force of 4 lb stretches it 1 ft. Find the work done in stretching the spring **(a)** from 6 ft to 9 ft; **(b)** from 6.5 ft to 10 ft.

3. A force of 12 lb stretches a spring 2 ft. If the spring has a natural length of 8 ft, find the work done in **(a)** compressing it from 8 ft to 6 ft; **(b)** stretching it from 10 ft to 13 ft.

4. A spring has a natural length of 7 ft. A force of 4 lb stretches it 1/3 ft. Find the work done in compressing the spring from 6 ft to 4 ft.

5. A chain 20 m long and weighing 3 N/m is hanging from a winch. Find the work done in winding it up.

6. A chain weighing 2 lb/ft is 8 ft long and has a 12-lb weight attached to the end. Find the work done in winding up the chain.

7. A cable weighing 10 lb/ft is 10 ft long. If a 20-lb weight is attached to the end, find the work done in winding up the chain.

8. If only half the chain in Exercise 5 is pulled up, show that the work done is 450 J.

9. A tank full of water is in the shape of a box 3 m deep, 4 m long, and 3 m wide. Find the work done in pumping out all the water over one of its sides.

10. Repeat Exercise 9, assuming now that the tank is filled to a level of 2 m.

11. A tank in the shape of a rectangular solid is filled with water to a level of 3.5 m. Find the work done in pumping out all of the water over the top, given that the tank is 10 m long, 6 m wide, and 5 m deep.

12. A cylindrical tank full of water has a radius of 2 m and a height of 3 m. Find the work done in pumping out all of the water over the top.

13. A cylindrical tank is buried in the ground with the top at the surface. The tank has a radius of 3 m and a depth of 10 m. If the tank is half full of water, find the work done in pumping all the water to ground level.

14. A cylindrical tank has a radius of 5 m and a depth of 8 m. If the tank is filled with water to a level of 6 m, how much work is done in pumping the water out over the top?

15. A trough is 12 m long and has a cross-sectional area in the shape of an isosceles triangle 4 m across the top and 3 m high. If the trough is filled with water, how much work is done in pumping the water to a level 2 m above the trough? (Use similar triangles.)

16. A hemispherical tank full of water has a radius of 3 m. Find the work done in pumping the water out over the top.

17. A conical tank (vertex downward) has a radius of 3 m and a height of 5 m. If the tank is filled with water, find the work done in pumping the water out over the top.

18. A trough 6 m long has a semicircular cross-section of radius 1 m. If the trough is full of oil with density 800 kg/m^3, find the work done in pumping the oil out over the top.

19. A tank in the shape of a cube measures 4 m on the side. Find the work required to fill the tank through a hole in the base.

20. A tank in the shape of a rectangular solid is 5 m long, 3 m wide, and 2 m deep. Find the work required to fill the tank through a hole in the base.

21. A cylindrical tank having a radius of 3 m and a height of 5 m is filled to a level of 2 m through an opening in the bottom of the tank. Find the work done.

22. A conical tank with vertex down has a radius of 3 m and a height of 6 m. Find the work done in filling the tank to a level of 4 m through a hole in the bottom.

23. Repeat Exercise 19 if the bottom of the tank is 2 m above ground level and parallel to the ground. (See Figure 5.52.)

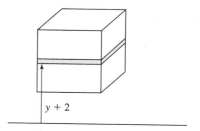

Figure 5.52

24. Repeat Exercise 20 if the bottom of the tank is 5 m above ground level.

25. In an adiabatic expansion the pressure P and the volume V are related by $PV^{1.4} = k$. For a certain steam engine, $k = 1600$. Find the work done (in foot-pounds) by the engine during a cycle in which the volume increases from 0.059 ft^3 to 0.417 ft^3, given that $W = \int_{V_1}^{V_2} P \, dV$.

26. The gravitational force between two objects r units apart is $F = Gm_1m_2/r^2$, where m_1 and m_2 are the respective masses and G is the gravitational constant whose value

depends on the units. Determine the formula for the work required to separate two objects that are r_a units apart to r_b units apart.

27. If two charged particles have opposite charges, they attract each other with a force that is inversely proportional to the square of the distance between them. (If F is the force, then $F = k/s^2$, where s is the distance between the particles.) If the force of attraction is 20.0 dynes when the particles are 1.0 cm apart, find the work done (in ergs) in separating them from **(a)** a distance of 1.0 cm to 10.0 cm; **(b)** a distance of 1.0 cm to an infinite separation.

28. If two particles have like charges, they repel each other with a force that is inversely proportional to the square of the distance between them. If the force of separation is 15 dynes at a distance of 0.50 cm, find the work done (in ergs) in bringing the particles together from an infinite separation to a separation of 1.0 cm.

29. A vertical gate on a dam has a rectangular shape 4 m long and 2 m deep. If the surface of the water is level with the top, find the force against the gate.

30. A vertical gate on a dam is a rectangle 8 m long and 3 m deep. The surface of the water is level with the gate. Find the force against the gate.

31. A rectangular tank is 5 m wide and 4 m deep. If the tank is filled with water, find the force against one end.

32. Find the force against the end of the tank in Exercise 31 if the tank is only half full.

33. The vertical gate on a dam is a rectangle 5 m long and 2 m high. If the top of the gate is horizontal and 3 m below the surface, find the force against the gate.

34. The vertical gate on a dam is a rectangle 10 m long and 4 m high. If the top of the gate is horizontal and 1 m below the surface, find the force against the gate.

35. A rectangular tank is 3 m wide across its top and is 2 m deep. If the tank is filled with water, find the force against the lower half of one end.

36. A half-filled cylindrical tank of water is lying on its side. If the diameter of the tank is 4 m, find the force against the end.

37. A trough full of water has vertical ends in the shape of an isosceles triangle 4 m across the top and 2 m deep. Find the force against the end.

38. Find the force against the end of the trough in Exercise 37, assuming now that the water is only 1 m deep in the center.

39. A thin plate has the shape of an isosceles triangle 2 m wide at the bottom and 3 m high. If the plate is submerged vertically in water with the top vertex 5 m below the surface and the bottom edge horizontal, find the total force against the (two sides of) the plate.

40. A gate on a vertical dam is an isosceles trapezoid with upper base 6 m, lower base 8 m, and height 3 m. Find the force against the gate if the upper base is 5 m below the surface.

41. A swimming pool is 9 m long. Its bottom is flat but not horizontal. If the depth ranges from 0 m to 3 m, find the total force against one side, assumed vertical.

42. The deep end of a swimming pool is vertical and has the shape of an isosceles trapezoid 6 m across the top, 4 m across the bottom, and 3 m deep. Find the force against the end.

43. Find the force on a vertical dam if the dam is in the shape of a parabola 8 m across the top and 10 m deep.

44. Suppose the face of a dam is inclined at an angle of 30° from the vertical. If the face of the dam is a rectangle 40 m wide and slant height 30 m, what is the force against the dam?

45. Suppose the end of the trough in Exercise 37 is tilted so that it makes an angle of 30° with the vertical. What is the force against the end?

■ Review Exercises / Chapter 5

1. A stone is dropped from a height of 256 ft. Find its mean velocity. ($g = 32$ ft/s²)

2. For a short time interval the current in a certain circuit is $i = 2.0t^{1/3}$ A. Find the mean current from $t = 1.0$ s to $t = 8.0$ s.

3. A current $i = 2.1 - 0.18t^{5/2}$ A is flowing through a 10.0-Ω resistor. Find the mean power from $t = 0.0$ s to $t = 4.0$ s.

4. Derive the formula for the volume of a sphere by the disk and shell methods.

5. Find the volume obtained by revolving the region bounded by $y = x^{3/2}$, $x = 4$, and the x-axis about:
   a. the x-axis   **b.** $x = 4$ **c.** $y = 8$ **d.** the y-axis

6. Find the volume of the solid generated by rotating the region bounded by $y = x^3$, $y = 1$, and $x = 0$ about the line $y = 1$.

7. Find the volume obtained by revolving the following region about the y-axis: bounded by $x + y = 2$, $x = \sqrt{y}$, and the y-axis.

8. Find the volume obtained by revolving the region in Exercise 7 about the x-axis.

9. Find the volume of the solid generated by revolving the region bounded by $y = 2x^2$, $x = 1$, and the x-axis about the line $x = 2$.

10. Find the volume of the solid generated by revolving the region bounded by $y = \sqrt{x - 2}$, $x = 6$, and the x-axis about the x-axis.

11. Find the centroid of the region bounded by $y = 4 - x^2$ and the coordinate axes (first quadrant).

12. Find I_y for the region in Exercise 11.

13. Find the centroid of the region bounded by $y = x^2 - x^3$ and the x-axis.

14. Find the centroid of the region bounded by $y = x$ and $y = 2 - x^2$.

15. Find the centroid of the solid obtained by rotating the region bounded by $x^2 - y^2 = 1$ and $x = 2$ about the x-axis.

16. Find the centroid of the solid obtained by revolving the region bounded by $y = x^2$, $x + y = 2$, and $y = 0$ about the y-axis.

17. Find I_x and R_x of the region bounded by $x = y - y^2$ and the y-axis.

18. Find R_y and R_x of the region bounded by $y = \sqrt{x}$, $x = 4$, and the x-axis.

19. Find the radius of gyration with respect to its axis of the solid obtained by revolving the region bounded by $y = \sqrt{x}$, $x = 4$, and the x-axis about the y-axis.

20. Find the moment of inertia with respect to its axis of the solid obtained by revolving the region bounded by $y = x^3$, $x = 1$, and the x-axis about the x-axis.

21. A spring has a natural length of 4 ft. A force of 2 lb stretches the spring 1/2 ft. Find the work done in stretching the spring from:
    a. its natural length to a length of 6 ft
    b. a length of 5 ft to 7 ft

22. A cable 10 m long has a density of 2 kg/m. If the cable hangs from a winch, find the work done in winding it up.

23. A cylindrical reservoir half full of water is 30 m high and has a radius of 10 m. Find the work done in pumping the water out over the top.

24. A conical tank with vertex down has a radius of 5 m and a height of 8 m. If the tank is filled with water, find the work done in pumping the water out over the top.

25. A square plate 3 m on the side is submerged vertically in water with one side parallel to and 5 m below the surface. Find the total force against one side of the plate.

26. A cylindrical tank half full of water is lying on its side. If the radius of the tank is 5 m, find the force against one end.

6

Derivatives of Transcendental Functions

Review of Trigonometry

This section consists of a brief review of trigonometry. Many of the ideas discussed here will be needed in the remainder of this chapter.

Standard Position of an Angle

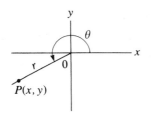

Figure 6.1

We recall that an angle θ is said to be in **standard position** if its vertex is at the origin and its initial side on the positive x-axis (Figure 6.1). Such an angle θ is considered positive if measured in the counterclockwise direction and negative if measured in the clockwise direction.

The Trigonometric Functions

We define the six trigonometric functions by selecting an arbitrary point $P(x, y)$ on the terminal side of θ, θ being in standard position (Figure 6.1). Let r be the distance from P to the origin. Then the following relationships hold:

$$\sin \theta = \frac{y}{r} \qquad \csc \theta = \frac{r}{y}$$

$$\cos \theta = \frac{x}{r} \qquad \sec \theta = \frac{r}{x}$$

$$\tan \theta = \frac{y}{x} \qquad \cot \theta = \frac{x}{y}$$

We will illustrate these functions by making use of some special angles ($0°$, $30°$, $45°$, $60°$, and so on). These angles occur so often in examples and problems that you should be thoroughly familiar with them. From Figure 6.2 we see that $\sin 30° = 1/2$, from Figure 6.3 that $\tan 45° = 1$, and so on. The terminal side may not be in the first quadrant. For example, to evaluate $\sin 210°$, we draw the angle as shown in Figure 6.4 and let P be the point $(-\sqrt{3}, -1)$. From the definition, $\sin 210° = -1/2$. Similarly, $\cot 210° = -\sqrt{3}/(-1) = \sqrt{3}$, $\cos 210° = -\sqrt{3}/2$, and so forth.

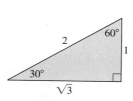

Figure 6.2

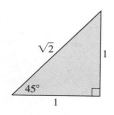

Figure 6.3

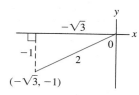

Figure 6.4

Example 1

Find $\cos 300°$, $\csc 300°$, and $\tan 300°$.

Solution. From Figure 6.5, we have

$$\cos 300° = \frac{x}{r} = \frac{1}{2}$$

$$\csc 300° = \frac{r}{y} = \frac{2}{-\sqrt{3}} = -\frac{2\sqrt{3}}{3}$$

$$\tan 300° = \frac{y}{x} = \frac{-\sqrt{3}}{1} = -\sqrt{3}$$

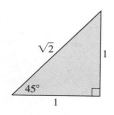

Figure 6.5

Radian Measure

Recall the following relationship between radian measure and degree measure:

$$\pi \text{ rad} = 180°$$

As a result, $\pi/2 = 90°$, $\pi/3 = 60°$, $\pi/4 = 45°$, and $\pi/6 = 30°$. Furthermore,

$$1 \text{ rad} = \frac{180°}{\pi} \approx 57.3°$$

and

$$1° = \frac{\pi}{180} \approx 0.01745 \text{ rad}$$

From the relationship $\pi \text{ rad} = 180°$, we obtain the following rules for converting from one system of measurement to the other:

To convert from degree measure to radian measure, multiply by $\pi/180°$.
To convert from radian measure to degree measure, multiply by $180°/\pi$.

Example 2

Convert:

a. $72°$ to radians   **b.** $\dfrac{5\pi}{36}$ to degrees

Solution.

a. $72° = 72° \cdot \dfrac{\pi}{180°} = \dfrac{2\pi}{5}$ **b.** $\dfrac{5\pi}{36} = \dfrac{5\pi}{36} \cdot \dfrac{180°}{\pi} = 25°$

It also follows from the definition of radian measure that the arc length s of a circular sector is given by $s = r\theta$.

Formula for the Arc Length of a Circle

$$s = r\theta \qquad \text{where } \theta \text{ is in radians} \qquad\qquad (6.1)$$

Now we will obtain a useful formula for the area of a circular sector. We know from elementary geometry that the area of a circular sector is always proportional to its central angle. In the special case where the central angle is the whole angle (2π radians), the area is πr^2. Denote the area of the sector by A and the central angle by θ. Since the ratios of the areas to their central angles are equal, we get

$$\frac{A}{\theta} = \frac{\pi r^2}{2\pi} \qquad \text{or} \qquad A = \frac{1}{2}r^2\theta$$

Formula for the Area of a Circular Sector

$$A = \frac{1}{2}r^2\theta \qquad \text{where } \theta \text{ is in radians} \qquad\qquad (6.2)$$

Graphs

The relationship between an angle and its sine or cosine can be studied conveniently from the graphs of these functions. The functions $y = \sin x$ (Figure 6.6) and $y = \cos x$ (Figure 6.7) are periodic (repeating) with **period** 2π; they have a maximum value, or **amplitude,** of 1 unit. In general, the functions

$$y = a \sin bx \qquad \text{and} \qquad y = a \cos bx$$

have amplitude a and period

$$P = \frac{2\pi}{b}$$

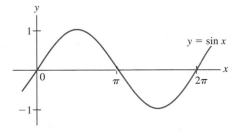

Figure 6.6

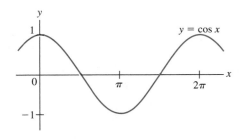

Figure 6.7

Example 3   Sketch the graph of $y = 3 \sin 2x$.

Solution. The amplitude is 3 and the period is $2\pi/2 = \pi$. Noting the basic shape of the sine function, we obtain the sketch directly (Figure 6.8). ∎

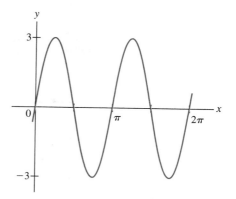

Figure 6.8

Trigonometric Identities

Let us now consider some trigonometric identities of the type that frequently occur in calculus, starting with the basic identities.

Basic Identities

$$\sec\theta = \frac{1}{\cos\theta} \qquad\qquad \csc\theta = \frac{1}{\sin\theta}$$

$$\cot\theta = \frac{1}{\tan\theta} \qquad\qquad \tan\theta = \frac{\sin\theta}{\cos\theta}$$

$$\sin^2\theta + \cos^2\theta = 1 \qquad \cot\theta = \frac{\cos\theta}{\sin\theta}$$

$$1 + \tan^2\theta = \sec^2\theta \qquad 1 + \cot^2\theta = \csc^2\theta$$

Note especially that all trigonometric functions can be expressed in terms of sines and cosines.

Example 4   Change $\cot\theta + \tan\theta$ to sines and cosines and simplify.

Solution.

$$\cot\theta + \tan\theta = \frac{\cos\theta}{\sin\theta} + \frac{\sin\theta}{\cos\theta}$$

$$= \frac{\cos\theta}{\sin\theta}\cdot\frac{\cos\theta}{\cos\theta} + \frac{\sin\theta}{\cos\theta}\cdot\frac{\sin\theta}{\sin\theta}$$

$$= \frac{\cos^2\theta + \sin^2\theta}{\sin\theta\cos\theta}$$

$$= \frac{1}{\sin\theta\cos\theta} \qquad\qquad\blacksquare$$

Identities are frequently used to convert one form to another.

Example 5   Convert:

    a. $\cos^2 3x$ to a sine function
    b. $1 + \cot^2 4x$ to a cosecant function

Solution.

    a. Since $\sin^2 \theta + \cos^2 \theta = 1$, we have

$$\cos^2 3x = 1 - \sin^2 3x$$

    b. Since $1 + \cot^2 \theta = \csc^2 \theta$, we get

$$1 + \cot^2 4x = \csc^2 4x \qquad \blacksquare$$

Sum and Difference Identities

$$\sin(\theta \pm \phi) = \sin \theta \cos \phi \pm \cos \theta \sin \phi$$
$$\cos(\theta \pm \phi) = \cos \theta \cos \phi \mp \sin \theta \sin \phi$$

Example 6   Show that $\cos\left(\dfrac{\pi}{2} - x\right) = \sin x$.

Solution. From the second identity,

$$\cos\left(\frac{\pi}{2} - x\right) = \cos\frac{\pi}{2}\cos x + \sin\frac{\pi}{2}\sin x$$
$$= 0 \cdot \cos x + 1 \cdot \sin x$$
$$= \sin x \qquad \blacksquare$$

Double-Angle Identities

$$\sin 2\theta = 2\sin\theta\cos\theta$$
$$\cos 2\theta = \cos^2\theta - \sin^2\theta = 1 - 2\sin^2\theta = 2\cos^2\theta - 1$$

Example 7   Write each expression as a single trigonometric function:

    a. $\sin 2x \cos 2x$ **b.** $2\cos^2 4x - 1$

Solution.

    a. From the identity $\sin 2\theta = 2\sin\theta\cos\theta$, we get

$$\sin 2x \cos 2x = \frac{1}{2}(2\sin 2x \cos 2x)$$
$$= \frac{1}{2}\sin 4x$$

    b. From the identity $\cos 2\theta = 2\cos^2\theta - 1$, we get

$$2\cos^2 4x - 1 = \cos 8x \qquad \blacksquare$$

Half-Angle Identities	Alternate Half-Angle Identities
$\sin\dfrac{\theta}{2} = \pm\sqrt{\dfrac{1-\cos\theta}{2}}$	$\sin^2\theta = \dfrac{1}{2}(1-\cos 2\theta)$
$\cos\dfrac{\theta}{2} = \pm\sqrt{\dfrac{1+\cos\theta}{2}}$	$\cos^2\theta = \dfrac{1}{2}(1+\cos 2\theta)$

Example 8   Write $\sin^2 6x$ without the square.

Solution. By the first alternate half-angle identity,

$$\sin^2 6x = \frac{1}{2}(1-\cos 12x)$$ ∎

■ Exercises / Section 6.1

In Exercises 1–24, evaluate the given expressions without tables or calculators.

1. $\sin 30°$

2. $\cos 120°$

3. $\tan(-45°)$

4. $\csc 240°$

5. $\sec 150°$

6. $\cot 300°$

7. $\sin(-150°)$

8. $\cos 210°$

9. $\csc(-30°)$

10. $\sec 225°$

11. $\tan 390°$

12. $\sin 0°$

13. $\tan 90°$

14. $\sec 180°$

15. $\cot 270°$

16. $\csc 180°$

17. $\sin 135°$

18. $\cos 225°$

19. $\sec 330°$

20. $\csc 240°$

21. $\cot 120°$

22. $\sin 315°$

23. $\cos 0°$

24. $\sin 270°$

In Exercises 25–33, convert the degree measures to radian measures.

25. $60°$

26. $45°$

27. $150°$

28. $-30°$

29. $135°$

30. $32°$

31. $144°$

32. $15°$

33. $20°$

In Exercises 34–43, express the radian measures in degree measures.

34. 0

35. $\dfrac{\pi}{4}$

36. $\dfrac{5\pi}{4}$

37. $\dfrac{\pi}{6}$

38. $\dfrac{\pi}{15}$

39. $\dfrac{5\pi}{3}$

40. $-\dfrac{5\pi}{4}$

41. $\dfrac{11\pi}{10}$

42. $\dfrac{17\pi}{18}$

43. $-\dfrac{5\pi}{9}$

In Exercises 44–48, find the period and amplitude, and sketch.

44. $y = 2\sin x$

45. $y = \dfrac{1}{3}\sin 2x$

46. $y = \dfrac{1}{2}\sin 3x$

47. $y = 3\cos\dfrac{1}{2}x$

48. $y = \dfrac{1}{2}\cos 4x$

In Exercises 49–62, change each expression to an expression involving sines and cosines, and simplify. (See Example 4.)

49. $\cos\theta\tan\theta$

50. $1 - \cot\theta\sin\theta$

51. $\tan\theta + \sec\theta$

52. $\dfrac{\cot\theta}{\csc\theta}$

53. $\dfrac{\tan^2\theta - \sec^2\theta}{\sec\theta}$

54. $\tan^2\theta - \dfrac{\sec^2\theta}{\csc^2\theta}$

55. $\csc^2\theta - \cot^2\theta$

56. $\dfrac{\sec\theta}{\tan\theta + \cot\theta}$

57. $\dfrac{1}{\sin^2\theta + \tan^2\theta + \cos^2\theta}$

58. $(1 + \tan^2\theta)\cos^2\theta$

59. $\sec^2\theta + \csc^2\theta$

60. $\dfrac{1 - \tan^2\theta}{1 + \tan^2\theta}$

61. $\cos\theta\cot\theta + \sin\theta$

62. $\sin^2\theta\sec^2\theta + \sin^2\theta\csc^2\theta$

In Exercises 63–70, convert sines to cosines and cosines to sines. (See Example 5.)

63. $1 - \cos^2 4x$

64. $1 - \cos^2 3x$

65. $1 - \sin^2 2x$

66. $1 - \sin^2 7x$

67. $\cos^2 5x$

68. $\sin^2 3x$

69. $\sin^2 6x$

70. $\cos^2 8x$

In Exercises 71–78, convert tangents to secants and secants to tangents.

71. $1 + \tan^2 6x$

72. $1 + \tan^2 2x$

73. $\sec^2 2x - 1$

74. $1 - \sec^2 4x$

75. $\tan^2 5x$

76. $\tan^2 9x$

77. $\sec^2 7x$

78. $\sec^2 8x$

79. Change $\cot^2 3x$ to a cosecant function.

80. Change $\csc^2 2x - 1$ to a cotangent function.

In Exercises 81–84, verify the given identities. (See Example 6.)

81. $\cos\left(x - \dfrac{\pi}{2}\right) = \sin x$

82. $\cos(\pi + x) = -\cos x$

83. $\sin(x + \pi) = -\sin x$

84. $\sin\left(x - \dfrac{\pi}{2}\right) = -\cos x$

In Exercises 85–92, write each expression as a single trigonometric function. (See Example 7.)

85. $2 \sin 5x \cos 5x$

86. $2 \sin 3x \cos 3x$

87. $\sin \dfrac{x}{2} \cos \dfrac{x}{2}$

88. $\sin 4x \cos 4x$

89. $\cos^2 3x - \sin^2 3x$

90. $\cos^2 4x - \sin^2 4x$

91. $2 \cos^2 8x - 1$

92. $1 - 2 \sin^2 5x$

In Exercises 93–98, write each expression without the square. (See Example 8.)

93. $\sin^2 3x$

94. $\cos^2 4x$

95. $\cos^2 2x$

96. $\sin^2 5x$

97. $\sin^2 \dfrac{1}{2}x$

98. $\cos^2 \dfrac{1}{2}x$

6.2   Derivatives of Sine and Cosine Functions

In Chapter 2 we developed a number of formulas for differentiating algebraic functions, thereby avoiding the tedious four-step process. In this chapter we will learn how to differentiate **transcendental functions**—functions that are not algebraic. For such functions our earlier formulas offer no help, and we need to return to the definition of derivative.

Two Special Limits

To obtain the derivative of $y = \sin x$, we need to develop two special limits. The first is

$$\lim_{\theta \to 0} \frac{\sin \theta}{\theta} \qquad (\theta \text{ in radians})$$

Since the substitution of zero for θ yields the undefined expression $0/0$, we need to fall back on a geometric argument.

Figure 6.9

Let θ be the acute angle in Figure 6.9. BD is the altitude of triangle OAB and CA is the altitude of triangle OAC. Observe that $\sin \theta = BD/1 = BD$ and $\tan \theta = CA/1 = CA$. We can get an inequality between various areas:

area of triangle OAB < area of sector OAB < area of triangle OAC

Since the area of a triangle is $(1/2)$base $\times$ height and that of a sector $(1/2)r^2\theta$ (Formula 6.2), we get

$$\frac{1}{2} \sin \theta < \frac{1}{2}\theta < \frac{1}{2} \tan \theta$$

Dividing by $(1/2)\sin \theta$, we have

$$1 < \frac{\theta}{\sin \theta} < \tan \theta \cdot \frac{1}{\sin \theta} = \frac{1}{\cos \theta}$$

Taking reciprocals reverses the sense of the inequalities:

$$1 > \frac{\sin \theta}{\theta} > \cos \theta$$

Finally, since $\cos \theta$ is a continuous function, $\lim_{\theta \to 0} \cos \theta = 1$.

It follows that the middle expression $(\sin \theta)/\theta$ also approaches 1:

$$\lim_{\theta \to 0} \frac{\sin \theta}{\theta} = 1 \tag{6.3}$$

If θ is negative, let $h = -\theta$, so that h is positive. Then

$$\frac{\sin \theta}{\theta} = \frac{\sin(-h)}{-h} = \frac{-\sin h}{-h} = \frac{\sin h}{h}$$

Hence Formula (6.3) is valid if $\theta \to 0$ through negative values.

Formula (6.3) can be used to obtain the other limit:

$$\lim_{\theta \to 0} \frac{\cos \theta - 1}{\theta} = 0 \tag{6.4}$$

Note that

$$\lim_{\theta \to 0} \frac{\cos \theta - 1}{\theta} = \lim_{\theta \to 0} \frac{(\cos \theta - 1)(\cos \theta + 1)}{\theta(\cos \theta + 1)}$$

$$= \lim_{\theta \to 0} \frac{\cos^2 \theta - 1}{\theta(\cos \theta + 1)} = \lim_{\theta \to 0} \frac{-\sin^2 \theta}{\theta(\cos \theta + 1)} \qquad \textbf{sin}^2\, \boldsymbol{\theta} + \textbf{cos}^2\, \boldsymbol{\theta} = \textbf{1}$$

$$= \lim_{\theta \to 0} \frac{\sin \theta}{\theta} \cdot \frac{-\sin \theta}{\cos \theta + 1}$$

$$= \lim_{\theta \to 0} \frac{\sin \theta}{\theta} \lim_{\theta \to 0} \frac{-\sin \theta}{\cos \theta + 1} \qquad \textbf{Theorem B on limits}$$

$$= 1 \cdot \frac{0}{1 + 1} = 0 \qquad \textbf{by Formula (6.3)}$$

The Derivative of $y = \sin u$ and $y = \cos u$

The limits (6.3) and (6.4) can now be used to find the derivative of $y = \sin x$ by the four-step process:

Step 1. $f(x + \Delta x) = \sin(x + \Delta x)$

Step 2. $\Delta y = \sin(x + \Delta x) - \sin x$

By the sum formula for the sine function,

$$\Delta y = (\sin x \cos \Delta x + \cos x \sin \Delta x) - \sin x$$

$$= \sin x \cos \Delta x - \sin x + \cos x \sin \Delta x \qquad \textbf{rearranging}$$

$$= \sin x(\cos \Delta x - 1) + \cos x \sin \Delta x \qquad \textbf{factoring sin } x$$

Step 3. $\dfrac{\Delta y}{\Delta x} = \dfrac{\sin x(\cos \Delta x - 1) + \cos x \sin \Delta x}{\Delta x}$

$$= \dfrac{\sin x(\cos \Delta x - 1)}{\Delta x} + \dfrac{\cos x \sin \Delta x}{\Delta x}$$

$$= \sin x \dfrac{\cos \Delta x - 1}{\Delta x} + \cos x \dfrac{\sin \Delta x}{\Delta x}$$

Step 4. $\displaystyle\lim_{\Delta x \to 0} \dfrac{\Delta y}{\Delta x} = \sin x \cdot 0 + \cos x \cdot 1$ **by (6.3) and (6.4)**

$$= \cos x$$

We conclude that

$$\frac{d}{dx}(\sin x) = \cos x$$

Remark: Note that we made use of Formula (6.3), which depends on Formula (6.2), but (6.2) is valid only for angles in radians. Since the derivatives of the other trigonometric functions will be obtained from the derivative of $y = \sin x$, we are going to be using mostly radian measure from now on.

A general form of the derivative of the sine function can be obtained from the chain rule (Formula 2.10):

$$\frac{dy}{dx} = \frac{dy}{du}\frac{du}{dx}$$

So if u is a function of x, then

$$\frac{d}{dx}(\sin u) = \frac{d}{du}\sin u \frac{du}{dx}$$

or

$$\frac{d}{dx}(\sin u) = \cos u \frac{du}{dx} \tag{6.5}$$

Example 1   Find the derivative of $y = \sin \sqrt{x^2 + 1}$.

Solution. Since $y = \sin(x^2 + 1)^{1/2}$, from Formula (6.5) we get

$$\frac{dy}{dx} = \cos(x^2 + 1)^{1/2} \frac{d}{dx}(x^2 + 1)^{1/2} \qquad \frac{d}{dx}\sin u = \cos u \frac{du}{dx}$$

$$= (\cos \sqrt{x^2 + 1}) \cdot \frac{1}{2}(x^2 + 1)^{-1/2} \cdot 2x = \frac{x \cos \sqrt{x^2 + 1}}{\sqrt{x^2 + 1}}$$

(Don't forget to multiply by du/dx.)   ■

Example 2   Differentiate $y = \cos x$.

Solution. From the difference identity for the sine function we get

$$\sin\left(\frac{\pi}{2} - x\right) = \sin\frac{\pi}{2}\cos x - \cos\frac{\pi}{2}\sin x$$
$$= (1)(\cos x) - (0)(\sin x) = \cos x$$

Hence

$$\frac{d}{dx}(\cos x) = \frac{d}{dx}\left[\sin\left(\frac{\pi}{2} - x\right)\right]$$
$$= \cos\left(\frac{\pi}{2} - x\right)\frac{d}{dx}\left(\frac{\pi}{2} - x\right) = \cos\left(\frac{\pi}{2} - x\right)(-1)$$
$$= \left(\cos\frac{\pi}{2}\cos x + \sin\frac{\pi}{2}\sin x\right)(-1) = -\sin x$$ ∎

By Example 2 and the chain rule,

$$\frac{d}{dx}(\cos u) = \frac{d}{du}(\cos u)\frac{du}{dx}$$

or

$$\frac{d}{dx}(\cos u) = -\sin u\frac{du}{dx} \tag{6.6}$$

Example 3   Find the derivative of $y = x^2\cos x^3$.

Solution. By the **product rule**,

Product rule

$$\frac{d}{dx}(uv) = u\frac{dv}{dx} + v\frac{du}{dx}$$

we get

$$\frac{dy}{dx} = x^2\frac{d}{dx}\cos x^3 + \cos x^3\frac{d}{dx}x^2$$
$$= x^2(-\sin x^3)(3x^2) + (\cos x^3)(2x) = -3x^4\sin x^3 + 2x\cos x^3$$ ∎

Example 4   Find the derivative of $y = \dfrac{\sin^2 x}{\sqrt{x}}$.

Solution. Since $\sin^2 x = (\sin x)^2$, we need both the power and quotient rules. From the **quotient rule**,

Quotient rule

$$\frac{d}{dx}\left(\frac{u}{v}\right) = \frac{v\dfrac{du}{dx} - u\dfrac{dv}{dx}}{v^2}$$

we get

$$\frac{dy}{dx} = \frac{x^{1/2}\frac{d}{dx}(\sin x)^2 - (\sin x)^2\frac{d}{dx}x^{1/2}}{(\sqrt{x})^2} \qquad \textbf{quotient rule}$$

Generalized power rule

By the **generalized power rule,**

$$\frac{d}{dx}u^n = nu^{n-1}\frac{du}{dx}$$

we have

$$\frac{d}{dx}(\sin x)^2 = 2(\sin x)\frac{d}{dx}(\sin x) = (2\sin x)(\cos x)$$

It follows that

$$\frac{dy}{dx} = \frac{x^{1/2}(2\sin x)(\cos x) - (\sin x)^2(\frac{1}{2})x^{-1/2}}{x}$$

$$= \frac{x^{1/2}(2\sin x \cos x) - (\sin x)^2(\frac{1}{2})x^{-1/2}}{x} \cdot \frac{2x^{1/2}}{2x^{1/2}}$$

$$= \frac{4x\sin x \cos x - \sin^2 x}{2x\sqrt{x}} \qquad\blacksquare$$

■ Exercises / Section 6.2

In Exercises 1–42, differentiate the given functions.

1. $y = \cos 5x$ **2.** $y = 2\sin 3x$ **3.** $y = 2\cos 4x$

4. $y = 3\sin 5x$ **5.** $y = \sin x^2$ **6.** $y = \cos x^2$

7. $s = 3\cos t^3$ **8.** $s = 4\sin t^3$ **9.** $y = \sin 3x$

10. $y = \cos 4x$ **11.** $y = x\sin x$

12. $y = x\sin x^2$ **13.** $y = \sin^2 x$

14. $y = \cos^2 3x$ **15.** $w = \cos^2 4v$

16. $z = 2\sin^3 2y$ **17.** $y = \dfrac{\sin x}{x}$

18. $y = \cos\sqrt{x}$ **19.** $w = \cos(v^2 + 3)$

20. $r = \sin(\theta + 1)$ **21.** $y = x\cos 2x$

22. $y = x^2\sin 4x$ **23.** $y = 2x\sin(2x + 2)$

24. $y = 3x\cos(4x - 3)$ **25.** $y = \sin\dfrac{1}{x}$

26. $y = \dfrac{\cos\sqrt{x}}{\sqrt{x}}$ **27.** $y = \dfrac{x}{\cos x}$

28. $y = \dfrac{x^2}{\sin x}$ **29.** $y = \dfrac{x}{\sin 4x}$

30. $y = \dfrac{3x}{\cos 5x}$ **31.** $N = \dfrac{\cos 2\theta}{3\theta}$

32. $r = \dfrac{\sin 5\omega}{\omega^3}$ **33.** $y = \sqrt{x}\sin x$

34. $y = \sin^2 x$ **35.** $y = \cos^2 x^3$

36. $y = x\sin^2 x$ **37.** $y = \sin x\cos x$

38. $y = \cos^2 x\sin x^2$ **39.** $y = \dfrac{\sin^3 x}{x}$

40. $y = \dfrac{\cos^3 x}{x^3}$ **41.** $y = x\cos^2 3x$

42. $y = \dfrac{x}{\sin^2 2x}$

43. Show that $\dfrac{d^2}{dx^2}(\sin x) = -\sin x$.

44. Show that $\dfrac{d^4}{dx^4}(\cos x) = \cos x$.

45. Show that $\dfrac{d^4}{dx^4}(\sin x) = \sin x$.

46. Show that $\dfrac{d^2}{dx^2}(\cos 3x) = -9\cos 3x$.

47. Find the slope of the tangent line to the curve $y = x\sin 2x$ at $x = \pi/4$.

48. The charge on a capacitor is $q(t) = 0.25\cos(t - 1.20)$. Find the current to the capacitor at $t = 3.60$ s. (Set your calculator in radian mode.)

49. The current in a circuit is given by $i = 20.0\sin 4.0t$. Find the voltage across an inductor of 0.0050 H when $t = 0.20$ s. (Recall that $v = L\,di/dt$.)

50. The position of a weight on a vertical spring as a function of time is $y(t) = A\sin\omega t$. Find the velocity and acceleration as functions of time.

51. The displacement s of a point on a certain vibrating string is

$$s(t) = \frac{1}{8}\sin(20\pi t)$$

where s is measured in centimeters and t in seconds. Find the velocity of the point at $t = 0.1$ s.

52. The temperature T (in degrees Fahrenheit) in a certain city for a twelve-hour period from 10 AM to 10 PM was found to be $T = 71 + 12\sin 0.26t$, where $t = 0$ corresponds to 10 AM. Find the rate of change of T at 2 PM.

53. Use a symbolic differentiation utility to check the derivatives in Exercises 29–37. Do you get the same form for the answer?

6.3   Other Trigonometric Functions

Using the formulas from the last section, we can readily obtain the derivatives of the remaining trigonometric functions. From the identity $\tan u = \sin u/\cos u$ and the quotient rule,

$$\frac{d}{dx}\tan u = \frac{\cos u\dfrac{d}{dx}\sin u - \sin u\dfrac{d}{dx}\cos u}{\cos^2 u}$$

$$= \frac{\cos^2 u + \sin^2 u}{\cos^2 u}\frac{du}{dx} = \frac{1}{\cos^2 u}\frac{du}{dx} \qquad \cos^2 u + \sin^2 u = 1$$

$$= \sec^2 u\frac{du}{dx} \qquad\qquad \sec u = 1/\cos u$$

Hence

$$\frac{d}{dx}(\tan u) = \sec^2 u\frac{du}{dx} \qquad\qquad (6.7)$$

Similarly, from $\csc u = 1/\sin u$ we find that

$$\frac{d}{dx}\csc u = \frac{d}{dx}(\sin u)^{-1} = -(\sin u)^{-2}\frac{d}{dx}\sin u \qquad \textbf{power rule}$$

$$= -\frac{1}{\sin^2 u}\cos u\frac{du}{dx} = -\frac{1}{\sin u}\frac{\cos u}{\sin u}\frac{du}{dx}$$

Hence

$$\frac{d}{dx}(\csc u) = -\csc u\cot u\frac{du}{dx} \qquad\qquad (6.8)$$

The remaining two derivations are similar and will be left as exercises:

$$\frac{d}{dx}(\cot u) = -\csc^2 u \frac{du}{dx} \tag{6.9}$$

$$\frac{d}{dx}(\sec u) = \sec u \tan u \frac{du}{dx} \tag{6.10}$$

Let us now summarize the derivative formulas for the trigonometric functions.

Derivatives of the Trigonometric Functions

$$\frac{d}{dx}(\sin u) = \cos u \frac{du}{dx} \qquad \frac{d}{dx}(\cos u) = -\sin u \frac{du}{dx}$$

$$\frac{d}{dx}(\tan u) = \sec^2 u \frac{du}{dx} \qquad \frac{d}{dx}(\cot u) = -\csc^2 u \frac{du}{dx}$$

$$\frac{d}{dx}(\sec u) = \sec u \tan u \frac{du}{dx} \qquad \frac{d}{dx}(\csc u) = -\csc u \cot u \frac{du}{dx}$$

To help you remember these forms, note that the derivatives of the functions with prefix *co* (cosine, cotangent, and cosecant) are preceded by a negative sign, while the other three are not.

Example 1   Differentiate the function $y = \sqrt{\tan x}$.

Solution.   We write $y = (\tan x)^{1/2}$ and use the power rule:

$$\frac{dy}{dx} = \frac{1}{2}(\tan x)^{-1/2}\frac{d}{dx}\tan x \qquad \frac{du^n}{dx} = nu^{n-1}\frac{du}{dx}$$

$$= \frac{1}{2}(\tan x)^{-1/2}(\sec^2 x)$$

by Formula (6.7), so that

$$\frac{dy}{dx} = \frac{\sec^2 x}{2\sqrt{\tan x}}$$ ∎

Example 2   Differentiate $y = x \sec x^2$.

Solution.   By the product rule,

$$\frac{d}{dx}(uv) = u\frac{dv}{dx} + v\frac{du}{dx}$$

$$\frac{dy}{dx} = x\frac{d}{dx}\sec x^2 + \sec x^2 \frac{dx}{dx}$$

$$= x(\sec x^2 \tan x^2)(2x) + \sec x^2$$

by Formula (6.10). Hence

$$\frac{dy}{dx} = 2x^2 \sec x^2 \tan x^2 + \sec x^2$$

∎

Example 3 Differentiate $y = \sin 2x \cot x^2$.

Solution. We use the product rule and Formulas (6.5) and (6.9):

$$\frac{dy}{dx} = \sin 2x \frac{d}{dx} \cot x^2 + \cot x^2 \frac{d}{dx} \sin 2x$$

$$= \sin 2x (-\csc^2 x^2)(2x) + \cot x^2 (\cos 2x)(2)$$

$$= -2x \sin 2x \csc^2 x^2 + 2 \cos 2x \cot x^2$$

∎

Example 4 Differentiate $z = \sqrt{\omega + \csc \omega^3}$.

Solution. By the power rule

$$\frac{dz}{d\omega} = \frac{1}{2}(\omega + \csc \omega^3)^{-1/2} \frac{d}{d\omega}(\omega + \csc \omega^3)$$

and by Formula (6.8)

$$\frac{dz}{d\omega} = \frac{1}{2}(\omega + \csc \omega^3)^{-1/2}[1 - \csc \omega^3 \cot \omega^3 (3\omega^2)]$$

$$= \frac{1 - 3\omega^2 \csc \omega^3 \cot \omega^3}{2\sqrt{\omega + \csc \omega^3}}$$

∎

Example 5 Find dy/dx by implicit differentiation: $y^2 = \tan y + x$.

Solution. By Formula (6.7),

$$2y \frac{dy}{dx} = \sec^2 y \frac{dy}{dx} + 1 \qquad \frac{d}{dx} \tan u = \sec^2 u \frac{du}{dx}$$

Thus

$$2y \frac{dy}{dx} - \sec^2 y \frac{dy}{dx} = 1 \qquad \text{or} \qquad (2y - \sec^2 y)\frac{dy}{dx} = 1$$

and

$$\frac{dy}{dx} = \frac{1}{2y - \sec^2 y}$$

∎

■ Exercises / Section 6.3

In Exercises 1–39, differentiate each of the functions.

1. $y = \sec 5x$

2. $y = 3\tan 4x$

3. $y = 2\csc 3t$

4. $y = 2\cot 6t$

5. $y = 3\cot 4x$

6. $y = 2\csc 3x$

7. $z = 2\csc w^2$

8. $W = \tan R^2$

9. $s = \tan 2t$

10. $P = \sec 3t$

11. $y = x\cot 2x$

12. $y = 2x\csc 4x$

13. $y = \sec(x^3 + 1)$

14. $y = \sqrt{\cot x}$

15. $r = \dfrac{\sec\theta}{\theta}$

16. $y = \dfrac{\tan x}{x}$

17. $y = \cot\sqrt{3x}$

18. $y = \sec^3 x$

19. $y = \sqrt{\tan 2x}$

20. $y = \cot^{1/3} x$

21. $y = 2\tan^4 4x$

22. $y = 4\cot^6 7x$

23. $r = \sqrt{\csc\omega^2}$

24. $r = \sqrt[4]{\sec 3\theta}$

25. $T_1 = T_2{}^2\csc T_2$

26. $L_1 = \sin L_2\tan L_2$

27. $y = \cos^2 x\cot x$

28. $y = \sqrt{x}\cos x$

29. $y = \dfrac{1 + \tan x}{\sin x}$

30. $y = \cot^2\sqrt{x}$

31. $y = x\sin(1 - x)^2$

32. $y = \dfrac{\sec 4x}{x^2}$

33. $y = \dfrac{x^3}{\tan 3x}$

34. $y = \dfrac{x}{\sec^2 2x}$

35. $y = \dfrac{x}{\csc^2 5x}$

36. $y = \dfrac{x^2}{\cot 4x + 1}$

37. $y = \dfrac{\csc 2x^2}{4x}$

38. $y = \dfrac{\sin 2x}{1 - \sec x}$

39. $y = \dfrac{\cos 3x}{1 - \cot x^2}$

40. Show that $(d/dx)(\tan x - x) = \tan^2 x$.

41. Show that $(d/dx)((1/3)\tan^3 x + \tan x) = \sec^4 x$.

In Exercises 42–46, find the second derivative of each function.

42. $y = \cot 2x$

43. $y = \sec 4x$

44. $y = \cos 5x$

45. $y = x\tan x$

46. $y = \dfrac{\sin x}{x}$

In Exercises 47–55, find dy/dx by implicit differentiation.

47. $y^2 = \tan x$

48. $y^2 = \sec 4x$

49. $y^2 = x\sec x$

50. $y = \cos(x - y)$

51. $y^2 = \sin(x + y^2)$

52. $x^2 y^2 = \csc 2x$

53. $y = x\cot y^2$

54. $y^2 = y + \sec x^2 - 2$

55. $\cos y = x^2 y - 2x$

56. Derive Formulas (6.9) and (6.10).

57. Find the slope of the line normal to the curve $y = 2\cot 2x$ at $x = \pi/8$.

58. A particle is moving along the x-axis so that its distance x (in feet) from the origin is $x = 2.0\sqrt{\tan 4.0t}$. Find the velocity at $t = 0.96$ s.

6.4 Inverse Trigonometric Functions

The purpose of this section is to review the inverse trigonometric functions. We can recall from trigonometry that the equality $\sin(\pi/6) = 1/2$ can be written $\arcsin 1/2 = \pi/6$, or "$\pi/6$ is the number (or angle) whose sine is $1/2$." The problem is that if we start with the expression $\arcsin 1/2 = y$, then there exist many possible choices for y—namely, $\pi/6$, $5\pi/6$, $13\pi/6$, $17\pi/6$, and so on. It would be highly desirable to restrict the values so that y is unique. In other words, we wish $y = \arcsin x$ to represent a *function,* denoted by

$$y = \text{Arcsin } x$$

Notice that the "A" in "Arcsin" is capitalized. (The notation $y = \sin^{-1} x$ is also commonly used.)

We can accomplish such a restriction in several ways. From the graph of $y = $ Arcsin x (Figure 6.10), we see that if we restrict y to the interval $[-\pi/2, \pi/2]$, then every value of x yields a unique value for y, although many other intervals could have been chosen. The advantage of our choice is that whenever x is positive, the terminal side of angle y will lie in the first quadrant. The function $y = $ Arcsin x is called the **inverse sine of x.**

Inverses can also be defined for the other trigonometric functions by suitable restriction of the domain of the trigonometric function. The most important cases are listed next.

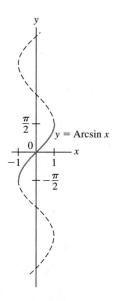

Figure 6.10

Inverse Trigonometric Functions		
$y = $ Arcsin x	$-\dfrac{\pi}{2} \le y \le \dfrac{\pi}{2}$	(Figure 6.10)
$y = $ Arctan x	$-\dfrac{\pi}{2} < y < \dfrac{\pi}{2}$	(Figure 6.11)
$y = $ Arccos x	$0 \le y \le \pi$	(Figure 6.12)

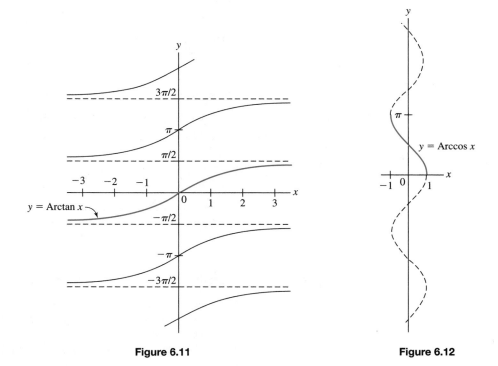

Figure 6.11   **Figure 6.12**

Although the remaining inverse trigonometric functions occur less often, let us list the ranges:

$$y = \text{Arccsc } x \qquad -\frac{\pi}{2} \le y \le \frac{\pi}{2} \qquad (y \ne 0)$$

$$y = \text{Arccot } x \qquad 0 < y < \pi$$

$$y = \text{Arcsec } x \qquad 0 \leq y \leq \pi \qquad \left(y \neq \frac{\pi}{2} \right)$$

In the following example and the corresponding exercises we are going to rely on our special angles again.

Example 1   Evaluate:

    a. $\text{Arcsin}\dfrac{1}{\sqrt{2}}$ **b.** $\text{Arccos}\left(-\dfrac{\sqrt{3}}{2}\right)$ **c.** $\text{Arctan}(-\sqrt{3})$

Solution.

    a. We are looking for an angle whose sine is $1/\sqrt{2}$. Inasmuch as y is restricted to the range $[-\pi/2, \pi/2]$, the only admissible value is $\pi/4$.

    b. An angle whose cosine is $-\sqrt{3}/2$ cannot lie in the first quadrant. Hence $5\pi/6$ is the only value that lies in the defined range $[0, \pi]$.

    c. In the range $(-\pi/2, \pi/2)$ the only admissible value lies in the interval $(-\pi/2, 0]$. Thus

$$\text{Arctan}(-\sqrt{3}) = -\frac{\pi}{3} \qquad\blacksquare$$

Example 2   Solve the equation $y = 1 + \tan 2x$ for x.

Solution.   We write the equation in the form $\tan 2x = y - 1$ and use the definition of inverse tangent:

$$2x = \text{Arctan}(y - 1) \qquad \text{or} \qquad x = \frac{1}{2}\text{Arctan}(y - 1) \qquad\blacksquare$$

The simplification of certain expressions involving inverse trigonometric functions will be encountered again in Section 7.6. Study the next example.

Example 3   Find an algebraic expression for $\tan(\text{Arcsin } 2x)$.

Solution.   Since $\text{Arcsin } 2x$ is an angle, we let $\theta = \text{Arcsin } 2x$. Then $\sin \theta = 2x = 2x/1$. We now construct the diagram in Figure 6.13. Since $\sin \theta$ is the opposite side over the hypotenuse, we let $2x$ be the length of the opposite side and let 1 be the length of the hypotenuse. The radical is obtained from the Pythagorean theorem. The desired expression can now be read from the diagram:

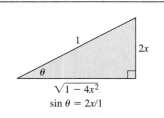

$\sin \theta = 2x/1$

Figure 6.13

$$\tan(\text{Arcsin } 2x) = \tan \theta = \frac{2x}{\sqrt{1 - 4x^2}} \qquad \tan \theta = \frac{\text{opposite}}{\text{adjacent}} \qquad\blacksquare$$

Numerical expressions can be evaluated the same way, without tables or calculators.

Example 4   Evaluate sin[Arccos(−3/4)].

Solution. Let $\theta = \text{Arccos}(-3/4)$, so that $\cos\theta = -3/4$. Since $0 \leq \text{Arccos } x \leq \pi$, θ is in the second quadrant. We now use the definition of the cosine function ($\cos\theta = x/r$) to construct the diagram in Figure 6.14. It follows that

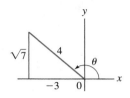

Figure 6.14

$$\sin\left[\text{Arccos}\left(-\frac{3}{4}\right)\right] = \frac{\sqrt{7}}{4} \qquad \sin\theta = \frac{\textbf{opposite}}{\textbf{hypotenuse}}$$ ■

■ Exercises / Section 6.4

In Exercises 1–28, evaluate each expression.

1. Arctan 1

2. $\text{Arcsin}\dfrac{\sqrt{3}}{2}$

3. $\text{Arcsin}\left(-\dfrac{1}{2}\right)$

4. $\text{Arccos}\dfrac{1}{\sqrt{2}}$

5. Arctan 0

6. $\text{Arccos}\left(-\dfrac{1}{2}\right)$

7. $\text{Arctan}\dfrac{1}{\sqrt{3}}$

8. Arcsin(−1)

9. Arccos(−1)

10. $\text{Arcsin}\left(-\dfrac{\sqrt{3}}{2}\right)$

11. Arctan(−1)

12. $\text{Arctan}\left(-\dfrac{1}{\sqrt{3}}\right)$

13. $\text{Arcsin}\left(-\dfrac{1}{\sqrt{2}}\right)$

14. Arcsin 0

15. $\text{Arccos}\left(-\dfrac{1}{\sqrt{2}}\right)$

16. $\text{Arctan}(\sqrt{3})$

17. $\text{Arccos}\left(\dfrac{1}{2}\right)$

18. $\text{Arccos}\left(\dfrac{\sqrt{3}}{2}\right)$

19. sin(Arctan 2)

20. $\cos\left(\text{Arcsin}\dfrac{1}{3}\right)$

21. cos(Arctan 6)

22. $\tan\left(\text{Arccos}\dfrac{2}{3}\right)$

23. $\tan\left(\text{Arcsin}\dfrac{2}{3}\right)$

24. $\sin\left(\text{Arccos}\dfrac{4}{5}\right)$

25. $\sec\left[\text{Arcsin}\left(-\dfrac{1}{3}\right)\right]$

26. cos [Arctan(−2)]

27. $\tan\left[\text{Arccos}\left(-\dfrac{3}{4}\right)\right]$

28. $\cot\left[\text{Arcsin}\left(-\dfrac{2}{5}\right)\right]$

In Exercises 29–37, write each expression as an algebraic expression.

29. sin(Arccos x)

30. sec(Arctan x)

31. csc(Arcsin x)

32. sec(Arcsin x)

33. cos(Arctan 2x)

34. sin(Arccos 3x)

35. cot(Arcsin 2x)

36. cos(Arctan 4x)

37. sin(Arccos 3x)

In Exercises 38–47, solve each equation for x.

38. $y = \text{Arcsin } 2x$

39. $y = 2\,\text{Arctan }\dfrac{x}{2}$

40. $y = \text{Arcsec } x^2$

41. $y = 1 + \sin x$

42. $y = 2 - \csc x$

43. $y = 1 + \cos 3x$

44. $y = 1 + 2\sin 2x$

45. $y = 3\tan 4x + 1$

46. $y = 2\cos 3x - 4$

47. $y = \sin\dfrac{1}{2}x - 2$

6.5 Derivatives of Inverse Trigonometric Functions

To find the derivative of $y = \text{Arcsin } u$, where u is a function of x, we write the expression in the form

$$u = \sin y \qquad \left(-\frac{\pi}{2} < y < \frac{\pi}{2} \right) \tag{6.11}$$

Since y is a function of x, Formula (6.5) applies:

$$\frac{du}{dx} = \cos y \frac{dy}{dx} \qquad \frac{d}{dx}(\sin u) = \cos u \frac{du}{dx}$$

or

$$\frac{dy}{dx} = \frac{1}{\cos y} \frac{du}{dx} \tag{6.12}$$

We recall that $\sin^2 y + \cos^2 y = 1$, so that

$$\cos y = \pm\sqrt{1 - \sin^2 y}$$

Note, however, that for $-\pi/2 < y < \pi/2$, we must choose the positive square root, since $\cos y > 0$ in this range. Hence Formula (6.12) becomes.

$$\frac{dy}{dx} = \frac{1}{\sqrt{1 - \sin^2 y}} \frac{du}{dx} \qquad \left(|y| < \frac{\pi}{2} \right)$$

By (6.11), we have $u = \sin y$, and the last formula reduces to

$$\frac{d}{dx}(\text{Arcsin } u) = \frac{1}{\sqrt{1 - u^2}} \frac{du}{dx} \qquad (|u| < 1) \tag{6.13}$$

The derivation of

$$\frac{d}{dx}(\text{Arccos } u) = -\frac{1}{\sqrt{1 - u^2}} \frac{du}{dx} \qquad (|u| < 1) \tag{6.14}$$

is similar and will be left as an exercise.

For $y = \text{Arctan } u$ we have $u = \tan y$, so that

$$\frac{du}{dx} = \sec^2 y \frac{dy}{dx}$$

and, since $\sec^2 y = 1 + \tan^2 y$,

$$\frac{dy}{dx} = \frac{1}{\sec^2 y} \frac{du}{dx} = \frac{1}{1 + \tan^2 y} \frac{du}{dx}$$

Since $u = \tan y$, we get

$$\frac{d}{dx}(\operatorname{Arctan} u) = \frac{1}{1+u^2}\frac{du}{dx} \qquad\qquad (6.15)$$

Example 1   Differentiate $y = \operatorname{Arcsin} 2x^3$.

Solution. By Formula (6.13) with $u = 2x^3$,

$$\frac{dy}{dx} = \frac{1}{\sqrt{1-(2x^3)^2}}\frac{d}{dx}(2x^3) = \frac{6x^2}{\sqrt{1-4x^6}}$$

∎

Example 2   Differentiate $v = (\operatorname{Arctan} t)^2$.

Solution. By the generalized power rule,

$$\frac{dv}{dt} = 2(\operatorname{Arctan} t)\frac{d}{dt}\operatorname{Arctan} t = 2(\operatorname{Arctan} t)\frac{1}{1+t^2} = \frac{2\operatorname{Arctan} t}{1+t^2}$$

by Formula (6.15).

∎

Example 3   Differentiate

$$y = \frac{\operatorname{Arccos} 2x}{x}$$

Solution. By the quotient rule,

$$\frac{d}{dx}\left(\frac{u}{v}\right) = \frac{v\dfrac{du}{dx} - u\dfrac{dv}{dx}}{v^2}$$

we have

$$\frac{dy}{dx} = \frac{x\dfrac{d}{dx}\operatorname{Arccos} 2x - (\operatorname{Arccos} 2x)\dfrac{dx}{dx}}{x^2}$$

$$= \frac{x\left(-\dfrac{2}{\sqrt{1-4x^2}}\right) - \operatorname{Arccos} 2x}{x^2} \qquad \text{Formula (6.14)}$$

$$= \frac{x\left(-\dfrac{2}{\sqrt{1-4x^2}}\right) - \operatorname{Arccos} 2x}{x^2}\cdot\frac{\sqrt{1-4x^2}}{\sqrt{1-4x^2}}$$

$$= \frac{x\left(-\dfrac{2}{\sqrt{1-4x^2}}\right)\sqrt{1-4x^2} - (\operatorname{Arccos} 2x)\sqrt{1-4x^2}}{x^2\sqrt{1-4x^2}}$$

$$= -\frac{2x + \sqrt{1-4x^2}\,\operatorname{Arccos} 2x}{x^2\sqrt{1-4x^2}}$$

∎

■ Exercises / Section 6.5

In Exercises 1–37, differentiate each function.

1. $y = \text{Arctan } 3x$

2. $y = \text{Arcsin } 3x$

3. $y = \text{Arccos } 5x$

4. $y = \text{Arctan } x^2$

5. $s = \text{Arctan } 2t^2$

6. $r = \text{Arctan } \sqrt{w}$

7. $u = \text{Arcsin } 3v^2$

8. $P = \text{Arccos } 9Q$

9. $y = \text{Arcsin } 2w$

10. $y = \text{Arctan } 4s$

11. $v_1 = \text{Arccos } v_2{}^2$

12. $s_1 = \text{Arcsin}\sqrt{s_2}$

13. $y = \text{Arcsin } 2x^2$

14. $y = \text{Arctan } 5x$

15. $y = \text{Arctan } 7x$

16. $y = \text{Arccos } 2x$

17. $y = x\,\text{Arctan } x$

18. $y = x^2\text{Arcsin } x$

19. $y = x\,\text{Arccos } x^2$

20. $y = \text{Arccos}(1 - x^2)$

21. $r = \theta\,\text{Arcsin } 3\theta$

22. $\omega = \theta\,\text{Arccos } \theta^2$

23. $R = 2V\,\text{Arctan } 3V$

24. $S = v\,\text{Arctan}\sqrt{v}$

25. $y = \dfrac{\text{Arcsin } x}{x}$

26. $y = \text{Arctan}\dfrac{1}{x}$

27. $y = \text{Arccos}\sqrt{1 - x}$

28. $y = \text{Arctan}\sqrt{1 + x^2}$

29. $y = \sqrt{x}\,\text{Arcsin } x$

30. $y = \dfrac{x}{\text{Arccos } x}$

31. $y = (\text{Arcsin } x)^2$

32. $y = x(\text{Arcsin } x)^2$

33. $y = \sqrt{\text{Arccos } x}$

34. $y = \sqrt[3]{\text{Arccos } x}$

35. $y = \dfrac{\text{Arctan } x}{x}$

36. $y = \dfrac{x}{\text{Arctan } x}$

37. $y = \dfrac{2x}{\text{Arcsin } x^2}$

38. A balloon is rising from the ground at the rate of 5.0 ft/s from a point 190 ft from an observer. Determine the rate of increase of the angle of inclination of the observer's line of sight when the balloon is 65 ft high. (*Hint:* If θ is the angle and h the altitude of the balloon, then $\tan\theta = h/190$ and $\theta = \text{Arctan}(h/190)$.)

39. A tractor is moving at 12.4 ft/s toward a building 80.4 ft high. Determine the rate at which the angle of elevation of the top is increasing at the instant when the tractor is 20.6 ft from the base of the building.

In Exercises 40 and 41, derive the given formulas.

40.
$$\frac{d}{dx}(\text{Arccos } u) = -\frac{1}{\sqrt{1 - u^2}}\frac{du}{dx}$$

41.
$$\frac{d}{dx}(\text{Arccot } u) = -\frac{1}{1 + u^2}\frac{du}{dx}$$

6.6   Exponential and Logarithmic Functions

In Chapter 2 we introduced powers with rational exponents. Since discussion of irrational powers is beyond the scope of this book, we will assume that if x is an irrational number and $a > 0$, then a^x is defined and the usual laws of exponents hold.

Exponential Function

Example 1   Discuss and sketch the graph of $y = 2^x$.

Solution. Since $2^x > 0$ for all x, there is no x-intercept; in fact the graph lies entirely above the x-axis. The y-intercept is $(0, 1)$. If $x > 0$ and increasing, then y increases rapidly, but if $x \to -\infty$, then $y \to 0$. Plotting a few points, we obtain the graph in Figure 6.15.

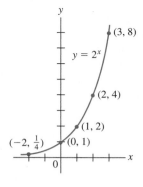

Figure 6.15 ■

x:	-2	0	1	2	3
y:	$\dfrac{1}{4}$	1	2	4	8

The function in Example 1 is typical of $y = a^x$, $a > 1$, in that $(0, 1)$ is the y-intercept and

$$\lim_{x \to -\infty} a^x = 0$$

Logarithmic Function

The function $y = a^x$, $a > 0$, $a \neq 1$, can be expressed in a different form by means of **logarithms.** We define x to be the logarithm of y to the base a and write

$$x = \log_a y \qquad (a > 0, a \neq 1)$$

Let us be very clear about this: the expressions $y = a^x$ and $x = \log_a y$ mean exactly the same thing; only the notation is different. Since $x = \log_a y$, we conclude that logarithms are exponents. For example, in the equality $2^3 = 8$ the exponent 3 is the logarithm and we would now say "3 is the logarithm of 8 to the base 2," written as $\log_2 8 = 3$.

Definition of $\log_a y$

If $a > 0$ and $a \neq 1$, then $\log_a y = x$ if, and only if, $a^x = y$.

Example 2

a. Change $3^{-2} = 1/9$ to logarithmic form and $\log_{32} 2 = 1/5$ to exponential form.
b. Show that $\log_a 1 = 0$ and $\log_a a = 1$, $a > 0$.

Solution.

a. By definition, since -2 is the exponent and 3 the base, $3^{-2} = 1/9$ is equivalent to $\log_3 1/9 = -2$.

In the expression $\log_{32} 2 = 1/5$, the value $1/5$, being the logarithm, is an exponent. Since 32 is the base, we get $(32)^{1/5} = 2$.

b. For these expressions we simply note that $a^0 = 1$ and $a^1 = a$. ■

Returning now to the logarithmic function $x = \log_a y$, suppose we employ the usual functional form with x as the independent variable. Then the function becomes $y = \log_a x$.

Example 3   Sketch the graph of $y = \log_2 x$.

Solution. The simplest way to sketch this graph is to write the function in its equivalent exponential form: since 2 is the base and y the exponent, the function becomes $x = 2^y$. We plot a few points by assigning values to y and calculating the corresponding values for x. The graph is shown in Figure 6.16. (Note that the equation is identical to the equation in Example 1 with x and y interchanged.)

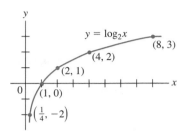

y:	-2	0	1	2	3
x:	$\dfrac{1}{4}$	1	2	4	8

Figure 6.16 ∎

Figure 6.16 is a typical graph of $y = \log_a x$, $a > 1$, in the sense that the x-intercept is always $(1, 0)$, the negative y-axis the asymptote, and the domain the open interval $(0, \infty)$.

Properties of Logarithms

Let us now recall the three basic laws of logarithms:

Properties of Logarithms	
$\log_a MN = \log_a M + \log_a N$	(6.16)
$\log_a \dfrac{M}{N} = \log_a M - \log_a N$	(6.17)
$\log_a M^k = k \log_a M$	(6.18)

We will check only Formula (6.16); the proofs of the other two are similar. Let $u = \log_a M$ and $v = \log_a N$; the equivalent exponential forms are $M = a^u$ and $N = a^v$. Now

$$MN = a^u a^v = a^{u+v}$$

and the equivalent logarithmic form is

$$\log_a MN = u + v$$

Substituting back, we get

$$\log_a MN = \log_a M + \log_a N$$

Example 4   Write

$$2 \log_a x + 3 \log_a y - \log_a z$$

as a single logarithm.

Solution. We have

$$\begin{aligned}
2 \log_a x + 3 \log_a y - \log_a z &= \log_a x^2 + \log_a y^3 - \log_a z &\textbf{by (6.18)} \\
&= \log_a x^2 y^3 - \log_a z &\textbf{by (6.16)} \\
&= \log_a \frac{x^2 y^3}{z} &\textbf{by (6.17)} \quad \blacksquare
\end{aligned}$$

In the next section we are going to use the laws of logarithms to change certain logarithmic expressions into simpler expressions. The technique is illustrated in the next example.

Example 5   Break up the following logarithms into sums, differences, or multiples of logarithms. (Use the relationships $\log_a 1 = 0$ and $\log_a a = 1$.)

    a. $\log_4 4x^2$ **b.** $\log_2 \dfrac{1}{4\sqrt{x}}$

Solution.

    a. $\begin{aligned}[t]
\log_4 4x^2 &= \log_4 4 + \log_4 x^2 &\textbf{Property (6.16)} \\
&= 1 + \log_4 x^2 &\textbf{log}_a\ \textbf{\textit{a}} = \textbf{1} \\
&= 1 + \log_4 x^2 \\
&= 1 + 2 \log_4 x &\textbf{Property (6.18)}
\end{aligned}$

    b. $\begin{aligned}[t]
\log_2 \frac{1}{4\sqrt{x}} &= \log_2 1 - \log_2 4\sqrt{x} &\textbf{Property (6.17)} \\
&= 0 - \log_2 4\sqrt{x} &\textbf{log}_a\ \textbf{1} = \textbf{0} \\
&= -(\log_2 4 + \log_2 \sqrt{x}) &\textbf{Property (6.16)} \\
&= -\log_2 2^2 - \log_2 x^{1/2} \\
&= -2 \log_2 2 - \frac{1}{2} \log_2 x &\textbf{Property (6.18)} \\
&= -2 - \frac{1}{2} \log_2 x &\textbf{log}_a\ \textbf{\textit{a}} = \textbf{1} \quad \blacksquare
\end{aligned}$

Logarithms can be used to solve equations in which the unknown is an exponent.

Example 6   Solve the equation

$$(2.79)^x = 4.68$$

Solution. We take the common logarithm (base 10) of both sides and use Property (6.18):

$$(2.79)^x = 4.68 \qquad \text{given equation}$$

$$\log (2.79)^x = \log 4.68 \qquad \text{taking logarithms}$$

$$x \log 2.79 = \log 4.68 \qquad \log_a M^k = k \log_a M$$

$$x = \frac{\log 4.68}{\log 2.79} \qquad \text{solving for } x$$

Using a calculator, we get $x = 1.50$. ■

As a final comment, note that the logarithmic function is continuous; that is,

$$\lim_{x \to x_0} \log_a x = \log_a (\lim_{x \to x_0} x) = \log_a x_0 \qquad (6.19)$$

■ Exercises / Section 6.6

In Exercises 1–6, write each of the expressions in logarithmic form.

1. $3^3 = 27$

2. $2^{-4} = \dfrac{1}{16}$

3. $9^0 = 1$

4. $(27)^{1/3} = 3$

5. $(32)^{-1/5} = \dfrac{1}{2}$

6. $5^{-2} = \dfrac{1}{25}$

In Exercises 7–12, write each of the expressions in exponential form.

7. $\log_3 243 = 5$

8. $\log_{10} 1000 = 3$

9. $\log_{1/4} \dfrac{1}{16} = 2$

10. $\log_{1/25} \dfrac{1}{5} = \dfrac{1}{2}$

11. $\log_9 \dfrac{1}{3} = -\dfrac{1}{2}$

12. $\log_5 \dfrac{1}{125} = -3$

In Exercises 13–20, determine the value of x.

13. $\log_{1/3} x = 2$

14. $\log_x 4 = \dfrac{1}{2}$

15. $\log_3 81 = x$

16. $\log_{16} \dfrac{1}{4} = x$

17. $\log_x \dfrac{1}{3} = -\dfrac{1}{3}$

18. $\log_x \dfrac{1}{7} = -\dfrac{1}{2}$

19. $\log_3 x = -2$

20. $\log_5 x = -1$

In Exercises 21–26, sketch the curves.

21. $y = 3^x$

22. $y = 4^x$

23. $y = 2^{-x}$

24. $y = 3^{-x}$

25. $y = \log_3 x$

26. $y = \log_4 x$

In Exercises 27–34, write each of the expressions as a single logarithm.

27. $\log_3 4 + \log_3 6$

28. $\log_5 7 - \log_5 6$

29. $5 \log_5 2 - 3 \log_5 2$

30. $5 \log_a x + 3 \log_a y - \log_a y^2$

31. $\dfrac{1}{2} \log_b 3 - \dfrac{1}{2} \log_b 9$

32. $\dfrac{1}{2} \log_2 4 - 2 \log_2 x$

33. $2 \log_3 y + \dfrac{1}{3} \log_3 8 - 2 \log_3 5$

34. $5 \log_b x + \dfrac{1}{3} \log_b y - 2 \log_b 4$

In Exercises 35–48, write each expression as a sum, difference, or multiple of logarithms. Whenever possible, use $\log_a 1 = 0$ and $\log_a a = 1$. (See Example 5.)

35. $\log_3 27$

36. $\log_2 16$

37. $\log_6 \sqrt{6x}$

38. $\log_2 4x^2$

39. $\log_5 \dfrac{1}{25x^2}$

40. $\log_3 \dfrac{1}{\sqrt{3x}}$

41. $\log_3 \dfrac{1}{\sqrt[3]{3x}}$

42. $\log_a \dfrac{1}{\sqrt{a}}$

43. $\log_5 \dfrac{1}{\sqrt{y-2}}$

44. $\log_{10} \dfrac{1}{\sqrt{x+1}}$

45. $\log_{10} \dfrac{x}{\sqrt{x+2}}$

46. $\log_{10} x\sqrt{x^2+1}$

47. $\log_{10} \dfrac{\sqrt{x}}{x+1}$

48. $\log_{10} \dfrac{\sqrt[3]{x}}{1-x}$

In Exercises 49–54, solve each equation for x. (See Example 6.)

49. $(3.62)^x = 12.4$

50. $(15.3)^x = 2.30$

51. $(8.04)^x = 2.85$

52. $(2.37)^x = 14.4$

53. $(36.4)^x = 0.147$

54. $(24.7)^x = 0.254$

6.7	**Derivative of the Logarithmic Function**

Some time before studying calculus you probably studied logarithms to base 10, called *common logarithms*. Historically, common logarithms were particularly useful in computational work. Logarithms were first used in this way by the Scottish mathematician John Napier (1550–1617), who published his first treatment on the subject in 1614. His work was enthusiastically received since it did much to simplify tedious computations.

Given the usefulness of the base 10, there seems to be little reason to consider any other base. Yet in obtaining the derivative of the logarithmic function, a different base suggests itself quite naturally (hence "natural logarithm").

To see how natural logarithms arise, let us find the derivative of $y = \log_a x$ by the four-step process.

Step 1. $f(x + \Delta x) = \log_a(x + \Delta x)$

Step 2. $\Delta y = \log_a(x + \Delta x) - \log_a x$

$$= \log_a \frac{x + \Delta x}{x} = \log_a\left(1 + \frac{\Delta x}{x}\right) \qquad \log_a \frac{M}{N} = \log_a M - \log_a N$$

Step 3. $\dfrac{\Delta y}{\Delta x} = \dfrac{1}{\Delta x} \log_a\left(1 + \dfrac{\Delta x}{x}\right)$

Step 4. $\displaystyle\lim_{\Delta x \to 0} \frac{\Delta y}{\Delta x} = \lim_{\Delta x \to 0} \frac{1}{\Delta x} \log_a\left(1 + \frac{\Delta x}{x}\right)$

Since the last expression cannot be simplified, there is no direct way to proceed. The limit on the right has to be studied more closely.

$$\lim_{\Delta x \to 0} \frac{1}{\Delta x} \log_a\left(1 + \frac{\Delta x}{x}\right) = \frac{x}{x} \lim_{\Delta x \to 0} \frac{1}{\Delta x} \log_a\left(1 + \frac{\Delta x}{x}\right) \qquad \text{multiplying by } \frac{x}{x}$$

$$= \lim_{\Delta x \to 0} \frac{x}{\Delta x} \frac{\log_a\left(1 + \dfrac{\Delta x}{x}\right)}{x}$$

$$= \lim_{\Delta x \to 0} \frac{\log_a\left(1 + \dfrac{\Delta x}{x}\right)^{x/\Delta x}}{x} \qquad k \log_a M = \log_a M^k$$

$$= \frac{1}{x} \lim_{\Delta x \to 0} \log_a\left(1 + \frac{\Delta x}{x}\right)^{x/\Delta x}$$

$$= \frac{1}{x} \log_a \lim_{\Delta x \to 0}\left(1 + \frac{\Delta x}{x}\right)^{x/\Delta x} \qquad \text{continuity}$$

by Formula (6.19). Now we have a chance. The last limit has the form

$$\lim_{h \to 0}(1 + h)^{1/h}$$

whose value can be estimated by using a calculator. When $h = 0$, the function $f(h) = (1 + h)^{1/h}$ is undefined, but the function is continuous at all other points. The following point charts give some of the values near $h = 0$:

h:	-0.1	-0.01	-0.001	-0.0001	-0.000001
$f(h)$:	2.8680	2.7320	2.7196	2.71842	2.71828

h:	0.000001	0.0001	0.001	0.01	0.1
$f(h)$:	2.71828	2.71815	2.7169	2.7048	2.5937

Based on our calculations, the limit appears to exist and to be approximately equal to 2.71828. We can get a rough confirmation of this limit by graphing $y = (1 + x)^{1/x}$ with a graphing utility (Figure 6.17) and using the trace feature. It is customary to denote this limit by the letter e. (We will see in Chapter 10 that e is indeed correct to the number of decimal places given.)

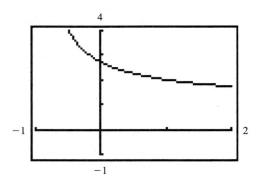

Figure 6.17

Definition of e

$$e = \lim_{h \to 0} (1 + h)^{1/h} \qquad\qquad (6.20)$$

It turns out that e is a *transcendental number,* which means that it is not the root of a polynomial equation. (Proving that a number is transcendental can be extremely difficult. For e this was first shown by the French mathematician C. Hermite in 1873.)

Returning now to Step 4, we see that

$$\frac{d}{dx} \log_a x = \frac{1}{x} \log_a \lim_{\Delta x \to 0} \left(1 + \frac{\Delta x}{x} \right)^{x/\Delta x} = \frac{1}{x} \log_a e$$

Finally, since the number e cannot be avoided in the expression for the derivative, it was decided in the early days of calculus to adopt e for the base. Since $\log_e e = 1$, we get the striking formula

$$\frac{d}{dx} \log_e x = \frac{1}{x}$$

Natural logarithm

Logarithms to base e are referred to as **natural logarithms.** To avoid having to write the base each time, we adopt the notation

$$\ln x = \log_e x$$

read "natural log of x" or, if there is no ambiguity, just "log x"; $\ln x$ should not be read "l.n. of x." The last derivative formula is now written

$$\frac{d}{dx}(\ln\ x) = \frac{1}{x}$$

Furthermore, by the chain rule,

$$\frac{d}{dx}(\log_a u) = \frac{1}{u}(\log_a e)\frac{du}{dx}$$

and

$$\frac{d}{dx}(\ln u) = \frac{1}{u}\frac{du}{dx}$$

These ideas will be summarized next.

Logarithms to base e are called **natural logarithms** and are denoted by

$$\log_e x = \ln x$$

The derivative of the **logarithmic function** is given by

$$\frac{d}{dx}(\log_a u) = \frac{1}{u}(\log_a e)\frac{du}{dx} \qquad (6.21)$$

The derivative of the **natural logarithmic function** is given by

$$\frac{d}{dx}(\ln u) = \frac{1}{u}\frac{du}{dx} \qquad (6.22)$$

In particular,

$$\frac{d}{dx}(\ln x) = \frac{1}{x} \qquad (6.23)$$

Example 1

Differentiate $y = \log_2 x^2$.

Solution. By Formula (6.21) with $u = x^2$ we get

$$\frac{dy}{dx} = \frac{1}{x^2}\log_2 e\left(\frac{d}{dx}x^2\right) = \frac{2}{x}\log_2 e$$

Alternatively, since $y = 2\log_2 x$,

$$\frac{dy}{dx} = \frac{2}{x}\log_2 e$$

directly.   ∎

Example 2   Differentiate $T = \log_{10}(v^2 + v)$.

Solution.

$$\frac{dT}{dv} = \frac{1}{v^2 + v} \log_{10} e(2v + 1)$$

$$= \frac{2v + 1}{v^2 + v} \log_{10} e \qquad (\log_{10} e \approx 0.43429) \qquad \blacksquare$$

Example 3   Differentiate $y = \ln \sec x$.

Solution. By Formula (6.22),

$$\frac{dy}{dx} = \frac{1}{\sec x} \frac{d}{dx}(\sec x) = \frac{1}{\sec x}(\sec x \tan x) = \tan x \qquad \blacksquare$$

In more complicated cases it is advisable to take advantage of the laws of logarithms.

Example 4   Find dy/dx if $y = \ln \sqrt[3]{x^2 + 1}$.

Solution. Since $y = \ln(x^2 + 1)^{1/3}$, we have, by Formula (6.18),

$$y = \frac{1}{3} \ln(x^2 + 1) \qquad \textbf{log}_a \textbf{M}^k = \textbf{k log}_a \textbf{M}$$

Hence

$$\frac{dy}{dx} = \frac{1}{3}\frac{1}{x^2 + 1}\frac{d}{dx}(x^2 + 1) = \frac{1}{3}\frac{1}{x^2 + 1}(2x) = \frac{2x}{3(x^2 + 1)}$$

by Formula (6.22). $\blacksquare$

Example 5   Find the derivative of $y = \ln(\sin^2 x /x)$.

Solution. The function can be written in the form

$$y = \ln(\sin x)^2 - \ln x \qquad \textbf{log}_a \frac{\textbf{M}}{\textbf{N}} = \textbf{log}_a \textbf{M} - \textbf{log}_a \textbf{N}$$

$$= 2 \ln \sin x - \ln x \qquad \textbf{log}_a \textbf{M}^k = \textbf{k log}_a \textbf{M}$$

It follows that

$$\frac{dy}{dx} = \frac{2}{\sin x}\cos x - \frac{1}{x}$$

$$= 2 \cot x - \frac{1}{x} \qquad \textbf{cot } \textbf{x} = \frac{\textbf{cos } \textbf{x}}{\textbf{sin } \textbf{x}}$$

$$= \frac{2x \cot x - 1}{x} \qquad \blacksquare$$

■ Exercises / Section 6.7

Differentiate each of the following functions. (Whenever possible, break down the function as in Examples 4 and 5.)

1. $y = \ln 2x$

2. $y = \ln 3x$

3. $y = 4 \ln 3x$

4. $y = 3 \ln 4x$

5. $R = \ln s^2$

6. $P = \ln t^3$

7. $y = 2 \ln x^3$

8. $y = 3 \ln x^4$

9. $y = \log_{10} x^3$

10. $y = \log_5 \sqrt{x}$

11. $R_1 = \ln \sin R_2$

12. $V_1 = \ln \cos V_2$

13. $y = x \ln x$

14. $y = \ln(x^2 - 1)^2$

15. $y = \ln \sqrt{x - 1}$

16. $y = \ln \dfrac{1}{\sqrt{x}}$

17. $y = \ln \dfrac{1}{\sqrt{x + 2}}$

18. $y = \ln \sqrt[3]{1 - x}$

19. $z = 3 \ln \sqrt[3]{t^2 + 2}$

20. $N = \ln \dfrac{1}{\sqrt{1 - T^2}}$

21. $y = \ln \dfrac{x^2}{x + 1}$

22. $y = \ln \dfrac{x}{2x - 1}$

23. $y = \ln \dfrac{2x}{x^2 + 1}$

24. $y = \ln \dfrac{\sqrt{x}}{1 - x}$

25. $y = \ln \dfrac{\sqrt{x}}{2 - x^2}$

26. $y = \ln \dfrac{\sqrt[3]{x}}{x + 2}$

27. $y = \ln \dfrac{\cos x}{\sqrt{x}}$

28. $y = \ln \dfrac{\sqrt{x}}{\sec x}$

29. $y = \ln \dfrac{\sec^2 x}{\sqrt{x + 1}}$

30. $y = \ln \dfrac{\sqrt{x - 1}}{\csc^2 x}$

31. $y = \ln^2 x$

32. $y = \sqrt{\ln x}$

33. $y = \sqrt{x + 1} \ln x$

34. $y = \tan \ln x$

35. $s = (\sec \theta) \ln \theta$

36. $v = (\ln \omega^2) \csc \omega$

37. $y = \ln(1 + \sqrt{x^2 - 1})$

38. $y = \ln \sqrt{\sec x}$

39. $y = (\ln^2 x) \ln x^2$

40. $y = \tan(\ln \tan x)$

41. $y = \ln(\ln x)$

42. $y = \ln^3 x$

43. $v_1 = \dfrac{\ln v_2}{v_2}$

44. $r_1 = \dfrac{r_2}{\ln r_2}$

6.8   **Derivative of the Exponential Function**

Using logarithms to the base e, we can examine the function $y = e^x$. This function has the surprising property of being equal to its derivative. To check this statement, we will obtain a general formula, the derivative of $y = a^u$, $a > 0$, $a \neq 1$.

From $y = a^u$ we have $u = \log_a y$. Hence

$$\frac{du}{dx} = \frac{1}{y} \log_a e \frac{dy}{dx}$$

by Formula (6.21). Thus

$$\frac{dy}{dx} = \frac{1}{\log_a e} y \frac{du}{dx}$$

and, since $y = a^u$,

$$\frac{d}{dx}(a^u) = \frac{1}{\log_a e} a^u \frac{du}{dx} \qquad (6.24)$$

It is readily shown that $1/\log_a e = \ln a$. Let $N = \log_a e$. Then $a^N = e$, and, taking natural logarithms of both sides,

$$\log_e a^N = \log_e e$$

$$N \log_e a = 1 \qquad \qquad \textbf{log}_a\ \textbf{M}^k = \textbf{k} \ \textbf{log}_a\ \textbf{M}; \ \textbf{log}_e\ \textbf{e} = \textbf{1}$$

$$N = \frac{1}{\log_e a} = \frac{1}{\ln a} \qquad \textbf{notation for natural logarithm}$$

But $N = \log_a e$, so $\log_a e = 1/\ln a$ and $1/\log_a e = \ln a$. Consequently, Formula (6.24) can also be written

$$\frac{d}{dx}(a^u) = a^u(\ln a)\frac{du}{dx}$$

In the special case where $a = e$, this formula reduces to

$$\frac{d}{dx}(e^u) = e^u\frac{du}{dx}$$

(since $\log_e e = 1$). Finally, if $u = x$,

$$\frac{d}{dx}(e^x) = e^x$$

as claimed. These rules are summarized next.

Derivatives of Exponential Functions

$$\frac{d}{dx}(a^u) = a^u(\ln a)\frac{du}{dx} = \frac{1}{\log_a e}a^u\frac{du}{dx} \qquad (6.25)$$

$$\frac{d}{dx}(e^u) = e^u\frac{du}{dx} \qquad (6.26)$$

$$\frac{d}{dx}(e^x) = e^x \qquad (6.27)$$

Example 1

Differentiate $y = 2^{x^2}$

Solution. By Formula (6.25),

$$\frac{d}{dx}(a^u) = a^u(\ln a)\frac{du}{dx}$$

we get

$$\frac{dy}{dx} = 2^{x^2}(\ln 2)\frac{d}{dx}(x^2) = 2^{x^2}(\ln 2)(2x) = 2x(\ln 2)2^{x^2} \qquad \blacksquare$$

Example 2 Find dy/dx if $y = e^{\sin^2 x}$.

Solution. By Formula (6.26), we have

$$\frac{dy}{dx} = e^{\sin^2 x}\frac{d}{dx}(\sin x)^2 \qquad \frac{d}{dx}e^u = e^u\frac{du}{dx}$$

$$= e^{\sin^2 x}(2\sin x \cos x) \qquad \text{power rule}$$

$$= e^{\sin^2 x}\sin 2x \qquad \text{double-angle identity} \qquad \blacksquare$$

Example 3 Differentiate $y = \ln(\tan e^{3x})$.

Solution. From the derivative of the logarithm,

$$\frac{dy}{dx} = \frac{1}{\tan e^{3x}}\frac{d}{dx}\tan e^{3x} \qquad \frac{d}{dx}\ln u = \frac{1}{u}\frac{du}{dx}$$

$$= \frac{1}{\tan e^{3x}}\sec^2 e^{3x}\frac{d}{dx}e^{3x} \qquad \frac{d}{dx}\tan u = \sec^2 u\frac{du}{dx}$$

$$= \frac{1}{\tan e^{3x}}\sec^2 e^{3x}(e^{3x}\cdot 3) \qquad \frac{d}{dx}e^u = e^u\frac{du}{dx}$$

$$= \frac{3e^{3x}\sec^2 e^{3x}}{\tan e^{3x}} \qquad\qquad \blacksquare$$

Our methods can be adapted to differentiate functions of the form $y = u^v$, where u and v are both functions of x. The simplest way is to take the natural logarithm of both sides and differentiate.

Example 4 Differentiate $y = x^{\tan x}$.

Solution. Since $\ln y = \tan x(\ln x)$, we get

$$\frac{1}{y}\frac{dy}{dx} = \tan x \cdot \frac{1}{x} + \sec^2 x(\ln x) \qquad \text{product rule}$$

so that

$$\frac{dy}{dx} = y\left[\frac{1}{x}\tan x + \sec^2 x(\ln x)\right]$$

Substituting back,

$$\frac{dy}{dx} = x^{\tan x}\left[\frac{1}{x}\tan x + \sec^2 x(\ln x)\right] \qquad y = x^{\tan x}$$

$$= x^{(\tan x - 1)}[\tan x + x\sec^2 x(\ln x)] \qquad \text{factoring } 1/x \qquad \blacksquare$$

Certain combinations of exponential functions occur so often that they have been given special names; these are **hyperbolic functions.** Two of these functions are referred to in Exercise 31:

$$\sinh x = \frac{1}{2}(e^x - e^{-x}) \qquad \text{called the } \textit{hyperbolic sine of } x \text{ (Figure 6.18)}$$

and

$$\cosh x = \frac{1}{2}(e^x + e^{-x}) \qquad \text{called the } \textit{hyperbolic cosine of } x \text{ (Figure 6.19)}$$

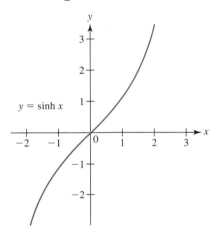

Figure 6.18   **Figure 6.19**

■ Exercises / Section 6.8

In Exercises 1–30, differentiate the given functions.

1. $y = e^{4x}$

2. $y = e^{-2x}$

3. $y = e^{x^2}$

4. $y = e^{-x^3}$

5. $y = 2e^{-t^2}$

6. $y = e^{2t^2}$

7. $y = e^{\tan x}$

8. $y = e^{\cot x}$

9. $y = 3^{x^3}$

10. $y = 2^{6x}$

11. $y = 4^{x^2}$

12. $y = e^{\sqrt{x}}$

13. $C = 2re^r$

14. $y = 2^{\sqrt{x}}$

15. $y = e^{\sin x}$

16. $y = e^{(1-\cos x)}$

17. $y = \sin e^x$

18. $y = \dfrac{e^x}{x^2}$

19. $S = e^{2\omega} \sin \omega$

20. $M = e^{\theta^2} \cos \theta$

21. $y = \dfrac{\tan x}{e^{x^2}}$

22. $y = \text{Arcsin } e^x$

23. $y = (\ln x)e^{\sec x}$

24. $y = e^{\sin x} \ln \sqrt{x}$

25. $y = \ln(\sin e^{2x})$

26. $y = e^{\text{Arctan } x}$

27. $y = \dfrac{x}{e^{2x}}$

28. $y = \dfrac{x^2}{e^{4x}}$

29. $y = \dfrac{\sin 2x}{e^x + 1}$

30. $y = \dfrac{\ln x}{e^x + 1}$

31. Two of the hyperbolic functions defined earlier are

$$\sinh x = \frac{1}{2}(e^x - e^{-x}) \quad \text{and} \quad \cosh x = \frac{1}{2}(e^x + e^{-x})$$

Show that $d(\sinh x)/dx = \cosh x$ and $d(\cosh x)/dx = \sinh x$.

In Exercises 32–37, use the results from Exercise 31 and the chain rule to differentiate the given functions.

32. $y = \sinh 3x$

33. $y = \sinh 5x$

34. $y = \cosh x^2$

35. $y = \cosh 2x^3$

36. $y = x \sinh 2x$

37. $y = x \cosh 3x$

In Exercises 38–43, find the derivative of each of the given functions. (See Example 4.)

38. $y = x^x$

39. $y = x^{\sin x}$

40. $y = (\sin x)^x$

41. $y = (\ln x)^x$

42. $y = (\sin x)^{\cos x}$

43. $y = (\tan x)^x$

44. Show that $\dfrac{d}{dx}u^v = vu^{v-1}\dfrac{du}{dx} + u^v(\ln u)\dfrac{dv}{dx}$.

45. The charge on a certain capacitor is $q(t) = 2.0e^{-0.60t} \times \cos 10\pi t$. Find the current to the capacitor at $t = 0.20$ s.

46. A radioactive substance decays according to $N = 2.0e^{-0.10t}$, where N is in kilograms and t in minutes. Find the time rate of change of N when $t = 3.4$ min.

47. Some diseases spread according to the law

$$p(t) = \frac{1}{1 + ae^{-kt}}$$

where $p(t)$ is the proportion of the population with the disease. Find an expression for the time-rate of change of $p(t)$.

6.9   L'Hospital's Rule

In this section we are going to use our differentiation techniques to obtain certain difficult limits in a simple way. Such limits can be found by means of L'Hospital's rule.

L'Hospital's Rule

If $f(x)$ and $g(x)$ are differentiable for every x other than a in some interval, and if $\lim_{x\to a} f(x) = \lim_{x\to a} g(x) = 0$ or if $\lim_{x\to a} f(x) = \lim_{x\to a} g(x) = \infty$ or $-\infty$, then

$$\lim_{x\to a}\frac{f(x)}{g(x)} = \lim_{x\to a}\frac{f'(x)}{g'(x)} \qquad (6.28)$$

provided that the latter limit exists.

Example 1   Evaluate

$$\lim_{x\to 0}\frac{1 - \cos x}{x}$$

Solution.   Since both numerator and denominator are equal to zero when $x = 0$, we say that $(1 - \cos x)/x$ tends to the indeterminate form $0/0$. So by L'Hospital's rule,

$$\lim_{x\to 0}\frac{1 - \cos x}{x} = \lim_{x\to 0}\frac{\dfrac{d}{dx}(1 - \cos x)}{\dfrac{d}{dx}(x)}$$

$$= \lim_{x\to 0}\frac{\sin x}{1} = 0 \qquad \blacksquare$$

Example 2   Evaluate

$$\lim_{x\to\infty}\frac{e^x}{x^3}$$

Solution. Here we say that the expression tends to the indeterminate form ∞/∞. Again by L'Hospital's rule,

$$\lim_{x\to\infty} \frac{e^x}{x^3} = \lim_{x\to\infty} \frac{e^x}{3x^2}$$

The last limit also tends to the form ∞/∞. So we apply the rule two more times to get

$$\lim_{x\to\infty} \frac{e^x}{6x} = \lim_{x\to\infty} \frac{e^x}{6} = \infty$$

That is, the limit does not exist.   ∎

L'Hospital's rule can also be applied to the indeterminate form $0 \cdot \infty$, as illustrated in the next example.

Example 3   Evaluate

$$\lim_{x\to\pi/2} (\tan x) \ln \sin x$$

Solution. Since $\tan \pi/2$ is undefined, the function takes on the indeterminate form $0 \cdot \infty$. L'Hospital's rule applies if the expression is written as a quotient:

$$\lim_{x\to\pi/2} (\tan x) \ln \sin x = \lim_{x\to\pi/2} \frac{\ln \sin x}{\cot x} \qquad \textbf{tan x = 1/cot x}$$

which tends to $0/0$. Thus

$$\lim_{x\to\pi/2} \frac{\ln \sin x}{\cot x} = \lim_{x\to\pi/2} \frac{\cos x / \sin x}{-\csc^2 x} = \lim_{x\to\pi/2} (-\cos x \sin x) = 0 \qquad ∎$$

L'Hospital's rule is named after the French marquis G. F. A. de L'Hospital (1661–1704), who was a pupil of the great Swiss mathematician Johann Bernoulli (1667–1748). The rule was discovered by Bernoulli in 1694.

■ Exercises / Section 6.9

Evaluate each of the following limits.

1. $\displaystyle\lim_{x\to-2} \frac{x^2 - 4}{x + 2}$

2. $\displaystyle\lim_{x\to\infty} \frac{x^2 + 2x + 1}{2x^2 + 3}$

3. $\displaystyle\lim_{x\to\infty} \frac{3x^2 - 4x}{2x^2 + 1}$

4. $\displaystyle\lim_{x\to\infty} \frac{4x^2 - 5x + 4}{3x^2 - 6x + 1}$

5. $\displaystyle\lim_{t\to3} \frac{t^2 + t - 12}{t - 3}$

6. $\displaystyle\lim_{s\to-1} \frac{s^2 - 3s - 4}{s + 1}$

7. $\displaystyle\lim_{m\to0} \frac{\sin 6m}{m}$

8. $\displaystyle\lim_{n\to0} \frac{\tan 2n}{\sin n}$

9. $\displaystyle\lim_{x\to0} \frac{\tan 3x}{1 - \cos x}$

10. $\displaystyle\lim_{x\to\pi/2} \frac{1 - \sin x}{\cos x}$

11. $\displaystyle\lim_{x\to0} \frac{e^x - e^{-x}}{\sin x}$

12. $\displaystyle\lim_{x\to\pi/2} \frac{\ln \sin x}{1 - \sin x}$

13. $\displaystyle\lim_{x\to0} \frac{1 - e^x}{2x}$

14. $\displaystyle\lim_{x\to0} \frac{x - \sin x}{x}$

15. $\displaystyle\lim_{x\to\infty} \frac{x + \ln x}{x \ln x}$

16. $\displaystyle\lim_{x\to\pi/2} \frac{\cos x}{\sin 2x}$

17. $\displaystyle\lim_{x\to0} \frac{x + \sin 2x}{x - \sin 2x}$

18. $\displaystyle\lim_{x\to0} \frac{\ln \sec x}{x^2}$

19. $\displaystyle\lim_{x\to\pi/2} \frac{\cos x}{\pi - 2x}$

20. $\displaystyle\lim_{x\to+\infty} \frac{\ln^2 x}{x}$

21. $\lim\limits_{x \to 0+} (\sin x) \ln \sin x$

22. $\lim\limits_{x \to 0} x \cot x$

23. $\lim\limits_{x \to 0+} x \ln x$

24. $\lim\limits_{x \to \infty} x \tan \dfrac{1}{x}$

25. $\lim\limits_{x \to \infty} (1 + 1/x)^x$ *Hint:* denote the given limit by y and evaluate

$$\ln y = \lim_{x \to \infty} \ln \left(1 + \frac{1}{x}\right)^x$$

26. $\lim\limits_{x \to 0+} (2x)^x$

6.10 Applications

Transcendental functions arise in many situations in science and technology. Some of these applications are considered in the following examples and exercises.

Example 1

If air resistance is neglected, then the range R (in meters) of a projectile fired from a gun that is inclined at an angle θ from the ground is known to be

$$R = \frac{v_0{}^2}{5} \sin \theta \cos \theta$$

where v_0 is the initial velocity. (This formula is derived in Exercise 29 of Section 8.1.) Find the angle for which the range will be as large as possible.

Solution. To find the critical value, we differentiate R with respect to θ. By the product rule,

$$\frac{dR}{d\theta} = \frac{v_0{}^2}{5}[\sin \theta(-\sin \theta) + \cos \theta(\cos \theta)]$$

$$= \frac{v_0{}^2}{5}(-\sin^2 \theta + \cos^2 \theta) = \frac{v_0{}^2}{5}(-\sin^2 \theta + 1 - \sin^2 \theta)$$

$$= \frac{v_0{}^2}{5}(1 - 2\sin^2 \theta) = 0$$

Solving for θ, we get

$$2\sin^2 \theta = 1 \qquad \text{or} \qquad \sin \theta = \frac{1}{\sqrt{2}}$$

whence $\theta = 45° = \pi/4$. To check this value by the second-derivative test, we need to find

$$\frac{d^2 R}{d\theta^2} = \frac{v_0{}^2}{5}(-4\sin \theta \cos \theta)$$

When $\theta = \pi/4$,

$$\frac{d^2 R}{d\theta^2} = -\frac{v_0{}^2}{5}\left(4\sin \frac{\pi}{4}\cos \frac{\pi}{4}\right) = -\frac{v_0{}^2}{5}(4)\left(\frac{1}{\sqrt{2}}\right)\left(\frac{1}{\sqrt{2}}\right)$$

$$= -\frac{2v_0{}^2}{5} < 0$$

Consequently, the range is maximal when the angle is $\pi/4$. ∎

Example 2

One of the pioneers in the study of polarized light was the French scientist E. L. Malus, who discovered the following law, called *Malus' law,* experimentally in 1809:

$$I = I_{max} \cos^2 \theta$$

where I_{max} is the maximum amount of light transmitted and I is the amount transmitted at angle θ. Suppose $I = 20$ lumens when $\theta = \pi/4$. If θ changes at the rate of 0.50 rad/min, how fast is I changing at that instant?

Solution. First we need to calculate I_{max} from the given information:

$$20 = I_{max} \cos^2 \frac{\pi}{4} = I_{max} \cdot \frac{1}{2} \qquad \textbf{I = 20, } \boldsymbol{\theta} \textbf{ = } \boldsymbol{\pi}\textbf{/4}$$

or $I_{max} = 40$ lumens. Thus

$$I = 40 \cos^2 \theta$$

We are given that $d\theta/dt = 0.50$ and wish to find dI/dt when $\theta = \pi/4$. Recall that both I and θ are treated as functions of time in a related-rates problem. Differentiating with respect to t, we get

$$\frac{dI}{dt} = 40(2 \cos \theta)(-\sin \theta) \frac{d\theta}{dt} \qquad \textbf{power rule}$$

Now substitute the given values:

$$\frac{dI}{dt} = 40 \left(2 \cos \frac{\pi}{4}\right)\left(-\sin \frac{\pi}{4}\right)(0.50)$$

$$= -20 \text{ lumens/min} \qquad \blacksquare$$

Example 3

The number of bacteria in a culture as a function of time is given by

$$N = Ae^{kt} \qquad \text{(where } A \text{ and } k \text{ are positive constants)}$$

under ideal conditions. Find the rate of change of N with respect to time.

Solution. Differentiation yields directly

$$\frac{dN}{dt} = Ae^{kt} \cdot k = kN \qquad \textbf{since } \boldsymbol{N = Ae^{kt}}$$

This equation says effectively that "the bigger it is, the faster it grows." $\qquad \blacksquare$

Example 4

The retarding force on a falling body owing to air resistance is proportional to the velocity; that is, the force is kv. If air resistance is neglected, then $k = 0$. We will see in Chapter 11 that the motion of a falling body is given by

$$v = \frac{mg}{k}(1 - e^{-kt/m})$$

Show that as $k \to 0$, we get $v = gt$, the usual formula without air resistance.

Solution. This problem is an application of L'Hospital's rule. Thus

$$\lim_{k \to 0} v = \lim_{k \to 0} mg \left[\frac{1 - e^{-kt/m}}{k} \right] = \lim_{k \to 0} mg \left[\frac{-e^{-kt/m} \left(-\dfrac{t}{m} \right)}{1} \right] = gt$$

(Note that the derivatives were taken with respect to k.) ∎

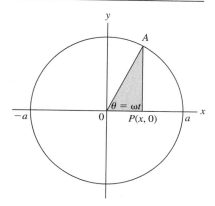

Figure 6.20

Example 5 Consider the motion of particle A moving counterclockwise at a constant rate around a circle of radius a in Figure 6.20. We wish to study the motion of point $P(x, 0)$, the projection of A on the x-axis. To this end we denote angle AOP by $\theta = \omega t$, ω a constant, so that the angular velocity is $d\theta/dt = \omega$. Then

$$x = a \cos \theta = a \cos \omega t$$

As θ ranges from 0 to 2π, t will range from 0 to $2\pi/\omega$. The following table gives some of the other values:

$\theta = \omega t$:	0	$\dfrac{\pi}{2}$	π	$\dfrac{3\pi}{2}$	2π
x:	a	0	$-a$	0	a

We conclude that the point moves back and forth along the x-axis between $x = a$ and $x = -a$.

The motion of point P, given by $x = a \cos \omega t$, is called **simple harmonic motion** and will be studied in greater detail in Chapter 12. In particular, the weight on a spring moves in simple harmonic motion; a condition similar to this mechanical motion can be given for alternating electric current.

Returning to the motion of P, the velocity and acceleration are readily found to be

$$\frac{dx}{dt} = -a\omega \sin \omega t \qquad \text{and} \qquad \frac{d^2x}{dt^2} = -a\omega^2 \cos \omega t$$

respectively. Note especially that

$$\frac{d^2x}{dt^2} = -\omega^2 x \tag{6.29}$$

This *differential equation* is sometimes used to define simple harmonic motion. The equation states that the acceleration is proportional to the displacement with direction opposite to the direction of motion. (See also Exercise 19.) ∎

Example 6 Find the minima and maxima and points of inflection of the function $y = x^2 e^{-x}$; sketch the graph.

Solution. The derivatives are

$$f'(x) = -x^2 e^{-x} + 2xe^{-x} = xe^{-x}(2 - x)$$

and

$$f''(x) = x^2 e^{-x} - 4xe^{-x} + 2e^{-x} = e^{-x}(x^2 - 4x + 2)$$

a. Critical points:

$$f'(x) = xe^{-x}(2-x) = 0$$

whence $x = 0, 2$. So the critical points are $(0, 0)$ and $(2, 4/e^2)$.

b. Test of critical points:

$$f''(0) = 2; \quad (0, 0) \text{ is a minimum}$$

$$f''(2) = -2e^{-2} < 0; (2, 4/e^2) \text{ is a maximum}$$

c. Concavity and points of inflection:

$$f''(x) = 0 \text{ whenever } x^2 - 4x + 2 = 0$$

so that

$$x = 2 \pm \sqrt{2}$$

Since $f''(x) \neq 0$ at all other points, we substitute arbitrary values such as $x = 0, 2$, and 4. For example, since $f''(0) > 0$, $f''(x) > 0$ on the interval $(-\infty, 2 - \sqrt{2}]$. So f is concave up on this interval. Similarly, f is concave down on $[2 - \sqrt{2}, 2 + \sqrt{2}]$ and concave up on $[2 + \sqrt{2}, \infty)$. Hence $(2 - \sqrt{2}, 0.19)$ and $(2 + \sqrt{2}, 0.38)$ are points of inflection.

d. Other: since

$$\lim_{x \to +\infty} x^2 e^{-x} = \lim_{x \to +\infty} \frac{x^2}{e^x}$$

takes on the indeterminate form ∞/∞, we may apply L'Hospital's rule:

$$\lim_{x \to +\infty} \frac{x^2}{e^x} = \lim_{x \to +\infty} \frac{2x}{e^x} = \lim_{x \to +\infty} \frac{2}{e^x} = 0$$

We conclude that the positive x-axis is an asymptote. The graph is shown in Figure 6.21.

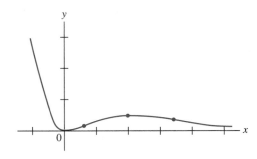

Figure 6.21

Example 7 A narrow hallway 1 m wide runs perpendicularly into another hallway. A thin pole 8 m long is carried horizontally along the narrow hallway and around the corner. How wide must the other hallway be to permit this?

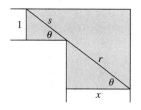

Figure 6.22

Solution. First we draw Figure 6.22. Referring to the figure, the problem is to find the largest distance x, given that the pole must touch the corner and both walls. Put another way, we find θ for which x is a maximum.

To be able to do so, we write x as a function of θ, making use of r and s. From the right triangles in the figure,

$$x = r\cos\theta \qquad \text{and} \qquad s = \csc\theta$$

Since $r + s = 8$, the length of the pole, we now have

$$r = 8 - s \qquad \text{or} \qquad r = 8 - \csc\theta$$

and since $x = r\cos\theta$, it follows that

$$x = (8 - \csc\theta)\cos\theta = 8\cos\theta - \csc\theta\cos\theta = 8\cos\theta - \cot\theta$$

To maximize x, we find the critical value:

$$x = 8\cos\theta - \cot\theta$$

$$\frac{dx}{d\theta} = -8\sin\theta + \csc^2\theta = 0$$

$$-8\sin\theta + \frac{1}{\sin^2\theta} = 0 \qquad \textbf{csc } \theta = \frac{1}{\textbf{sin } \theta}$$

$$8\sin^3\theta = 1 \qquad \textbf{multiplying by sin}^2\theta$$

and

$$\sin^3\theta = \frac{1}{8} \qquad \text{or} \qquad \sin\theta = \frac{1}{2}$$

Hence $\theta = 30°$ and from $x = 8\cos\theta - \cot\theta$, we get

$$x = 8\cos 30° - \cot 30° = 3\sqrt{3} \text{ m}$$

Since

$$\frac{d^2x}{d\theta^2} = -8\cos\theta - 2\csc^2\theta\cot\theta \bigg|_{\theta=\pi/6} < 0$$

it follows that $x = 3\sqrt{3}$ is indeed a maximum. ∎

■ Exercises / Section 6.10

In Exercises 1–5, find the minima and maxima, the points of inflection, and sketch the graph.

1. $y = xe^{-x}$

2. $y = \dfrac{\ln x}{x} (x > 0)$

3. $y = x\ln x (x > 0)$

4. $y = \sinh x$ (See Figure 6.18.)

5. $y = \cosh x$ (See Figure 6.19.)

6. Find the minimum point on the graph of $y = x^x$, $x > 0$, and sketch the curve.

7. If the current in an AC circuit is given by $i = \cos t + \sin t$, find the first maximum current after $t = 0$.

8. Repeat Exercise 7 for $i = \cos t + 2 \sin t$.

9. Find the voltage across a 0.0030-H inductor at $t = 0$ if the current is $i = 3.0 \sin 200t + 2.0 \cos 200t$.

10. The distance in meters from the origin of a certain object is given by $x = 2 \cos 2t + \sin 4t$. Find the velocity at $t = \pi/4$ s.

11. A certain radioactive substance decays according to the law $N = 12e^{-2.0t}$, where N (in kilograms) is the amount present and t is the time in years. Find an expression for the time rate of decrease of N as a function of time. Show that $dN/dt = -2.0N$.

12. A certain radioactive substance decays according to the law $N = 20(0.80)^{2t}$ where N(in grams) is the amount present. Find the rate of change of N with respect to t when $t = 5.0$ years.

13. A certain radioactive isotope decays according to the law $N = 6.0e^{-0.25t}$, where N is measured in grams. Find the time rate of change of N when $t = 8.5$ min.

14. At $t = 0$, the number of bacteria in a culture is P_0. For $t \geq 0$, the number is $P = P_0e^{kt}$. Show that the time rate of change of the number of bacteria is proportional to the number present.

15. The number N of bacteria in a culture as a function of time is $N = 200e^{(1/2)t}$, where t is measured in hours. Find the rate of increase after 3 h.

16. A warm object is placed in a refrigerator to cool. If the temperature in degrees Celsius at any time is given by $T = 10 + 20e^{-0.0933t}$, find the rate of change of the temperature when $t = 10$ min.

17. An underwater telegraph cable consists of a circular core and a layer of insulation. The speed S of a signal is given by $S = kR^2 \ln(1/R)$, where R is the ratio of the radius of the core to the thickness of the covering and k is a constant. For what value of R is the speed the greatest?

18. Since the human ear is sensitive to a large range of intensities, a logarithmic scale is most convenient. The *intensity level* ß (in decibels) of a sound is given by the equation $ß = 10 \log(I/I_0)$, where $I_0 = 10^{-16}$ W/cm^2, the intensity corresponding roughly to the faintest sound that can be heard. In testing the noise made by a quiet automobile, the intensity is measured to be $1.1 \times 10^{-11.2}$ W/cm^2, with a possible error of 4.0×10^{-12} W/cm^2. Use differentials to determine the approximate error in ß.

19. Consider the simple harmonic motion $x = 5 \cos 2t$. Show that the magnitude of the velocity is a maximum whenever the particle passes the origin.

20. Show that for the straight-line motion described by $x = a \sin 2\pi kt + b \cos 2\pi kt$, where x gives the position as a function of time, the acceleration is proportional to the displacement from the origin and oppositely directed.

21. A balloon is rising from the ground at the rate of 6.0 m/s from a point 100 m from an observer, also on the ground. Use inverse trigonometric functions to determine how fast the angle of inclination of the observer's line of sight is increasing when the balloon is at an altitude of 150 m.

22. A fugitive is running along a wall at 4.0 m/s. A searchlight 20 m from the wall is trained on him. How fast is the searchlight rotating at the instant when he is 10 m from the point on the wall nearest the searchlight?

23. A man is walking toward a building 80.0 m high at the rate of 1.5 m/s. How fast is the angle of elevation of the top increasing when he is 40.0 m away from the building?

24. The largest weight W that can be pulled up a plane inclined at an angle θ with the horizontal by a force F is $W = (F/\mu)(\cos \theta + \mu \sin \theta)$, where μ is the coefficient of friction. Find θ so that W is a maximum.

25. A gutter is to be made from a long piece of metal 24 cm wide by turning up strips 8 cm wide along each side in such a way that they make equal angles θ with the vertical. For what value of θ will the cross-sectional area be the greatest?

26. A wall is 3 m high and 2 m from a building. Find the length of the shortest ladder that touches the ground and building and just clears the wall. (*Hint:* If x is the length of the ladder and θ is the angle made with the ground, then $x = 2 \sec \theta + 3 \csc \theta$.)

27. A hallway is 2 m wide and runs perpendicularly into another hallway 5 m wide. What is the length of the longest thin pole that can be moved horizontally around the corner?

28. An object of weight W is dragged along a horizontal plane by a force F whose line of action makes an angle θ with the plane. The force is given by $F = (\mu W)/(\mu \sin \theta + \cos \theta)$, where μ is the coefficient of friction. Show that the pull is minimal when $\mu = \tan \theta$.

29. The range R of a projectile fired from a gun along an inclined plane making an angle α with the horizontal is
$$R = \frac{2v_0^2 \cos \theta \sin(\theta - \alpha)}{g \cos^2 \alpha}$$
where θ is the angle of elevation of the gun and v_0 is the initial velocity. For what value of θ is the range up the plane a maximum? (*Suggestion:* Write the derivative in the form $K \cos(A + B)$.)

30. The following law (derived in Chapter 11) is called *logistic growth* and gives the size of a population as a function of time:
$$N(t) = \frac{MN_0}{N_0 + (M - N_0)e^{-Mkt}}$$
where N_0 is the initial size of the population and k a positive constant. **(a)** Show that $\lim_{t\to\infty} N(t) = M$. **(b)** Use $N'(t)$ to show that $N(t)$ is increasing if $N_0 < M$ and $N(t)$ is decreasing if $N_0 > M$.

31. The profit P on a certain commodity as a function of the number x of units sold is $P = Axe^{-x/n}$, where A and n are constants. What value of x yields the maximum profit?

6.11   Newton's Method

Newton's method is an application of calculus to the solution of equations.

To solve the equation $f(x) = 0$ for x, we let $y = f(x)$ and use Newton's method to find the x-intercept. (See Figure 6.23.) Let x_0 be a rough estimate of the intercept on the right and consider the tangent line to the graph at $(x_0, f(x_0))$. The point $(x_1, 0)$, where the tangent line crosses the x-axis, is an approximation of the root. We can find the value of x_1 by noting that the slope of the tangent line is

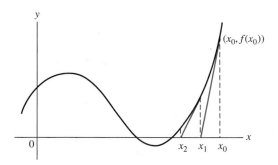

Figure 6.23

$$\frac{f(x_0) - 0}{x_0 - x_1} = f'(x_0)$$

It follows that

$$f(x_0) = f'(x_0)(x_0 - x_1)$$

Solving for x_1, we get the following formula:

Newton's Method

$$x_1 = x_0 - \frac{f(x_0)}{f'(x_0)}$$

The procedure is now repeated: we use x_1 for our new estimate and find

$$x_2 = x_1 - \frac{f(x_1)}{f'(x_1)}$$

and so on. (See Figure 6.23.)

Example 1   Find the root of the cubic equation $x^3 - 3x - 4 = 0$.

Solution. To obtain the first estimate x_0 of the root, we need to graph the equation by the methods of Chapter 3 or by using a graphing utility. (The graph is shown in

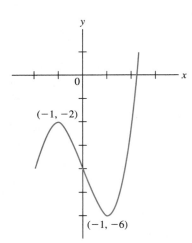

Figure 6.24

Figure 6.24.) As a rough first guess, $x = 2$ seems to do. Using a calculator, we quickly find x_1, x_2, and so on. If we decide to carry out the solution to six decimal places, we continue the iteration until the first six decimal places no longer change:

$$x_1 = x_0 - \frac{f(x_0)}{f'(x_0)} = x_0 - \frac{x_0^3 - 3x_0 - 4}{3x_0^2 - 3} \qquad f(x) = x^3 - 3x - 4$$

or

$$x_1 = 2 - \frac{2^3 - 3(2) - 4}{3(2^2) - 3} = 2.222222 \qquad x_0 = 2$$

Using this value, we calculate x_2:

$$x_2 = x_1 - \frac{x_1^3 - 3x_1 - 4}{3x_1^2 - 3}$$

$$= 2.222222 - \frac{(2.222222)^3 - 3(2.222222) - 4}{3(2.222222)^2 - 3}$$

$$= 2.196215$$

For the next iteration, we get

$$x_3 = x_2 - \frac{x_2^3 - 3x_2 - 4}{3x_2^2 - 3}$$

$$= 2.196215 - \frac{(2.196215)^3 - 3(2.196215) - 4}{3(2.196215)^2 - 3}$$

$$= 2.195823$$

From this point on the values no longer change, so that $x = 2.195823$ must be the root accurate to six decimal places. ∎

■ Exercises / Section 6.11

Use Newton's method to find the root (or roots) of each of the following equations to six decimal places. In some of the exercises your calculator must be set in the **radian mode.**

1. $\cos x + x = 0$

2. $e^x + 2x - \dfrac{3}{2} = 0$

3. $4 \sin x - x = 0$; find the positive root

4. $\sin x = 1 - x$

5. $\sin x = x^2$; find the positive root

6. $\sin x = e^x - 2$

7. $e^x - \dfrac{5}{x} = 0$

8. $\ln x - \tan x = 0$; find the smallest positive root

9. $\tan x + x = 2$; find the smallest positive root

10. $e^x = 2 \cos x$; find the positive root

11. $x^2 - 2x - 12 = 0$; find all the roots

12. $x^2 + 3x - 11 = 0$; find all the roots

13. $x^3 - 6x^2 - 4x + 6 = 0$; find all the roots

14. $x^3 + 3x - 6 = 0$

15. $2x^3 - 4x^2 - x + 4 = 0$

16. $x^3 - 4x + 2 = 0$; find all the roots

■ Review Exercises / Chapter 6

In Exercises 1–6, evaluate each expression.

1. $\text{Arcsin}(-1)$ **2.** $\text{Arctan}(-1)$

3. $\text{Arccos}\left(-\dfrac{1}{2}\right)$ **4.** $\text{Arcsin } 0$

5. $\cos(\text{Arctan } 3)$ **6.** $\sin\left(\text{Arccos }\dfrac{1}{3}\right)$

In Exercises 7 and 8, write each expression as an algebraic expression.

7. $\sin(\text{Arctan } x)$ **8.** $\tan(\text{Arccos } x)$

In Exercises 9–26, differentiate each of the given functions.

9. $y = x^2 \tan 3x$ **10.** $y = \dfrac{\sin 4x}{x}$

11. $v = \dfrac{e^{2t}}{t}$ **12.** $V_1 = \cos V_2 \ln V_2$

13. $y = \ln \dfrac{1}{\sqrt{x+3}}$ [Use the properties of logarithms.]

14. $y = \ln \dfrac{\sqrt{x}}{1-x}$ **15.** $y = \ln \dfrac{\sqrt{2x^2+1}}{x}$

16. $y = \sin e^{-x}$ **17.** $y = \sqrt{\ln 2x}$

18. $y = e^{2 \tan x}$ **19.** $y = \cos(\ln x)$

20. $y = 5^{x^2}$ **21.** $y = e^{2x} \cot x$

22. $y = x \text{ Arcsin } x$ **23.** $y = e^{\text{Arccos } 3x}$

24. $y = \ln \sec^2 x$ **25.** $y = (\cot x)^x$

26. $y = (\sec x)^x$

27. Find dy/dx implicitly: $e^{\sin y} + \csc x = 1$.

28. Find dy/dx implicitly: $\ln y + \tan y + x = 3$.

In Exercises 29–33, evaluate the given limits.

29. $\lim\limits_{x \to 0} \dfrac{\sin 4x}{x}$ **30.** $\lim\limits_{x \to \infty} \dfrac{\ln x}{x}$

31. $\lim\limits_{x \to \pi/4} (1 - \tan x) \sec 2x$ **32.** $\lim\limits_{x \to 0} \dfrac{1 - \cos x}{x^2}$

33. $\lim\limits_{x \to 0+} \dfrac{\sin x - x}{x \sin x}$

34. Find the equation of the line normal to the curve $y = x \ln x$ at $(1, 0)$. (Recall that a normal line is perpendicular to a tangent line.)

35. Find the dimensions of the largest rectangle in the fourth quadrant whose sides lie along the axes with one vertex on the curve $y = \ln x$.

36. The velocity in meters per second of a certain falling object was determined to be $v = 20(1 - e^{-0.1t})$. Find the limiting velocity.

37. A deposit of S dollars that earns $100r\%$ annual interest compounded continuously leaves a balance of $P = Se^{rt}$ dollars after t years. **(a)** What will an amount of $5000 grow to after 15 years at 10% annual interest compounded continuously? **(b)** Determine the rate at which P is growing.

38. The atmospheric pressure at an altitude of h meters above sea level is $P = 10^4 e^{-0.00012h}$ kg/m^2. If the altitude increases at the rate of 25 m/min, what is the rate of change of pressure at the instant when $h = 1000.0$ m?

39. The vapor pressure p, in millimeters of mercury, of carbon tetrachloride is given by

$$\log p = -\dfrac{1706.4}{T} - 7.7760$$

where T is in kelvin (K). If T increases at the rate of 2.00 K/min, find the rate at which p increases when $T = 300.0$ K.

40. An inductance L and a resistance R are in series with a battery of E volts. We will see in Chapter 11 that the current is given by

$$I = \dfrac{E}{R}(1 - e^{-Rt/L})$$

Find an expression for the time-rate of change of the current.

41. A light is hung above the center of a round table of unit radius. Let d be the length of the line segment joining the light to a point on the edge and θ the angle that this line makes with the table. Then the illumination at the edge is given by $I = k(\sin \theta)/d^2$, where k is a constant. Find θ so that the illumination at the edge is a maximum.

42. A tracking device 1 mi from a launching pad is kept trained on a missile traveling along the parabolic path $y^2 = x$ (x and y are measured in miles). By placing the launching pad at the origin and the tracking device at $(-1, 0)$, determine the largest angle of elevation that the device must be capable of.

43. Use Newton's method to solve the following equation, accurate to six decimal places:

$$\cos x - x = \dfrac{1}{2}$$

7   Integration Techniques

7.1   The Power Formula Again

In the last chapter we obtained the derivatives of various transcendental functions. As a result we will now be able to integrate a much larger class of functions, both algebraic and transcendental. For our first integration formula, recall the general power formula from Chapter 4.

General Power Formula

$$\int u^n du = \frac{u^{n+1}}{n+1} + C \qquad (n \neq -1) \tag{7.1}$$

Since we are no longer confined to algebraic functions, we may apply Formula (7.1) to integrals involving transcendental functions; thus integrals of this form may vary greatly in appearance. Consequently, though the power formula has a very simple structure, integrals of this form are among the hardest to recognize. You must be thoroughly familiar with the differentiation formulas of the last chapter to be able to pick out the function u and its differential du. *Recognizing the type of integral is an important part of the integration process.*

Example 1   Integrate $\int \sqrt{\cos 4x} \, \sin 4x \, dx$.

Solution. Written as

$$\int (\cos 4x)^{1/2} \sin 4x \, dx$$

the integral seems to fit Formula (7.1). If we let $u = \cos 4x$, then $du = -4 \sin 4x \, dx$. As a consequence, we must introduce -4 to complete the differential and place $-1/4$

in front of the integral. Thus

$$-\frac{1}{4}\int (\cos 4x)^{1/2} (-4\sin 4x)\,dx = -\frac{1}{4}\int u^{1/2}\,du \qquad \left[\begin{array}{l} u = \cos 4x \\ du = -4\sin 4x\,dx \end{array}\right]$$

$$= -\frac{1}{4}\cdot\frac{2}{3}u^{3/2} + C = -\frac{1}{6}\cos^{3/2} 4x + C \qquad \blacksquare$$

Example 2   Find

$$\int \frac{e^x\,dx}{(1+2e^x)^3}$$

Solution. We try $u = 1 + 2e^x$; then $du = 2e^x\,dx$. For the integral to fit the proper form, we need to introduce a 2 and place $1/2$ in front of the integral. Thus

$$\frac{1}{2}\int \frac{2e^x\,dx}{(1+2e^x)^3} = \frac{1}{2}\int \frac{du}{u^3} = \frac{1}{2}\int u^{-3}\,du \qquad \left[\begin{array}{l} u = 1 + 2e^x \\ du = 2e^x dx \end{array}\right]$$

$$= \frac{1}{2}\frac{u^{-2}}{-2} + C = -\frac{1}{4}\frac{1}{u^2} + C$$

$$= -\frac{1}{4(1+2e^x)^2} + C \qquad \blacksquare$$

Example 3   Integrate

$$\int \frac{\ln^2 x\,dx}{x}$$

Solution. At first glance this integral appears to be quite different from those above. However, recalling that $d\ln x = (1/x)\,dx$, we see that the substitution $u = \ln x$ leads quite simply to

$$\int (\ln x)^2\,\frac{dx}{x} = \int u^2\,du = \frac{1}{3}u^3 + C = \frac{1}{3}\ln^3 x + C \qquad \left[\begin{array}{l} u = \ln x \\ du = \dfrac{1}{x}\,dx \end{array}\right] \qquad \blacksquare$$

Example 4   Evaluate

$$\int_0^1 \frac{\text{Arctan } v\,dv}{1+v^2}$$

Solution. Let $u = \text{Arctan } v$; then $du = dv/(1+v^2)$ and

$$\int_0^1 \frac{\text{Arctan } v\,dv}{1+v^2} = \frac{1}{2}(\text{Arctan } v)^2\Big|_0^1 \qquad \int u\,du = \frac{1}{2}u^2 + C$$

$$= \frac{1}{2}[(\text{Arctan } 1)^2 - (\text{Arctan } 0)^2] = \frac{1}{2}\left(\frac{\pi}{4}\right)^2 = \frac{\pi^2}{32} \qquad \blacksquare$$

■ Exercises / Section 7.1

Perform the integrations indicated. Check your answers by differentiation.

1. $\int x\sqrt{x^2 + 1}\, dx$

2. $\int x^2\sqrt{1 - x^3}\, dx$

3. $\int \dfrac{dx}{\sqrt{1 - x}}$

4. $\int \dfrac{dx}{\sqrt{x + 2}}$

5. $\int \sin^2 x \cos x\, dx$

6. $\int \cos^3 x \sin x\, dx$

7. $\int \tan^2 2x \sec^2 2x\, dx$

8. $\int \cot^3 4x \csc^2 4x\, dx$

9. $\int (1 + \tan 3t)^3 \sec^2 3t\, dt$

10. $\int \dfrac{V\, dV}{\sqrt{1 - 3V^2}}$

11. $\int (1 + e^{4r})^3 e^{4r}\, dr$

12. $\int e^{2\theta}\sqrt{1 + e^{2\theta}}\, d\theta$

13. $\int (1 - \cos 5x)^3 \sin 5x\, dx$

14. $\int (\cot 4x - 1)^2 \csc^2 4x\, dx$

15. $\int \dfrac{\sec^2 3x}{(\tan 3x + 1)^3}\, dx$

16. $\int \dfrac{\cos 2x}{\sqrt{\sin 2x + 4}}\, dx$

17. $\int \dfrac{\ln x\, dx}{x}$

18. $\int \dfrac{1 + 2\ln x}{x}\, dx$

19. $\int \dfrac{dx}{x\sqrt{\ln x}}$

20. $\int \dfrac{\text{Arcsin}\, 3x\, dx}{\sqrt{1 - 9x^2}}$

21. $\int \dfrac{\text{Arccos}\, x\, dx}{\sqrt{1 - x^2}}$

22. $\int \dfrac{1 + \cos x}{(x + \sin x)^2}\, dx$

23. $\int_{\pi/6}^{\pi/2} \cos^2 x \sin x\, dx$

24. $\int_0^1 \dfrac{x\, dx}{\sqrt{x^2 + 1}}$

25. $\int_1^e \dfrac{\sqrt{\ln x}}{x}\, dx$

26. $\int \dfrac{e^x\, dx}{\sqrt{2 - e^x}}$

27. $\int \dfrac{\cot 2x\, dx}{\sin^2 2x}$

28. $\int \dfrac{dx}{\cot^3 x \cos^2 x}$

29. $\int \sqrt{\tan x} \sec^2 x\, dx$

30. $\int \dfrac{\sqrt[3]{\ln x}}{x}\, dx$

31. $\int \dfrac{\text{Arctan}\, 4R}{1 + 16R^2}\, dR$

32. $\int (1 + \sin z)^3 \cos z\, dz$

33. $\int \dfrac{(1 + \ln x)^2}{x}\, dx$

34. $\int_1^e \dfrac{\ln^3 x}{x}\, dx$

7.2   The Logarithmic and Exponential Forms

The general power Formula (7.1) is valid for all n except $n = -1$. We can fill this gap by making use of the derivative of the logarithmic function. We know from Chapter 6 that for $u > 0$,

$$\frac{d}{dx} \ln u = \frac{1}{u}\frac{du}{dx}$$

or, in differential form,

$$d \ln u = \frac{du}{u}$$

It follows that

$$\int \frac{du}{u} = \ln u + C \qquad (u > 0)$$

This formula can be extended. First recall the definition of absolute value:

$$|x| = \begin{cases} x & x \geq 0 \\ -x & x < 0 \end{cases}$$

For example,

$$|2| = |+2| = 2 \qquad \textbf{since } x \textbf{ is positive}$$

and

$$|-2| = -(-2) = 2 \qquad \textbf{since } x \textbf{ is negative}$$

Thus

$$\ln|x| = \begin{cases} \ln x & x > 0 \\ \ln(-x) & x < 0 \end{cases}$$

Differentiating, we get

$$\frac{d}{dx}\ln|x| = \begin{cases} \dfrac{d}{dx}\ln x & x > 0 \\ \dfrac{d}{dx}\ln(-x) & x < 0 \end{cases} = \begin{cases} \dfrac{1}{x} & x > 0 \\ \dfrac{-1}{-x} = \dfrac{1}{x} & x < 0 \end{cases}$$

In either case,

$$\frac{d}{dx}\ln|x| = \frac{1}{x} \qquad (x \neq 0)$$

It now follows from the chain rule that

$$\frac{d}{dx}\ln|u| = \frac{1}{u}\frac{du}{dx} \qquad (u \neq 0)$$

By reversing this formula, we get the following integration form:

Logarithmic Form

$$\int \frac{du}{u} = \ln|u| + C \qquad (7.2)$$

This formula fills the gap left by the Power Formula (7.1).

From $de^u/dx = e^u \, du/dx$, we get the differential form

$$de^u = e^u \, du$$

By reversing this formula, we get the corresponding integration form:

Exponential Form

$$\int e^u \, du = e^u + C \qquad (7.3)$$

For bases other than e we have, from Formula (6.25),

$$\int a^u \, du = a^u \log_a e + C = \frac{a^u}{\ln a} + C$$

Example 1   Integrate

$$\int \frac{x\,dx}{x^2 + 1}$$

Solution. If we let $u = x^2 + 1$, then $du = 2x\,dx$; since $n = -1$, Formula (7.1) does not apply. But by Formula (7.2) we get

$$\frac{1}{2} \int \frac{2x\,dx}{x^2 + 1} = \frac{1}{2} \int \frac{du}{u} = \frac{1}{2} \ln|u| + C \qquad \left[\begin{matrix} u = x^2 + 1 \\ du = 2x\,dx \end{matrix} \right]$$

$$= \frac{1}{2} \ln|x^2 + 1| + C = \frac{1}{2} \ln(x^2 + 1) + C$$

since $x^2 + 1 > 0$.

Remark: If the quantity $x^2 + 1$ were raised to *any* power besides 1, we would have to use the power formula. For example, to integrate

$$\int \frac{x\,dx}{(x^2 + 1)^2}$$

we let $u = x^2 + 1$, so that $du = 2x\,dx$. The integral then becomes

$$\frac{1}{2} \int \frac{2x\,dx}{(x^2 + 1)^2} = \frac{1}{2} \int \frac{du}{u^2} = \frac{1}{2} \int u^{-2}\,du = \frac{1}{2} \frac{u^{-1}}{-1} + C$$

$$= -\frac{u^{-1}}{2} + C = -\frac{1}{2u} + C = -\frac{1}{2(x^2 + 1)} + C \qquad ∎$$

Example 2   Find

$$\int \frac{dx}{x(\ln x + 1)}$$

Solution. Since the derivative of $\ln x$ is $1/x$, the derivative of $\ln x + 1$ is also $1/x$. So we try the substitution $u = \ln x + 1$. Then $du = dx/x$, and we obtain

$$\int \frac{1}{\ln x + 1} \frac{dx}{x} = \int \frac{du}{u} = \ln|u| + C = \ln|\ln x + 1| + C \qquad \left[\begin{matrix} u = \ln x + 1 \\ du = \frac{1}{x}\,dx \end{matrix} \right]$$

∎

Example 3   Integrate $\int e^{\cos 5\theta} \sin 5\theta \, d\theta$.

Solution. This integral seems to fit Formula (7.3). Let $u = \cos 5\theta$; then $du = -5 \sin 5\theta \, d\theta$ and

$$\int e^{\cos 5\theta} \sin 5\theta \, d\theta = -\frac{1}{5} \int e^{\cos 5\theta} (-5 \sin 5\theta) \, d\theta$$

$$= -\frac{1}{5} \int e^u \, du = -\frac{1}{5} e^u + C = -\frac{1}{5} e^{\cos 5\theta} + C \qquad ∎$$

Example 4   Integrate

$$\int \frac{e^{\text{Arcsin}\,2x}}{\sqrt{1 - 4x^2}} \, dx$$

Solution. The form fits Formula (7.3):

$$\int \frac{e^{\text{Arcsin }2x}}{\sqrt{1-4x^2}}\,dx = \frac{1}{2}\int e^{\text{Arcsin }2x}\frac{2\,dx}{\sqrt{1-4x^2}} \qquad \left[\begin{array}{l} u = \text{Arcsin }2x \\[2mm] du = \dfrac{2\,dx}{\sqrt{1-4x^2}} \end{array}\right]$$

$$= \frac{1}{2}\int e^u\,du = \frac{1}{2}e^u + C$$

$$= \frac{1}{2}e^{\text{Arcsin }2x} + C$$

Example 5   Evaluate

$$\int_0^1 \frac{e^x\,dx}{1+e^x}$$

Solution. Even though the integrand contains e^x, the integral is not an exponential form. But since $d(1+e^x) = e^x\,dx$, we see that Formula (7.2) applies. Hence

$$\int_0^1 \frac{e^x\,dx}{1+e^x} = \ln(1+e^x)\Big|_0^1 = \ln(1+e) - \ln 2 = \ln\frac{1+e}{2}$$

■ Exercises / Section 7.2

In Exercises 1–47, perform the designated integrations.

1. $\displaystyle\int \frac{dx}{x-1}$

2. $\displaystyle\int \frac{dx}{1-4x}$

3. $\displaystyle\int \frac{dx}{2+3x}$

4. $\displaystyle\int \frac{x\,dx}{1-2x^2}$

5. $\displaystyle\int \frac{ds}{1-3s}$

6. $\displaystyle\int \frac{d\omega}{7\omega+6}$

7. $\displaystyle\int_0^1 \frac{x\,dx}{x^2+1}$

8. $\displaystyle\int \frac{x\,dx}{2x^2+4}$

9. $\displaystyle\int e^{-x}\,dx$

10. $\displaystyle\int e^{6x}\,dx$

11. $\displaystyle\int_0^2 2e^{3x}\,dx$

12. $\displaystyle\int 2e^{-4x}\,dx$

13. $\displaystyle\int e^{4x}\,dx$

14. $\displaystyle\int_{-1}^0 3e^{-3x}\,dx$

15. $\displaystyle\int te^{t^2}\,dt$

16. $\displaystyle\int t^2 e^{t^3}\,dt$

17. $\displaystyle\int e^{\sin R}\cos R\,dR$

18. $\displaystyle\int \frac{\cos\theta}{1+\sin\theta}\,d\theta$

19. $\displaystyle\int \frac{\sec^2 2x}{1+\tan 2x}\,dx$

20. $\displaystyle\int e^{\tan 2x}\sec^2 2x\,dx$

21. $\displaystyle\int \frac{e^{\text{Arctan }x}}{1+x^2}\,dx$

22. $\displaystyle\int \frac{e^{\sqrt{x}}\,dx}{\sqrt{x}}$

23. $\displaystyle\int \frac{dx}{x\ln x}$ (See Example 2.)

24. $\displaystyle\int \frac{dx}{x(1-\ln x)}$

25. $\displaystyle\int \frac{e^x+1}{e^x}\,dx$

26. $\displaystyle\int \frac{\sin 4z\,dz}{(1+\cos 4z)^2}$

27. $\displaystyle\int \frac{e^{-3W}}{1-e^{-3W}}\,dW$

28. $\displaystyle\int xe^{(1+x^2)}\,dx$

29. $\displaystyle\int \frac{2x}{(1+x^2)^2}\,dx$

30. $\displaystyle\int (1+e^x)^2 e^x\,dx$

31. $\displaystyle\int (1+e^x)^2\,dx$

32. $\displaystyle\int \frac{\sec 3x\tan 3x}{1+\sec 3x}\,dx$

33. $\displaystyle\int \frac{x+1}{x+2}\,dx$ (*Hint:* Divide numerator by denominator.)

34. $\displaystyle\int \frac{x^2+x-1}{x+1}\,dx$

35. $\displaystyle\int \frac{\cos x}{\sin^2 x}\,dx$

36. $\displaystyle\int \frac{\sec^2 x\,dx}{1-2\tan x}$

37. $\displaystyle\int e^{\sin^2 x}\sin 2x\,dx$

38. $\displaystyle\int_0^1 xe^{-x^2}\,dx$

39. $\displaystyle\int_0^{\pi/3} \frac{\sin x\,dx}{\cos x}$

40. $\displaystyle\int \frac{\sec^2 3x}{4-\tan 3x}\,dx$

41. $\displaystyle\int \frac{\cos 2x}{1+\sin 2x}\,dx$

42. $\displaystyle\int \frac{\sec^2 3x}{(4-\tan 3x)^2}\,dx$

43. $\displaystyle\int \frac{\cos 2x}{(1+\sin 2x)^2}\,dx$

44. $\int 2^x \, dx$

45. $\int 5^{3x} \, dx$

46. $\int x \, 3^{x^2} \, dx$

47. $\int 2^{\cot x} \csc^2 x \, dx$

48. Find the area of the region bounded by $y = 1/(x+1)$, $x = 1$, and the coordinate axes.

49. Find I_y, the moment of inertia about the y-axis, of the region bounded by $y = 1/(1+x^3)$, $x = 2$, and the coordinate axes.

50. Find the first-quadrant area of the region bounded by $y = e^{-x}$ and the coordinate axes.

51. Find the average value of the current $i = e^{(-1/3)t}$ in a certain circuit from $t = 0$ to $t = 2$.

52. Show that the root mean square of the function $(\ln x)/\sqrt{x}$ from $x = 1$ to $x = e$ is $1/\sqrt{3(e-1)}$.

53. Find the volume generated by rotating the region bounded by $y = e^{-x^2}$, $x = 1$, and the coordinate axes about the y-axis.

54. Find the coordinates of the centroid of the region bounded by $y = 1/\sqrt{x}$, $x = 1$, $x = 4$, and the x-axis.

55. Find the volume of the solid generated by rotating the region bounded by $y = 4/x^2$, $x = 1$, $x = 2$, and the x-axis about the y-axis.

56. The number N ($N > 0$) of bacteria in a culture at any time satisfies the relation $dN/N = k \, dt$, $k > 0$. Show that $N = N_0 e^{kt}$ if $N_0 > 0$ denotes the number of bacteria when $t = 0$.

57. The velocity in meters per second of a falling object of mass 1 kg experiencing a retarding force due to air resistance satisfies the relation $dv/(10 - kv) = dt$, $k > 0$, $10 - kv > 0$. Find v as a function of time if the body is dropped from rest.

58. A small squirrel population in a new area increases at the rate $r = 150 - 20e^{t/100}$, where t is measured in years. Determine the population after 4 years. (For convenience assume that the population started at zero.)

59. Find the mass of a rod 6.00 m long whose density (in kilograms per meter) is given by $\rho(x) = 3.00 + 0.0300e^{0.0100x}$ where the distance x (in meters) is measured from one end of the rod.

60. Find the approximate value of

$$\int_0^2 e^{-x^2} \, dx$$

by Simpson's rule with $n = 6$.

7.3 Trigonometric Forms

The purpose of this section is to discuss the basic trigonometric integrals. These forms include the integrals of the six trigonometric functions, as well as the forms obtained by reversing the derivative formulas. Other trigonometric forms will be taken up in the next section.

Since integration is the inverse of differentiation, integrating the derivatives of the six trigonometric functions yields the following formulas:

$$\int \cos u \, du = \sin u + C \tag{7.4}$$

$$\int \sin u \, du = -\cos u + C \tag{7.5}$$

$$\int \sec^2 u \, du = \tan u + C \tag{7.6}$$

$$\int \csc^2 u \, du = -\cot u + C \tag{7.7}$$

$$\int \sec u \, \tan u \, du = \sec u + C \tag{7.8}$$

$$\int \csc u \, \cot u \, du = -\csc u + C \tag{7.9}$$

The integrals of the remaining trigonometric functions are:

$$\int \tan u \, du = -\ln|\cos u| + C = \ln|\sec u| + C \tag{7.10}$$

$$\int \cot u \, du = \ln|\sin u| + C = -\ln|\csc u| + C \tag{7.11}$$

$$\int \sec u \, du = \ln|\sec u + \tan u| + C \tag{7.12}$$

$$\int \csc u \, du = \ln|\csc u - \cot u| + C \tag{7.13}$$

Formula (7.10) may be obtained by integrating $\int [(\sin u)/(\cos u)] \, du$ and is left as an exercise. Formula (7.11) is obtained similarly. The derivation of Formula (7.13) requires a trick:

$$\int \csc u \, du = \int \frac{\csc u (\csc u - \cot u)}{\csc u - \cot u} \, du$$

$$= \int \frac{\csc^2 u - \csc u \cot u}{\csc u - \cot u} \, du$$

This is recognized to be a logarithmic form; Formula (7.13) follows. The derivation of Formula (7.12) is similar and will also be left as an exercise.

Example 1 | Integrate $\int \sec 4x \tan 4x \, dx$.

Solution. Let $u = 4x$; then $du = 4 \, dx$, and we get

$$\frac{1}{4} \int \sec 4x \tan 4x (4 \, dx) = \frac{1}{4} \int \sec u \tan u \, du$$

$$= \frac{1}{4} \sec u + C \qquad \text{by (7.8)}$$

$$= \frac{1}{4} \sec 4x + C \qquad \blacksquare$$

Example 2 | Integrate $\int x^2 \cot x^3 \, dx$.

Solution. Let $u = x^3$, so that $du = 3x^2 dx$. The integral now reduces to

$$\frac{1}{3} \int \cot x^3 (3x^2 dx) = \frac{1}{3} \int \cot u \, du = \frac{1}{3} \ln|\sin u| + C \qquad \text{by (7.11)}$$

$$= \frac{1}{3} \ln|\sin x^3| + C \qquad \blacksquare$$

Example 3 Find $\int e^{2x} \sec e^{2x} \, dx$.

Solution. Try $u = e^{2x}$; then $du = 2e^{2x} \, dx$. The integral becomes

$$\frac{1}{2} \int (\sec e^{2x})(2e^{2x}) \, dx = \frac{1}{2} \int \sec u \, du \qquad \left[\begin{array}{l} u = e^{2x} \\ du = 2e^{2x} \, dx \end{array} \right]$$

$$= \frac{1}{2} \ln|\sec u + \tan u| + C \qquad \textbf{by (7.12)}$$

$$= \frac{1}{2} \ln|\sec e^{2x} + \tan e^{2x}| + C \qquad \blacksquare$$

■ Exercises / Section 7.3

In Exercises 1–35, perform the designated integrations.

1. $\int \sec^2 2x \, dx$

2. $\int \sin 3x \, dx$

3. $\int \sec 3x \tan 3x \, dx$

4. $\int \csc 2x \, dx$

5. $\int \csc 4x \cot 4x \, dx$

6. $\int \sec 2x \, dx$

7. $\int \tan \frac{1}{2} x \, dx$

8. $\int r \tan r^2 \, dr$

9. $\int \cos 2t \, dt$

10. $\int \sec^2 5t \, dt$

11. $\int y \csc y^2 \, dy$

12. $\int \cot y \csc^2 y \, dy$

13. $\int \frac{\csc^2 y}{1 + \cot y} \, dy$

14. $\int \tan^2 5t \sec^2 5t \, dt$

15. $\int \frac{\sec^2 5t}{4 - \tan 5t} \, dt$

16. $\int \csc \frac{1}{2} x \cot \frac{1}{2} x \, dx$

17. $\int x \sin x^2 \, dx$

18. $\int x^2 \sec x^3 \tan x^3 \, dx$

19. $\int T \cot \frac{1}{2} T^2 \, dT$

20. $\int \cos^4 V \sin V \, dV$

21. $\int \frac{\cos \sqrt{x}}{\sqrt{x}} \, dx$

22. $\int e^x \tan e^x \, dx$

23. $\int \tan^3 4x \sec^2 4x \, dx$

24. $\int x^3 \sec x^4 \, dx$

25. $\int e^{3x} \csc e^{3x} \, dx$

26. $\int \frac{\sin \ln x}{x} \, dx$

27. $\int \frac{\csc^2 \ln x}{x} \, dx$

28. $\int \frac{\cos x \, dx}{\sin^3 x}$

29. $\int \frac{\sec^2 x \, dx}{\tan^2 x}$

30. $\int (1 + \sec x)^2 \, dx$

31. $\int \frac{\sin x \, dx}{1 + 2 \cos x}$

32. $\int_{\pi/4}^{\pi/2} \frac{d\theta}{\sin^2 \theta}$

33. $\int_0^{\sqrt{\pi/2}} \omega \cos \omega^2 \, d\omega$

34. $\int_0^{\pi/16} \sec^2 4x \, dx$

35. $\int_0^{\pi/6} \frac{\cos x}{1 - \sin x} \, dx$

36. The current in a circuit is given by $i = 6.0 \sin 3.0t$. How many coulombs of charge pass a point in the first 2.0 s if $q = 1.5$ C when $t = 0$?

37. The current in a circuit is $i = 2.00 \cos 100t$. Find the voltage across a 100-μF capacitor after 0.200 s, if the initial voltage is zero (one microfarad (μF) $= 10^{-6}$ F).

38. Find the volume of the solid of revolution obtained by rotating the region bounded by $y = \sqrt{\sin x}$, $x = 0$, $x = \pi$, and the x-axis about the x-axis.

39. Find the volume of the solid of revolution obtained by rotating the region bounded by $y = \cos x^2$, $x = 0$, $x = \sqrt{\pi/2}$, and $y = 0$ about the y-axis.

40. Derive Formulas (7.10) and (7.12).

41. Find M_y, the moment about the y-axis, of the region bounded by $y = \tan x^2$, $x = 0$, $x = \sqrt{\pi}/2$, and the x-axis. (Assume the weight per unit area to be equal to 1.)

42. A particle is moving along the x-axis in simple harmonic motion. If the velocity is $v = 10 \cos 2t$, find the position of the particle after 4 s if $x = 0$ when $t = 0$.

43. The voltage in a certain circuit is given by $e(t) = E \sin \omega t$. Find the mean voltage over half a period.

44. A certain circuit consists of an inductor connected to a generator with electromotive force $e(t) = E \sin \omega t$. Recall that if i is the current, then $L(di/dt) = E \sin \omega t$. Show that $i = [E/(L\omega)](1 - \cos \omega t)$ if $i = 0$ when $t = 0$.

45. (Set calculator in radian mode.) Use the trapezoidal rule with $n = 8$ to find the approximate value of

$$\int_1^2 \frac{\sin x}{x} \, dx$$

46. Repeat Exercise 45 using Simpson's rule for

$$\int_0^\pi \sqrt{x} \sin x \, dx \qquad (n = 6)$$

Further Trigonometric Forms

The techniques of the last section can be extended to more complex trigonometric integrals by proper use of certain trigonometric identities in Section 6.1. For convenience these will now be repeated.

Trigonometric Identities Needed in This Section

$$\sin^2 x + \cos^2 x = 1 \tag{7.14}$$

$$1 + \tan^2 x = \sec^2 x \tag{7.15}$$

$$1 + \cot^2 x = \csc^2 x \tag{7.16}$$

$$\cos^2 x = \frac{1}{2}(1 + \cos 2x) \tag{7.17}$$

$$\sin^2 x = \frac{1}{2}(1 - \cos 2x) \tag{7.18}$$

The integrals discussed in this section are essentially of three types:

Types of Integrals

Type 1: $\displaystyle\int \sin^n u \cos^m u \, du$

Type 2: $\displaystyle\int \tan^n u \sec^m u \, du$

Type 3: $\displaystyle\int \cot^n u \csc^m u \, du$

For each type of integral the method of integration depends on whether the exponents are even or odd. In each case one of the exponents may be zero.

Identity (7.14) can be used to integrate products of powers of sines and cosines, provided that one of the powers is odd. The integral is transformed by means of Identity (7.14), so that it consists of powers of sines with a cosine term set aside for the differential du (or powers of cosines with a sine term set aside for du). Although the procedure can best be seen from an example, let us state the general case first.

Type 1: *n* Odd or *m* Odd
Use Identity (7.14) to integrate

$$\int \sin^n u \cos^m u \, du$$

Example 1 Integrate $\int \sin^2 x \cos^3 x \, dx$.

Solution. The trick in this integration is to write $\cos^3 x \, dx = \cos^2 x (\cos x \, dx)$ and then change $\cos^2 x$ to $1 - \sin^2 x$. Thus

$$\int \sin^2 x \cos^3 x \, dx = \int \sin^2 x \, \cos^2 x (\cos x \, dx)$$

$$= \int \sin^2 x (1 - \sin^2 x)(\cos x \, dx)$$

The reason for this step is now clear: if we let $u = \sin x$, then $du = \cos x \, dx$, and the integral becomes

$$\int u^2 (1 - u^2) \, du = \int (u^2 - u^4) \, du \qquad \left[\begin{matrix} u = \sin x \\ du = \cos x \, dx \end{matrix}\right]$$

$$= \frac{1}{3} u^3 - \frac{1}{5} u^5 + C = \frac{1}{3} \sin^3 x - \frac{1}{5} \sin^5 x + C \qquad \blacksquare$$

Note that the integration cannot be performed by the method of Example 1 if the power of the cosine is even (while the power of the sine is also even). If the power of the sine is odd, then the procedure is similar, as can be seen from the next example.

Example 2 Integrate $\int \sin^5 x \cos^4 x \, dx$.

Solution. We save $\sin x \, dx$ for the differential du and change the remaining sines to cosines:

$$\int \sin^5 x \cos^4 x \, dx = \int \sin^4 x \cos^4 x (\sin x \, dx)$$

$$= \int (\sin^2 x)^2 \cos^4 x (\sin x \, dx)$$

$$= \int (1 - \cos^2 x)^2 \cos^4 x (\sin x \, dx)$$

Now let $u = \cos x$; then $du = -\sin x \, dx$, and the integral can be written as

$$- \int (1 - \cos^2 x)^2 \cos^4 x (-\sin x \, dx)$$

$$= - \int (1 - u^2)^2 u^4 \, du \qquad \left[\begin{matrix} u = \cos x \\ du = -\sin x \, dx \end{matrix}\right]$$

$$= - \int (u^4 - 2u^6 + u^8) \, du$$

$$= -\frac{1}{5} u^5 + \frac{2}{7} u^7 - \frac{1}{9} u^9 + C$$

$$= -\frac{1}{5} \cos^5 x + \frac{2}{7} \cos^7 x - \frac{1}{9} \cos^9 x + C \qquad \blacksquare$$

The method of Examples 1 and 2 can be used even if one of the exponents is zero.

Example 3   Integrate $\int \cos^3 7x \, dx$.

Solution.

$$\int \cos^3 7x \, dx = \int \cos^2 7x (\cos 7x \, dx)$$

$$= \int (1 - \sin^2 7x)(\cos 7x \, dx)$$

$$= \frac{1}{7} \int (1 - \sin^2 7x)(7 \cos 7x \, dx) \qquad \left[\begin{array}{l} u = \sin 7x \\ du = 7 \cos 7x \, dx \end{array} \right]$$

$$= \frac{1}{7} \int (1 - u^2) \, du$$

$$= \frac{1}{7} \left(u - \frac{1}{3} u^3 \right) + C$$

$$= \frac{1}{7} \sin 7x - \frac{1}{21} \sin^3 7x + C \qquad\qquad \blacksquare$$

To integrate even powers of sines and cosines we may use the Half-Angle Identities (7.17) and (7.18) to reduce the powers, as illustrated in Example 4.

> **Type 1: *n* and *m* Even**
> Use Identities (7.17) and (7.18) to integrate
> $$\int \sin^n u \, du \qquad \int \cos^m u \, du \qquad \int \sin^n u \cos^m u \, du$$

Example 4   Integrate $\int \sin^4 2x \, dx$.

Solution. By Identity (7.18),

$$\int \sin^4 2x \, dx = \int (\sin^2 2x)^2 dx = \int \left(\frac{1 - \cos 4x}{2} \right)^2 dx$$

$$= \frac{1}{4} \int (1 - 2 \cos 4x + \cos^2 4x) \, dx$$

We now use Identity (7.17) to obtain

$$\frac{1}{4} \left(\int dx - 2 \int \cos 4x \, dx + \int \frac{1 + \cos 8x}{2} \, dx \right)$$

The remaining integrations can be performed readily:

$$\int \sin^4 2x \, dx = \frac{1}{4} \int dx - \frac{1}{2} \int \cos 4x \, dx + \frac{1}{8} \int dx + \frac{1}{8} \int \cos 8x \, dx$$

$$= \frac{1}{4} \int dx - \frac{1}{2} \cdot \frac{1}{4} \int \cos 4x (4 \, dx) + \frac{1}{8} \int dx + \frac{1}{8} \cdot \frac{1}{8} \int \cos 8x (8 \, dx)$$

Thus

$$\int \sin^4 2x \, dx = \frac{1}{4}x - \frac{1}{8}\sin 4x + \frac{1}{8}x + \frac{1}{64}\sin 8x + C$$

$$= \frac{3}{8}x - \frac{1}{8}\sin 4x + \frac{1}{64}\sin 8x + C \qquad \blacksquare$$

The procedure for integrating powers of sines and cosines is summarized next:

1. If the power of the sine function is odd and positive, save one sine function for the differential and change the remaining sines to cosines:

$$\int \overset{\text{odd}}{\sin^n} u \cos^m u \, du = \int \sin^{n-1} u \cos^m u \overset{\text{save}}{(\sin u)} \, du$$

$\qquad\qquad\qquad\qquad\qquad\qquad$ change to cosines

2. If the power of the cosine function is odd and positive, save one cosine function for the differential and change the remaining cosines to sines:

$$\int \sin^n u \overset{\text{odd}}{\cos^m} u \, du = \int \sin^n u \cos^{m-1} u \overset{\text{save}}{(\cos u)} \, du$$

$\qquad\qquad\qquad\qquad\qquad\qquad$ change to sines

3. If both powers are even, make repeated use of the half-angle identities to eliminate the even powers.

Many integrals of type 2 can be broken down in analogous fashion by means of Identity (7.15).

Type 2: *n* Odd or *m* Even
Use Identity (7.15) to integrate

$$\int \tan^n u \sec^m u \, du$$

Example 5   Integrate $\int \tan^4 x \sec^4 x \, dx$.

Solution. The even power of the secant suggests saving $\sec^2 x \, dx$ for the term du and using Identity (7.15) to change the remaining secants to tangents:

$$\int \tan^4 x \sec^4 x \, dx = \int \tan^4 x \, \mathbf{sec^2}x \, (\sec^2 x) \, dx$$

$$= \int \tan^4 x (1 + \mathbf{tan^2} x)(\sec^2 x \, dx)$$

Let $u = \tan x$, so that $du = \sec^2 x \, dx$. We now have

$$\int u^4(1 + u^2) \, du = \int (u^4 + u^6) \, du = \frac{1}{5}u^5 + \frac{1}{7}u^7 + C \qquad \left[\begin{array}{l} u = \tan x \\ du = \sec^2 x \, dx \end{array} \right]$$

$$= \frac{1}{5}\tan^5 x + \frac{1}{7}\tan^7 x + C \qquad \blacksquare$$

Example 6   Integrate $\int \tan^3 3x \sec 3x \, dx$.

Solution. Since the power of the secant is odd, the method of Example 5 does not work. But since $d(\sec 3x) = 3 \sec 3x \tan 3x \, dx$, the odd power of the tangent suggests the following breakdown:

$$\int \tan^3 3x \sec 3x \, dx = \int \tan^2 3x(\sec 3x \tan 3x \, dx)$$

$$= \int (\sec^2 3x - 1)(\sec 3x \tan 3x) \, dx$$

by Identity (7.15). Now let $u = \sec 3x$, so that $du = 3 \sec 3x \tan 3x \, dx$; we get

$$\frac{1}{3} \int (\sec^2 3x - 1)(3 \sec 3x \tan 3x \, dx) = \frac{1}{3} \int (u^2 - 1) \, du$$

$$= \frac{1}{3} \left(\frac{1}{3} u^3 - u \right) + C$$

$$= \frac{1}{9} \sec^3 3x - \frac{1}{3} \sec 3x + C \qquad \blacksquare$$

The procedure for integrating powers of tangents and secants is summarized next:

1. If the power of the secant function is even and positive, save $\sec^2 u$ for the differential and change the remaining secants to tangents:

$$\int \tan^n u \overbrace{\sec^m}^{\text{even}} u \, du = \int \tan^n u \sec^{m-2} u \overbrace{(\sec^2 u)}^{\text{save}} \, du$$
$$\text{change to tangents}$$

2. If the power of the tangent function is odd and positive, save $\sec u \tan u$ for the differential and change the remaining tangents to secants:

$$\int \overbrace{\tan^n}^{\text{odd}} u \sec^m u \, du = \int \tan^{n-1} u \sec^{m-1} u \overbrace{(\sec u \tan u)}^{\text{save}} \, du$$
$$\text{change to secants}$$

Powers of tangents (type 2 with $m = 0$) and even powers of secants (type 2 with m even and $n = 0$) can be integrated with the help of Identity (7.15).

Type 2: $m = 0$, n Even or Odd

Use Identity (7.15) to integrate

$$\int \tan^n u \, du$$

Type 2: $n = 0$, m Even

Use Identity (7.15) to integrate

$$\int \sec^m u \, du$$

Odd powers of secants cannot be integrated by present techniques. We will return to this case in Section 7.7.

Example 7   Find $\int \tan^4 x \, dx$.

Solution. By Identity (7.15),

$$\int \tan^4 x \, dx = \int \tan^2 x \, \tan^2 x \, dx$$

$$= \int \tan^2 x (\sec^2 x - 1) \, dx$$

$$= \int \tan^2 x \sec^2 x \, dx - \int \tan^2 x \, dx$$

$$= \int \tan^2 x \sec^2 x \, dx - \int (\sec^2 x - 1) \, dx$$

$$= \int \tan^2 x \sec^2 x \, dx - \tan x + x + C$$

$$= \frac{1}{3} \tan^3 x - \tan x + x + C \qquad \left[\begin{array}{l} u = \tan x \\ du = \sec^2 x \, dx \end{array} \right] \qquad \blacksquare$$

Integrals of type 3 can be integrated with the help of Identity (7.16). Otherwise the cases are identical to those for integrals of type 2. For example, if m is even, we set aside $\csc^2 u \, du$ for the differential.

Example 8   Integrate $\int \cot x \, \csc^4 x \, dx$.

Solution.

$$\int \cot x \, \csc^2 x \, \csc^2 x \, dx = \int \cot x (1 + \cot^2 x)(\csc^2 x \, dx)$$

$$= -\int u(1 + u^2) \, du \qquad \left[\begin{array}{l} u = \cot x \\ du = -\csc^2 x \, dx \end{array} \right]$$

$$= -\frac{1}{2} u^2 - \frac{1}{4} u^4 + C$$

$$= -\frac{1}{2} \cot^2 x - \frac{1}{4} \cot^4 x + C \qquad \blacksquare$$

■ Exercises / Section 7.4

In Exercises 1–36, perform the designated integrations.

1. $\int \sin^2 2x \cos^3 2x \, dx$

2. $\int \sin^3 x \cos^2 x \, dx$

3. $\int \sin^3 x \, dx$

4. $\int \cos^3 2x \, dx$

5. $\int \sin^3 x \cos^4 x \, dx$

6. $\int \sin^4 3x \cos^3 3x \, dx$

7. $\int \sin^3 x \cos^3 x \, dx$

8. $\int \sin^5 x \cos^6 x \, dx$

9. $\int \cos^2 4x \, dx$

10. $\int \sin^2 x \, dx$

11. $\int \sin^2 x \cos^2 x \, dx$

12. $\int \cos^4 x \, dx$

13. $\int \sin^3 2t \cos^2 2t \, dt$

14. $\int \sin^3 3\theta \, d\theta$

15. $\int \sin^4 4x \cos^3 4x \, dx$

16. $\int \sin^3 3x \cos^4 3x \, dx$

17. $\int \tan^3 x \, dx$

18. $\int \sec^4 2x \, dx$

19. $\int \tan^2 x \sec^4 x \, dx$

20. $\int \tan^4 3x \sec^4 3x \, dx$

21. $\int \tan y \sec^3 y \, dy$

22. $\int \tan^5 z \sec^4 z \, dz$

23. $\int \tan^3 x \sec^3 x \, dx$

24. $\int \csc^4 x \, dx$

25. $\int \cot^6 2x \csc^4 2x \, dx$

26. $\int \cot x \csc^3 x \, dx$

27. $\int \csc^6 x \, dx$

28. $\int \cot^4 2x \, dx$

29. $\int_0^{\pi/4} \sqrt{\tan x} \sec^4 x \, dx$

30. $\int_0^{\pi/2} \cos^2 x \, dx$

31. $\int_0^{\pi} (1 + \sin x)^2 \, dx$

32. $\int_0^{\pi/3} \tan^3 x \sec x \, dx$

33. $\int_0^{\pi/4} \frac{\sec^2 x}{1 + \tan x} \, dx$

34. $\int \tan^3 2x \sec^3 2x \, dx$

35. $\int \tan^5 4x \sec^4 4x \, dx$

36. $\int \sqrt{\sin x} \cos^3 x \, dx$

37. Find the volume of the solid of revolution obtained by rotating the region bounded by $y = \sin x$, $y = 0$, $x = 0$, and $x = \pi$ about the x-axis.

38. Find the effective current (root mean square value) for $i = 3 \sin 2t$ for one period.

39. Repeat Exercise 38 for $i = 20 \cos 100\pi t$.

40. The mean power supplied to an electrical device in which there is an alternating current is given by

$$P = \frac{1}{T} \int_0^T ei \, dt$$

where T is the common period of e and i. If $e = 5 \sin \omega t$ and $i = 3 \sin(\omega t - \pi/2) = -3 \cos \omega t$, show that the average power over one period is zero. (During part of the cycle the device returns energy to the circuit.)

41. (Refer to Exercise 40.) Determine the mean power delivered to an electrical device if $e = 3 \sin 2t$ and $i = 5 \sin(2t - \pi/3)$. Note that $T = 2\pi/2 = \pi$.

42. Integrate $\int \sin x \cos x \, dx$ in two ways: first let $u = \sin x$, and then let $u = \cos x$. Explain the result.

43. (Set calculator in radian mode.) Use Simpson's rule with $n = 8$ to find the approximate value of

$$\int_0^{\pi/2} \frac{dx}{\sqrt{1 + \cos x}}$$

44. Repeat Exercise 43 for

$$\int_0^{\pi/2} \sqrt{1 + \sin x} \, dx$$

using the trapezoidal rule with $n = 10$.

45. Another application of the root mean square is the determination of the effective voltage of regular household current. This is an alternating current that varies from -155 V to $+155$ V with a frequency of 60 cycles per second: $f = 155 \sin(120\pi t)$. Voltmeters read the root mean square voltage over one cycle. Use the integration capability of your graphing utility to determine this value.

7.5 Inverse Trigonometric Forms

In this section we are going to discuss the integrals of certain algebraic functions that lead to inverse trigonometric forms. The simplest of these are

$$\int \frac{du}{\sqrt{a^2 - u^2}} \quad \text{and} \quad \int \frac{du}{a^2 + u^2}$$

Both integrals may be obtained from the derivatives of inverse trigonometric functions. In particular,

$$\frac{d}{dx} \operatorname{Arcsin} \frac{u}{a} = \frac{\frac{1}{a}}{\sqrt{1 - \frac{u^2}{a^2}}} \frac{du}{dx} = \frac{\frac{1}{a}}{\sqrt{\frac{a^2 - u^2}{a^2}}} \frac{du}{dx} = \frac{\frac{1}{a}}{\frac{\sqrt{a^2 - u^2}}{a}} \frac{du}{dx}$$

$$= \frac{1}{\sqrt{a^2 - u^2}} \frac{du}{dx}$$

Hence

$$\int \frac{du}{\sqrt{a^2 - u^2}} = \text{Arcsin}\, \frac{u}{a} + C \qquad (7.19)$$

Starting with Arctan(u/a), we obtain in similar manner the integration formula

$$\int \frac{du}{a^2 + u^2} = \frac{1}{a}\, \text{Arctan}\, \frac{u}{a} + C \qquad (7.20)$$

Example 1 Integrate

$$\int \frac{x\, dx}{\sqrt{4 - x^4}}$$

Solution. Since the method of substitution is one of trial and error, we might try $u = 4 - x^4$, based on earlier experience. Unfortunately, $du = -4x^3 dx$, which does not match $x\, dx$. Recognizing the proper form is more than half the battle! Instead, we write

$$\int \frac{x\, dx}{\sqrt{4 - x^4}} = \int \frac{x\, dx}{\sqrt{2^2 - (x^2)^2}}$$

which may be reduced to Formula (7.19) by the substitution $u = x^2$, so that $du = 2x\, dx$. Thus

$$\frac{1}{2} \int \frac{2x\, dx}{\sqrt{2^2 - (x^2)^2}} = \frac{1}{2} \int \frac{du}{\sqrt{2^2 - u^2}} = \frac{1}{2}\, \text{Arcsin}\, \frac{u}{2} + C = \frac{1}{2}\, \text{Arcsin}\, \frac{x^2}{2} + C$$

by Formula (7.19) with $a = 2$. ∎

Example 2 Integrate

$$\int \frac{t^2\, dt}{9 + 4t^6}$$

Solution. The integral can be written

$$\int \frac{t^2\, dt}{3^2 + (2t^3)^2}$$

Let $u = 2t^3$; then $du = 6t^2\, dt$, and we get

$$\frac{1}{6} \int \frac{6t^2\, dt}{3^2 + (2t^3)^2} = \frac{1}{6} \int \frac{du}{3^2 + u^2} = \frac{1}{6} \cdot \frac{1}{3}\, \text{Arctan}\, \frac{u}{3} + C \qquad \left[\begin{array}{l} u = 2t^3 \\ du = 6t^2\, dt \end{array} \right]$$

$$= \frac{1}{18}\, \text{Arctan}\, \frac{2t^3}{3} + C$$

by Formula (7.20) with $a = 3$. ∎

More general quadratic forms can often be handled by first completing the square and then substituting, as shown in the following example.

Example 3   Find

$$\int \frac{dx}{x^2 - 4x + 5}$$

Solution. To complete the square, we add and subtract the square of one-half the coefficient of x: $[(1/2)(-4)]^2 = 4$. Thus

$$x^2 - 4x + 4 - 4 + 5 = (x^2 - 4x + 4) + 1 = (x - 2)^2 + 1$$

This form suggests the substitution $u = x - 2$, so that $du = dx$:

$$\int \frac{dx}{x^2 - 4x + 5} = \int \frac{dx}{(x - 2)^2 + 1} = \int \frac{du}{u^2 + 1} \qquad \begin{bmatrix} \textbf{\textit{u} = \textit{x} - 2} \\ \textbf{\textit{du} = \textit{dx}} \end{bmatrix}$$

$$= \text{Arctan}\, u + C = \text{Arctan}(x - 2) + C \qquad \blacksquare$$

Example 4   Find

$$\int \frac{dx}{\sqrt{2 - 2x - x^2}}$$

Solution. The quadratic expression can be written as follows:

$$2 - x^2 - 2x = 2 - (x^2 + 2x) = 2 - (x^2 + 2x + 1 - 1)$$

$$= 3 - (x^2 + 2x + 1) = 3 - (x + 1)^2$$

Hence

$$\int \frac{dx}{\sqrt{2 - 2x - x^2}} = \int \frac{dx}{\sqrt{3 - (x + 1)^2}}$$

Let $u = x + 1$; then $du = dx$ and

$$\int \frac{dx}{\sqrt{3 - (x + 1)^2}} = \int \frac{du}{\sqrt{3 - u^2}} = \text{Arcsin}\,\frac{u}{\sqrt{3}} + C = \text{Arcsin}\,\frac{x + 1}{\sqrt{3}} + C$$

by Formula (7.19) with $a = \sqrt{3}$. $\qquad \blacksquare$

■ Exercises / Section 7.5

In Exercises 1–38, perform the designated integrations.

1. $\displaystyle \int \frac{dx}{\sqrt{1 - x^2}}$

2. $\displaystyle \int \frac{dx}{\sqrt{4 - 4x^2}}$

3. $\displaystyle \int \frac{dx}{9 + 4x^2}$

4. $\displaystyle \int \frac{dx}{4 + 3x^2}$

5. $\displaystyle \int \frac{x\,dx}{\sqrt{1 - x^2}}$

6. $\displaystyle \int \frac{z\,dz}{\sqrt{1 - 4z^2}}$

7. $\displaystyle \int \frac{x\,dx}{16 + 9x^2}$

8. $\displaystyle \int \frac{dx}{16 + 9x^2}$

9. $\displaystyle \int \frac{t\,dt}{4 + t^4}$

10. $\displaystyle \int \frac{P\,dP}{4 + P^2}$

11. $\displaystyle \int \frac{\csc^2 x\,dx}{\sqrt{4 - \cot^2 x}}$

12. $\displaystyle \int \frac{\sec^2 x\,dx}{\sqrt{9 - \tan^2 x}}$

13. $\displaystyle\int \frac{dx}{\sqrt{5-3x^2}}$

14. $\displaystyle\int \frac{x\,dx}{\sqrt{1-5x^4}}$

15. $\displaystyle\int \frac{\cos y\,dy}{2+\sin y}$

16. $\displaystyle\int \frac{x^2\,dx}{\sqrt{1-x^6}}$

17. $\displaystyle\int_3^{3\sqrt{3}} \frac{3\,dx}{9+x^2}$

18. $\displaystyle\int_1^2 \frac{dx}{\sqrt{4-x^2}}$

19. $\displaystyle\int \frac{dx}{\sqrt{4-3x^2}}$

20. $\displaystyle\int \frac{dx}{9+7x^2}$

21. $\displaystyle\int \frac{x}{\sqrt{4-3x^2}}\,dx$

22. $\displaystyle\int x\sqrt{4-x^2}\,dx$

23. $\displaystyle\int \frac{dx}{x^2-6x+9}$

24. $\displaystyle\int \frac{dx}{\sqrt{4-x^2}}$

25. $\displaystyle\int \frac{dx}{x^2-6x+10}$

26. $\displaystyle\int \frac{dx}{\sqrt{4x-x^2}}$

27. $\displaystyle\int \frac{dx}{\sqrt{1-x^2-4x}}$

28. $\displaystyle\int \frac{dx}{x^2+2x+3}$

29. $\displaystyle\int \frac{dx}{x^2+3x+3}$

30. $\displaystyle\int \frac{1-x}{\sqrt{1-x^2}}\,dx$

31. $\displaystyle\int \frac{x+4}{x^2+16}\,dx$

32. $\displaystyle\int \frac{x\,dx}{\sqrt{9-x^2}}$

33. $\displaystyle\int_0^{\pi/2} \frac{\cos x}{1+\sin^2 x}\,dx$

34. $\displaystyle\int \frac{e^x}{1+e^{2x}}\,dx$

35. $\displaystyle\int \frac{e^{2x}}{1+e^{2x}}\,dx$

36. $\displaystyle\int \frac{x^3}{4+x^4}\,dx$

37. $\displaystyle\int \frac{e^\theta}{\sqrt{1-e^{2\theta}}}\,d\theta$

38. $\displaystyle\int_0^{\pi/2} \frac{\sin\omega}{1+\cos^2\omega}\,d\omega$

39. Find the area "bounded" by $y=1/(x^2+1)$, the y-axis, and the positive x-axis.

40. Find M_y, the moment about the y-axis, of the region bounded by $y=1/(1+x^4)$, $y=0$, $x=0$, and $x=1$. (Assume the weight per unit area to be 1.)

41. Find the volume of the solid of revolution obtained by rotating the first-quadrant region "bounded" by $y=1/(4+x^4)$ about the y-axis.

42. Find the volume of the solid generated by revolving about the x-axis the region bounded by $y=2/\sqrt{x^2+9}$, $y=0$, $x=0$, and $x=3$.

7.6   Integration by Trigonometric Substitution

So far all of our substitutions in performing integration have been of a certain type: we let u be some function appearing in the integrand. More specifically, if the integral has the form

$$\int f(g(x))g'(x)\,dx$$

then we let $u=g(x)$, so that $du=g'(x)\,dx$. Then, if the resulting integral

$$\int f(u)\,du$$

is in recognizable form, we perform the integration and substitute back.

It is also possible in some cases to make a substitution of a totally different kind: let x be some function of a new variable θ. For example, x can be replaced by some trigonometric function, although other possibilities exist. Suppose we look at the familiar integral

$$\int \frac{dx}{\sqrt{1-x^2}}$$

If we let $x=\sin\theta$, then $dx=\cos\theta\,d\theta$, and we get

$$\int \frac{dx}{\sqrt{1-x^2}} = \int \frac{\cos\theta\,d\theta}{\sqrt{1-\sin^2\theta}} = \int \frac{\cos\theta\,d\theta}{\sqrt{\cos^2\theta}} = \int \frac{\cos\theta}{\cos\theta}\,d\theta$$

$$= \int d\theta = \theta + C$$

Since $\sin\theta = x$, it follows that $\theta = \text{Arcsin}\,x$, so that

$$\int \frac{dx}{\sqrt{1-x^2}} = \text{Arcsin}\,x + C$$

which we know to be correct. (See Formula (7.19).)

This example reveals the reason why such a substitution works: certain trigonometric forms will collapse in a way that the corresponding algebraic forms will not. As we saw, the radicand $1 - \sin^2\theta$ can be combined into a single squared term, so that the troublesome radical disappears. Identities (7.14) and (7.15) in Section 7.4 will suffice to eliminate radicals containing $a^2 - x^2$, $a^2 + x^2$, and $x^2 - a^2$. The procedure is summarized in Table 7.1.

Table 7.1 *Rules for trigonometric substitution*

For integrals containing . . .	Use . . .	To obtain . . .	
$\sqrt{a^2 - x^2}$	$x = a\sin\theta$	$\sqrt{a^2 - x^2} = a\cos\theta$	$(a > 0)$
$\sqrt{x^2 + a^2}$	$x = a\tan\theta$	$\sqrt{x^2 + a^2} = a\sec\theta$	$(a > 0)$
$\sqrt{x^2 - a^2}$	$x = a\sec\theta$	$\sqrt{x^2 - a^2} = a\tan\theta$	$(a > 0)$

One could also employ the cofunctions ($\cos\theta$, $\cot\theta$, or $\csc\theta$) in the trigonometric substitution. For example, letting $x = a\cos\theta$,

$$\sqrt{a^2 - x^2} = \sqrt{a^2 - a^2\cos^2\theta} = a\sin\theta$$

Avoiding the cofunctions altogether has the advantage of giving us a one-to-one correspondence between the radicals and the functions to be substituted. Some students claim that they never know what to substitute; according to our scheme, there is really nothing to decide!

Example 1   Integrate

$$\int \frac{\sqrt{x^2 - 4}}{x}\,dx$$

Solution. To eliminate the radical, we must let $x = 2\sec\theta$, so that $dx = 2\sec\theta\tan\theta\,d\theta$. (A common error is forgetting to substitute the expression for dx.) Then

$$\sqrt{x^2 - 4} = \sqrt{4\sec^2\theta - 4} = \sqrt{4(\sec^2\theta - 1)} = 2\sqrt{\sec^2\theta - 1}$$
$$= 2\sqrt{\tan^2\theta} = 2\tan\theta$$

and

$$\int \frac{\sqrt{x^2 - 4}}{x}\,dx = \int \frac{(2\tan\theta)(2\sec\theta\tan\theta)\,d\theta}{2\sec\theta}$$
$$= 2\int \tan^2\theta\,d\theta = 2\int (\sec^2\theta - 1)\,d\theta$$
$$= 2(\tan\theta - \theta) + C$$

It remains to rewrite the last expression in terms of x. As long as θ stands alone, the substituted expression $x/2 = \sec\theta$ gives us

$$\theta = \operatorname{Arcsec} \frac{x}{2}$$

directly. For a function of θ we need to go a step further and appeal to the definition of the trigonometric function. Since

$$\sec\theta = \frac{x}{2}$$

we may construct the triangle in Figure 7.1 with x the length of the hypotenuse and 2 the length of the adjacent side. The length of the opposite side, $\sqrt{x^2 - 4}$, is computed by the Pythagorean theorem. Now we can read off any trigonometric function required. In particular, $\tan\theta = \sqrt{x^2 - 4}/2 = (1/2)\sqrt{x^2 - 4}$. So

$$\int \frac{\sqrt{x^2 - 4}}{x}\, dx = 2(\tan\theta - \theta) + C$$

$$= 2\left(\frac{1}{2}\sqrt{x^2 - 4} - \operatorname{Arcsec}\frac{x}{2}\right) + C$$

$$= \sqrt{x^2 - 4} - 2\operatorname{Arcsec}\frac{x}{2} + C$$

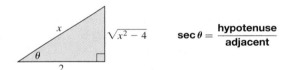

$$\sec\theta = \frac{\text{hypotenuse}}{\text{adjacent}}$$

Figure 7.1

Remark: According to Figure 7.1 the answer could have been written in other ways. In particular, if $\sec\theta = x/2$, then $\tan\theta = \sqrt{x^2 - 4}/2$, yielding

$$\sqrt{x^2 - 4} - 2\operatorname{Arctan}\left(\frac{1}{2}\sqrt{x^2 - 4}\right) + C \qquad\blacksquare$$

Example 2   Integrate $\int x^3 \sqrt{x^2 + 4}\, dx$.

Solution. In this case we need to substitute $x = 2\tan\theta$, for then

$$\sqrt{x^2 + 4} = \sqrt{4\tan^2\theta + 4} = \sqrt{4(\tan^2\theta + 1)} = 2\sqrt{\tan^2\theta + 1}$$

$$= 2\sqrt{\sec^2\theta} = 2\sec\theta$$

Also, $dx = 2 \sec^2 \theta \, d\theta$, and $x^3 = 8 \tan^3 \theta$. Substitution yields

$$\int x^3 \sqrt{x^2 + 4} \, dx = \int (8 \tan^3 \theta)(2 \sec \theta)(2 \sec^2 \theta) \, d\theta$$

$$= 32 \int \tan^3 \theta \sec^3 \theta \, d\theta$$

$$= 32 \int \tan^2 \theta \sec^2 \theta (\sec \theta \tan \theta) \, d\theta$$

$$= 32 \int (\sec^2 \theta - 1)(\sec^2 \theta)(\sec \theta \tan \theta) \, d\theta$$

$$= 32 \int (\sec^4 \theta - \sec^2 \theta)(\sec \theta \tan \theta \, d\theta)$$

Since $d(\sec \theta) = \sec \theta \tan \theta \, d\theta$, we get

$$32 \left(\frac{1}{5} \sec^5 \theta - \frac{1}{3} \sec^3 \theta \right) + C = \frac{32}{5} \sec^5 \theta - \frac{32}{3} \sec^3 \theta + C$$

We now use the substituted expression **$\tan \theta = x/2$** to construct the triangle in Figure 7.2. Note that

$$\sec \theta = \frac{\sqrt{x^2 + 4}}{2}$$

$\tan \theta = x/2$

Figure 7.2

so that

$$\int x^3 \sqrt{x^2 + 4} \, dx = \frac{32}{5} \left(\frac{\sqrt{x^2 + 4}}{2} \right)^5 - \frac{32}{3} \left(\frac{\sqrt{x^2 + 4}}{2} \right)^3 + C$$

$$= \frac{1}{5}(x^2 + 4)^{5/2} - \frac{4}{3}(x^2 + 4)^{3/2} + C \qquad \blacksquare$$

Example 3   Derive the following formula:

$$\int \frac{du}{\sqrt{u^2 + a^2}} = \ln \left| u + \sqrt{u^2 + a^2} \right| + C \qquad (7.21)$$

Solution. Let **$u = a \tan \theta$**; then $du = a \sec^2 \theta \, d\theta$. Thus

$$\int \frac{du}{\sqrt{u^2 + a^2}} = \int \frac{a \sec^2 \theta \, d\theta}{\sqrt{a^2 \tan^2 \theta + a^2}} = \int \frac{a \sec^2 \theta \, d\theta}{a \sec \theta}$$

$$= \int \sec \theta \, d\theta = \ln |\sec \theta + \tan \theta| + C'$$

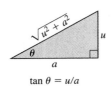

$\tan\theta = u/a$

Figure 7.3

From the substituted expression $\tan\theta = u/a$ and the corresponding triangle (Figure 7.3), we get

$$\int \frac{du}{\sqrt{u^2 + a^2}} = \ln\left|\frac{\sqrt{u^2 + a^2}}{a} + \frac{u}{a}\right| + C'$$

$$= \ln\left|\frac{\sqrt{u^2 + a^2} + u}{a}\right| + C'$$

$$= \ln\left|u + \sqrt{u^2 + a^2}\right| - \ln a + C'$$

Now let $C = C' - \ln a$, and Formula (7.21) follows. (If C' is arbitrary, so is $C' - \ln a$.) ∎

The derivation of

$$\int \frac{du}{\sqrt{u^2 - a^2}} = \ln\left|u + \sqrt{u^2 - a^2}\right| + C \qquad (7.22)$$

is left as an exercise.

■ Exercises / Section 7.6

In Exercises 1–20, perform the designated integrations.

1. $\displaystyle\int \frac{\sqrt{4 - x^2}}{x^2}\,dx$

2. $\displaystyle\int \frac{\sqrt{x^2 - 9}}{x}\,dx$

3. $\displaystyle\int x\sqrt{x^2 + 9}\,dx$

4. $\displaystyle\int \frac{dx}{x\sqrt{x^2 - 1}}$

5. $\displaystyle\int \frac{dx}{(x^2 + 25)^{3/2}}$

6. $\displaystyle\int \frac{dx}{x\sqrt{9 - x^2}}$

7. $\displaystyle\int \frac{dx}{x\sqrt{x^2 - 4}}$

8. $\displaystyle\int \frac{x^3\,dx}{\sqrt{16 - x^2}}$

9. $\displaystyle\int \frac{dx}{x^2\sqrt{x^2 + 16}}$

10. $\displaystyle\int \frac{dx}{(x^2 + 5)^{3/2}}$

11. $\displaystyle\int \frac{x\,dx}{(x^2 - 2)^{3/2}}$

12. $\displaystyle\int \frac{x\,dx}{(x^2 + 7)^{3/2}}$

13. $\displaystyle\int \frac{x^3\,dx}{\sqrt{x^2 - 3}}$

14. $\displaystyle\int x^3\sqrt{x^2 - 6}\,dx$

15. $\displaystyle\int \frac{\sqrt{x^2 + 1}}{x^2}\,dx$

16. $\displaystyle\int \frac{x\,dx}{\sqrt{x^2 + 1}}$

17. $\displaystyle\int_0^4 \frac{dx}{(16 + x^2)^{3/2}}$

18. $\displaystyle\int_1^2 \frac{dx}{x\sqrt{16 - x^2}}$

19. $\displaystyle\int_0^3 \sqrt{9 - x^2}\,dx$

20. $\displaystyle\int \frac{dx}{(x + 4)\sqrt{1 - (x + 4)^2}}$

21. A circular gate on a vertical dam has a diameter of 4 m, while its center is 20 m below the surface of the water. Find the force against the bottom half of the gate.

22. Derive the formula for the volume of a torus. (A torus is a doughnut-shaped solid formed by revolving the circular disk $(x - c)^2 + y^2 = r^2$ about the y-axis.)

23. Find the moment of inertia of a sphere with respect to its axis.

24. Derive Formula (7.22).

25. A cylindrical tank half full of water is lying on its side. If the tank has a radius of 3 m and a length of 8 m, find the work done in pumping the water out of the tank through an opening in the top.

26. Suppose the tank in Exercise 25 is full of water. Find the work done in pumping out half of the water through the opening in the top.

Use a symbolic integration utility to perform the following integrations:

27. $\displaystyle\int \sqrt{4 - x^2}\,dx$

28. $\displaystyle\int \sqrt{x^2 + 4}\,dx$

29. $\displaystyle\int x^3\sqrt{x^2 - 4}\,dx$

30. $\displaystyle\int \frac{\sqrt{x^2 - 1}}{x^3}\,dx$

Integration by Parts

The product rule from differentiation,

$$\frac{d}{dx}uv = u\frac{dv}{dx} + v\frac{du}{dx}$$

can also be written in differential form:

$$d(uv) = u\,dv + v\,du$$

or

$$u\,dv = d(uv) - v\,du$$

Integrating both sides, we get a formula known as the formula for **integration by parts.**

Formula for Integration by Parts

$$\int u\,dv = uv - \int v\,du \qquad\qquad (7.23)$$

This formula can be extremely useful. If the integral on the right in Formula (7.23) is easy to calculate, then the integral on the left can be obtained as well. Unfortunately, the formula does not tell us what part of the given integral $\int u\,dv$ to regard as the function u and which as dv. All we can do is try it out.

Example 1   Integrate $\int x \sin x\,dx$.

Solution. Note that this integral does not fit any of our earlier forms. Suppose we make some arbitrary assignment in the integral to the functions u and dv, say $u = x$ and $dv = \sin x\,dx$, and then arrange our work as follows:

$$u = x \qquad\qquad dv = \sin x\,dx \quad \textbf{Assign } u \textbf{ and } dv.$$
$$du = dx \qquad\qquad v = -\cos x \quad \textbf{Find } du \textbf{ and } v.$$

The expressions for du and v are obtained from u and dv, respectively. (The constant of integration will be included in the final result.) By Formula (7.23)

$$\int x \sin x\,dx = uv - \int v\,du$$

$$= x(-\cos x) - \int (-\cos x)\,dx = -x\cos x + \sin x + C$$

We conclude that

$$\int x \sin x\,dx = -x\cos x + \sin x + C$$

The procedure worked just fine. Suppose we now reverse the choices assigned to u and dv, as follows:

$$u = \sin x \qquad\qquad dv = x\,dx$$
$$du = \cos x\,dx \qquad\qquad v = \frac{1}{2}x^2$$

Now, by Formula (7.23) we have

$$\int x \sin x\,dx = \frac{1}{2}x^2 \sin x - \frac{1}{2}\int x^2 \cos x\,dx$$

This result is perfectly correct but useless, since the integral on the right is even more complicated than the given integral. ∎

We can see from Example 1 that we should try letting u be the factor whose derivative is a function simpler than u. As a result, the choice $u = x$ worked while the choice $u = \sin x$ did not.

Example 2   Integrate $\int x^2 \ln x\,dx$.

Solution. In this problem we actually have little choice since assigning $dv = \ln x\,dx$ leads nowhere. (We will see in Exercise 6 that $\int \ln x\,dx$ is itself integrated by parts.) Hence we choose $u = \ln x$ and $dv = x^2\,dx$:

$$u = \ln x \qquad\qquad dv = x^2\,dx \qquad \textbf{Assign } \boldsymbol{u} \textbf{ and } \boldsymbol{dv.}$$
$$du = \frac{1}{x}\,dx \qquad\qquad v = \frac{1}{3}x^3 \qquad \textbf{Find } \boldsymbol{du} \textbf{ and } \boldsymbol{v.}$$

We now obtain

$$\int x^2 \ln x\,dx = (\ln x)\left(\frac{1}{3}x^3\right) - \int \frac{1}{3}x^3 \cdot \frac{1}{x}\,dx \qquad \int u\,dv = uv - \int v\,du$$
$$= \frac{1}{3}x^3 \ln x - \frac{1}{3}\int x^2\,dx$$
$$= \frac{1}{3}x^3 \ln x - \frac{1}{9}x^3 + C \qquad\qquad\qquad ∎$$

Integration by parts can sometimes be used on integrals containing a single factor, such as $\int \ln x\,dx$ or $\int \text{Arcsin}\,x\,dx$. In such a case, we use $dv = dx$, as shown in the next example.

Example 3   Work out the integral $\int \text{Arctan}\,x\,dx$.

Solution. There is only one possible choice:

$$u = \text{Arctan}\,x \qquad\qquad dv = dx$$
$$du = \frac{dx}{1+x^2} \qquad\qquad v = x$$

Thus

$$\int \text{Arctan}\, x \, dx = x \, \text{Arctan}\, x - \int \frac{x\, dx}{1 + x^2}$$

$$= x \, \text{Arctan}\, x - \frac{1}{2} \ln(1 + x^2) + C \qquad \blacksquare$$

It is sometimes necessary to use integration by parts repeatedly, as shown in the remaining examples.

Example 4   Find $\int x^2 e^x \, dx$.

Solution. This integral is similar to the one in Example 1: $u = x^2$ is a good choice since the derivative of u is a simpler function.

$u = x^2$	$dv = e^x \, dx$
$du = 2x \, dx$	$v = e^x$

(By now the composition of the right-hand side of Formula (7.23) should be familiar.) Hence

$$\int x^2 e^x \, dx = x^2 e^x - 2 \int x e^x \, dx = x^2 e^x - 2 \left[\int x e^x \, dx \right]$$

The integral on the right can be obtained by using integration by parts again:

$u = x$	$dv = e^x \, dx$
$du = dx$	$v = e^x$

Hence

$$\int x^2 e^x \, dx = x^2 e^x - 2 \left[x e^x - \int e^x \, dx \right]$$

$$= x^2 e^x - 2x e^x + 2 \int e^x \, dx$$

$$= x^2 e^x - 2x e^x + 2 e^x + C \qquad \blacksquare$$

Example 5   Integrate $\int e^x \cos x \, dx$.

Solution. In this case one choice appears to be as good (or as bad) as another. Suppose we try

$u = e^x$	$dv = \cos x \, dx$
$du = e^x \, dx$	$v = \sin x$

Then

$$\int e^x \cos x \, dx = e^x \sin x - \int e^x \sin x \, dx$$

The resulting integral is of the same type as the given integral, so that no progress appears to have been made. But suppose we repeat this procedure on the resulting integral by letting

$$
\begin{array}{c|c}
u = e^x & dv = \sin x \, dx \\
\hline
du = e^x \, dx & v = -\cos x
\end{array}
$$

We now get

$$
\int e^x \cos x \, dx = e^x \sin x - \left[-e^x \cos x + \int e^x \cos x \, dx \right]
$$

$$
= e^x \sin x + e^x \cos x - \int e^x \cos x \, dx
$$

At this point we are right back where we started. Or are we? If we just transpose the last term, we end up with

$$
2 \int e^x \cos x \, dx = e^x \sin x + e^x \cos x
$$

and we obtain the integral after all! Thus

$$
\int e^x \cos x \, dx = \frac{1}{2}(e^x \sin x + e^x \cos x) + C
$$

∎

■ Exercises / Section 7.7

In Exercises 1–22, perform the designated integrations.

1. $\int x e^x \, dx$

2. $\int x \cos x \, dx$

3. $\int x \sin 2x \, dx$

4. $\int x e^{-x} \, dx$

5. $\int x \sec^2 x \, dx$

6. $\int \ln x \, dx$

7. $\int x \ln x \, dx$

8. $\int x e^{x^2} \, dx$

9. $\int \operatorname{Arcsin} x \, dx$

10. $\int \dfrac{\ln x}{x^2} \, dx$

11. $\int x \cos 3x \, dx$

12. $\int x^2 \sin x \, dx$

13. $\int x^2 e^{-x} \, dx$

14. $\int x^3 \ln x \, dx$

15. $\int \operatorname{Arccot} x \, dx$

16. $\int \operatorname{Arcsin} 4x \, dx$

17. $\int x \sin x^2 \, dx$

18. $\int x \operatorname{Arctan} x \, dx$

19. $\int e^x \sin x \, dx$

20. $\int e^{-x} \sin x \, dx$

21. $\int e^{-x} \cos \pi x \, dx$

22. $\int \sec^3 x \, dx$ (*Hint:* Let $u = \sec x$ and $dv = \sec^2 x \, dx$, and apply the method of Example 5.)

23. Evaluate $\int_0^1 x \, 2^x \, dx$.   24. Evaluate $\int_1^2 x^4 \ln x \, dx$.

25. Find the area of the region bounded by $y = x \sin x$, $x = 0$, $x = \pi$, and $y = 0$.

26. Find the volume of the solid of revolution obtained by rotating the area bounded by $y = \cos x$, $x = 0$, $x = \pi/2$, and $y = 0$ about the y-axis.

27. Find I_y, the moment of inertia about the y-axis, of the region in Exercise 26.

28. Find the average value of the current $i = e^{-2t} \cos 2t$ in a certain circuit on the time interval $0 \le t \le \pi/2$.

29. The current in a certain circuit is $i = e^{-t} \sin 4t$. Find the voltage across a 10-μF capacitor after 50 milliseconds if the initial voltage is zero. (One microfarad (μF) $= 10^{-6}$ F.)

Use a symbolic integration utility to perform the following integrations:

30. $\int x^3 e^x \, dx$

31. $\int x^3 e^{-4x} \, dx$

32. $\int x^2 \sin 3x \, dx$

33. $\int x^4 \cos x \, dx$

34. $\int x^4 \ln x \, dx$

35. $\int e^{-x} \cos 2x \, dx$

7.8 Integration of Rational Functions*

In this section we study integration of rational functions by means of partial fraction expansions. Partial fractions are taken up again in Chapter 13 in order to keep the presentation of Laplace transforms independent of this section. As a result, a study of the algebraic techniques in this section may be postponed.

A *rational function* has the form of a fraction in which both numerator and denominator are polynomials. Although the integrals of some special rational functions have already been considered, a general discussion is more involved. For example, the integral

$$\int \frac{x^3 + x}{x^2 - 1} \, dx$$

has an unfamiliar form. But suppose we divide numerator by denominator:

$$
\begin{array}{r}
x \\
x^2 - 1 \overline{\smash{)}\, x^3 + x} \\
\underline{x^3 - x} \\
2x
\end{array}
$$

Since the remainder is $2x$, we get

$$\int \frac{x^3 + x}{x^2 - 1} \, dx = \int \left(x + \frac{2x}{x^2 - 1} \right) dx$$

$$= \frac{1}{2}x^2 + \ln\left|x^2 - 1\right| + C$$

Since the remainder is always of lower degree than the divisor, long division reduces a rational function to either (1) a polynomial or (2) a polynomial plus a *proper* fraction (degree of numerator strictly less than degree of denominator).

Even proper fractions cannot always be integrated by the methods considered so far, as can be seen from the integral

$$\int \frac{2 \, dx}{x^2 - 1}$$

We can readily check, however, that the integrand can be written

$$\frac{2}{(x - 1)(x + 1)} = \frac{1}{x - 1} - \frac{1}{x + 1}$$

so that each term can be integrated separately. The fractions on the right are called *partial fractions*. Thus to integrate a rational function, we factor the denominator and split the whole fraction into a sum of simpler partial fractions.

To obtain the different cases, we are going to rely on the following fact from advanced algebra: a polynomial with real coefficients can be factored into *linear* and irreducible *quadratic* factors with real coefficients. Consequently, we do not have to consider factors of degree greater than 2.

---

*This section may be omitted without loss of continuity.

If the fraction is proper, it can be split into a sum of partial fractions according to the following rules:

Rule I. If a linear factor $ax + b$ occurs n times in the denominator, then there exist n partial fractions

$$\frac{A_1}{ax + b} + \frac{A_2}{(ax + b)^2} + \cdots + \frac{A_n}{(ax + b)^n}$$

where $A_1, A_2, \ldots, A_n$ are constants.

Rule II. If a quadratic factor $ax^2 + bx + c$ occurs n times in the denominator, then there exist n partial fractions

$$\frac{A_1 x + B_1}{ax^2 + bx + c} + \frac{A_2 x + B_2}{(ax^2 + bx + c)^2} + \cdots + \frac{A_n x + B_n}{(ax^2 + bx + c)^n}$$

where the A's and B's are constants. In all cases, n may be equal to 1.

Rules I and II will serve only as a general guide. How they are put to use will be illustrated in the examples. To make our job easier, we will classify the integrals according to whether the denominators have:

1. Distinct linear factors
2. Repeating linear factors
3. Distinct quadratic factors
4. Repeating quadratic factors

Case 1: Distinct Linear Factors

Example 1 Integrate

$$\int \frac{x + 8}{(x - 1)(x + 2)}\, dx$$

Solution. Note that the numerator is of first degree and the denominator of second degree, so that the integrand is a proper fraction. Furthermore, the factors are all distinct (each occurs only once). So by Rule I with $n = 1$, we have the following decomposition:

$$\frac{x + 8}{(x - 1)(x + 2)} = \frac{A}{x - 1} + \frac{B}{x + 2} \tag{7.24}$$

(Since there are only two constants, it is better to use A and B rather than subscripts.) The main task is to determine the constants A and B. To this end, we add the fractions on the right to obtain

$$\frac{A}{x - 1} + \frac{B}{x + 2} = \frac{A}{x - 1} \cdot \frac{x + 2}{x + 2} + \frac{B}{x + 2} \cdot \frac{x - 1}{x - 1}$$

$$= \frac{A(x + 2) + B(x - 1)}{(x - 1)(x + 2)}$$

It follows that the numerator of this fraction must be equal to the numerator of the left side of (7.24). In other words,

$$A(x + 2) + B(x - 1) = x + 8 \tag{7.25}$$

The constants A and B can be determined in two ways: (1) equating coefficients and (2) substituting convenient values for x.

1. *Equating coefficients.* If we multiply the expressions on the left side of (7.25) and combine similar terms, we get

$$A(x + 2) + B(x - 1) = x + 8$$
$$Ax + 2A + Bx - B = x + 8$$
$$(A + B)x + (2A - B) = x + 8$$

Now we equate corresponding coefficients to obtain the following system of equations:

$$
\begin{array}{ll}
A + B = 1 & \textbf{coefficients of } x \\
2A - B = 8 & \textbf{constants} \\
\hline
3A \phantom{{}- B} = 9 & \textbf{adding} \\
A = 3 &
\end{array}
$$

From $A + B = 1$, we get $3 + B = 1$, or $B = -2$.
Since $A = 3$ and $B = -2$, it follows from (7.24) that

$$\frac{x + 8}{(x - 1)(x + 2)} = \frac{3}{x - 1} - \frac{2}{x + 2}$$

2. *Substitution.* To see how A and B may be obtained by substitution, first note that (7.25),

$$A(x + 2) + B(x - 1) = x + 8$$

is an identity and is therefore valid for all values of x. If we let $x = -2$, for example, we obtain B at once:

$$x = -2: \qquad A(-2 + 2) + B(-2 - 1) = -2 + 8$$
$$-3B = 6$$
$$B = -2$$

Similarly, we may substitute 1 for x:

$$x = 1: \qquad A(1 + 2) + B(1 - 1) = 1 + 8$$
$$3A = 9$$
$$A = 3$$

We have shown again that

$$\frac{x + 8}{(x - 1)(x + 2)} = \frac{3}{x - 1} - \frac{2}{x + 2}$$

Finally, rewriting the given integral, we find that

$$\int \frac{x+8}{(x-1)(x+2)} \, dx = \int \left(\frac{3}{x-1} - \frac{2}{x+2} \right) dx$$

$$= 3 \ln|x-1| - 2 \ln|x+2| + C$$

$$= \ln|x-1|^3 - \ln|x+2|^2 + C$$

$$= \ln \left| \frac{(x-1)^3}{(x+2)^2} \right| + C \qquad \blacksquare$$

We can see from Example 1 that the method of substitution is the more efficient, at least for linear factors.

Example 2 Integrate

$$\int \frac{5x+10}{(x+1)(x-2)(x+3)} \, dx$$

Solution. As in Example 1, the factors are all distinct and linear. By Rule I with $n = 1$, the form is given by

$$\frac{5x+10}{(x+1)(x-2)(x+3)}$$

$$= \frac{A}{x+1} + \frac{B}{x-2} + \frac{C}{x+3}$$

$$= \frac{A(x-2)(x+3) + B(x+1)(x+3) + C(x+1)(x-2)}{(x+1)(x-2)(x+3)}$$

Equating numerators, we get

$$A(x-2)(x+3) + B(x+1)(x+3) + C(x+1)(x-2) = 5x+10$$

We now substitute 2, -3, and -1, respectively:

$$x = 2: \qquad 0 + B(2+1)(2+3) + 0 = 5(2) + 10$$

$$15B = 20$$

$$B = \frac{4}{3}$$

$$x = -3: \qquad 0 + 0 + C(-3+1)(-3-2) = 5(-3) + 10$$

$$10C = -5$$

$$C = -\frac{1}{2}$$

$$x = -1: \qquad A(-1-2)(-1+3) + 0 + 0 = 5(-1) + 10$$

$$-6A = 5$$

$$A = -\frac{5}{6}$$

Substituting the values of A, B, and C, we get

$$\int \frac{5x + 10}{(x + 1)(x - 2)(x + 3)} \, dx = \int \left(-\frac{5}{6} \frac{1}{x + 1} + \frac{4}{3} \frac{1}{x - 2} - \frac{1}{2} \frac{1}{x + 3} \right) dx$$

$$= -\frac{5}{6} \ln|x + 1| + \frac{4}{3} \ln|x - 2| - \frac{1}{2} \ln|x + 3| + C$$

This result can also be written

$$\frac{1}{6}(-5 \ln|x + 1| + 8 \ln|x - 2| - 3 \ln|x + 3|) + C$$

$$= \frac{1}{6}(-\ln|x + 1|^5 + \ln|x - 2|^8 - \ln|x + 3|^3) + C$$

$$= \frac{1}{6} \ln \left| \frac{(x - 2)^8}{(x + 1)^5(x + 3)^3} \right| + C \qquad \blacksquare$$

Case 2: Repeating Linear Factors

Example 3   Integrate

$$\int \frac{x - \frac{1}{2}}{(x - 3)^2} \, dx$$

Solution. By Rule I with $n = 2$, the integrand has the form

$$\frac{x - \frac{1}{2}}{(x - 3)^2} = \frac{A}{x - 3} + \frac{B}{(x - 3)^2} \qquad \textbf{repeating linear factors}$$

Adding the fractions, we get

$$\frac{A}{x - 3} \frac{x - 3}{x - 3} + \frac{B}{(x - 3)^2} = \frac{A(x - 3) + B}{(x - 3)^2}$$

Equating numerators, we obtain

$$A(x - 3) + B = x - \frac{1}{2}$$

$$\boldsymbol{x = 3}: \quad A(0) + B = 3 - \frac{1}{2} \qquad \textbf{substituting } x = 3$$

$$B = \frac{5}{2}$$

Because of the repeating factor, $x = 3$ is the only convenient value we can substitute. If we use the value of B already obtained, however, we can let x be equal to an arbitrary value (such as $x = 0$):

$$A(x - 3) + B = x - \frac{1}{2}$$

$$A(x - 3) + \frac{5}{2} = x - \frac{1}{2} \qquad \boldsymbol{B = \frac{5}{2}}$$

$$\boldsymbol{x = 0}: \quad A(0 - 3) + \frac{5}{2} = 0 - \frac{1}{2}$$

$$-3A = -\frac{6}{2}$$

$$A = 1$$

Substituting the values for A and B, we get

$$\int \frac{x - \frac{1}{2}}{(x-3)^2}\, dx = \int \left(\frac{1}{x-3} + \frac{5}{2} \frac{1}{(x-3)^2} \right) dx$$

$$= \int \frac{dx}{x-3} + \frac{5}{2} \int (x-3)^{-2}\, dx \qquad \begin{bmatrix} \boldsymbol{u = x - 3} \\ \boldsymbol{du = dx} \end{bmatrix}$$

$$= \ln|x-3| + \frac{5}{2} \frac{(x-3)^{-1}}{-1} + C$$

$$= \ln|x-3| - \frac{5}{2} \frac{1}{x-3} + C \qquad\qquad \blacksquare$$

Example 4   Integrate

$$\int \frac{x^5 - x^4 - 8x^3 + 14x^2 - 5x - 8}{(x-2)^2(x+3)}\, dx$$

Solution. Since the numerator is of higher degree than the denominator, we first perform the long division. (To be able to do so, we need to multiply out the denominator.)

$$x^3 - x^2 - 8x + 12 \overline{\smash{\big)}\, x^5 - x^4 - 8x^3 + 14x^2 - 5x - 8}$$

with quotient x^2, subtracting $x^5 - x^4 - 8x^3 + 12x^2$, leaving $2x^2 - 5x - 8$.

The integrand may now be written as

$$x^2 + \frac{2x^2 - 5x - 8}{(x-2)^2(x+3)}$$

We need to split up the fraction. Now note that the denominator contains only linear factors, the repeating factor $x - 2$ and the single factor $x + 3$. So by Rule I:

$$\frac{2x^2 - 5x - 8}{(x-2)^2(x+3)} = \frac{A}{x-2} + \frac{B}{(x-2)^2} + \frac{C}{x+3}$$

As before, we combine the fractions on the right to obtain

$$\frac{A(x-2)(x+3) + B(x+3) + C(x-2)^2}{(x-2)^2(x+3)}$$

Equating numerators, we get

$$A(x-2)(x+3) + B(x+3) + C(x-2)^2 = 2x^2 - 5x - 8$$

$x = 2$: $0 + 5B + 0 = -10$ and $B = -2$

$x = -3$: $0 + 0 + 25C = 25$ and $C = 1$

We now substitute the values already obtained for B and C:

$$A(x-2)(x+3) - 2(x+3) + 1(x-2)^2 = 2x^2 - 5x - 8$$

To find A, we can let x be equal to any value, say $x = 0$:

$$x = 0: \qquad A(-2)(3) - 2(3) + 1(-2)^2 = -8$$

It follows that $A = 1$.

The integral now becomes

$$\int \left(x^2 + \frac{1}{x-2} - \frac{2}{(x-2)^2} + \frac{1}{x+3} \right) dx$$

$$= \frac{1}{3}x^3 + \ln|x - 2| + \frac{2}{x-2} + \ln|x + 3| + C \qquad \blacksquare$$

Case 3: Distinct Quadratic Factors

Example 5   Integrate

$$\int \frac{3x^2 + 2x + 4}{(x^2 + 4)(x + 1)} \, dx$$

Solution. Since one of the factors is quadratic, Rule II applies. (The linear factor $x + 1$ leads to the usual form by Rule I.) Thus

$$\frac{3x^2 + 2x + 4}{(x^2 + 4)(x + 1)} = \frac{Ax + B}{x^2 + 4} + \frac{C}{x + 1}$$

After adding the fractions on the right and equating numerators, we get

$$(Ax + B)(x + 1) + C(x^2 + 4) = 3x^2 + 2x + 4$$

Proceeding as we did before, we let $x = -1$.

$$x = -1: \qquad 0 + C(1 + 4) = 3 - 2 + 4$$
$$5C = 5$$
$$C = 1$$

At this point we seem to have run out of values to substitute. By using the value of C already obtained, however, we can let $x = 0$ and solve for B:

$$(Ax + B)(x + 1) + 1(x^2 + 4) = 3x^2 + 2x + 4 \qquad \boldsymbol{C = 1}$$
$$x = 0: \qquad (0 + B)(1) + 1(4) = 4$$
$$B = 0$$

We now have

$$(Ax)(x + 1) + 1(x^2 + 4) = 3x^2 + 2x + 4 \qquad \boldsymbol{B = 0}$$

Finally, we let x be any value (say $x = 1$) and solve for A:

$$x = 1: \qquad A(2) + 5 = 9$$
$$A = 2$$

After substituting the values of A, B, and C, the integral becomes

$$\int \left(\frac{2x}{x^2+4} + \frac{1}{x+1} \right) dx = \ln|x^2+4| + \ln|x+1| + C$$

$$= \ln|(x^2+4)(x+1)| + C \qquad \blacksquare$$

In some cases involving quadratic factors, it is best to return to the method of comparing coefficients, as in Example 1.

Example 6   (*Trinomial factor.*) Integrate

$$\int \frac{2x^2+10x+10}{x(x^2+4x+5)} \, dx$$

Solution. By Rules I and II,

$$\frac{2x^2+10x+10}{x(x^2+4x+5)} = \frac{A}{x} + \frac{Bx+C}{x^2+4x+5}$$

It follows that

$$A(x^2+4x+5) + (Bx+C)x = 2x^2+10x+10$$

or

$$(A+B)x^2 + (4A+C)x + 5A = 2x^2+10x+10$$

Equating coefficients, we get the following system of equations:

$A+B=2$ **coefficients of x^2**

$4A+C=10$ **coefficients of x**

$5A=10$ **constants**

From the last equation, $A=2$. Substituting in the first equation, we get $2+B=2$, or $B=0$. From the second equation, $4(2)+C=10$, or $C=2$. Consequently,

$$\int \frac{2x^2+10x+10}{x(x^2+4x+5)} \, dx = \int \left(\frac{2}{x} + \frac{2}{x^2+4x+5} \right) dx$$

$$= \int \frac{2}{x} \, dx + \int \frac{2}{(x+2)^2+1} \, dx$$

$$= 2\ln|x| + 2\operatorname{Arctan}(x+2) + C \qquad \blacksquare$$

Case 4: Repeating Quadratic Factors

Example 7   Integrate

$$\int \frac{x^3 \, dx}{(x^2+1)^2}$$

Solution. By Rule II,

$$\frac{x^3}{(x^2 + 1)^2} = \frac{Ax + B}{x^2 + 1} + \frac{Cx + D}{(x^2 + 1)^2}$$

Adding fractions again and equating numerators, we get

$$(Ax + B)(x^2 + 1) + (Cx + D) = x^3$$

or, collecting similar terms,

$$Ax^3 + Bx^2 + (A + C)x + (B + D) = x^3$$

The system of equations resulting from comparing coefficients is particularly simple in this case. Note that

$$A = 1 \qquad B = 0 \qquad A + C = 0 \qquad B + D = 0$$

whence $C = -1$ and $D = 0$. Thus

$$\int \frac{x^3\,dx}{(x^2 + 1)^2} = \int \left(\frac{x}{x^2 + 1} - \frac{x}{(x^2 + 1)^2}\right) dx$$

Now let $u = x^2 + 1$, so that $du = 2x\,dx$. Then we get

$$\frac{1}{2}\int \left(\frac{1}{x^2 + 1} - \frac{1}{(x^2 + 1)^2}\right) 2x\,dx = \frac{1}{2}\int (u^{-1} - u^{-2})\,du$$

$$= \frac{1}{2}\left(\ln|u| + \frac{1}{u}\right) + C$$

$$= \frac{1}{2}\ln(x^2 + 1) + \frac{1}{2(x^2 + 1)} + C \qquad \blacksquare$$

The partial fraction technique is due to German mathematician Carl Jacobi (1804–1851) and was developed in his Berlin dissertation of 1825. In spite of this modest start, Jacobi soon established himself as a mathematician of first rank through his work on elliptic functions, a difficult branch of the theory of functions of complex variables. The theory of determinants, in the form now taught in algebra, is also due to Jacobi.

■ Exercises / Section 7.8

Perform the following integrations.

1. $\displaystyle\int \frac{dx}{x^2 - 4}$

2. $\displaystyle\int \frac{x - 5}{x + 2}\,dx$

3. $\displaystyle\int \frac{dx}{x - x^2}$

4. $\displaystyle\int \frac{2x + 1}{x^2 + x}\,dx$

5. $\displaystyle\int \frac{5x - 4}{(x - 2)(x + 1)}\,dx$

6. $\displaystyle\int \frac{x^3 - 3x^2 + 5x - 4}{x^2 - 3x + 2}\,dx$

7. $\displaystyle\int \frac{x^3\,dx}{x^2 - 2x - 3}$

8. $\displaystyle\int \frac{3x^2 + 2x - 6}{x(x - 2)(x + 3)}\,dx$

9. $\displaystyle\int \frac{x^2 + 10x - 20}{(x - 4)(x - 1)(x + 2)}\,dx$

10. $\displaystyle\int \frac{5 - 2x}{(x - 2)^2}\,dx$

11. $\displaystyle\int \frac{3x + 5}{(x + 1)^2}\,dx$

12. $\displaystyle\int \frac{4\,dx}{(x + 2)^2}$

13. $\displaystyle\int \frac{-2x^2 + 9x - 7}{(x - 2)^2(x + 1)}\,dx$

14. $\displaystyle\int \frac{4x^2 - x - 7}{(x - 1)^2(x + 3)}\,dx$

15. $\int \dfrac{2x^2 + 1}{(x-2)^3} \, dx$ $\left(\textit{Hint:}\ \text{By Rule I,}\ \dfrac{2x^2 + 1}{(x-2)^3} \right.$

$\left. = \dfrac{A}{x-2} + \dfrac{B}{(x-2)^2} + \dfrac{C}{(x-2)^3} \right)$

16. $\int \dfrac{x \, dx}{(x-1)^3}$

17. $\int \dfrac{x^2 - 3x - 2}{(x+2)(x^2+4)} \, dx$ (See Example 5.)

18. $\int \dfrac{2x^2 + 3x + 9}{(x-3)(x^2+9)} \, dx$

19. $\int \dfrac{3x^2 + 4x + 3}{(x+1)(x^2+1)} \, dx$

20. $\int \dfrac{x^3 + 6x^2 + 2x + 3}{(x-1)(x+2)(x^2+1)} \, dx$

21. $\int \dfrac{x^5 \, dx}{(x^2+4)^2}$

22. $\int \dfrac{x^3 + 3x}{(x^2+1)^2} \, dx$

23. $\int \dfrac{x^2 - 3x + 5}{x(x^2 - 2x + 5)} \, dx$

24. $\int \dfrac{3x^2 - 4x - 3}{(x-4)(x^2 + 2x + 5)} \, dx$

25. $\int \dfrac{dx}{x(x^2 + 2x + 2)}$

7.9   Integration by Use of Tables

It should not be surprising that many basic integrals have been tabulated over the years and collected in tables. (Table 2, Appendix A, contains a short list of such integrals.) Although integration tables may be highly useful, they do have certain natural limitations. To see why, pretend that you have forgotten most of the integration formulas and decided to rely on the table. You are faced with the integral

$$\int \frac{\sin e^{-x}}{e^x} \, dx$$

Neither Table 2 nor any other table contains this particular case. What can be done? Based on our earlier experience, it is not difficult to see that by letting $u = e^{-x} [du = -e^{-x} \, dx = (-1/e^x) \, dx]$, the integral assumes the form

$$-\int \sin u \, du = \cos u + C = \cos e^{-x} + C$$

In other words, the table lists only the general form, which in this case ($\int \sin u \, du$) you would probably have remembered anyway. Consequently, *you have to be familiar with the integration techniques in order to convert a given integral to a recognizable form listed in the table.*

Fortunately, many integrals can be obtained from a table without any special tricks. All that may be required is a little care in identifying the constants.

Example 1   Use Table 2 to integrate

$$\int \frac{dx}{x\sqrt{5 + 2x}}$$

Solution. We hunt up the form in the second section of the table ("Forms Containing $\sqrt{a+bu}$"). Note that in Formula 12, $a = 5$ and $b = 2$. By direct substitution we now get

$$\int \frac{dx}{x\sqrt{5+2x}} = \frac{1}{\sqrt{5}} \ln\left|\frac{\sqrt{5+2x}-\sqrt{5}}{\sqrt{5+2x}+\sqrt{5}}\right| + C \qquad \blacksquare$$

Example 2 Integrate $\int \sin 4x \cos 2x\, dx$.

Solution. We look in the table under trigonometric forms and find that Formula 63 fits our case. Since $m = 4$ and $n = 2$, we get

$$\int \sin 4x \cos 2x\, dx = -\frac{\cos(4+2)x}{2(4+2)} - \frac{\cos(4-2)x}{2(4-2)} + C$$

$$= -\frac{1}{12}\cos 6x - \frac{1}{4}\cos 2x + C \qquad \blacksquare$$

Example 3 Integrate $\int \sec^4 x\, dx$.

Solution. Note that Formula 71 fits our form. Since the right-hand side contains an integral, such a form is referred to as a *reduction formula*. With $n = 4$, we now get directly

$$\int \sec^4 x\, dx = \frac{\sec^2 x \tan x}{4-1} + \frac{4-2}{4-1}\int \sec^2 x\, dx$$

$$= \frac{1}{3}\sec^2 x \tan x + \frac{2}{3}\tan x + C$$

by Formula 53 in Table 2 or directly from memory. $\qquad \blacksquare$

Example 4 Integrate

$$\int \frac{dx}{x\sqrt{2x^2+4}}$$

Solution. As indicated earlier, the use of tables depends on the proper recognition of the form. With $u^2 = 2x^2$, the form can be made to fit Formula 34 by adjusting the constants. Since $u = \sqrt{2}x$, we get $du = \sqrt{2}\, dx$ and write the integral as

$$\int \frac{\sqrt{2}\, dx}{\sqrt{2}x\sqrt{2x^2+4}}$$

Now Formula 34 in the table fits precisely, so that

$$\int \frac{\sqrt{2}\, dx}{\sqrt{2}x\sqrt{2x^2+4}} = \int \frac{du}{u\sqrt{u^2+a^2}} = \frac{1}{a}\ln\left|\frac{u}{a+\sqrt{u^2+a^2}}\right| + C$$

$$= \frac{1}{2}\ln\left|\frac{\sqrt{2}x}{2+\sqrt{2x^2+4}}\right| + C \qquad \blacksquare$$

■ Exercises / Section 7.9

Integrate each of the given functions by using Table 2 (Appendix A).

1. $\displaystyle\int \frac{dx}{x(2+x)}$

2. $\displaystyle\int x\sqrt{1+2x}\,dx$

3. $\displaystyle\int \sqrt{x^2-7}\,dx$

4. $\displaystyle\int \frac{dx}{x^2-9}$

5. $\displaystyle\int \frac{dx}{5-x^2}$

6. $\displaystyle\int \frac{dx}{(x^2+2)^{3/2}}$

7. $\displaystyle\int \frac{dx}{x^2\sqrt{5x^2+4}}$

8. $\displaystyle\int xe^x\,dx$

9. $\displaystyle\int \sin 2x \sin x\,dx$

10. $\displaystyle\int e^{-x}\sin 2x\,dx$

11. $\displaystyle\int \frac{dx}{\sqrt{3x^2+5}}$

12. $\displaystyle\int \frac{\sqrt{x^2+8}}{x}\,dx$

13. $\displaystyle\int x^2 e^{2x}\,dx$

14. $\displaystyle\int \sin^4 x\,dx$

15. $\displaystyle\int \tan^6 x\,dx$

16. $\displaystyle\int x\,\text{Arcsin}\,x^2\,dx$

17. $\displaystyle\int \frac{dx}{4x^2-9}$

18. $\displaystyle\int \cos 3x \cos x\,dx$

19. $\displaystyle\int \frac{dx}{x\sqrt{3+x}}$

20. $\displaystyle\int \frac{dx}{3x^2+5}$

21. $\displaystyle\int \frac{dx}{3x^2-5}$

22. $\displaystyle\int x\sin x\,dx$

23. $\displaystyle\int \frac{\sqrt{x^2-10}}{x}\,dx$

24. $\displaystyle\int \frac{dx}{\sqrt{4x^2-9}}$

25. $\displaystyle\int \sin 3x \cos 2x\,dx$

26. $\displaystyle\int \sqrt{5-x^2}\,dx$

27. $\displaystyle\int \frac{dx}{(4x^2+5)^{3/2}}$

28. $\displaystyle\int \sin^2 2x\,dx$

7.10   Additional Remarks

Elementary function

After studying the various integration techniques, you could easily get the impression that any function can be integrated by finding its antiderivative. Such is not the case in the following sense. Suppose we define a function to be **elementary** if it is an algebraic, trigonometric, exponential, logarithmic, inverse trigonometric, or hyperbolic function. It turns out that even apparently simple integrals such as

$$\int e^{x^3}\,dx \qquad \int \sqrt{x}\,\sin x\,dx \qquad \int \sqrt{a^2\cos^2 x + b^2\sin^2 x}\,dx$$

do not lead to elementary antiderivatives. (For example, there is no elementary function whose derivative is e^{x^3}.) Certain definite integrals will be evaluated in Chapter 10 by means of infinite series, which yield nonelementary forms. Other well-known definite integrals such as

$$\int_0^\infty \frac{\sin x}{x}\,dx = \frac{\pi}{2} \qquad \text{and} \qquad \frac{2}{\sqrt{\pi}}\int_0^\infty e^{-x^2}\,dx = 1$$

can only be handled by more advanced methods. Fortunately, numerical techniques have been developed for evaluating many definite integrals. Two such methods were discussed in Chapter 4 (Section 4.9).

■ Review Exercises / Chapter 7

One of the problems in integration is recognizing the type of integral. In the previous sections the type was usually known and the idea was to apply the technique correctly. In the following exercises the main task is to decide what technique to apply. *DO NOT USE TABLE 2 OF APPENDIX A.*

1. $\int \dfrac{x \, dx}{x^2 + 1}$

2. $\int \dfrac{x \, dx}{(x^2 + 1)^2}$

3. $\int \dfrac{2 \, dx}{x^2 + 1}$

4. $\int \dfrac{e^{\operatorname{Arctan} x}}{x^2 + 1} \, dx$

5. $\int \dfrac{t \, dt}{\sqrt{9 - t^2}}$

6. $\int \dfrac{d\theta}{\sqrt{9 - \theta^2}}$

7. $\int x \cos 2x^2 \, dx$

8. $\int x \cos 2x \, dx$

9. $\int \dfrac{e^x \, dx}{4 + e^{2x}}$

10. $\int \dfrac{e^x \, dx}{4 + e^x}$

11. $\int \dfrac{e^x \, dx}{(4 + e^x)^2}$

12. $\int x \ln x \, dx$

13. $\int \dfrac{\ln x}{x} \, dx$

14. $\int \dfrac{\cos(\ln x)}{x} \, dx$

15. $\int \dfrac{x + 2}{x^2 + 4x + 5} \, dx$

16. $\int \dfrac{dx}{x^2 + 4x + 5}$

17. $\int \dfrac{dx}{x^2 + 4x + 4}$

18. $\int \dfrac{dx}{x(1 + \ln^2 x)}$

19. $\int \ln^2 x \, dx$

20. $\int \dfrac{dx}{\sqrt{4 - 5x^2}}$

21. $\int \dfrac{dx}{x\sqrt{4 - x^2}}$

22. $\int x\sqrt{4 - x^2} \, dx$

23. $\int \dfrac{\sin^2 2x \cos 2x \, dx}{1 + \sin^3 2x}$

24. $\int \sin^3 2x \cos^2 2x \, dx$

25. $\int \sin^2 2x \, dx$

26. $\int x e^{2x^2} \, dx$

27. $\int x e^{2x} \, dx$

28. $\int \dfrac{\operatorname{Arctan} 2x \, dx}{1 + 4x^2}$

29. $\int \operatorname{Arctan} 2x \, dx$

30. $\int \dfrac{\sin 2x \, dx}{e^{\sin^2 x}}$

31. $\int \dfrac{e^{\tan x}}{\cos^2 x} \, dx$

32. $\int \dfrac{\sin(1/x) \, dx}{x^2}$

33. $\int \dfrac{x^2 - 1}{x^2 + 3} \, dx$

34. $\int \dfrac{\sin \omega}{1 - \cos \omega} \, d\omega$

35. $\int \dfrac{1 - \cos \omega}{\sin \omega} \, d\omega$

36. $\int \dfrac{dx}{x \sqrt[3]{\ln x}}$

37. $\int \cot^4 x \csc^4 x \, dx$

38. $\int \dfrac{dx}{\sqrt{4x - 2 - x^2}}$

39. $\int \sec^3 2x \tan^3 2x \, dx$

40. $\int e^{\tan x} \sec^2 x \tan x \, dx$

41. $\int \tan^2 x \sec^2 x \, dx$

42. $\int \dfrac{dx}{x(2 - \ln x)}$

43. $\int \sec^4 3x \, dx$

44. $\int \csc^3 4x \cot^3 4x \, dx$

45. $\int e^x \cos 4x \, dx$

46. $\int e^{\cos 4x} \sin 4x \, dx$

47. $\int \dfrac{dx}{5x^2 + 4}$

48. $\int \dfrac{3x \, dx}{5x^2 + 4}$

49. $\int \dfrac{x + 1}{x^2 + 2x - 8} \, dx$

50. $\int \dfrac{x + 2}{x^2 + 2x - 8} \, dx$

51. $\int \dfrac{3x^2 - 4x + 9}{(x - 2)(x^2 + 9)} \, dx$

8

Parametric Equations, Vectors, and Polar Coordinates

Vectors and Parametric Equations

In this section we will make a brief study of vectors and parametric equations. Although it had its origins in antiquity, the vector concept was developed by Galileo through his discovery of the parallelogram law of forces. Later developments correlate with the discovery of coordinate geometry and the notion of complex numbers. More advanced areas of vector analysis date from about the middle of the nineteenth century, beginning with the work of the Irish mathematician William Rowan Hamilton (1805–1865) and Hermann Grassmann (German, 1809–1877). Hamilton was the first to represent a complex number $a + bj$ as an ordered pair (a, b), as is done in many algebra books. Out of this representation grew his famous concept of "quaternions," of which modern vector analysis is an offshoot. (A quaternion is something like a four-dimensional complex number.)

Since Grassmann's approach to vectors was more philosophical than physical, the merit of his work was not recognized by his contemporaries—partly because he was far ahead of his time but also because his book on the subject was considered essentially unreadable. Frustrated in his efforts to obtain a university position, Grassmann turned to linguistics, a field in which he finally achieved distinction.

Although vectors have many applications, we will restrict their use to the study of motion in a plane. To this end we need to examine a commonly used method for describing plane curves.

Parametric Equations

So far all our functions have been of the form $y = F(x)$. Functions, and their corresponding graphs, can also be expressed as *parametric equations*.

Parametric Equations

Parametric equations represent the x and y coordinates of a curve as functions of a third variable t, called the **parameter:**

$$\left. \begin{array}{l} x = f(t) \\ y = g(t) \end{array} \right\} \tag{8.1}$$

Consider the set of parametric equations

$$\left.\begin{array}{l} x = t \\ y = t^2 \end{array}\right\}$$

Both coordinates are expressed as functions of t, so that each value of t determines the coordinates of a point. A few of these coordinates are listed in the following chart:

t:	-2	-1	0	1	2
x:	-2	-1	0	1	2
y:	4	1	0	1	4

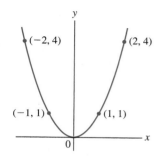

Figure 8.1

Plotting these points and connecting them by a smooth curve, we obtain the graph in Figure 8.1. There is something familiar about the shape of this curve. Indeed, we may eliminate t by algebraic means: square both sides of the first equation to obtain

$$\left.\begin{array}{l} x^2 = t^2 \\ y = t^2 \end{array}\right\}$$

It follows that $y = x^2$, the equation of a *parabola*.

Most graphing utilities have a parametric graphing mode. Try using it to obtain the graph in Figure 8.1.

Example 1   Show that

$$\left.\begin{array}{l} x = r \cos t \\ y = r \sin t \end{array}\right\}$$

are the parametric equations of a circle centered at the origin.

Solution. Since the parametric equations involve trigonometric functions, t cannot be eliminated algebraically. If we square both sides of these equations and divide by r^2, however, we get

$$\frac{x^2}{r^2} = \cos^2 t$$

$$\frac{y^2}{r^2} = \sin^2 t$$

Adding, we obtain

$$\frac{x^2}{r^2} + \frac{y^2}{r^2} = \cos^2 t + \sin^2 t = 1$$

or

$$x^2 + y^2 = r^2 \qquad\qquad ■$$

It is easy to find the slope of a tangent line to a curve represented by parametric equations: the differentials of Equations (8.1) are

$$\left.\begin{array}{l} dx = f'(t)\,dt \\ dy = g'(t)\,dt \end{array}\right\}$$

so that

$$\frac{dy}{dx} = \frac{g'(t)\,dt}{f'(t)\,dt} = \frac{g'(t)}{f'(t)} \tag{8.2}$$

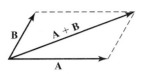

Figure 8.2

provided, of course, that f and g are differentiable.

Vectors

Formula (8.2) is particularly useful in the study of motion by means of vectors. A **vector** is an entity that has both magnitude and direction. Examples of vectors are velocity, force, and magnetic field intensity. It is customary to represent a vector by an arrow indicating the direction; the length of the arrow represents the magnitude. A boldface letter is normally used to denote a vector; the magnitude of vector **A** is denoted by $|\mathbf{A}|$ or A.

Vectors are added by the "parallelogram law," as shown in Figure 8.2: vectors **A** and **B** determine a parallelogram and the sum $\mathbf{A} + \mathbf{B}$ is the indicated diagonal, also called the **resultant.** The vector $\mathbf{A} - \mathbf{B}$ is shown in Figure 8.3.

Resultant

It is convenient to place a vector with initial point at the origin, so that any vector can be expressed as a sum of its **components.** For example, the vector **V** in Figure 8.4 extending from the origin to $(-1, \sqrt{3})$ is the sum of the vectors along the coordinate axes. Suppose we now let **i** denote the unit vector (magnitude equal to 1) along the positive x-axis and let **j** denote the unit vector along the positive y-axis (Figure 8.5). Then

Components

$$\mathbf{V} = -\mathbf{i} + \sqrt{3}\mathbf{j}$$

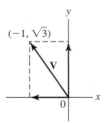

Figure 8.3

In fact, any vector can be written uniquely in the form

$$\mathbf{A} = A_x \mathbf{i} + A_y \mathbf{j}$$

for suitable choices of A_x and A_y. It is understood that the arrow representing the vector can be placed anywhere in the plane (Figure 8.6).

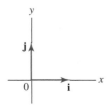

Figure 8.4

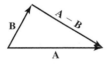

Figure 8.5

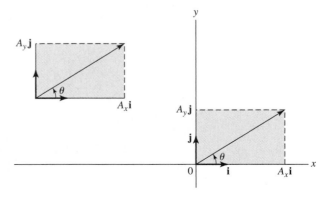

Figure 8.6

It now follows that the magnitude of **A** is

$$|\mathbf{A}| = \sqrt{A_x{}^2 + A_y{}^2}$$

by the Pythagorean theorem, while the direction θ is an angle such that $\tan\theta = A_y/A_x$.

Magnitude of Vector **A** is given by

$$A = |\mathbf{A}| = \sqrt{A_x{}^2 + A_y{}^2}$$

Direction θ of Vector **A** is determined from

$$\tan\theta = \frac{A_y}{A_x}$$

Also, addition (as well as subtraction) may be performed componentwise: if $\mathbf{A} = A_x\mathbf{i} + A_y\mathbf{j}$ and $\mathbf{B} = B_x\mathbf{i} + B_y\mathbf{j}$, then

$$\mathbf{A} \pm \mathbf{B} = (A_x \pm B_x)\mathbf{i} + (A_y \pm B_y)\mathbf{j}$$

Returning now to Equations (8.1), if the parameter t is time, then the parametric equations may be interpreted as the curvilinear motion of a particle. In that case the value of t not only specifies the point on the curve but also the position of the particle at any time and, consequently, the **direction** in which the particle is moving.

We can now find the velocity and acceleration vectors at any point by means of Formula (8.2). We will illustrate the method by two examples.

Example 2   The motion of a particle is described by the parametric equations

$$\left.\begin{array}{l} x = 1 - \dfrac{1}{2}t^3 \\[2mm] y = \dfrac{1}{\sqrt{t}} \end{array}\right\}$$

where x and y are in meters and t is in seconds. Find the velocity vector **v** when $t = 1$.

Solution. Since x and y are the coordinates of the position of the particle at any time, dx/dt and dy/dt are the components of the velocity. We denote these components by v_x and v_y, respectively. Thus

$$\left.\begin{array}{l} v_x = \dfrac{dx}{dt} = -\dfrac{3}{2}t^2 \\[2mm] v_y = \dfrac{dy}{dt} = -\dfrac{1}{2t^{3/2}} \end{array}\right\}$$

so that

$$\mathbf{v} = -\frac{3}{2}t^2\mathbf{i} - \frac{1}{2t^{3/2}}\mathbf{j}$$

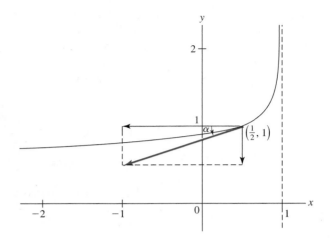

Figure 8.7

By Formula (8.2)

$$\frac{dy}{dx} = \frac{v_y}{v_x}$$

which implies that the velocity vector is tangent to the curve. (The line determined by the velocity vector has the same slope as the tangent line to the curve). At $t = 1$, $v_x = -3/2$ and $v_y = -1/2$, or

$$\mathbf{v} = -\frac{3}{2}\mathbf{i} - \frac{1}{2}\mathbf{j}$$

When $t = 1$ the particle is located at $(1/2, 1)$. (The curve and the velocity vector $\mathbf{v}$ are shown in Figure 8.7.) The magnitude is now found to be

$$|\mathbf{v}| = \sqrt{\left(-\frac{3}{2}\right)^2 + \left(-\frac{1}{2}\right)^2} \approx 1.58 \text{ m/s}$$

For the angle α in the figure we have

$$\tan \alpha = \frac{v_y}{v_x} = \frac{-\frac{1}{2}}{-\frac{3}{2}} = \frac{1}{3}$$

or $\alpha = 18.4°$. Since both components are negative, θ is in the third quadrant; so the direction is given by $\theta = 18.4° + 180° = 198.4°$.   ∎

Example 3   Find the acceleration vector $\mathbf{a}$ at $t = 2\pi/3$ if the motion of the particle is described by the parametric equations

$$\left. \begin{array}{l} x = 2\cos t \\ y = \sin t \end{array} \right\}$$

with x and y in meters and t in seconds.

Solution. We let

$$a_x = \frac{dv_x}{dt} = -2\cos t \left.\vphantom{\frac{dv_x}{dt}}\right\} \quad \text{since } v_x = -2\sin t$$

$$a_y = \frac{dv_y}{dt} = -\sin t \quad \text{since } v_y = \cos t$$

so that

$$\mathbf{a} = a_x\mathbf{i} + a_y\mathbf{j} = -2\cos t\,\mathbf{i} - \sin t\,\mathbf{j}$$

At $t = 2\pi/3$, $a_x = 1$ and $a_y = -\sqrt{3}/2$. (See Figure 8.8; the curve is the ellipse $x^2/4 + y^2 = 1$.) So the magnitude is

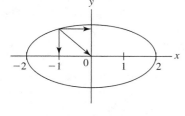

Figure 8.8

$$\sqrt{1^2 + \left(-\frac{\sqrt{3}}{2}\right)^2} \approx 1.32 \text{ m/s}^2$$

while the direction is given by

$$\theta = \text{Arctan}\left(\frac{a_y}{a_x}\right) = \text{Arctan}\left(-\frac{\sqrt{3}}{2}\right) = -40.9° \qquad \blacksquare$$

If the parametric equations

$$\left. \begin{array}{l} x = f(t) \\ y = g(t) \end{array} \right\}$$

describe the position of a particle moving on a curvilinear path, the position may also be expressed in vector form:

$$\mathbf{P}(t) = f(t)\mathbf{i} + g(t)\mathbf{j}$$

It can be seen from Examples 2 and 3 that

$$\mathbf{v} = f'(t)\mathbf{i} + g'(t)\mathbf{j}$$

and

$$\mathbf{a} = f''(t)\mathbf{i} + g''(t)\mathbf{j}$$

According to Example 3 the acceleration vector will not, in general, be tangent to the curve. To see why, consider the special case of a particle moving around a circle. By Newton's second law the acceleration is proportional to the force and is in the same direction as the force. But the centripetal force acts toward the center, which must also be the direction of the acceleration. (See also Exercise 30.)

■ Exercises / Section 8.1

In Exercises 1–10, eliminate the parameter. Use a graphing utility to graph the given equations.

1. $x = 3t$, $y = t + 1$
2. $x = 2t^2$, $y = 1 - 2t$
3. $x = t^2 + 1$, $y = t$
4. $x = -t^2$, $y = t - 1$
5. $x = \cos\theta$, $y = \sin^2\theta$
6. $x = 2 + \sin\theta$, $y = 3 - \cos\theta$
7. $x = 3\tan\theta$, $y = \sec^2\theta$
8. $x = 2 - 2\cos\theta$, $y = 3 + 4\sin\theta$
9. $x = \ln t$, $y = t + 2$
10. $x = e^{-t}$, $y = e^t$

In Exercises 11–20, find the velocity vector **v** and its magnitude and direction at the indicated value of t.

11. $x = t^2, y = 2 - t$ $(t = 2)$

12. $x = 2t + 4, y = 1 - t^2$ $(t = 2)$

13. $x = t^2 - 2, y = \frac{1}{3}t^3 + 2t + 1$ $(t = -1)$

14. $x = 1 - 2\sqrt{t}, y = \frac{1}{4}t^2 - 1$ $(t = 4)$

15. $x = (t - 3)^2, y = 2t$ $(t = 2)$

16. $x = \frac{2}{3}t^{3/2}, y = \frac{1}{3(t - 2)^3}$ $(t = 3)$

17. $x = e^t, y = e^{-t}$ $(t = 0)$

18. $x = 4 \ln t, y = 2\sqrt{5 - t}$ $(t = 4)$

19. $x = \sec t, y = 2 \tan t$ $(t = \pi/6)$

20. $x = \cot t, y = \csc t$ $(t = \pi/3)$

In Exercises 21–25, find the magnitude and direction of the velocity and acceleration vectors at the indicated value of t.

21. $x = 2t^2, y = \frac{1}{3}t^3$ $(t = -1)$

22. $x = \frac{1}{2}t^2 - t, y = \frac{2}{3}t^3$ $(t = -2)$

23. $x = \frac{1}{4}t^4 - 2t^2 + t, y = \frac{1}{3}t^3 - t$ $(t = 0)$

24. $x = 4\sqrt{t}, y = \frac{1}{4}t^2$ $(t = 4)$

25. $x = 4 \cos t, y = 4 \sin t$ $\left(t = \frac{3\pi}{4}\right)$

26. The motion of a particle is given by

$$\left.\begin{array}{l} x = \cos \pi t \\ y = \sqrt{8t - 1} \end{array}\right\}$$

where x and y are in meters and t is in seconds. Find the magnitude and direction of the acceleration vector at $t = 1/4$.

27. If the motion of a particle is described by $x = \tan^2 t$ and $y = \csc t$, find the magnitude and direction of the velocity vector at $t = \pi/3$. (Assume x and y to be in meters and t in seconds.)

28. Show that the velocity vector of a particle moving according to the equations $x = a \cos t + b$ and $y = a \sin t + c$ has a constant magnitude.

29. If a projectile is hurled from the ground with velocity v_0, where v_0 makes an angle θ with the ground, then the horizontal and vertical motions are completely independent of each other, a principle discovered by Galileo.

   a. Show that the motion of the projectile is given by $x = v_0 t \cos \theta$, $y = v_0 t \sin \theta - 5t^2$ (x and y in meters; upward direction positive). Here $v_0 = |\mathbf{v_0}|$.

   b. If $v_0 = 40$ m/s and $\theta = 30°$, find the magnitude and direction of the velocity and acceleration vectors at the end of the first second.

   c. Show by eliminating t that the range R of the projectile is given by $R = (v_0^2/5) \sin \theta \cos \theta$.

30. The motion of a particle is given by

$$\left.\begin{array}{l} x = a \sin t \\ y = b \cos t \end{array}\right\}$$

Show that the path is an ellipse and that the acceleration vector always points toward the center.

8.2   Arc Length

The formula for the *length of an arc* of an arbitrary curve $y = f(x)$ from $x = a$ to $x = b$ (Figure 8.9) involves an integral. Suppose we obtain the formula by an informal argument using the shortcut method. In Figure 8.9, let Δs be the length of the small arc from A to B. If Δx is very small, then Δs will be approximately equal to the chord AB. So by the Pythagorean theorem,

$$(\Delta s)^2 \approx (\Delta x)^2 + (\Delta y)^2 \tag{8.3}$$

Also,

$$\left(\frac{\Delta s}{\Delta x}\right)^2 \approx 1 + \left(\frac{\Delta y}{\Delta x}\right)^2 \qquad \text{and} \qquad \frac{\Delta s}{\Delta x} \approx \sqrt{1 + \left(\frac{\Delta y}{\Delta x}\right)^2}$$

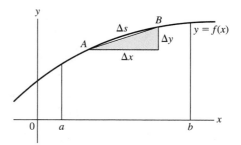

Figure 8.9

As $\Delta x \to 0$, we get

$$\frac{ds}{dx} = \sqrt{1 + \left(\frac{dy}{dx}\right)^2}$$

and, in differential form,

$$ds = \sqrt{1 + \left(\frac{dy}{dx}\right)^2}\, dx$$

Summing from $x = a$ to $x = b$ (Figure 8.9), we get

$$s = \int_a^b \sqrt{1 + \left(\frac{dy}{dx}\right)^2}\, dx$$

which is the formula for the arc length of the curve $y = f(x)$ from $x = a$ to $x = b$.
    Returning to Formula (8.3), we also have

$$\left(\frac{\Delta s}{\Delta t}\right)^2 \approx \left(\frac{\Delta x}{\Delta t}\right)^2 + \left(\frac{\Delta y}{\Delta t}\right)^2$$

whence

$$\frac{ds}{dt} = \sqrt{\left(\frac{dx}{dt}\right)^2 + \left(\frac{dy}{dt}\right)^2} \qquad \text{or} \qquad ds = \sqrt{\left(\frac{dx}{dt}\right)^2 + \left(\frac{dy}{dt}\right)^2}\, dt$$

Summing from $t = t_1$ to $t = t_2$, we have

$$s = \int_{t_1}^{t_2} \sqrt{\left(\frac{dx}{dt}\right)^2 + \left(\frac{dy}{dt}\right)^2}\, dt$$

which is the arc length formula if the curve is represented by parametric equations.

Formula for the Arc Length of $y = f(x)$ from a to b

$$s = \int_a^b \sqrt{1 + \left(\frac{dy}{dx}\right)^2}\, dx \tag{8.4}$$

Formula for the Arc Length of a Curve in Parametric Form from $t = t_1$ to $t = t_2$

$$s = \int_{t_1}^{t_2} \sqrt{\left(\frac{dx}{dt}\right)^2 + \left(\frac{dy}{dt}\right)^2}\, dt \tag{8.5}$$

Example 1 Find the arc length of the curve $y = x^{2/3}$ from $x = 0$ to $x = 8$.

Solution. From

$$y' = \frac{2}{3}x^{-1/3} \qquad \text{we get} \qquad (y')^2 = \frac{4}{9}x^{-2/3}$$

So, by Formula (8.4),

$$s = \int_0^8 \sqrt{1 + \frac{4}{9}x^{-2/3}}\, dx = \int_0^8 \sqrt{1 + \frac{4}{9x^{2/3}}}\, dx = \int_0^8 \sqrt{\frac{9x^{2/3} + 4}{9x^{2/3}}}\, dx$$

$$= \int_0^8 \frac{\sqrt{9x^{2/3} + 4}}{3x^{1/3}}\, dx = \frac{1}{3}\int_0^8 x^{-1/3}\sqrt{9x^{2/3} + 4}\, dx$$

Now let $u = 9x^{2/3} + 4$, so that $du = 6x^{-1/3}\, dx$. It follows that

$$\frac{1}{3}\int x^{-1/3}\sqrt{9x^{2/3} + 4}\, dx = \frac{1}{3} \cdot \frac{1}{6}\int 6x^{-1/3}\sqrt{9x^{2/3} + 4}\, dx$$

$$= \frac{1}{18}\int u^{1/2}\, du = \frac{1}{18} \cdot \frac{2}{3}u^{3/2}$$

$$= \frac{1}{27}(9x^{2/3} + 4)^{3/2}\Big|_0^8$$

$$= \frac{1}{27}[(9 \cdot 8^{2/3} + 4)^{3/2} - (4)^{3/2}]$$

$$= \frac{1}{27}[(40)^{3/2} - (4)^{3/2}]$$

$$= \frac{1}{27}(40\sqrt{40} - 8) \approx 9.07 \qquad \blacksquare$$

Example 2 Find the arc length of the *cycloid*

$$x = a(\theta - \sin\theta) \qquad y = a(1 - \cos\theta)$$

from $\theta = 0$ to $\theta = 2\pi$. (See Figure 8.10.)

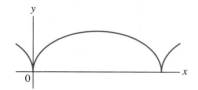

Figure 8.10

Solution. We first note that $y = 0$ when $\theta = 0$ and 2π. Thus one arc of this curve is traced out as θ ranges from 0 to 2π. Since $dx/d\theta = a(1 - \cos\theta)$ and $dy/d\theta = a\sin\theta$, we have, by Formula (8.5),

$$s = \int_0^{2\pi} \sqrt{a^2(1 - \cos\theta)^2 + a^2 \sin^2\theta}\, d\theta = a \int_0^{2\pi} \sqrt{2 - 2\cos\theta}\, d\theta$$

Making use of the half-angle identity

$$\sin\frac{\theta}{2} = \sqrt{\frac{1 - \cos\theta}{2}} \qquad (0 \le \theta \le 2\pi)$$

the last integral can be written

$$s = a \int_0^{2\pi} \sqrt{\frac{2(2 - 2\cos\theta)}{2}}\, d\theta$$

$$= 2a \int_0^{2\pi} \sqrt{\frac{1 - \cos\theta}{2}}\, d\theta$$

$$= 2a \int_0^{2\pi} \sin\frac{\theta}{2}\, d\theta = -4a \cos\frac{\theta}{2}\Big|_0^{2\pi} = 8a \qquad \left[\begin{array}{c} u = \dfrac{\theta}{2} \\[2mm] du = \dfrac{1}{2}d\theta \end{array}\right] \qquad \blacksquare$$

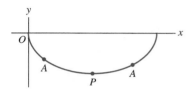

Figure 8.11

Remark: The cycloid is the curve of quickest descent. Suppose P is the lowest point on the inverted cycloid in Figure 8.11. Then a particle acted on by gravity, sliding without friction from O, will reach the lowest point P in less time than it would take the particle to slide from O to P along any other curve. Also, the time taken to reach the bottom of the cycloid is always the same, regardless of the starting point of the particle. In other words, it takes the same length of time to travel from any point A to P as from O to P.

The cycloid was the first transcendental (nonalgebraic) curve whose arc length was found by mathematical means. This was done by Sir Christopher Wren (1632–1723) in 1658. Wren was professor of astronomy at Oxford and later turned to architecture. He was the architect of the magnificent St. Paul's Cathedral in London.

■ Exercises / Section 8.2

In Exercises 1–12, find the arc length of the curves on the intervals indicated.

1. $y = \dfrac{2}{3}x^{3/2}, 0 \le x \le 3$

2. $y = \ln\cos x, 0 \le x \le \dfrac{\pi}{4}$

3. $y = \dfrac{1}{2}(e^x + e^{-x}), 0 \le x \le 1$

4. $y = \dfrac{1}{3}x^3 + \dfrac{1}{4x}, 1 \le x \le 2$

5. $y = \dfrac{1}{6}x^3 + \dfrac{1}{2x}, 1 \le x \le 3$

6. $x = 2\cos\theta, y = 2\sin\theta, \dfrac{\pi}{6} \le \theta \le \dfrac{\pi}{4}$

7. $x = 3t^2, y = t^3, 0 \le t \le \sqrt{5}$

8. $y = \dfrac{x^2}{4} - \dfrac{\ln x}{2}, 1 \le x \le 2$

9. $x = \cos^3\theta, y = \sin^3\theta, 0 \le \theta \le \dfrac{\pi}{2}$

10. $y = \ln x, 2 \le x \le 6$ (Use the table of integrals in Appendix A.)

11. $x = e^t \sin t, y = e^t \cos t, 0 \le t \le 1$

12. $x = 8t^3, y = 4t^2, 0 \le t \le 1$

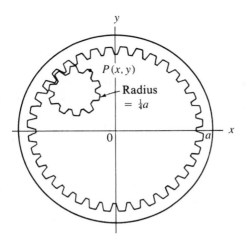

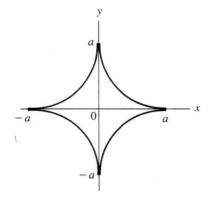

Figure 8.12

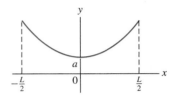

Figure 8.13

14. A perfectly flexible cable suspended from two points at the same height forms a curve called a *catenary*. The equation of a catenary is a hyperbolic cosine function. (See Exercise 31, Section 6.8.) For example, the cable suspended from two points L feet apart (Figure 8.14) is described by the equation

$$y = a \cosh \frac{x}{a}$$

a. Show that $1 + \sinh^2 A = \cosh^2 A$.
b. Use the formula in part (a) to find the length of the cable.

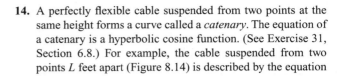

Figure 8.14

Use the integration capability of a graphing utility to find the arc length of the curves on the indicated intervals.

15. $y = 2x^2$, $[0, 3]$

16. $y = \sin x$, $[0, \pi]$

17. $y = \ln x$, $[1, 4]$

18. $y = x^3$, $[2, 3]$

19. $y = e^{-x}$, $[1, 5]$

20. $y = \dfrac{1}{1 + x^2}$, $[-2, 2]$

13. A gear of radius $(1/4)a$ is rolling inside a gear of radius a. (See Figure 8.12.) The point $P(x, y)$ on the inside gear traces out a curve called a *hypocycloid*, whose equation is $x^{2/3} + y^{2/3} = a^{2/3}$ (Figure 8.13). Find its arc length. Note that the equation of the upper half of the curve is $y = (a^{2/3} - x^{2/3})^{3/2}$.

8.3   Polar Coordinates

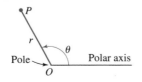

Figure 8.15

So far we have consistently used the rectangular coordinate system. For certain types of curves another coordinate system may be more convenient. Suppose we start with a point O, called the **pole**, and draw a ray with O as the endpoint, called the **polar axis** (Figure 8.15). Using this ray as the initial side, we may generate angles just as we do in trigonometry. Let θ be such an angle and P a point on the terminal side. If r is the distance from P to the pole, then we say that (r, θ) are the **polar coordinates** of P.

> **Polar Coordinates**
>
> Let P be a point in the plane, r the distance from P to a fixed point O (the pole), and θ the angle between the ray OP and a fixed ray (the polar axis). Then P is said to be represented by the polar coordinates (r, θ).

Recall that in trigonometry we always assumed that r is positive. In the polar coordinate system, however, we agree to plot a negative value for r by extending the terminal side of the angle in the opposite direction through the pole and marking off r units on the extended side. For example, the point $(2, 150°)$ is located on the terminal side of the angle $\theta = 150°$, 2 units from the pole, as shown in Figure 8.16. To plot $(-2, 150°)$, we first extend the terminal side through the pole and then locate the point two units from the pole. (See Figure 8.16.) Consider another example.

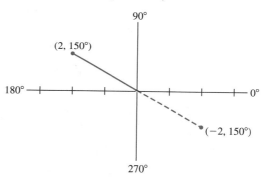

Figure 8.16

Example 1   Plot the points whose polar coordinates are $(2, 30°)$, $(3, -\pi/4)$, $(4, 5\pi/6)$, $(-3, 240°)$, $(-1, 120°)$, $(3, 180°)$, and $(4, 0°)$.

Solution.   To plot these points we need only the polar axis, but for convenience we keep the vertical axis as a line of reference and also extend the polar axis to the left (Figure 8.17).

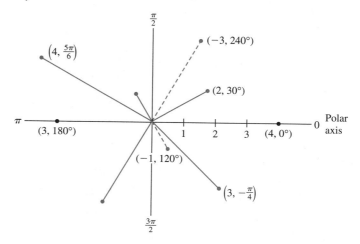

Figure 8.17

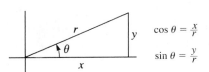

$\cos\theta = \frac{x}{r}$

$\sin\theta = \frac{y}{r}$

Figure 8.18

We can see from Example 1 that the pair of values (r, θ) determines a unique point. The converse is not true, however. The point $(2, 30°)$, for example, can also be represented by $(2, 390°)$, $(2, -330°)$, or $(-2, 210°)$; in fact, there are infinitely many possibilities.

The relationships between the rectangular and polar coordinate systems can be readily seen from the triangle in Figure 8.18.

Conversion Formulas

$$x = r\cos\theta \qquad\qquad y = r\sin\theta \qquad\qquad\qquad (8.6)$$

$$r^2 = x^2 + y^2 \qquad \tan\theta = \frac{y}{x} \qquad\qquad\qquad (8.7)$$

Example 2   Express

    a. $(-2, -\pi/3)$ in rectangular coordinates;

    b. $(-2, 2)$ in polar coordinates.

Solution.

    a. By Formula (8.6),

$$x = r\cos\theta = -2\cos\left(-\frac{\pi}{3}\right) = -2\left(\frac{1}{2}\right) = -1$$

$$y = r\sin\theta = -2\sin\left(-\frac{\pi}{3}\right) = -2\left(-\frac{\sqrt{3}}{2}\right) = \sqrt{3}$$

So the point is $(-1, \sqrt{3})$ in rectangular coordinates.

    b. Since $r^2 = x^2 + y^2$, we obtain at once

$$r = \pm\sqrt{(-2)^2 + 2^2} = \pm\sqrt{4+4} = \pm 2\sqrt{2}$$

and

$$\tan\theta = 2/(-2) = -1$$

Hence $\theta = 135° + n\cdot 360°$ or $315° + n\cdot 360°$. If we choose the positive root for r, then the coordinates are $(2\sqrt{2}, 135°)$ and, more generally, $(2\sqrt{2}, 135° + n\cdot 360°)$. For the negative root we have $(-2\sqrt{2}, 315° + n\cdot 360°)$. ∎

Relationships (8.6) and (8.7) can also be used to convert equations from one system to the other.

Example 3   Write $y^2 = x$ in polar coordinates.

Solution. By direct substitution,

$$(r\sin\theta)^2 = r\cos\theta \qquad\qquad y = r\sin\theta, \; x = r\cos\theta$$

$$r^2\sin^2\theta - r\cos\theta = 0$$

and

$$r(r \sin^2 \theta - \cos \theta) = 0 \qquad \textbf{factoring } r$$

from which it follows that $r = 0$ or

$$r \sin^2 \theta - \cos \theta = 0$$

$$r = \frac{\cos \theta}{\sin^2 \theta} = \frac{\cos \theta}{\sin \theta} \frac{1}{\sin \theta} \qquad \textbf{solving for } r$$

or

$$r = \cot \theta \csc \theta$$

Notice that $r = 0$ may be dropped since it represents only the pole, which is included in the equation $r = \cot \theta \csc \theta$. (For example, $r = 0$ when $\theta = 90°$.)   ■

Example 4   Write the equation $r = 2 \cos 2\theta$ in rectangular form.

Solution. By the double-angle identity,

$$r = 2(\cos^2 \theta - \sin^2 \theta) \qquad \textbf{cos } 2\theta = \textbf{cos}^2 \, \theta - \textbf{sin}^2 \, \theta$$

Since $\cos \theta = x/r$ and $\sin \theta = y/r$ by (8.6), we get

$$\pm\sqrt{x^2 + y^2} = 2 \left(\frac{x^2}{r^2} - \frac{y^2}{r^2} \right)$$

$$\pm\sqrt{x^2 + y^2} = 2 \left(\frac{x^2}{x^2 + y^2} - \frac{y^2}{x^2 + y^2} \right) = 2 \frac{x^2 - y^2}{x^2 + y^2} \qquad r^2 = x^2 + y^2$$

and, squaring both sides,

$$x^2 + y^2 = 4 \frac{(x^2 - y^2)^2}{(x^2 + y^2)^2}$$

or

$$(x^2 + y^2)^3 = 4(x^2 - y^2)^2 \qquad \textbf{multiplying by } (x^2 + y^2)^2$$

A simple alternative is to multiply both sides of the equation $r = 2(\cos^2 \theta - \sin^2 \theta)$ by r^2 to get

$$r^2 r = r^2 \cdot 2(\cos^2 \theta - \sin^2 \theta)$$

or

$$r^3 = 2(r^2 \cos^2 \theta - r^2 \sin^2 \theta)$$

In this form the equation converts directly to

$$\pm(x^2 + y^2)^{3/2} = 2(x^2 - y^2) \qquad \begin{array}{l} \textbf{r cos } \theta = \textbf{x, r sin } \theta = \textbf{y} \\ \textbf{r} = \pm\sqrt{x^2 + y^2} \end{array}$$

or

$$(x^2 + y^2)^3 = 4(x^2 - y^2)^2 \qquad \textbf{squaring both sides}$$ ■

Remark: The alternative method in Example 4 works particularly well with simple equations. For example, if $r = 4 \sin \theta$, then

$$r \cdot r = 4r \sin \theta$$

From $r^2 = 4r \sin \theta$, we get $x^2 + y^2 = 4y$.

Example 5   Convert the equation $r = 3 \cot \theta$ to rectangular form.

Solution. Since

$$r = 3 \cot \theta = 3 \frac{\cos \theta}{\sin \theta}$$

there is an easy way to insert r:

$$r = 3 \frac{r \cos \theta}{r \sin \theta} \qquad \textbf{multiplying numerator and denominator by } r$$

This equation converts directly to

$$\pm\sqrt{x^2 + y^2} = 3\frac{x}{y} \qquad r \cos \theta = x, r \sin \theta = y$$

or

$$x^2 + y^2 = \frac{9x^2}{y^2} \qquad \textbf{squaring both sides}$$

■ Exercises / Section 8.3

1. Plot the following points: $(1, 60°)$, $(3, 3\pi/4)$, $(2, -50°)$, $(-2, 90°)$, $(-1, -\pi/4)$, $(-2, 270°)$, $(-4, 0°)$.

In Exercises 2–7, express the points in rectangular coordinates.

2. $\left(\sqrt{2}, \frac{\pi}{4}\right)$ **3.** $(2, 120°)$

4. $\left(4, -\frac{\pi}{6}\right)$ **5.** $\left(-6, \frac{\pi}{3}\right)$

6. $(1, 50°)$ **7.** $(-2, 170°)$

In Exercises 8–12, find a set of polar coordinates for each of the following points in rectangular coordinates.

8. $(\sqrt{3}, 1)$ **9.** $(1, -1)$

10. $(-2, -2\sqrt{3})$ **11.** $(3, 4)$

12. $(-2, 5)$

In Exercises 13–24, express the equations in polar coordinates.

13. $x = 2$ **14.** $y = 4$

15. $x^2 + y^2 = 2$ **16.** $x^2 + y^2 + 3x - 2y = 0$

17. $2x - 4y = 5$ **18.** $xy = 4$

19. $x^2 - 2x + y = 2$ **20.** $x + y^2 - 1 = 0$

21. $x^2 - y^2 = 1$ **22.** $y = 2x^2$

23. $2x^2 + 4y^2 = 1$ **24.** $y = x^3$

In Exercises 25–49, express the equations in rectangular coordinates.

25. $r \sin \theta = 2$ **26.** $r = 3 \sec \theta$

27. $\theta = \frac{\pi}{4}$ **28.** $r = 5$

29. $r = \cos \theta$ **30.** $r = \sin \theta$

31. $r = 1 + \cos \theta$ **32.** $r = 1 - \sin \theta$

33. $r = 1 - \cos \theta$ **34.** $r = 2 + \cos \theta$

35. $r = 1 - 2 \sin \theta$ **36.** $r = 3 - 2 \cos \theta$

37. $r = -2 + 4 \sin \theta$

38. $r = 3 \sin 2\theta$ (Recall that $\sin 2\theta = 2 \sin \theta \cos \theta$.)

39. $r^2 = \sin 2\theta$ **40.** $r = \tan \theta$

41. $r = \dfrac{2}{1 - \cos\theta}$

42. $r^2 = \cot\theta$

43. $r = \dfrac{4}{1 - \sin\theta}$

44. $r^2 \cos 2\theta = 4$

45. $r^2 = 3\cos 2\theta$

46. $r = a\cos 3\theta$ (Use $\cos 3\theta = \cos(2\theta + \theta)$.)

47. $r = a\sin 3\theta$

48. $r = 4\sec\theta\tan\theta$

49. $r = 1 - 2\sin\theta$

8.4   Curves in Polar Coordinates

In this section we are going to study curves in polar coordinates. Without prior experience we are essentially reduced to the point-plotting method. We will soon find out, however, that a number of polar curves can be classified according to type. Once the type is identified, drawing the curve becomes relatively easy.

Example 1   Draw the graph of $r = 4\sin\theta$.

Solution. First we make a table of values. Since we are merely plotting points, let us use degree measure:

θ:	0°	30°	45°	60°	90°	120°	135°	150°	180°	225°	270°	315°
r:	0	2	$2\sqrt{2}$	$2\sqrt{3}$	4	$2\sqrt{3}$	$2\sqrt{2}$	2	0	$-2\sqrt{2}$	-4	$-2\sqrt{2}$

Since $r = 4\sin\theta$ is a periodic function with period 2π, we would expect to trace out the entire curve in the interval $\theta = 0°$ to $\theta = 360°$, and we do. There is a surprise, however: when θ passes 180°, the r-values become negative, and the same curve is being traced out again. For example, $(2\sqrt{2}, 45°)$ represents the same point as $(-2\sqrt{2}, 225°)$. So the entire curve is actually traced out in the interval $\theta = 0°$ to $\theta = 180°$. The graph is shown in Figure 8.19.

(A convenient way to graph a polar equation by hand is to use polar graph paper.)

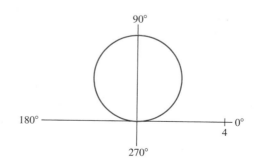

Figure 8.19

Example 2   Draw the graph of $r = 4(1 + \sin\theta)$.

Solution. Once again making a table of values:

θ:	0°	30°	45°	60°	90°	120°	135°	150°	180°	225°	270°	315°
r:	4	6	6.8	7.5	8	7.5	6.8	6	4	1.2	0	1.2

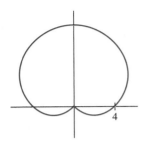

Figure 8.20

In comparing $r = 4(1 + \sin\theta)$ to the curve $r = 4\sin\theta$ in Example 1, observe that the quantity $1 + \sin\theta$ can never become negative. This results in a very different behavior: As θ ranges from 0° to 90°, r ranges from 4 to 8; from 90° to 180° the values of r repeat in inverse order. Continuing from 180° to 270°, we see that $\sin\theta$ ranges from 0 to -1; hence r decreases from 4 to 0. Finally, as θ ranges from 270° to 360°, r increases from 0 to 4. Once we have a complete circuit, the values of r repeat and it becomes unnecessary to find additional points. The graph is shown in Figure 8.20. ■

It is easy to check that the curve $r = 4(1 + \cos\theta)$ is the curve in Example 2 rotated by 90° in the clockwise direction.

Figure 8.20 is an example of a curve called a *cardioid* (meaning "heart-shaped").

Cardioid

A cardioid is a curve whose general equation is

$$r = a(1 \pm \sin\theta) \qquad \text{or} \qquad r = a(1 \pm \cos\theta) \qquad (8.8)$$

The coefficient a may be either positive or negative.

Example 3   Sketch the graph of $r = 5 + 3\sin\theta$.

Solution. Because of the similarity to Example 2, we locate only the four points analogous to the intercepts: (5, 0°), (8, 90°), (5, 180°), and (2, 270°). The graph is shown in Figure 8.21. Note that at 270°, r is still positive, so that the curve (unlike the cardioid) does not pass through the pole.

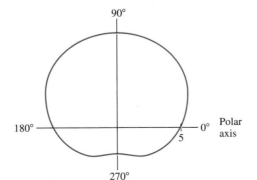

θ:	0°	90°	180°	270°
r:	5	8	5	2

Figure 8.21

■

The curve in Example 3 is called a *limaçon of Pascal.*

Limaçon

A limaçon is a curve whose general equation is

$$r = a + b\sin\theta \quad \text{or} \quad r = a + b\cos\theta \tag{8.9}$$

If $|a| > |b|$ and $a > 0$, r is always positive (as in Example 3). If $|a| = |b|$, then $r = 0$ for one value of θ between $0°$ and $360°$, so that the curve is actually a cardioid. If $|a| < |b|$ and $a > 0$, r is negative for some values of θ, and we obtain a "limaçon with a loop." The three cases are shown in Figure 8.22. (If a or b are negative, the conclusion is the same in the sense that the limaçon has a loop if, and only if, $|a| < |b|$.)

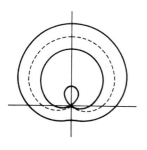

Figure 8.22

Example 4 Sketch the limaçon $r = 1 - 2\cos\theta$.

Solution. Since $|a| < |b|$, this limaçon has a loop. To find the values of θ tracing out the loop, we must first see where the loop starts and ends; that is, we need to find the values of θ for which the curve passes through the pole. To this end we let $r = 0$ to get

$$1 - 2\cos\theta = 0 \quad \text{and} \quad \cos\theta = \frac{1}{2}$$

so that $\theta = \pm 60°$. If $-60° < \theta < 60°$, then $r < 0$. To sketch the rest of the graph, we find the four "intercepts": $(-1, 0°)$, $(1, 90°)$, $(3, 180°)$, and $(1, 270°)$. The graph is shown in Figure 8.23.

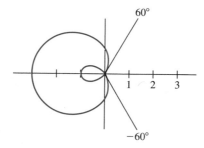

θ:	$0°$	$90°$	$180°$	$270°$
r:	-1	1	3	1

Figure 8.23

∎

Example 5 Sketch the graph of $r = 2\sin 3\theta$.

Solution. Instead of plotting points, we will discuss the general behavior of this function: r is zero whenever 3θ is coterminal with $0°$ or $180°$. This will occur whenever $\theta = 0°$, $60°$, $120°$, $180°$, $240°$, and $300°$. Numerically, r attains its maximum value of 2 whenever $\theta = 30°$, $90°$, $150°$, $210°$, $270°$, and $330°$. We note especially that when θ ranges from $0°$ to $30°$, 3θ ranges from $0°$ to $90°$, so that $\sin 3\theta$ ranges

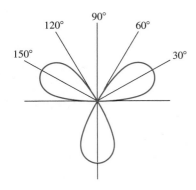

Figure 8.24

from 0 to 1; at $\theta = 60°$, r is back to zero. As θ ranges from 60° to 120°, the values of r are negative. The graph is shown in Figure 8.24. ∎

The curve in Figure 8.24 is an example of a *rose*.

Rose

A rose is a curve whose general equation is

$$r = a \sin n\theta \qquad \text{or} \qquad r = a \cos n\theta \qquad\qquad (8.10)$$

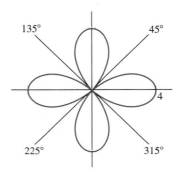

Figure 8.25

Number of leaves

The coefficient n gives the **number of "leaves":** the rose has **n** leaves if **n is odd** and **$2n$** leaves if **n is even.** For example, $r = 4 \cos 2\theta$ is a four-leaf rose (Figure 8.25).

Another type of curve, called a *lemniscate of Bernoulli,* has the form given next.

Lemniscate

A lemniscate is a curve whose general equation is

$$r^2 = a \sin 2\theta \qquad \text{or} \qquad r^2 = a \cos 2\theta \qquad\qquad (8.11)$$

Example 6   Sketch the lemniscate $r^2 = 4 \cos 2\theta$.

Solution. A brief table of values is

θ	0°	10°	20°	30°	45°
r	±2	±1.9	±1.75	±1.4	0

We get the same values for r if θ starts at $0°$ and goes to $-45°$. In the range $\theta = 45°$ to $\theta = 135°$, $\cos 2\theta < 0$, so that the values of r become imaginary. The graph is shown in Figure 8.26. (A lemniscate always has two "leaves".) ■

Circle
Straight line
Spiral of Archimedes

Certain special cases may also occur. The curve $r = a$ is a **circle** centered at the pole, while $\theta = b$ is a **straight line** through the pole. The curve $r = a\theta$ is called a **spiral of Archimedes.** Other spirals will be taken up in the exercises.

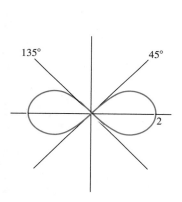

Figure 8.26

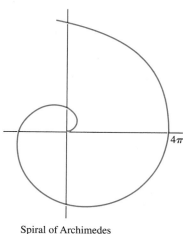

Spiral of Archimedes

Figure 8.27

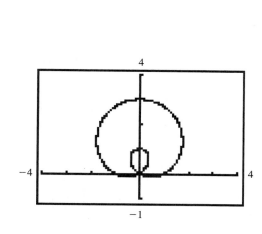

Figure 8.28

Example 7   Draw the spiral $r = 2\theta$, $\theta \geq 0$.

Solution. In this case the values of θ must be in radian measure. Since r is directly proportional to θ, the points recede farther and farther from the pole (Figure 8.27). ■

Graphing utilities having a parametric but no polar mode will handle polar curves by using the parametric mode in conjunction with Formula (8.6), where $r = f(\theta)$:

$$\left.\begin{array}{l} x = f(\theta)\cos\theta \\ y = f(\theta)\sin\theta \end{array}\right\}$$

For example, the graph of the limaçon $r = 1 + 2\sin\theta$ in Figure 8.28 results from

$$\left.\begin{array}{l} x = (1 + 2\sin\theta)\cos\theta \\ y = (1 + 2\sin\theta)\sin\theta \end{array}\right\}$$

■ Exercises / Section 8.4

Sketch the following curves. Whenever possible, identify the curve before attempting the sketch.

1. $r = 2$

2. $\theta = \dfrac{\pi}{3}$

3. $r = 2\cos\theta$

4. $r = 3\sin\theta$

5. $r\cos\theta = 4$

6. $r = 2\csc\theta$

7. $r = 2(1 + \cos\theta)$

8. $r = -4(1 + \sin\theta)$

9. $r = 9 + 5\cos\theta$

10. $r = 4 - 3\sin\theta$

11. $r = 3 - 2\cos\theta$

12. $r = -1 + \cos\theta$

13. $r = 1 - 2\sin\theta$

14. $r = 1 + 2\cos\theta$

15. $r = -1 - 2\sin\theta$

16. $r = 2\cos 3\theta$

17. $r = 4\sin 2\theta$

18. $r = 3\sin 5\theta$

19. $r = \cos 3\theta$

20. $r^2 = 9\cos 2\theta$

21. $r^2 = 4\sin 2\theta$

22. $r = \sqrt{\sin\theta}$

23. $r = 3\sqrt{\sin 2\theta},\ 0° \le \theta \le 90°$

24. $r = 4\cos 2\theta$

25. $r = \cos 3\left(\theta - \dfrac{\pi}{6}\right)$

26. $r^2 = 2\sin 2\theta$

27. $r = \sin 3\theta$

28. $r = \theta$ (spiral of Archimedes)

29. $r\theta = 2,\ \theta > 0$ (hyperbolic spiral)

30. $r = e^{\theta/3},\ \theta \ge 0$ (logarithmic spiral)

31. $r = \dfrac{3}{2 - \cos\theta}$ (ellipse)

32. $r = \dfrac{2}{1 - \cos\theta}$ (parabola)

33. $r = \dfrac{6}{2 + 3\sin\theta}$ (hyperbola)

In the remaining exercises, use your graphing utility to graph the given curves in polar coordinates.

34. $r = 1 + 4\cos 2\theta$

35. $r = 6\sin^2\theta\cos\theta$

36. $r = 3\sec\theta + 4$

37. $r = \tan^2\theta$

38. $r = 2\sin\left(\dfrac{2\theta}{3}\right)$

39. $r = 2\cos\left(\dfrac{3\theta}{2}\right)$

40. $r = 10^{\cos\theta}$

41. $r = 1 + \cos 2\theta$

42. $r = 1 + \sin 4\theta$

43. $r = 1 + 4\sin 3\theta$

44. $r = 1 + 4\cos 4\theta$

8.5 Areas in Polar Coordinates

The area of a region bounded by curves in polar coordinates is found by integration. To obtain the formula, we will return to the method of Riemann sums, which is preferable to the shortcut in this case. Consider the region bounded by $r = f(\theta)$ and rays $\theta = \alpha$ and $\theta = \beta$ in Figure 8.29. Since rectangular partitions are unsuitable in the polar coordinate system, we subdivide the region into n circular sectors as follows. Let

$$\alpha = \theta_0' < \theta_1' < \cdots < \theta_i' < \cdots < \theta_n' = \beta$$

and let θ_i be any ray between θ_i' and θ_{i-1}' $(i = 1, 2, \ldots, n)$. Each θ_i determines a circular sector of radius $r_i = f(\theta_i)$ and central angle

$$\Delta\theta_i = \theta_i' - \theta_{i-1}' \qquad (i = 1, 2, \ldots, n)$$

(See Figure 8.29.) Recall from Section 6.1, Formula (6.2), that the area of a circular sector of radius r and central angle θ is given by $(1/2)r^2\theta$. The area of the ith sector is therefore

$$\frac{1}{2}r_i^2\Delta\theta_i = \frac{1}{2}[f(\theta_i)]^2\,\Delta\theta_i \qquad r_i = f(\theta_i)$$

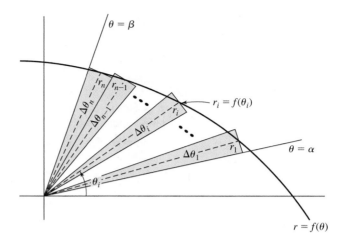

Figure 8.29

The total area A then becomes approximately

$$\frac{1}{2} \sum_{i=1}^{n} [f(\theta_i)]^2 \, \Delta\theta_i$$

which has the form of a Riemann sum. By the definition of the definite integral, we now get the formula for the area.

Area in Polar Coordinates

$$A = \lim_{n \to \infty} \frac{1}{2} \sum_{i=1}^{n} [f(\theta_i)]^2 \Delta\theta_i$$

$$= \frac{1}{2} \int_{\alpha}^{\beta} [f(\theta)]^2 \, d\theta = \frac{1}{2} \int_{\alpha}^{\beta} r^2 d\theta \qquad (8.12)$$

(As always, it is understood that upon taking the limit, all the subdivisions tend to zero.)

Since differentiation and integration formulas are valid only for angles in radians, we need to use radian measure for the limits of integration.

Example 1   Find the area of the region bounded by the spiral $r = \theta$ ($\theta \geq 0$) and the rays $\theta = 0$ and $\theta = \pi/2$.

Solution. By Formula (8.12),

$$A = \frac{1}{2} \int_{0}^{\pi/2} \theta^2 d\theta = \frac{\pi^3}{48} \qquad \blacksquare$$

The real difficulty in calculating areas of given regions is to determine the limits of integration since, in most problems, these are not explicitly given.

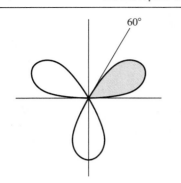

Figure 8.30

Example 2   Find the area of one leaf of the rose $r = 4 \sin 3\theta$.

Solution. From the general form we see that the equation represents a three-leaf rose. The first leaf is traced out starting at $\theta = 0°$. It reaches its maximum value at $\theta = 30°$, since $\sin 3 \cdot 30° = \sin 90° = 1$, and gets back to the pole at $\theta = 60°$ (since $\sin 3 \cdot 60° = \sin 180° = 0$). We conclude that the leaf is traced out in the interval $0° \leq \theta \leq 60°$, so that the limits of integration are $\theta = 0$ and $\theta = \pi/3$. (The limits of integration, incidentally, also give us the curve in Figure 8.30, since the leaves are all equally spaced.) By Formula (8.12),

$$A = \frac{1}{2} \int_0^{\pi/3} (4 \sin 3\theta)^2 \, d\theta = \frac{1}{2} \int_0^{\pi/3} 16 \sin^2 3\theta \, d\theta$$

$$= 8 \int_0^{\pi/3} \sin^2 3\theta \, d\theta$$

Recall from Section 7.4 that the even power on the sine may be reduced by the half-angle identity $\sin^2 x = (1/2)(1 - \cos 2x)$. Thus

$$\sin^2 3\theta = \frac{1 - \cos 6\theta}{2}$$

and

$$A = 8 \int_0^{\pi/3} \frac{1 - \cos 6\theta}{2} \, d\theta$$

$$= 4 \int_0^{\pi/3} (1 - \cos 6\theta) \, d\theta$$

$$= 4 \left[\int_0^{\pi/3} d\theta - \int_0^{\pi/3} \cos 6\theta \, d\theta \right]$$

$$= 4 \left[\int_0^{\pi/3} d\theta - \frac{1}{6} \int_0^{\pi/3} \cos 6\theta (6 \, d\theta) \right] \qquad \begin{bmatrix} u = 6\theta \\ du = 6 \, d\theta \end{bmatrix}$$

$$= 4 \left(\theta - \frac{1}{6} \sin 6\theta \right) \Big|_0^{\pi/3} = \frac{4\pi}{3} \qquad \blacksquare$$

Example 3   Find the area enclosed by the lemniscate $r^2 = 4 \cos 2\theta$.

Solution. The graph has already been obtained in Example 6 of Section 8.4 (Figure 8.31). Suppose we take advantage of the symmetry by finding the area of the shaded region and multiplying by 4. Since the given equation is not a function, we solve for r to obtain

$$r = \pm 2\sqrt{\cos 2\theta}$$

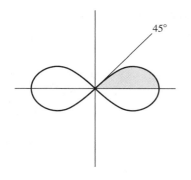

Figure 8.31

The positive branch $r = 2\sqrt{\cos 2\theta}$ traces out only the loop on the right (for $-45° \le \theta \le 45°$). Observe that the portion of the curve enclosing the shaded region begins at $\theta = 0°$ and reaches the pole at $\theta = 45°$ (since $\cos 2 \cdot 45° = \cos 90° = 0$). Hence the limits of integration are $\theta = 0$ and $\theta = \pi/4$:

$$\frac{1}{4}A = \frac{1}{2}\int_0^{\pi/4}\left(2\sqrt{\cos 2\theta}\right)^2 d\theta$$

$$= \int_0^{\pi/4} 2\cos 2\theta\, d\theta = \int_0^{\pi/4}(\cos 2\theta)\cdot 2\, d\theta \qquad \begin{bmatrix} u = 2\theta \\ du = 2\,d\theta \end{bmatrix}$$

$$= \sin 2\theta\Big|_0^{\pi/4} = 1$$

Consequently, the total area is 4 square units. ■

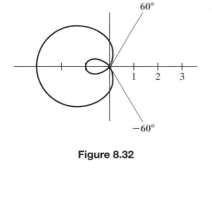

Figure 8.32

Example 4   Find the area of the inner loop of the limaçon $r = 1 - 2\cos\theta$ shown in Figure 8.32.

Solution.   To obtain the limits of integration, we let $r = 0$ and solve for θ:

$$1 - 2\cos\theta = 0$$

$$\cos\theta = \frac{1}{2}$$

$$\theta = \pm 60°$$

So the loop begins at $\theta = -\pi/3$ and ends at $\pi/3$. We can also find the area by integrating from $\theta = 0$ to $\theta = \pi/3$ and doubling the result:

$$A = 2\cdot\frac{1}{2}\int_0^{\pi/3}(1 - 2\cos\theta)^2 d\theta$$

$$= \int_0^{\pi/3}(1 - 4\cos\theta + 4\cos^2\theta)\, d\theta$$

$$= \int_0^{\pi/3}\left(1 - 4\cos\theta + 4\frac{1 + \cos 2\theta}{2}\right) d\theta \qquad \cos^2\theta = \frac{1}{2}(1 + \cos 2\theta)$$

$$= \int_0^{\pi/3}(1 - 4\cos\theta + 2 + 2\cos 2\theta)\, d\theta$$

$$= \int_0^{\pi/3}(3 - 4\cos\theta + 2\cos 2\theta)\, d\theta$$

$$= 3\theta - 4\sin\theta + \sin 2\theta\Big|_0^{\pi/3} \qquad \text{(See Example 3.)}$$

$$= \left[\left(3\cdot\frac{\pi}{3} - 4\sin\frac{\pi}{3} + \sin 2\left(\frac{\pi}{3}\right)\right)\right] - 0$$

$$= \pi - 4\left(\frac{\sqrt{3}}{2}\right) + \frac{\sqrt{3}}{2} = \pi - \frac{3\sqrt{3}}{2} \approx 0.54$$ ■

For the cardioid and the limaçon without a loop, we need to make a complete circuit from $0°$ to $360°$ to trace out the whole curve. For example, the integral for finding the area enclosed by the limaçon $r = 3 + 2\cos\theta$ is

$$\frac{1}{2}\int_0^{2\pi}(3 + 2\cos\theta)^2 d\theta$$

Area of a Region Bounded by Two Curves

In the next example the region is bounded by two curves.

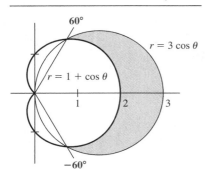

Figure 8.33

Example 5   Find the area of the region outside the cardioid $r = 1 + \cos\theta$ and inside the circle $r = 3\cos\theta$.

Solution. The graphs are shown in Figure 8.33. The desired area is the shaded portion. To get the limits of integration, we find the θ values of the points of intersection of the curves. To do so, we must eliminate r between the given equations. Substituting $r = 3\cos\theta$ in $r = 1 + \cos\theta$, we have

$$3\cos\theta = 1 + \cos\theta \qquad \text{or} \qquad \cos\theta = \frac{1}{2}$$

so that $\theta = \pm 60°$. Hence the shaded region lies between the rays $\theta = -\pi/3$ and $\theta = \pi/3$.

The region inside the circle between the rays $\theta = -\pi/3$ and $\theta = \pi/3$ is given by

$$\frac{1}{2}\int_{-\pi/3}^{\pi/3} 9\cos^2\theta\, d\theta$$

and the region inside the cardioid between $\theta = -\pi/3$ and $\theta = \pi/3$ by

$$\frac{1}{2}\int_{-\pi/3}^{\pi/3} (1 + \cos\theta)^2\, d\theta$$

To find the shaded area we subtract the smaller region from the larger. To simplify the calculation, we integrate from 0 to $\pi/3$ and double the result. Thus

$$A = 2\cdot\frac{1}{2}\left[\int_0^{\pi/3} 9\cos^2\theta\, d\theta - \int_0^{\pi/3}(1 + 2\cos\theta + \cos^2\theta)\, d\theta\right]$$

$$= \int_0^{\pi/3}(8\cos^2\theta - 2\cos\theta - 1)\, d\theta$$

$$= \int_0^{\pi/3}\left[\frac{8}{2}(1 + \cos 2\theta) - 2\cos\theta - 1\right] d\theta$$

$$= 4\left(\theta + \frac{1}{2}\sin 2\theta\right) - 2\sin\theta - \theta\,\Big|_0^{\pi/3} = \pi \qquad \blacksquare$$

■ Exercises / Section 8.5

In Exercises 1–30, find each area bounded by the given curves.

1. $r = 2, \theta = 0, \theta = \dfrac{\pi}{6}$

2. $r = 2, \theta = \dfrac{\pi}{6}, \theta = \dfrac{\pi}{3}$

3. $r = \sec\theta, \theta = 0, \theta = \dfrac{\pi}{4}$

4. $r = \csc\theta, \theta = \dfrac{\pi}{4}, \theta = \dfrac{\pi}{2}$

5. $r = \theta\ (\theta \geq 0), \theta = 0, \theta = 2$

6. $r = \tan\theta, \theta = 0, \theta = \dfrac{\pi}{4}$

7. $r = 2\sin\theta$

8. $r = \cos\theta$

9. $r = 4\cos 3\theta$

10. $r = 2\cos 2\theta$

11. $r = 3\sin 2\theta$

12. $r = 2\sin 3\theta$

13. $r = 6 \sin 3\theta$

14. $r = 2 \cos 3\theta$

15. $r = \sqrt{\sin \theta}$

16. $r = \sqrt{\cos \theta}$

17. $r = 2 + \cos \theta$

18. $r = 3 - \sin \theta$

19. $r = 2(1 - \cos \theta)$

20. $r = 4 + 4 \sin \theta$

21. $r = 2 - \sin \theta$

22. $r = a(1 - \cos \theta)$

23. $r = 3 + 3 \cos \theta$

24. $r = 1 + \sin \theta$

25. $r^2 = 4 \sin 2\theta$

26. $r^2 = 9 \cos 2\theta$

27. $r^2 = 2 \cos 2\theta$

28. $r^2 = 2 \sin 2\theta$

29. $r^2 = 6 \sin 2\theta$

30. $r^2 = 7 \cos 2\theta$

In Exercises 31–42, find the area of each region.

31. The inner loop of $r = 1 - 2 \sin \theta$

32. The inner loop of $r = 1 + 2 \cos \theta$

33. Inside $r = 2 + \cos \theta$, outside $r = 2$

34. Inside $r = 1 + \sin \theta$, outside $r = 1$

35. Inside $r = 4 \cos \theta$, outside $r = 2$

36. Inside $r^2 = 8 \cos 2\theta$, outside $r = 2$

37. Inside $r = \sin \theta$, outside $r = 1 + \cos \theta$

38. Inside $r = 5 \sin \theta$, outside $r = 2 + \sin \theta$

39. Inside $r = 4(1 + \sin \theta)$, outside $r = 8 \sin \theta$

40. The region common to $r = \cos \theta$ and $r = \sin \theta$

41. The region common to $r = 3 \cos \theta$ and $r = 1 + \cos \theta$

42. The region common to $r = 1 - \sin \theta$ and $r = 1$

■ Review Exercises / Chapter 8

1. Eliminate the parameter: $x = 4 \cos^2 \theta$, $y = 4 \sin \theta$.

2. Eliminate the parameter: $x = 1 - \tan \theta$, $y = 2 \sec \theta$.

3. A particle is traveling according to the equations
   a. $x = 5 \cos t$, $y = 5 \sin t$
   b. $x = 5 \sin t$, $y = 5 \cos t$

   (x and y are measured in meters and t in seconds). Find the velocity and acceleration vectors at $t = \pi/4$.

4. a. Find the acceleration vector of the particle whose motion is described by $x = 1 + 2 \cos t$ and $y = 2 + \sin t$ at $t = \pi/3$; x and y are in meters and t is in seconds.
   b. Show that the path is an ellipse centered at (1, 2) and that the acceleration vector points toward the center.

5. Find the arc length of the curve $y = \ln \sec x$ from $x = 0$ to $x = \pi/4$.

6. Find the arc length of the curve whose parametric equations are $x = t^3$, $y = t^2$ from $t = 0$ to $t = 1$.

7. Express the equation $y^2 = 3x$ in polar coordinates.

In Exercises 8–10, express the equations in rectangular coordinates.

8. $r = 4 \csc \theta$

9. $r^2 = 4 \tan \theta$

10. $r = 2 - 3 \cos \theta$

In Exercises 11–16, identify the curves and sketch them.

11. $r = \sin \theta$

12. $r = 3(1 - \cos \theta)$

13. $r = 2 \cos 2\theta$

14. $r = 2 - 5 \sin \theta$

15. $r^2 = 9 \sin 2\theta$

16. $r = \sin 3\theta$

In Exercises 17–24, find the areas bounded by the given curves.

17. $r = \tan 2\theta$, $\theta = 0$, $\theta = \dfrac{\pi}{8}$

18. Inside the cardioid $r = 2(1 - \sin \theta)$

19. Inside the lemniscate $r^2 = a^2 \cos 2\theta$

20. Inside the limaçon $r = 3 + \cos \theta$

21. The inner loop of $r = 1 - 2 \cos \theta$

22. Inside the cardioid $r = 2 + 2 \sin \theta$, outside $r = 3$

23. Inside $r^2 = 8 \cos 2\theta$, outside $r = 2$

24. One leaf of the rose $r = \sin 2\theta$

25. If r is replaced by $f(\theta)$ in the transformation formulas, we obtain $x = f(\theta) \cos \theta$ and $y = f(\theta) \sin \theta$. By treating these formulas as a set of parametric equations, show that the slope of a line tangent to a curve in polar coordinates is

$$\frac{dy}{dx} = \frac{f'(\theta) \sin \theta + f(\theta) \cos \theta}{f'(\theta) \cos \theta - f(\theta) \sin \theta}$$

26. Find the slope of the tangent line to the curve $r = 2(1 + \sin \theta)$ at $(2, \pi)$. (Refer to Exercise 25.)

27. Find the slope of the line tangent to the spiral $r = 2\theta$ at $(\pi, \pi/2)$.

28. Show that the cardioid $r = 1 + \sin \theta$ has a vertical tangent line at the pole. (*Hint:* Use L'Hospital's rule.)

9 Three-Dimensional Space; Partial Derivatives; Multiple Integrals

Surfaces in Three Dimensions

Up to now all graphs discussed have been located in a plane. We are now going to extend the idea of a graph to three dimensions; such graphs require three coordinate axes.

To locate a point in three dimensions, we give the direction and distance from three mutually perpendicular planes. The intersections of these planes define the x-axis, y-axis, and z-axis, as shown in Figure 9.1. Each point P has three coordinates, denoted by the ordered triple (x, y, z), where x, y, and z are the distances (with proper signs) from the respective planes. These **coordinate planes,** called the xy-, yz-, and xz-planes, divide space into eight *octants;* the one in which all the coordinates are positive, called the **first octant,** is shown in Figure 9.1. (The other octants are not usually numbered.)

Now recall from Chapter 1 that the slope of the line $x = a$ is undefined, so that the equation does not fit the point-slope form $y - y_1 = m(x - x_1)$. However, we may argue instead that the set of points having a fixed abscissa $x = a$ is necessarily a vertical line through $(a, 0)$. An analogous statement can be made in three-space: the set of points having a fixed x-coordinate a is a plane through $(a, 0, 0)$ parallel to the yz-plane (Figure 9.2).

A similar argument applies to the set of points satisfying the condition $z = y^2$, which is independent of x. Suppose we first sketch the curve $z = y^2$ in the yz-plane. Now, since the coordinates of $(0, 1, 1)$ satisfy the condition, the point lies on the surface and, consequently, so does any point of the form $(a, 1, 1)$. (See Figure 9.3.) More generally, if $(0, y_1, z_1)$ lies on the surface, then so does (a, y_1, z_1) since $z = y^2$ is independent of x. Figure 9.3 is an example of a **cylindrical surface,** defined to be a surface generated by a moving line that remains parallel to its original position and passes through a plane curve. For example, in Figure 9.3 the line l could serve as the moving line and the parabola $z = y^2$ in the yz-plane as the curve. Moreover, *any equation having only two variables in three-space represents a cylindrical surface.*

Coordinate planes

First octant

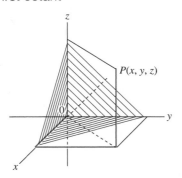

Figure 9.1

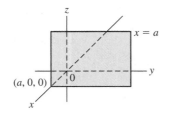

Figure 9.2

Cylindrical Surface

A cylindrical surface can be represented by an equation in two variables. (The moving line is parallel to the axis of the missing variable.)

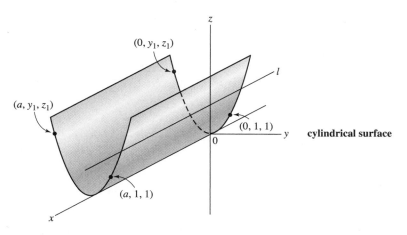

cylindrical surface

Figure 9.3

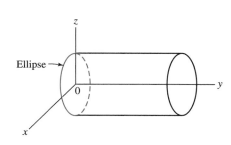

Figure 9.4

For example, the cylinder $2x^2 + z^2 = 2$ is shown in Figure 9.4. (Since the y-variable is missing, the surface extends indefinitely along the y-axis.)

Section

Trace

To sketch more complex surfaces, we proceed by studying various **sections,** defined to be the intersections of planes with the given surface. A section made by a coordinate plane is called a **trace.**

To find the trace in the xy-plane, let $z = 0$ in the equation. To find the trace in the xz-plane, let $y = 0$; to find the trace in the yz-plane, let $x = 0$. To facilitate this procedure, let us recall the equations of the conics.

Equations of the Conic Curves

The equation

$$Ax^2 + By^2 + Cx + Dy + E = 0$$

represents

1. An **ellipse** if A and B have like signs
2. A **hyperbola** if A and B have unlike signs
3. A **parabola** if either $A = 0$ or $B = 0$ (but not both)

If $A = B$ in (1), then the locus is a **circle.** Certain degenerate cases and imaginary loci may also occur.

Example 1

Sketch the surface $z = 4x^2 + y^2$ by finding the traces and cross-sections parallel to the coordinate planes.

Solution. To find the trace in the **xz-plane,** we set **$y = 0$** and draw the resulting parabola $z = 4x^2$ in this plane (Figure 9.5). The trace in the **yz-plane** is found by setting **$x = 0$** to obtain the parabola $z = y^2$ (Figure 9.5). Note that the trace in the xy-plane ($z = 0$) is a single point since $4x^2 + y^2 = 0$ only if $x = 0$ and $y = 0$. (The sum of two positive numbers cannot be zero.)

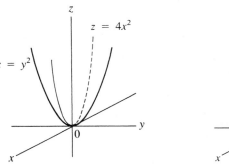

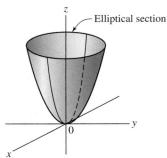

Figure 9.5   Figure 9.6

The sketch so far does not yet look like a surface. To finish the sketch we will take advantage of the fact that we can obtain as many additional cross-sections as we wish. For example, the plane $z = 4$ is parallel to the xy-plane and 4 units above it. The intersection of this plane with the given surface is found by substituting $z = 4$ in the equation $z = 4x^2 + y^2$. The resulting equation $4 = 4x^2 + y^2$ is that of an ellipse. This section, together with the traces already obtained, gives us the surface in Figure 9.6. ∎

Figure 9.6 is an example of a *paraboloid*.

> **Paraboloid**
>
> A paraboloid is a surface whose general equation is
>
> $$z = ax^2 + by^2 + c \qquad\qquad (9.1)$$

Example 2   Sketch the surface $4y^2 - 9x^2 - 4z^2 = 16$.

Solution. We first find the traces in the different coordinate planes.

1. xy-plane: let $z = 0$; the resulting curve $4y^2 - 9x^2 = 16$ is a hyperbola (Figure 9.7).

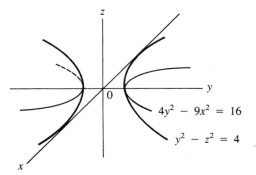

Figure 9.7

2. *yz*-plane: let $x = 0$; the trace is $4y^2 - 4z^2 = 16$ or $y^2 - z^2 = 4$, another hyperbola (Figure 9.7).
3. *xz*-plane: letting $y = 0$; we see that the resulting equation, $-9x^2 - 4z^2 = 16$, is an imaginary locus since the sum of two negative numbers cannot be positive. We conclude that the surface has no trace in the *xz*-plane.

As in Example 1, the two traces do not give us a surface. We can find additional cross-sections by letting *y* be constant. For example, the equation $y = 4$ is a plane parallel to the *xz*-plane and 4 units to the right. The intersection with the given surface $4y^2 - 9x^2 - 4z^2 = 16$ is

$$4(4)^2 - 9x^2 - 4z^2 = 16 \qquad \textbf{y = 4}$$

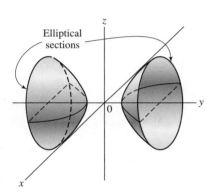

Elliptical sections

Figure 9.8

or

$$64 - 9x^2 - 4z^2 = 16$$

and

$$9x^2 + 4z^2 = 48 \qquad \textbf{elliptical section}$$

which represents an ellipse. The same equation is obtained by letting $y = -4$. (Note that we get real cross-sections only for $|y| \geq 2$.) The surface is shown in Figure 9.8. ∎

The surface in Example 2 is a special case of a *hyperboloid of two sheets*.

> **Hyperboloid of Two Sheets**
>
> A hyperboloid of two sheets is a surface whose general equation is
>
> $$\frac{y^2}{a^2} - \frac{x^2}{b^2} - \frac{z^2}{c^2} = 1 \tag{9.2}$$

Example 3   Sketch the surface $9x^2 + y^2 - 9z^2 = 9$.

Solution. As always, we first find the traces:

1. *xy*-plane: let $z = 0$; the trace is the ellipse $9x^2 + y^2 = 9$.
2. *xz*-plane: letting $y = 0$ and simplifying, we get the hyperbola $x^2 - z^2 = 1$.
3. *yz*-plane: if $x = 0$, the equation reduces to $y^2 - 9z^2 = 9$, representing a hyperbola.

In addition, we let $z = \pm 1$ to obtain two more elliptical sections. The resulting sketch is shown in Figure 9.9. ∎

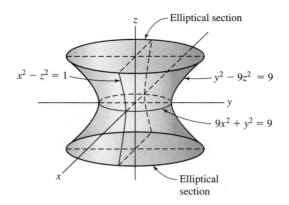

Figure 9.9

The surface in Figure 9.9 is called a *hyperboloid of one sheet*.

Hyperboloid of One Sheet

A hyperboloid of one sheet is a surface whose general equation is

$$\frac{x^2}{a^2} + \frac{y^2}{b^2} - \frac{z^2}{c^2} = 1 \qquad (9.3)$$

The next surface is called an *ellipsoid*.

Ellipsoid

An ellipsoid is a surface whose general equation is

$$\frac{x^2}{a^2} + \frac{y^2}{b^2} + \frac{z^2}{c^2} = 1 \qquad (9.4)$$

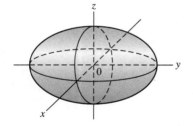

Figure 9.10

Sphere

For this surface all the traces exist and consist of ellipses (Figure 9.10). If $a = b = c$, the surface is a **sphere**. (See Exercise 32.)

The surface in Figure 9.11 is an example of a *hyperbolic paraboloid*.

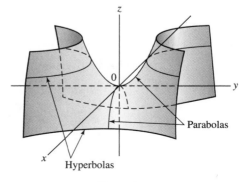

Figure 9.11

Hyperbolic Paraboloid

A hyperbolic paraboloid is a surface whose general equation is

$$\frac{y^2}{a^2} - \frac{x^2}{b^2} = 4cz \qquad (9.5)$$

The sections parallel to the xz- and yz-planes are parabolas, and sections parallel to the xy-plane are hyperbolas. Since the figure resembles a saddle, the origin is called a **saddle point.**

Saddle point

The surfaces in this section are collectively known as *quadric surfaces* since the equations are quadratics in three variables. When doing the sketches in the exercises, keep in mind that *the surfaces have different positions relative to the axes.* For example, the hyperboloid of two sheets

$$\frac{z^2}{c^2} - \frac{y^2}{a^2} - \frac{x^2}{b^2} = 1$$

has the z-axis for its axis of symmetry. (See Figure 9.12.)

A surface different from those described above is that of a *plane,* represented by a first-degree equation.

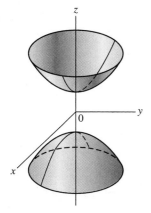

Figure 9.12

Plane

A plane is a surface represented by the first-degree equation

$$ax + by + cz = d \qquad (9.6)$$

If the plane does not pass through the origin, *the sketch can be obtained by locating the three intercepts* since three noncollinear points determine a plane.

Example 4   Sketch the plane $3x - 2y + 4z = 6$.

Solution. The intercepts are $x = 2$, $y = -3$, and $z = 3/2$ (Figure 9.13).

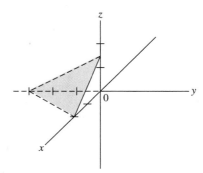

Figure 9.13

■ Exercises / Section 9.1

In Exercises 1–31, identify the surfaces and sketch them.

1. $x^2 + y^2 = 4$

2. $x^2 + z^2 = 9$

3. $y^2 = 9x$

4. $z^2 = 4y$

5. $y^2 + 4z^2 = 4$

6. $3x^2 + 2y^2 = 6$

7. $z = 3x - 3x^2$

8. $z = \sqrt{y}$

9. $y = e^x$

10. $2x + 4y + z = 4$

11. $2x - 3y + z = 6$

12. $-4x + 3y - z = 12$

13. $z = x$

14. $z = x - 2y$

15. $x^2 + y^2 + z^2 = 9$

16. $2x^2 + 2y^2 + z^2 = 8$

17. $9x^2 + 4y^2 + z^2 = 36$

18. $36x^2 + 9y^2 + 16z^2 = 144$

19. $z = x^2 + y^2$

20. $z = 4 - x^2 - y^2$

21. $z = 1 + 2x^2 + 4y^2$

22. $z = 2x^2 + y^2 - 2$

23. $4x^2 + y^2 - z^2 = 4$

24. $2x^2 + 2y^2 - z^2 = 6$

25. $-2x^2 - y^2 + z^2 = 6$

26. $9x^2 + 9y^2 - 4z^2 = 9$

27. $x^2 - 4y^2 + z^2 = 4$

28. $z^2 - x^2 - y^2 = 9$

29. $9y^2 - 36x^2 - 16z^2 = 144$

30. $z^2 - 4x^2 - y^2 = 0$ (cone)

31. $z = y^2 - x^2$

32. Use the distance formula to derive the equation of a sphere.

9.2   Partial Derivatives

Functions of Two Variables

In previous chapters all the functions we encountered contained a single independent variable. Yet some of the most common formulas in science and technology involve more than one independent variable. For example, the formula for the volume of a cylinder, $V = \pi r^2 h$, is a function of r and h, where r and h are totally independent of each other.

> **Definition of a Function of Two Variables**
>
> If for every pair of values of the independent variables x and y there exists a unique value for z, then we call z a function of x and y, denoted by
>
> $$z = f(x, y)$$

The definition can be extended to any number of independent variables; thus

$$z = g(x_1, x_2, \ldots, x_n)$$

denotes a function of n variables $x_1, x_2, \ldots, x_n$.

Example 1   If $f(x_1, x_2, x_3) = x_1{}^2 x_2 - 2x_3{}^2$, find $f(-2, 4, -3)$.

Solution. Substituting, we have

$$f(-2, 4, -3) = (-2)^2(4) - 2(-3)^2 = -2$$

■

Partial Derivatives

Next, suppose we have a function $z = f(x, y)$. If y is held fixed, then z will be a function of x alone and may be differentiated in the usual way. This is called the *partial derivative of f with respect to x,* denoted by $\partial f/\partial x$. Similarly, the *partial derivative of f with respect to y* (x held fixed) is denoted by $\partial f/\partial y$.

Definition of Partial Derivatives

If $z = f(x, y)$, then the partial derivatives are defined to be

$$\frac{\partial f}{\partial x} = \lim_{\Delta x \to 0} \frac{f(x + \Delta x, y) - f(x, y)}{\Delta x}$$

$$\frac{\partial f}{\partial y} = \lim_{\Delta y \to 0} \frac{f(x, y + \Delta y) - f(x, y)}{\Delta y}$$

Notation: If $z = f(x, y)$, then the partial derivative of f with respect to x is denoted by

$$\frac{\partial f}{\partial x} \quad \text{or} \quad \frac{\partial z}{\partial x} \quad \text{or} \quad f_x$$

and the partial derivative of f with respect to y is denoted by

$$\frac{\partial f}{\partial y} \quad \text{or} \quad \frac{\partial z}{\partial y} \quad \text{or} \quad f_y$$

Since the definition of partial derivative is similar to the definition of ordinary derivative (Section 2.4), the rules for finding partial derivatives are the same as the rules for finding ordinary derivatives. The only difference is that one of the variables must be treated as a constant. To find $\partial f/\partial x$, we *consider y constant* and differentiate with respect to x. To find $\partial f/\partial y$, we *consider x constant* and differentiate with respect to y.

As an example, if $f(x, y) = x^2 + 2y$, then $\partial f/\partial x = 2x + 0$ (since $2y$ is treated as a constant) and $\partial f/\partial y = 0 + 2$ (since x^2 is treated as a constant). In similar fashion, if $g(x, y) = 2xy^3$, then $\partial g/\partial y = 2x \cdot 3y^2 = 6xy^2$, since $2x$ is a constant coefficient.

Example 2   If $z = \ln(x^2 + y^3) + e^{2x} \sin y$, find $\partial z/\partial x$ and $\partial z/\partial y$.

Solution. To find $\partial z/\partial x$, we must remember to treat y as a constant. In particular, if y is constant, so is $\sin y$. Thus

$$\frac{\partial z}{\partial x} = \frac{1}{x^2 + y^3} \frac{\partial}{\partial x}(x^2 + y^3) + \sin y \frac{\partial}{\partial x}(e^{2x}) \qquad \textbf{sin y is constant}$$

$$= \frac{1}{x^2 + y^3}(2x + 0) + \sin y(2e^{2x}) \qquad \textbf{y}^3 \textbf{ is constant}$$

$$= \frac{2x}{x^2 + y^3} + 2e^{2x} \sin y$$

To find $\partial z / \partial y$, we must treat x as a constant:

$$\frac{\partial z}{\partial y} = \frac{1}{x^2 + y^3} \frac{\partial}{\partial y} (x^2 + y^3) + e^{2x} \frac{\partial}{\partial y} (\sin y) \qquad \textbf{e}^{2x} \text{ is constant}$$

$$= \frac{1}{x^2 + y^3} (0 + 3y^2) + e^{2x} \cos y \qquad \textbf{x}^2 \text{ is constant}$$

$$= \frac{3y^2}{x^2 + y^3} + e^{2x} \cos y \qquad\qquad\qquad ∎$$

Example 3   Find $\partial z / \partial x$ if $z = \tan x^2 y^2 + \text{Arctan } y$.

Solution. Since y is held fixed, the derivative of the second term is zero. We now have

$$\frac{\partial z}{\partial x} = (\sec^2 x^2 y^2) \frac{\partial}{\partial x} (x^2 y^2) + 0 \qquad \textbf{Arctan } \textbf{y} \text{ is constant}$$

$$= (\sec^2 x^2 y^2)(y^2) \frac{\partial}{\partial x} (x^2) \qquad \textbf{y}^2 \text{ is constant}$$

$$= 2xy^2 \sec^2 x^2 y^2 \qquad\qquad\qquad ∎$$

The geometric interpretation of the partial derivative is easily given. From our work on curve sketching we know that $y = y_0$ is a plane parallel to the xz-plane and passing through the point $(0, y_0, 0)$. Let $z = f(x, y)$ be the surface in Figure 9.14 and $P(x_0, y_0, z_0)$ be a point on the surface. Then $z = f(x, y_0)$ is the curve through P lying in the plane $y = y_0$. In this situation $\partial z / \partial x$ means exactly the same as dz/dx and is simply the slope of the tangent line to the curve at P (Figure 9.14).

Similarly, $\partial z / \partial y$ at P is the slope of the tangent line to the curve obtained by passing a plane through P parallel to the yz-plane. (See Figure 9.14.) If we pick a

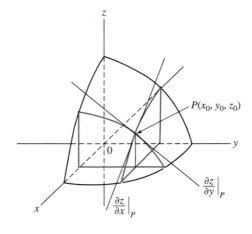

Figure 9.14

different point for P, we naturally get different curves and hence different slopes. The partial derivatives, then, give us the rate of change of the function in two special directions.

Partial derivatives can be defined in a natural way for any number of variables, although the geometric interpretation is lost when the number of independent variables exceeds two.

Example 4

The temperature at a point (x, y, z) in a solid is $T(x, y, z) = \sqrt{x^2 + y^2 + z^2}$ (in °C). Assuming the distance to be measured in centimeters, find the rate of change of the temperature with respect to the distance if we start at the point $(0, 0, 3)$ and move in the upward direction.

Solution. To find the rate of change in the z-direction, we hold x and y fixed. Thus

$$\frac{\partial T}{\partial z} = \frac{1}{2}(x^2 + y^2 + z^2)^{-1/2}(2z)\Big|_{(0, 0, 3)} = 1°\text{C/cm}$$ ∎

Higher Derivatives

In the case of higher partial derivatives we need to make some additional observations. Consider the function $z = \cos xy$. Differentiating with respect to x, we get

$$\frac{\partial z}{\partial x} = -y \sin xy$$

Differentiating a second time, we get the second derivative

$$\frac{\partial^2 z}{\partial x^2} = \frac{\partial}{\partial x}\left(\frac{\partial z}{\partial x}\right) = -y^2 \cos xy$$

Since there are two independent variables, why not differentiate the second time with respect to y? Then we get

$$\frac{\partial}{\partial y}\left(\frac{\partial z}{\partial x}\right) = \frac{\partial}{\partial y}(-y \sin xy) = -xy \cos xy - \sin xy$$

The operation

$$\frac{\partial}{\partial y}\left(\frac{\partial z}{\partial x}\right) \qquad \text{is denoted by} \qquad \frac{\partial^2 z}{\partial y \partial x}$$

Mixed partial derivative

or by f_{xy} if $z = f(x, y)$ and is called the **mixed partial derivative.**
If we reverse the order of differentiation, we get

$$\frac{\partial z}{\partial y} = -x \sin xy$$

and

$$\frac{\partial^2 z}{\partial x \partial y} = \frac{\partial}{\partial x}\left(\frac{\partial z}{\partial y}\right) = \frac{\partial}{\partial x}(-x \sin xy) = -xy \cos xy - \sin xy$$

which is the same expression we obtained before. It is true in general that the mixed partial derivatives are equal; that is,

$$\frac{\partial^2 z}{\partial x \partial y} = \frac{\partial^2 z}{\partial y \partial x}$$

provided that the function and the derivatives, as well as the mixed partial derivatives, are all continuous.

Example 5   Given $f(x, y) = 2x^2 y^3 + \sin x$, find

$$\frac{\partial^2 f}{\partial x \partial y}$$

Solution.

$$\frac{\partial f}{\partial y} = 2x^2 \cdot 3y^2 + 0 \qquad \textbf{\textit{x} held fixed}$$

$$= 6x^2 y^2$$

$$\frac{\partial^2 f}{\partial x \partial y} = \frac{\partial}{\partial x}\left(\frac{\partial f}{\partial y}\right) = \frac{\partial}{\partial x}(6x^2 y^2)$$

$$= 12xy^2 \qquad \textbf{\textit{y} held fixed}$$

We can also find the mixed partial derivative by differentiating with respect to x first and then with respect to y:

$$\frac{\partial f}{\partial x} = 4xy^3 + \cos x \qquad \textbf{\textit{y} held fixed}$$

$$\frac{\partial^2 f}{\partial x \partial y} = \frac{\partial^2 f}{\partial y \partial x} = \frac{\partial}{\partial y}\left(\frac{\partial f}{\partial x}\right)$$

$$= \frac{\partial}{\partial y}(4xy^3 + \cos x)$$

$$= 4x \cdot 3y^2 + 0 \qquad \textbf{\textit{x} held fixed}$$

$$= 12xy^2 \qquad \blacksquare$$

■ Exercises / Section 9.2

In Exercises 1–20, find **(a)** $\partial f/\partial x$; **(b)** $\partial f/\partial y$.

1. $f(x, y) = 2x^2 + 5y^2 + 1$

2. $f(x, y) = 4x^2 - 6y^2 - 3x + 2$

3. $f(x, y) = x + \sin 2y$

4. $f(x, y) = \dfrac{y}{x} + \operatorname{Arcsin} x$

5. $f(x, y) = \dfrac{1}{3}x^3 - 3\cos y$

6. $f(x, y) = 2x^2 + \tan 3y$

7. $f(x, y) = e^{2x} + \ln y$

8. $f(x, y) = 3\ln x - 5y$

9. $f(x, y) = x^2 \ln y$

10. $f(x, y) = ye^{2x}$

11. $f(x, y) = 3x^2 \tan 2y$

12. $f(x, y) = 3x \sec y^2$

13. $f(x, y) = x\sqrt{x + y}$

14. $f(x, y) = y\sqrt{y - x}$

15. $f(x, y) = y^2 \tan xy$

16. $f(x, y) = x \sec xy$

17. $f(x, y) = \dfrac{\cos x^2 y}{y}$

18. $f(x, y) = \dfrac{\sin 3xy}{x}$

19. $f(x, y) = \dfrac{\sqrt{x + y^2}}{x}$

20. $f(x, y) = \dfrac{\sqrt{x^2 + y^2}}{y}$

In Exercises 21–32, find **(a)** $\partial f/\partial x$; **(b)** $\partial f/\partial y$; **(c)** $\partial^2 f/\partial x^2$; **(d)** $\partial^2 f/\partial y^2$; **(e)** $\partial^2 f/\partial x\,\partial y$.

21. $f(x, y) = 5x - 2y - 3$

22. $f(x, y) = x^2 y$

23. $f(x, y) = 2x^2 + 3y^2 + 3x^2 y^2 + 5xy$

24. $f(x, y) = \dfrac{1}{xy}$

25. $f(x, y) = \sin(2x + y)$

26. $f(x, y) = (x - 2y)^6$

27. $f(x, y) = \ln(x^2 y) + \tan y$

28. $f(x, y) = x \sin(x + y)$

29. $f(x, y) = \operatorname{Arctan}(y/x)$

30. $f(x, y) = e^{x+y}\sin(x - y)$

31. $f(x, y, z) = \sqrt{xyz}$

32. $f(x, y, z) = e^{x+z^2}\sin(x + y + z)$

In Exercises 33–36, compute the slope of the line tangent to the curve determined by the intersection of the given surface with a vertical plane through the given point parallel to **(a)** the x-axis; **(b)** the y-axis. (See Figure 9.14.)

33. $z = ye^x$; $(1, -1, -e)$

34. $z = e^x \cos y$; $\left(0, \dfrac{\pi}{3}, \dfrac{1}{2}\right)$

35. $z = \ln(x^2 + y)$; $\left(\dfrac{1}{\sqrt{2}}, \dfrac{1}{2}, 0\right)$

36. $z = \sin xy$; $\left(1, \dfrac{\pi}{2}, 1\right)$

37. The impedance of a circuit consisting of a resistor and an inductor in parallel is

$$Z = \frac{RX}{R + X}$$

Find $\partial Z/\partial R$ when $R = 8.0\,\Omega$ and $X = 5.0\,\Omega$.

38. The current I (in amperes) in an AC circuit is given by

$$I = \frac{E}{\sqrt{R^2 + X^2}}$$

Find $\partial I/\partial R$ when $X = 3.0\,\Omega$, $R = 10.0\,\Omega$, and $E = 100\,\text{V}$.

39. The effective resistance R of two resistors R_1 and R_2 in parallel is such that

$$\frac{1}{R} = \frac{1}{R_1} + \frac{1}{R_2}$$

Find $\partial R/\partial R_1$ when $R_1 = 10.0\,\Omega$ and $R_2 = 20.0\,\Omega$.

40. The resonance frequency f in an AC circuit is given by

$$f = \frac{1}{2\pi\sqrt{LC}}$$

Find $\partial f/\partial L$.

41. The plate current i_p (milliamperes) in a vacuum tube is a function of the plate voltage v_p (volts) and the grid voltage v_g. If $i_p = 0.50(v_g + 0.1v_p)^{4/3}$, find the *mutual conductance* g_m, which is expressed as

$$g_m = \frac{\partial i_p}{\partial v_g}$$

and is measured in reciprocal ohms. (In practice, these relationships are usually described graphically since they are difficult to express by equations.)

42. If the plate voltage is given by $v_p = 27(i_p{}^{2/3} - v_g)$, find the *plate resistance*

$$r_p = \frac{\partial v_p}{\partial i_p}$$

when $i_p = 64\,\text{mA}$. (Refer to Exercise 41.)

43. The temperature of a metal plate at point (x, y) is given by $T = 1 + x^3 y$. If the temperature is measured in °C and the distance in meters, find the rate of change of the temperature with respect to distance at $(2, 1)$ in the direction parallel to the x-axis.

44. According to the ideal gas law the relationship between the pressure, temperature, and volume of a confined gas is given by $P = kT/V$, where k is a constant. Show that if the temperature remains constant, then the pressure and volume satisfy the condition

$$\frac{\partial P}{\partial V} = -\frac{P}{V}$$

(A process in which the temperature does not change is called *isothermal*.)

45. If a vibrating string is stretched along the *x*-axis, the displacement *y* at position *x* and time *t* is $y = A \cos \omega(t - x/v)$, where *A* is the amplitude of the wave and *v* is its velocity. Show that *y* satisfies the *wave equation*

$$\frac{\partial^2 y}{\partial x^2} = \frac{1}{v^2} \frac{\partial^2 y}{\partial t^2}$$

46. Given that the temperature of a long bar positioned along the *x*-axis with its left face at the origin is $T(x, t) = 2\sqrt{\pi} t e^{-x/t}$, find the *heat flux*

$$\frac{\partial T}{\partial x}\bigg|_{x=0}$$

applied to the left face.

<table>
<tr><td>**9.3**</td><td># Applications of Partial Derivatives</td></tr>
</table>

The applications of the ordinary derivative encountered in our earlier work will now be extended to functions of more than one variable by means of partial derivatives.

In Chapter 3 we learned that the differential $dy = f'(x)\,dx$ of a function of a single variable approximates the increment Δy if $\Delta x = dx$ is relatively small (Figure 9.15). Now let $z = f(x, y)$ be a function of two variables and let dx and dy be the usual increments (Figure 9.16). To get the analogous approximation for Δz, we need to consider a plane tangent to a surface, the exact analog of the line tangent to a curve. Assume that *PBGD* is a plane tangent to the surface at *P*, so that $EG = dz \approx \Delta z$. Now consider the box in Figure 9.16 with vertices at *A*, *E*, *C*, and *G*. The plane passes through *P* and *G* and intersects two of the edges at *B* and *D*, respectively. Note especially that the lines determined by *P* and *G* and by *B* and *D*

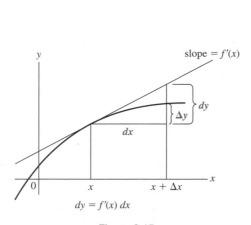

$$dy = f'(x)\,dx$$

Figure 9.15

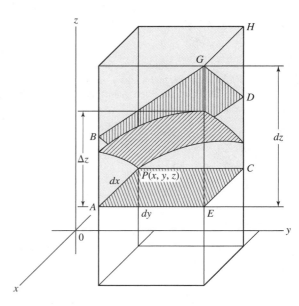

Figure 9.16

pass through the center of the box. With these facts in mind, it takes only a little reflection to see that $AB = DH$. It follows that $AB + CD = CH = EG$. If we now consider the approximate changes (in terms of differentials) in the x- and y-directions separately, then we see from Figure 9.16 that

$$AB = \frac{\partial f}{\partial x} dx = \text{approximate change in the } x\text{-direction}$$

and

$$CD = \frac{\partial f}{\partial y} dy = \text{approximate change in the } y\text{-direction}$$

But $AB + CD = EG$, the approximate total change dz. So we define the *total differential* as follows:

Total Differential of $z = f(x, y)$

$$dz = \frac{\partial f}{\partial x} dx + \frac{\partial f}{\partial y} dy \tag{9.7}$$

The definition can be extended to any number of variables. For example, if $w = F(x, y, z)$, then

$$dw = \frac{\partial F}{\partial x} dx + \frac{\partial F}{\partial y} dy + \frac{\partial F}{\partial z} dz \tag{9.8}$$

If x and y are both functions of time, then by Definition (9.7)

$$\frac{dz}{dt} = \frac{\partial f}{\partial x} \frac{dx}{dt} + \frac{\partial f}{\partial y} \frac{dy}{dt} \tag{9.9}$$

---

which can be applied to problems in related rates.

Example 1

The resistance R of a variable resistor is measured to be 20.00 Ω with a possible error of ± 0.04 Ω; the voltage across the resistor is measured to be 15.00 V with a possible error of ± 0.02 V. Find the approximate maximum possible error and percentage error in the current $i = V/R$.

Solution. Writing the expression for i in the form $i = VR^{-1}$, we obtain, by Definition (9.7),

$$di = \frac{\partial}{\partial V} (VR^{-1}) dV + \frac{\partial}{\partial R} (VR^{-1}) dR$$

$$= R^{-1} dV - VR^{-2} dR$$

$$= \frac{1}{R} dV - \frac{V}{R^2} dR$$

From the given information, we have

$$R = 20.00 \, \Omega \quad \text{and} \quad dR = \pm 0.04 \, \Omega$$

$$V = 15.00 \, \text{V} \quad \text{and} \quad dV = \pm 0.02 \, \text{V}$$

Substituting in di, we get

$$di = \frac{1}{20.00}(\pm 0.02) - \frac{15.00}{(20.00)^2}(\pm 0.04)$$

To get the *maximum* possible error, we need to consider the extreme cases:

$$di = \frac{1}{20.00}(+0.02) - \frac{15.00}{(20.00)^2}(-0.04)$$

or

$$di = \frac{1}{20.00}(-0.02) - \frac{15.00}{(20.00)^2}(+0.04)$$

Either way,

$$di = \pm 0.0025 \qquad \textbf{approximate maximum error}$$

The approximate percentage error (using $di = 0.0025$) is

$$\frac{di}{i} \times 100 = \frac{di}{V/R} \times 100$$

$$= \frac{0.0025}{15.00/20.00} \times 100 = 0.33\%$$ ∎

Example 2   The impedance of an RL circuit is given by $Z = \sqrt{R^2 + X^2}$. If R increases at the rate of $2.0 \, \Omega/\text{s}$ and X at $3.0 \, \Omega/\text{s}$, find the rate at which Z is changing at the instant when $R = 3.0 \, \Omega$ and $X = 4.0 \, \Omega$.

Solution. By Formula (9.9)

$$\frac{dZ}{dt} = \frac{1}{2}(R^2 + X^2)^{-1/2}(2R)\frac{dR}{dt} + \frac{1}{2}(R^2 + X^2)^{-1/2}(2X)\frac{dX}{dt}$$

Substituting $X = 4.0$, $R = 3.0$, $dX/dt = 3.0$, and $dR/dt = 2.0$, we get

$$\frac{dZ}{dt} = \frac{18}{5.0} = 3.6 \, \Omega/\text{s}$$ ∎

Our last topic in this section is the problem of finding a maximum or minimum point on a surface. Consider the surface $z = f(x, y)$ in Figure 9.17. If the point P is indeed a maximum point, then every tangent line at P is horizontal. In particular

$$\frac{\partial z}{\partial x} = 0 \quad \text{and} \quad \frac{\partial z}{\partial y} = 0 \tag{9.10}$$

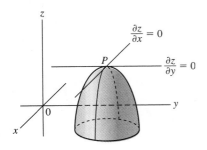

Figure 9.17

provided, of course, that the partial derivatives exist. The same statement holds if P is a minimum point. We see, then, that Statement (9.10) is a necessary condition for the existence of a minimum or maximum point and provides a method for finding it.

Unfortunately, Equations (9.10) yield only the critical values and do not guarantee the existence of a minimum or maximum. The following test, analogous to the second-derivative test for functions of one variable, is proved in many textbooks on advanced calculus.

Second-Derivative Test for Functions of Two Variables

Let

$$A = \frac{\partial^2 f}{\partial x^2}\frac{\partial^2 f}{\partial y^2} - \left(\frac{\partial^2 f}{\partial x \partial y}\right)^2$$

and suppose that for $x = a$ and $y = b$

$$\frac{\partial f}{\partial x} = 0 \quad \text{and} \quad \frac{\partial f}{\partial y} = 0$$

For these values:

1. If $A > 0$, then $f(a, b)$ is a maximum value whenever $\partial^2 f/\partial x^2 < 0$ and a minimum value whenever $\partial^2 f/\partial x^2 > 0$.
2. If $A < 0$, then $f(a, b)$ is neither a maximum nor a minimum. (The critical point is a saddle point.)
3. If $A = 0$, the test fails.

Example 3 Test the hyperbolic paraboloid

$$\frac{x^2}{a^2} - \frac{y^2}{b^2} = 4cz$$

for minima and maxima. (See Section 9.1.)

Solution. Solving the given equation for z, we get

$$z = \frac{1}{4c}\left(\frac{x^2}{a^2} - \frac{y^2}{b^2}\right)$$

Now we find all the partial derivatives needed:

$$\frac{\partial z}{\partial x} = \frac{1}{4c}\left(\frac{2x}{a^2}\right) = \frac{x}{2a^2c} \qquad \frac{\partial z}{\partial y} = \frac{1}{4c}\left(-\frac{2y}{b^2}\right) = -\frac{y}{2b^2c}$$

$$\frac{\partial^2 z}{\partial x^2} = \frac{1}{2a^2c} \qquad \frac{\partial^2 z}{\partial y^2} = -\frac{1}{2b^2c} \qquad \frac{\partial^2 z}{\partial x \partial y} = 0$$

To locate the critical points, we set the first partial derivatives to zero:

$$\frac{\partial z}{\partial x} = \frac{x}{2a^2c} = 0 \qquad \frac{\partial z}{\partial y} = -\frac{y}{2b^2c} = 0$$

so that $x = 0$ and $y = 0$. Hence $(0, 0, 0)$ is the critical point. Since $\partial^2 z/\partial x^2 = 1/(2a^2c)$ and $\partial^2 z/\partial y^2 = -1/(2b^2c)$, it follows that

$$A = \left(\frac{1}{2a^2c}\right)\left(\frac{-1}{2b^2c}\right) - 0^2 = -\frac{1}{4a^2b^2c^2} < 0$$

Consequently, the critical point is neither a minimum nor a maximum. As noted earlier, such a point is called a *saddle point*. (See Figure 9.11.) ∎

Example 4
Test the function $f(x, y) = x^3 + y^3 - 6xy$ for minima and maxima.

Solution. We first find the critical points:

$$\frac{\partial f}{\partial x} = 3x^2 - 6y = 0$$

$$\frac{\partial f}{\partial y} = 3y^2 - 6x = 0$$

From the first equation we get $x^2 = 2y$ and from the second $y^2 = 2x$. Eliminating x, we get

$$y^4 = 8y \qquad \text{or} \qquad y(y^3 - 8) = 0$$

It follows that $y = 0$ and $y = 2$ and that the corresponding x-values are $x = 0$ and $x = 2$. So the critical points are at $(0, 0)$ and $(2, 2)$. To test the critical points we compute A. Note that

$$\frac{\partial^2 f}{dx^2} = 6x \qquad \frac{\partial^2 f}{\partial y^2} = 6y \qquad \frac{\partial^2 f}{\partial x \partial y} = -6$$

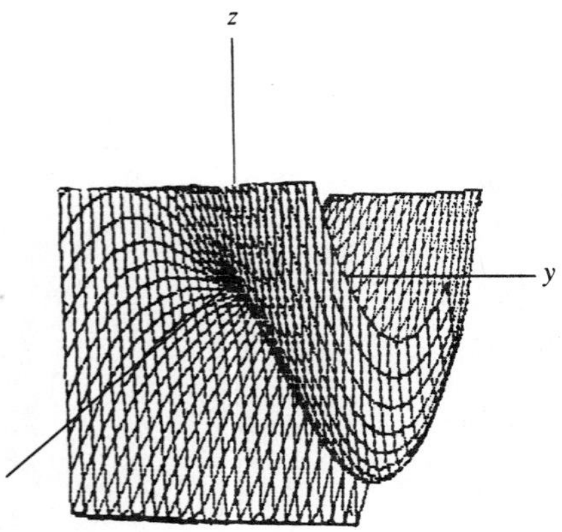

It follows that

$$A = (6x)(6y) - (-6)^2 = 36xy - 36$$

At $(2, 2)$, $A = 108 > 0$, while $\partial^2 f/\partial x^2 = 12 > 0$. We conclude that the function has a minimum at $(2, 2)$.

At $(0, 0)$, $A = -36 < 0$. We therefore have a saddle point at $(0, 0)$. (The surface is shown in Figure 9.18.) ∎

Figure 9.18

Remark: A particularly interesting application of minima and maxima is the derivation of the method for finding the best-fitting curve discussed in the next section.

■ Exercises / Section 9.3

In Exercises 1–8, find the differential of each given function.

1. $P = \dfrac{2L}{V^2}$

2. $z = 4r\sqrt{s}$

3. $L = \dfrac{1}{\sqrt{X^2 + Y^2}}$

4. $S = \dfrac{m}{\sqrt{n + 4}}$

5. $M = \dfrac{\sin\theta_1}{\sin\theta_2}$

6. $N = e^{-t/L}$

7. $f = r + \tan r\omega$

8. $X = \dfrac{\sec V}{Q^2}$

9. $V = L/P^2$; $L = 8.0$ with error of ± 0.5; $P = 3.0$ with error of ± 0.2; find the approximate maximum error in V.

10. $M = \sqrt{t_1}/t_2$; $t_1 = 5.00$ with error of ± 0.03; $t_2 = 8.00$ with error of ± 0.06; find the approximate maximum error in M.

11. A right circular cylinder is measured to be 10.0 cm high with a possible error of 0.1 cm. The radius of the base is measured to be 5.00 cm with a possible error of 0.08 cm. What is the approximate maximum error in the volume?

12. The radius and height of a right circular cone are measured to be 2.00 cm and 5.00 cm, respectively, with a possible error of ± 0.01 cm in measurement. What is the approximate maximum possible error and percentage error in the volume?

13. Assume that $i = 1 - e^{-R/L}$ at some instant. Suppose R is measured to be $1.20\ \Omega$ with a possible error of $\pm 0.05\ \Omega$ and L is measured to be 0.70 H with a possible error of ± 0.01 H. Determine the approximate maximum possible error and percentage error in i.

14. The resonance frequency f (cycles per second) in an LC circuit is given by

$$f = \frac{1}{2\pi\sqrt{LC}}$$

If L is measured to be 0.40 H with a possible error of ± 0.02 H and C is measured to be 1.30 C with a possible error of ± 0.01 C, determine the approximate maximum error and percentage error in f.

15. The period of a pendulum is given by $T = 2\pi\sqrt{L/g}$, where L is the length and g is the acceleration due to gravity. If L is measured to be 15.0 cm with an error of ± 0.2 cm and g is measured to be 980 cm/s^2 with an error of ± 6 cm/s^2, find the approximate maximum error and percentage error in the period.

16. The formula for the impedance in a circuit is $Z = \sqrt{X^2 + R^2}$. If X is measured to be $20.00\ \Omega$ with an error of $\pm 0.04\ \Omega$ and R is measured to be $30.00\ \Omega$ with an error of $\pm 0.06\ \Omega$, find the approximate maximum error and percentage error in Z.

17. At a certain instant the altitude of a right circular cone is 20 cm and the radius of the base is 10 cm. If both the base and height are increasing at the rate of 1.0 cm/min, how fast is the volume increasing at that instant?

18. If two legs of a right triangle are increasing at 2.0 cm/min and 3.0 cm/min, respectively, at what rate is the area increasing at the instant when both legs are 10 cm long?

19. Recall that the power delivered to a resistor is given by $P = i^2 R$. If i is increasing at the rate of 2.0 A/s and R at the rate of 3.0 Ω/s, how fast is the power increasing at the instant that $i = 10$ A and $R = 50\ \Omega$?

20. If $R = R_1 R_2/(R_1 + R_2)$ and R_1 and R_2 are both increasing at 4.0 Ω/s, find the rate of change of R when $R_1 = 10\ \Omega$ and $R_2 = 20\ \Omega$.

In Exercises 21–32, examine the functions for minima and maxima.

21. $f(x, y) = 3x^2 + 2y^2 + 4x - 4y - 1$

22. $f(x, y) = 2 + 2x - 8y - 2x^2 - 4y^2$

23. $f(x, y) = 2xy - x^2 - 2y^2 + 3x + 5$

24. $f(x, y) = x^2 + 2xy + 3y^2 + 2x + 10y$

25. $f(x, y) = y^2 - x^2 - 2xy - 4y$

26. $f(x, y) = 4x^2 - 3y^2 + 2$

27. $f(x, y) = x^2 + 2y^3 - x - 12y - 4$

28. $f(x, y) = y - x\ln y$

29. $f(x, y) = 3x^3 - xy^2 + x$

30. $f(x, y) = 2x^3 - 3xy + 2y^3 + 1$

31. $f(x, y) = 9xy - x^3 - y^3$

32. $f(x, y) = y^3 - y^2 + x^2 - y + 2x + 1$

33. The sum of three numbers is 60. Find the numbers if their product is a maximum.

34. Show that a box with an open top and a fixed volume requires the least material to build if it has a square base and a height equal to one-half the length of the base.

35. A company manufactures two types of widgets, W_1 and W_2, that sell for $\$P_1$ and $\$P_2$, respectively. Suppose that $C(x, y)$ is the cost of producing x units of W_1 and y units of W_2. Show that if the profit is a maximum at $x = a$ and $y = b$, then

$$\frac{\partial C(a, b)}{\partial x} = P_1 \quad \text{and} \quad \frac{\partial C(a, b)}{\partial y} = P_2$$

(Profit is revenue minus cost.)

36. Let N be the number of items produced by a manufacturer, M the number of worker hours required, and C the amount of capital. The relationship is denoted by $N = f(M, C)$. Show that N/M, the average production per worker-hour, is maximized when $\partial N / \partial M = N/M$. (Assume that a maximum exists.)

37. Consider the implicit function $f(x, y) = 0$. Show that the implicit derivative is given by

$$\frac{dy}{dx} = -\frac{\partial f / \partial x}{\partial f / \partial y}$$

(*Hint:* Find the differential of f and solve for dy/dx.)

In Exercises 38–49, use the formula in Exercise 37 to find the implicit derivative dy/dx in each case.

38. $x^5 - 3x^4 y^3 + 2x^2 y^2 + 2xy + 3y - 5 = 0$

39. $x^6 + 2x^5 y^2 - 6x^3 y^3 - 4y^4 + 10 = 0$

40. $x^7 - 4x^6 y^5 - 3x^5 y^2 + 3x^2 y + 1 = 0$

41. $2x^5 + 3x^4 y - 4x^3 y^2 + 7x^2 y^2 + 1 = 0$

42. $2x^7 y^6 - 2x^5 y^3 + 2x^3 - 7y^3 + 6 = 0$

43. $7x^4 y^8 + 16x^3 y^5 + 25x^2 - 7y^2 + 2 = 0$

44. $x \sin y + y = 1$

45. $y^2 + y \cos x = 2$

46. $xy^2 - \sin x^2 y = 0$

47. $\cot(x^2 - y^2) - 3xy + 7 = 0$

48. $e^{y^2} + y - 3x = 0$

49. $\ln y - x \sin y + 3y^2 = 0$

9.4 Curve Fitting*

In earlier chapters we studied a number of techniques for graphing a function $y = f(x)$. In this section we are going to reverse the procedure: given a set of points, find the function $y = f(x)$.

It is theoretically possible to find a polynomial that will pass through a given finite set of points, but is this really what we want? Suppose (x_1, y_1), (x_2, y_2), $\ldots$, (x_n, y_n) is a set of experimental data. If the underlying theory suggests that the relationship between x and y is linear, then it would make better sense to find a line that yields the "best fit," even though the line may not actually pass through any of these points. (See Figure 9.19).

A convenient way to obtain a best fit is by the method of *least squares*. Suppose the line has the form $y = ax + b$ for some a and b. Then for every x_i we get $y = ax_i + b$. The problem is that y_i will not usually lie on the line. Referring to Figure 9.19, suppose

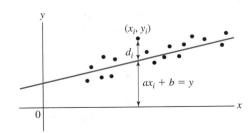

Figure 9.19

*This section may be omitted without loss of continuity.

we let d_i be the difference between y_i and the ordinate of the point on $y = ax + b$ corresponding to x_i; that is,

$$d_i = y_i - (ax_i + b) \qquad \mathbf{d_i = y_i - y}$$

Then we define the **best fit** to be the line for which the sum of the squares of the differences d_i is as small as possible. In other words, choose a and b so that

$$S = [y_1 - (ax_1 + b)]^2 + [y_2 - (ax_2 + b)]^2 + \cdots + [y_n - (ax_n + b)]^2$$

is a minimum. Treating S as a function of a and b, we can find the minimum by computing $\partial S/\partial a$ and $\partial S/\partial b$ and setting the partial derivatives equal to zero:

$$\frac{\partial S}{\partial a} = 2(y_1 - ax_1 - b)(-x_1) + 2(y_2 - ax_2 - b)(-x_2)$$

$$+ \cdots + 2(y_n - ax_n - b)(-x_n) = 0$$

and

$$\frac{\partial S}{\partial b} = 2(y_1 - ax_1 - b)(-1) + 2(y_2 - ax_2 - b)(-1)$$

$$+ \cdots + 2(y_n - ax_n - b)(-1) = 0$$

Dividing by 2 and collecting terms, we get

$$b(x_1 + x_2 + \cdots + x_n) + a\left(x_1{}^2 + x_2{}^2 + \cdots + x_n{}^2\right) = x_1 y_1 + x_2 y_2 + \cdots + x_n y_n$$

and

$$\overbrace{b + b + \cdots + b}^{n \text{ terms}} + a(x_1 + x_2 + \cdots + x_n) = y_1 + y_2 + \cdots + y_n$$

Using the sigma notation, the equations can be written

$$b \sum_{i=1}^{n} x_i + a \sum_{i=1}^{n} x_i{}^2 = \sum_{i=1}^{n} x_i y_i$$

and

$$nb + a \sum_{i=1}^{n} x_i = \sum_{i=1}^{n} y_i$$

Since this is a system of two equations in two unknowns a and b, we may write the solution conveniently in determinant form by using Cramer's rule. The resulting formulas are given next.

Line of Best Fit

The best-fitting (least-squares) line through the points (x_1, y_1), (x_2, y_2),...,
(x_n, y_n) is

$$y = ax + b$$

where

$$a = \frac{A}{D} \qquad b = \frac{B}{D} \tag{9.11}$$

and

$$D = \begin{vmatrix} \sum\limits_{i=1}^{n} x_i & \sum\limits_{i=1}^{n} x_i^2 \\ n & \sum\limits_{i=1}^{n} x_i \end{vmatrix} \qquad A = \begin{vmatrix} \sum\limits_{i=1}^{n} x_i & \sum\limits_{i=1}^{n} x_i y_i \\ n & \sum\limits_{i=1}^{n} y_i \end{vmatrix}$$

$$B = \begin{vmatrix} \sum\limits_{i=1}^{n} x_i y_i & \sum\limits_{i=1}^{n} x_i^2 \\ \sum\limits_{i=1}^{n} y_i & \sum\limits_{i=1}^{n} x_i \end{vmatrix} \tag{9.12}$$

Example 1   Find the best-fitting line $y = ax + b$ for the following set of data:

x:	0	5	10	15	20	25	30	35	40	45	50
y:	2.1	2.4	2.7	2.8	3.0	3.2	3.5	4.1	3.8	4.0	4.6

Solution. The most convenient way to perform the calculations is to write the data
in column form with x_i^2 and $x_i y_i$ in the third and fourth columns, respectively:

x_i	y_i	x_i^2	$x_i y_i$
0	2.1	0	0
5	2.4	25	12
10	2.7	100	27
15	2.8	225	42
20	3.0	400	60
25	3.2	625	80
30	3.5	900	105
35	4.1	1225	143.5
40	3.8	1600	152
45	4.0	2025	180
50	4.6	2500	230
Totals: 275	36.2	9625	1031.5

Using Formulas (9.11) and (9.12), we get

$$D = \begin{vmatrix} 275 & 9625 \\ 11 & 275 \end{vmatrix} = -30{,}250$$

$$A = \begin{vmatrix} 275 & 1031.5 \\ 11 & 36.2 \end{vmatrix} = -1391.5$$

and

$$B = \begin{vmatrix} 1031.5 & 9625 \\ 36.2 & 275 \end{vmatrix} = -64{,}762.5$$

from which

$$a = \frac{A}{D} = \frac{-1391.5}{-30{,}250} = 0.0460$$

and

$$b = \frac{B}{D} = \frac{-64{,}762.5}{-30{,}250} = 2.1409$$

Hence the best-fitting line $y = ax + b$ is

$$y = 0.0460x + 2.1409$$ ∎

The method of least squares can be extended to find the best curve of the form

$$y = [f(x)]a + b$$

If for theoretical reasons or after inspection of the data we decide that this form is more appropriate, we simply replace each x_i by $f(x_i)$ and proceed as in Example 1.

Example 2   Suppose for the set of data in the following table it is decided that the curve should be of the form $y = ae^{-x/2} + b$.

x:	3	6	9	12	15	18
y:	43.35	48.49	49.76	49.95	49.99	50.01

Find the best-fitting curve.

Solution. We let $x' = e^{-x/2}$; then $x_1' = e^{-3/2} = 0.2231$, $x_2' = e^{-6/2} = 0.0498$, and so on. The values are listed in the first column:

x_i'	y_i	$x_i'^2$	$x_i'y_i$
0.2231	43.35	0.04977	9.67139
0.0498	48.49	0.00248	2.41480
0.0111	49.76	0.00012	0.55234
0.0025	49.95	0.00001	0.12488
0.0006	49.99	0.00000	0.02999
0.0001	50.01	0.00000	0.00500
Totals: 0.2872	291.55	0.0524	12.7984

By (9.11) and (9.12) we now have

$$D = \begin{vmatrix} 0.2872 & 0.0524 \\ 6 & 0.2872 \end{vmatrix} = -0.2319$$

$$A = \begin{vmatrix} 0.2872 & 12.7984 \\ 6 & 291.55 \end{vmatrix} = 6.9428$$

$$B = \begin{vmatrix} 12.7984 & 0.0524 \\ 291.55 & 0.2872 \end{vmatrix} = -11.6015$$

and

$$a = \frac{A}{D} = -29.94 \qquad b = \frac{B}{D} = 50.03$$

So the desired curve is $y = -29.94x' + 50.03$, or

$$y = 50.03 - 29.94e^{-x/2} \qquad \blacksquare$$

■ Exercises / Section 9.4

In Exercises 1–2, use the method of least squares to fit a curve of the type indicated to the given data.

1. $y = ax + b$

x:	2	4	6	8	10
y:	14	36	53	78	92

2. $y = ax + b$

x:	1.0	1.5	2.0	2.5	3.0
y:	−12.5	−22.3	−33.3	−43.0	−54.1

x:	3.5	4.0	4.5	5.0	5.5
y:	−61.2	−73.9	−83.5	−94.1	−100.0

3. A tensile force is applied to a steel specimen. Suppose that x in the table is the force in metric tons and y is the elongation in millimeters. Find a linear law of the form $y = ax + b$ connecting x and y.

x:	1	2	3	4	5	6
y:	0.70	1.65	2.00	3.15	3.80	4.25

4. In the accompanying table T is the temperature of an iron bar (°C) and L is the length (cm). Find a law of the form $L = aT + b$ connecting length and temperature.

T:	15	30	45	60	75	90
L:	50.00	50.18	50.39	50.58	50.77	50.96

5. The resistance in a certain resistor as a function of temperature is known to have the form $R = aT^2 + b$. The resistance (ohms) at various temperatures (°C) is found to be as follows:

T:	20	30	40	50	60	70
R:	7	8	11	16	22	27

Find the empirical relationship.

6. The accompanying table gives the quantity N of a certain product sold at various prices P (in dollars). Find a relationship between N and P of the form $N = a(1/P) + b$.

P:	1.10	1.20	1.30	1.40	1.50
N:	56.6	53.7	50.9	48.8	46.9

P:	1.60	1.70	1.80	1.90	2.00
N:	45.4	43.0	42.1	39.8	39.4

7. Find a relationship between x and y of the form $y = a \ln x + b$.

x:	3	4	5	6	7
y:	−0.81	−0.22	0.22	0.57	0.88

9.5 Iterated Integrals

We found in Section 9.2 that we can differentiate a function with respect to one variable while holding the other variable fixed. It seems reasonable, therefore, to *integrate* a function with respect to one variable while holding the other variable fixed.

Consider, for example, the integral

$$\int_0^1 x^2 y^2 \, dy$$

The symbol dy indicates that y is the variable and x is a constant. If x (and therefore x^2) is held fixed, we obtain

$$\int_0^1 x^2 y^2 \, dy = x^2 \frac{y^3}{3} \Big|_0^1 = x^2 \left(\frac{1}{3} - 0 \right) = \frac{x^2}{3}$$

Since x is treated as a constant, even the limits of integration may involve x. For example,

$$\int_1^{2x} x^2 y^2 \, dy = x^2 \frac{y^3}{3} \Big|_1^{2x} = \frac{1}{3} x^2 y^3 \Big|_1^{2x} \qquad \textbf{variable } y$$

$$= \frac{1}{3} x^2 [(2x)^3 - 1^3] = \frac{1}{3} x^2 (8x^3 - 1)$$

This type of "partial integration" is usually part of an iterated integral. An iterated integral has the form

$$\int_a^b \left(\int_{g_1(x)}^{g_2(x)} F(x, y)\, dy \right) dx \qquad \text{or} \qquad \int_c^d \left(\int_{h_1(y)}^{h_2(y)} F(x, y)\, dx \right) dy$$

usually written without parentheses:

$$\int_a^b \int_{g_1(x)}^{g_2(x)} F(x, y)\, dy\, dx \qquad \text{or} \qquad \int_c^d \int_{h_1(y)}^{h_2(y)} F(x, y)\, dx\, dy$$

Consider the next example.

Example 1   Evaluate

$$\int_0^1 \int_{x^2}^{x} (x + y)\, dy\, dx$$

Solution. At this stage it may be convenient to insert the parentheses to help clarify the procedure:

$$\int_0^1 \left(\int_{x^2}^{x} (x + y)\, dy \right) dx$$

We now perform the integration on the inner integral in the usual way, remembering to hold x fixed. We have

$$\int_0^1 \left(xy + \frac{1}{2} y^2 \right) \Big|_{x^2}^{x} dx = \int_0^1 \left[\left(x^2 + \frac{1}{2} x^2 \right) - \left(x^3 + \frac{1}{2} x^4 \right) \right] dx$$

At this point we end up with an ordinary integral, which is easily evaluated:

$$\int_0^1 \left(\frac{3}{2} x^2 - x^3 - \frac{1}{2} x^4 \right) dx = \frac{3}{20} \qquad \blacksquare$$

Iterated integrals can be used to find the area of a region. To see why, consider the region in Figure 9.20, bounded by $x = a$, $x = b$, $y = g_1(x)$ and $y = g_2(x)$. Note that the area A of this region is

$$A = \int_a^b [g_2(x) - g_1(x)]\, dx$$

The following iterated integral leads to the same expression for A:

$$\int_a^b \int_{g_1(x)}^{g_2(x)} dy\, dx = \int_a^b \left(\int_{g_1(x)}^{g_2(x)} 1\, dy \right) dx$$

$$= \int_a^b y \Big|_{g_1(x)}^{g_2(x)} dx = \int_a^b [g_2(x) - g_1(x)]\, dx$$

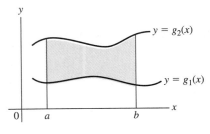

Figure 9.20

Example 2   Use an iterated integral to find the area of the region bounded by $y = x$ and $y = x^2$.

Solution. The region is shown in Figure 9.21.

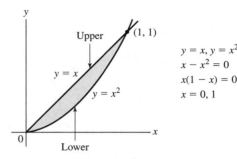

Figure 9.21

The curves $y = x$ and $y = x^2$ intersect at $(0, 0)$ and $(1, 1)$. For the iterated integral the x-limits are $x = 0$ and $x = 1$. The inner limits (y-limits) are the "lower" function $y = x^2$ and the "upper" function $y = x$. Thus

$$A = \int_0^1 \int_{x^2}^x dy\, dx = \int_0^1 y \Big|_{x^2}^x dx = \int_0^1 (x - x^2)\, dx$$

$$= \frac{x^2}{2} - \frac{x^3}{3} \Big|_0^1 = \frac{1}{2} - \frac{1}{3} = \frac{1}{6}$$

Note that the integral

$$\int_0^1 (x - x^2)\, dx$$

which we obtained after one integration, is the ordinary integral for finding the area by the method of Section 4.6.   ■

If the region is bounded by $y = c$, $y = d$, $x = h_1(y)$, and $x = h_2(y)$ we get the iterated integral

$$A = \int_c^d \int_{h_1(y)}^{h_2(y)} dx\, dy$$

$$= \int_c^d x \Big|_{h_1(y)}^{h_2(y)} dy \qquad \textbf{integrating with respect to x (y held fixed)}$$

$$= \int_c^d [h_2(y) - h_1(y)]\, dy$$

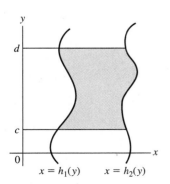

Figure 9.22

(See Figure 9.22.)

Note that the roles of x and y are interchanged, so that $x = h_1(y)$ and $x = h_2(y)$ are functions of y. In other words, viewed from the y-axis, the lower function is $x = h_1(y)$ and the upper function is $x = h_2(y)$.

Example 3   Find the area of the region in Example 2 by letting y be the outer variable.

Solution. If the outer variable is y, the iterated integral must have the form

$$\int_c^d \int_{h_1(y)}^{h_2(y)} dx\, dy$$

Thus we need to interchange the roles of x and y. When expressed as functions of y, the equations become

$$x = y \qquad \text{and} \qquad x = \sqrt{y}$$

as shown in Figure 9.23.

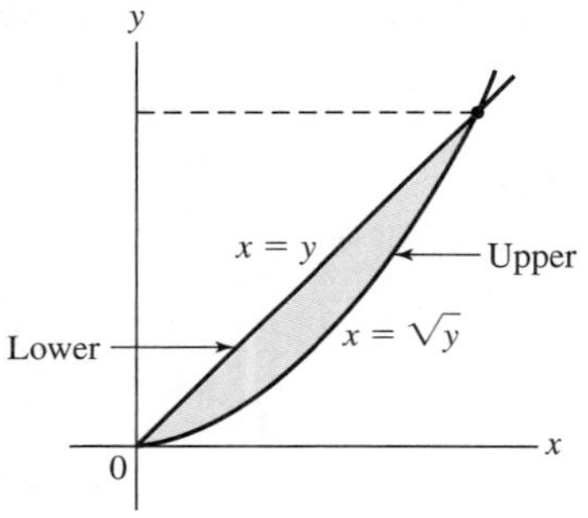

Observe that, viewed from the y-axis, the lower function is $x = y$ and the upper function is $x = \sqrt{y}$. For the outer limits (y-limits) we refer to the y-axis ($y = 0$ to $y = 1$). We now have

$$A = \int_0^1 \int_y^{\sqrt{y}} dx\, dy = \int_0^1 x \Big|_y^{\sqrt{y}} dy$$

$$= \int_0^1 \left(\sqrt{y} - y \right) dy = \frac{2}{3} y^{3/2} - \frac{1}{2} y^2 \Big|_0^1$$

$$= \frac{2}{3} - \frac{1}{2} = \frac{1}{6} \qquad \blacksquare$$

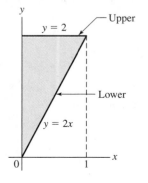

Figure 9.23

Example 4   Find the area of the region bounded by $y = 2x$, $y = 2$, and the y-axis in two ways:

a. using x for the outer variable ($dy\, dx$)
b. using y for the outer variable ($dx\, dy$)

Solution. The region is shown in Figure 9.24.

a. If $y = 2$ in the equation $y = 2x$, we get $x = 1$ (Figure 9.24). Note that $y = 2x$ is the lower function and $y = 2$ the upper function. Thus

$$A = \int_0^1 \int_{2x}^2 dy\, dx = \int_0^1 y \Big|_{2x}^2 dx$$

$$= \int_0^1 (2 - 2x)\, dx = 2x - x^2 \Big|_0^1 = 2 - 1 = 1$$

Figure 9.24

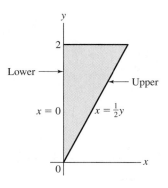

Figure 9.25

b. To reverse the order of integration, we must write the equations as functions of y. (See Figure 9.25.)

Viewed from the y-axis, $x = 0$ is the lower function and $x = (1/2)y$ is the upper function. The limits of integration along the y-axis are $y = 0$ and $y = 2$. We now have

$$A = \int_0^2 \int_0^{(1/2)y} dx\, dy = \int_0^2 x \Big|_0^{(1/2)y} dy = \int_0^2 \frac{1}{2}y\, dy = 1 \qquad \blacksquare$$

These examples show that we can find the area of a region by using either x or y for the outer variable. For many regions, however, one order of integration may lead to a simpler calculation than the opposite order. Consider the next example.

Example 5   Find the area of the region shown in Figure 9.26.

Solution. Viewed from the y-axis, the lower function is $x = (1/2)y$ and the upper function is $x = y$. Thus

$$A = \int_0^4 \int_{(1/2)y}^y dx\, dy = \int_0^4 x \Big|_{(1/2)y}^y dy = \int_0^4 \left(y - \frac{1}{2}y\right) dy$$

$$= \int_0^4 \frac{1}{2}y\, dy = 4$$

Interchanging the order of integration is not convenient in this problem. In fact, to use the order $dy\, dx$ we need to divide the region into two parts, as shown in Figure 9.27. The area is

$$A = \int_0^2 \int_x^{2x} dy\, dx + \int_2^4 \int_x^4 dy\, dx = 4$$

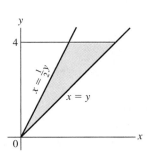

Figure 9.26

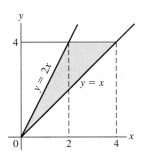

Figure 9.27 $\blacksquare$

Double integrals

The iterated integrals we have considered so far are called **double integrals.** Iterated integrals can be extended to **triple integrals** (Exercise 15) and even to *n-fold* integrals. Collectively, these are known as **multiple integrals.**

You may wonder why we even bother with iterated integrals, since we can find the area of a region using ordinary integrals. The need for iterated integrals will be seen in the next two sections, which deal with volumes, mass, centroids, and moments of inertia.

■ Exercises / Section 9.5

In Exercises 1–16, evaluate the multiple integrals.

1. $\displaystyle\int_0^1 \int_0^x x^2 y^2 \, dy \, dx$

2. $\displaystyle\int_0^3 \int_1^x 3x \, dy \, dx$

3. $\displaystyle\int_0^1 \int_{\sqrt{x}}^1 y \, dy \, dx$

4. $\displaystyle\int_0^1 \int_0^{\sqrt{4-y^2}} x \, dx \, dy$

5. $\displaystyle\int_0^2 \int_0^{\sqrt{y-1}} xy \, dx \, dy$

6. $\displaystyle\int_0^2 \int_{\sqrt{x}}^2 2x^2 y \, dy \, dx$

7. $\displaystyle\int_1^3 \int_0^{\sqrt{9-y^2}} y \, dx \, dy$

8. $\displaystyle\int_0^1 \int_0^x e^{x^2} \, dy \, dx$

9. $\displaystyle\int_0^{\sqrt{\pi/6}} \int_0^x \cos x^2 \, dy \, dx$

10. $\displaystyle\int_0^2 \int_{-\sqrt{4-x^2}}^{\sqrt{4-x^2}} x \, dy \, dx$

11. $\displaystyle\int_0^{\pi/4} \int_0^{\sec y} 2x \, dx \, dy$

12. $\displaystyle\int_{-\pi/4}^{\pi/4} \int_0^x \sec^2 y \, dy \, dx$

13. $\displaystyle\int_0^3 \int_0^x e^x \, dy \, dx$

14. $\displaystyle\int_0^{\pi/4} \int_0^{2y} y \cos x \, dx \, dy$

15. $\displaystyle\int_2^3 \int_1^x \int_0^{6y} xy \, dz \, dy \, dx$

16. $\displaystyle\int_0^a \int_0^x \int_0^y x^3 y^2 z \, dz \, dy \, dx$

In Exercises 17–34, find the area of the region bounded by the given curves using:

    a. x for the outer variable $(dy \, dx)$
    b. y for the outer variable $(dx \, dy)$

17. $x + 2y = 2$, coordinate axes

18. $4x + y = 4$, coordinate axes

19. $y = x, y = 2$, y-axis

20. $y = 2x, y = 2$, y-axis

21. $y = 2x, x = 4, y = 0$

22. $y = \frac{1}{2}x, x = 4, y = 0$

23. $y = x^2, y = 4$, y-axis (first quadrant)

24. $y = x^3, y = 1$, y-axis

25. $y = x, y = \sqrt{x}$

26. $y = x, y = x^3$

27. $x = 2\sqrt{y}, y = 9$, y-axis

28. $y = \sqrt{x}, x = 4$, x-axis

29. $y = 1 - x^2$, x-axis

30. $y = x^{2/3}, x = 8$, x-axis

31. $x = y^2, x = 9$

32. $y = 1 + x^2, y = 2$

33. $y = x^2, y^2 = 8x$

34. $y = 2x, y = x^2$

In Exercises 35–38, find the area of each given region by using the most convenient order of integration. (See Example 5.)

35. $y = x - x^2$, x-axis

36. $x = 2y - 2y^2$, y-axis

37. $y = 2x, y = \frac{1}{2}x, y = 2$

38. $y = 2x^2, y = x + 1$

9.6   Volumes by Double Integration

In this section we are going to use **double integrals** to find the volume of a solid. Let $z = F(x, y)$ be a continuous function whose surface is shown in Figure 9.28. Let R be a bounded region in the xy-plane. We wish to find the volume under the surface.

Suppose R is bounded by $x = a$, $x = b$, $y = g_1(x)$, and $y = g_2(x)$ as in Figure 9.28. First we divide the interval $[a, b]$ into small subintervals of length dx, thereby dividing the volume into thin slices. Consider the typical element in Figure 9.28, a slice of thickness dx with x temporarily held fixed. The volume of the typical slice is equal to the area of the front face multiplied by the thickness dx. Keeping in mind that x is still fixed, the area of the face is found by summing the areas $F(x, y) \, dy$ of the rectangles to get

$$\int_{g_1(x)}^{g_2(x)} F(x, y) \, dy \qquad \textbf{area of face}$$

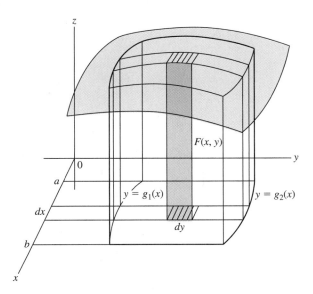

Figure 9.28

Although constant for this slice, the limits of integration depend on the extremities of the region and vary from one slice to another. The volume of the typical slice then becomes

$$\left(\int_{g_1(x)}^{g_2(x)} F(x, y)\, dy \right) dx \qquad \textbf{area of face} \cdot \textbf{thickness}$$

Finally, summing the slices from $x = a$ to $x = b$, we get

$$V = \int_a^b \left(\int_{g_1(x)}^{g_2(x)} F(x, y)\, dy \right) dx$$

which is usually written without parentheses:

$$V = \int_a^b \int_{g_1(x)}^{g_2(x)} F(x, y)\, dy\, dx$$

In some cases the roles of x and y are interchanged, in which case the integral will have the form

$$V = \int_c^d \int_{h_1(y)}^{h_2(y)} F(x, y)\, dx\, dy$$

(See Figure 9.29.)

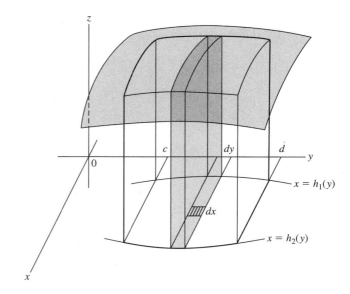

Figure 9.29

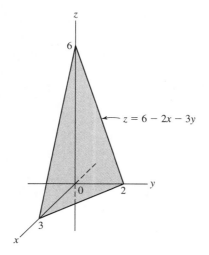

Figure 9.30

Volume of a Solid

If a region in the xy plane is bounded by $y = g_1(x)$, $y = g_2(x)$, $x = a$, and $x = b$, then the volume of the solid between the surface $z = F(x, y)$ and the region is given by

$$V = \int_a^b \int_{g_1(x)}^{g_2(x)} F(x, y)\, dy\, dx \qquad (9.13)$$

as in Figure 9.28.

If the region in the xy-plane is bounded by $x = h_1(y)$, $x = h_2(y)$, $y = c$, and $y = d$, then the integral has the form

$$V = \int_c^d \int_{h_1(y)}^{h_2(y)} F(x, y)\, dx\, dy \qquad (9.14)$$

as in Figure 9.29.

Example 1   Find the volume of the solid bounded by the plane $2x + 3y + z = 6$ and the coordinate planes (Figure 9.30).

Solution. In all problems involving volumes, the integrand is always the function representing the surface, in this case $F(x, y) = 6 - 2x - 3y$. The limits of integration are found by examining the region R in the xy-plane. This region is shown in Figure 9.31. In fact, if we are sure that the surface lies entirely above the region R,

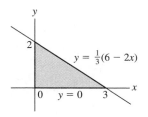

Figure 9.31

this is the only part that needs to be drawn. Of course, Figure 9.30 it not wasted since it tells us what the whole volume looks like. Note that the xy-trace in Figure 9.31 is the line $2x + 3y = 6$ or $y = (1/3)(6 - 2x)$. Referring to (9.13), $g_1(x) = 0$ and $g_2(x) = (1/3)(6 - 2x)$, while the outer limits are $x = 0$ and $x = 3$. The double integral now becomes

$$V = \int_0^3 \int_0^{(1/3)(6-2x)} (6 - 2x - 3y)\, dy\, dx$$

$$= \int_0^3 \left(6y - 2xy - \frac{3}{2}y^2 \right) \Big|_0^{(1/3)(6-2x)} dx$$

$$= \int_0^3 \left[6 \cdot \frac{1}{3}(6 - 2x) - 2x \cdot \frac{1}{3}(6 - 2x) - \frac{3}{2} \cdot \frac{1}{9}(6 - 2x)^2 \right] dx$$

$$= \int_0^3 \left(\frac{2}{3}x^2 - 4x + 6 \right) dx = 6 \qquad \blacksquare$$

Example 2   Reverse the order of integration in Example 1.

Solution.   We wish to write the integral in the form of Formula (9.14). To this end we draw the region R again and express the line as a function of y. (See Figure 9.32.)

Referring to Formula (9.14) and Figure 9.29, notice that the outer limits are along the y-axis: $y = 0$ and $y = 2$. In the x-direction, $h_1(y) = 0$ and $h_2(y) = (1/2)(6 - 3y)$. Hence

$$V = \int_0^2 \int_0^{(1/2)(6-3y)} (6 - 2x - 3y)\, dx\, dy = 6 \qquad \blacksquare$$

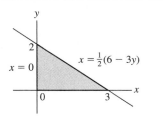

Figure 9.32

Example 3   Find the volume of the solid bounded by the surface $z = x + y$, the planes $y = x$ and $z = 0$, and the cylinder $y = x^2$.

Solution.   The region R is pictured in Figure 9.33. Since $z = x + y > 0$ for positive x and y, we know that the surface lies entirely above the region R. Consequently, a three-dimensional figure is not needed. Note that $g_1(x) = x^2$ is the lower curve and $g_2(x) = x$ the upper curve; x ranges from 0 to 1. Since the equation of the surface is $z = x + y$, the integrand is $x + y$. Thus

$$V = \int_0^1 \int_{x^2}^x (x + y)\, dy\, dx = \frac{3}{20} \qquad \blacksquare$$

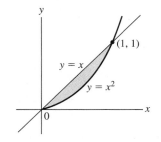

Figure 9.33

Example 4

Find the volume of the solid bounded by the paraboloid $z = x^2 + y^2$, the cylinder $x = y - y^2$, and the planes $z = 0$ and $x = 0$ (Figure 9.34).

Solution. Since $x^2 + y^2 \geq 0$, we know that the surface $z = x^2 + y^2$ lies above the xy-plane, except at the origin, so that the picture of the region R in Figure 9.35 would have sufficed. In this problem it is advantageous to integrate first with respect to x, referring to Formula (9.14). Thus $h_1(y) = 0$ and $h_2(y) = y - y^2$; that is, $h_1(y)$ is the lower function and $h_2(y)$ is the upper function when viewed from the y-axis. The outer limits are $y = 0$ and $y = 1$. We now get

$$V = \int_0^1 \int_0^{y-y^2} (x^2 + y^2)\, dx\, dy = \int_0^1 \left(\frac{1}{3}x^3 + xy^2\right)\Big|_0^{y-y^2} dy$$

$$= \int_0^1 \left(-\frac{1}{3}y^6 + y^5 - 2y^4 + \frac{4}{3}y^3\right) dy = \frac{11}{210}$$

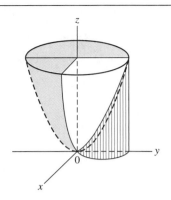

Figure 9.34

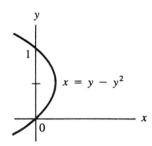

Figure 9.35

Remark: additional problems on volumes can be found in Section 9.8.

■ Exercises / Section 9.6

In Exercises 1–14, find the volumes of the solids bounded by the indicated surfaces.

1. The coordinate planes and the planes $x + y = 1$ and $z = 1$

2. The coordinate planes and the planes $y = 2 - x$ and $z = 2$

3. The cylinder $z = x^2$ and the planes $y = x$, $x = 1$, $y = 0$, and $z = 0$

4. The cylinder $z = y^2$ and the planes $y = x$, $y = 1$, $x = 0$, and $z = 0$

5. The plane $2x + y + 2z = 4$ and the coordinate planes

6. The planes $z = 1 - x$, $x = 0$, $y = 0$, $y = 2$, and $z = 0$

7. The cylinder $z = x^2$ and the planes $y = 2x$, $y = 2$, and $z = 0$

8. The cylinders $z = y^2$ and $y = x^2$ and the planes $x = 2$ and $z = 0$

9. First-octant volume: the plane $z = x$ and the cylinder $x^2 + y^2 = 9$

10. First-octant volume: the paraboloid $z = x^2 + y^2$ and the plane $x + y = 2$

11. First-octant volume: the plane $x + y = 4$ and the surface $z = xy$

12. The paraboloid $z = 4 - x^2 - y^2$; the cylinders $y = x^2$ and $y = \sqrt{x}$, above the xy-plane

13. The cylinder $y^2 = x$, the planes $x = 1$ and $z = 0$, and the paraboloid $z = x^2 + 3y^2$

14. The planes $z = y$ and $z = 0$ and the cylinder $y = \sqrt{1 - x^2}$

In Exercises 15–22, **set up, but do not evaluate,** each of the integrals to find the volume indicated.

15. The volume of a sphere of radius r

16. The volume above the xy-plane under the surface $z = 1 - x^2 - y^2$

17. The volume above the xy-plane, under $z = 9 - x^2 - y^2$, and inside $x^2 + y^2 = 4$

18. The volume bounded by $x^2 + y^2 = 1$ and $2x + y + z = 4$, above the xy-plane, inside the cylinder

19. The volume bounded by $x = \sqrt{y}$, $x + y = 2$, $y = 0$ and the ellipsoid $z^2 + 2x^2 + y^2 = 8$, above the xy-plane

20. First-octant volume bounded by $y = x^3$ and $y = 1$ and the sphere $x^2 + y^2 + z^2 = 4$

21. First-octant volume bounded by $z = xy$, $y^2 = x - 1$, $y = 1$, and $y = 2$

22. The smaller volume bounded by $x^2 + y^2 = 4$, $x + y = 2$, $z = 0$, and $z = 1 + x^2 + y^2$

In Exercises 23–31, graph each region and reverse the order of integration.

23. $\displaystyle\int_0^1 \int_0^x F(x, y)\, dy\, dx$ **24.** $\displaystyle\int_0^1 \int_{x^2+1}^2 F(x, y)\, dy\, dx$

25. $\displaystyle\int_0^4 \int_0^{\sqrt{x}} F(x, y)\, dy\, dx$ **26.** $\displaystyle\int_0^2 \int_{x^2}^{2x} F(x, y)\, dy\, dx$

27. $\displaystyle\int_{-2}^2 \int_{y^2}^4 F(x, y)\, dx\, dy$ **28.** $\displaystyle\int_0^1 \int_{2y}^2 F(x, y)\, dx\, dy$

29. $\displaystyle\int_0^2 \int_{4y^2}^{16} F(x, y)\, dx\, dy$ **30.** $\displaystyle\int_0^1 \int_y^{\sqrt{y}} F(x, y)\, dx\, dy$

31. $\displaystyle\int_0^1 \int_{y^2}^{\sqrt{y}} F(x, y)\, dx\, dy$

32. Set up the integral for finding the volume of the smaller solid bounded by the paraboloid $z = 9 - x^2 - y^2$ and the planes $z = 0$ and $x + y = 3$.

33. Find the volume of the solid bounded by $z = 4 - y^2$, $x = 2$, $x = 0$, $y = 0$, and $z = 0$ (first-octant).

34. Find the volume of the solid bounded by the following planes: $z = x$, $x = y$, $y = 1$, and $z = 0$.

35. Find the volume of the solid bounded by the plane $z = 0$ and the cylinders $x^2 + z = 1$ and $y^2 + z = 1$.

36. Find the volume of the solid bounded by $x + z = 2$, $y = x^3$, $y = 0$, and $z = 0$.

9.7   Mass, Centroids, and Moments of Inertia

Moments and mass of plane regions can be found readily by double integration. To see how, suppose we find the first-quadrant area bounded by $y = x^2$ and $y = 1$ (Figure 9.36) by the following scheme: let $z = 1$ be the surface above this region. Then the volume of the resulting solid is numerically equal to the area. Therefore

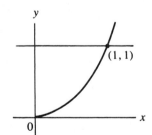

$$A = \int_0^1 \int_{x^2}^1 1\, dy\, dx = \int_0^1 y \Big|_{x^2}^1 dx$$

$$= \int_0^1 (1 - x^2)\, dx = \frac{2}{3}$$

Figure 9.36

Since this area could have been obtained by using a single integral, namely the last one, we seem to have done some extra work for nothing. Actually, this approach turns out to be extremely fruitful, as it leads to a different point of view regarding the composition of the area. To see why, consider the region in Figure 9.37, subdivided according to the scheme in Figure 9.28. Since the surface above the region is $z = 1$, the vertical columns (not shown) all have a height of 1 unit. As before, we fix x and sum $F(x, y)\, dy = 1 \cdot dy$ from $y = g_1(x)$ to $y = g_2(x)$ to get

$$\int_{g_1(x)}^{g_2(x)} dy$$

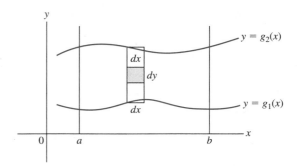

Figure 9.37

The volume of the typical slice is now given by

$$\left(\int_{g_1(x)}^{g_2(x)} dy \right) dx$$

Note that this is numerically equal to the area of the strip in Figure 9.37 and hence to the sum of the areas of the rectangles within the strip. Finally, we sum the strips from $x = a$ to $x = b$ to obtain

$$A = \int_a^b \int_{g_1(x)}^{g_2(x)} dy \, dx$$

which is the sum of all the little rectangles in the region. In other words, if the area is found by double integration, then the typical element is the small rectangle of dimensions dx by dy in Figure 9.37.

Typical Element for Double Integrals

The **typical element** in a plane region is a rectangle of dimensions dx by dy, as shown in Figure 9.37. If the area of the rectangle is denoted by dA, then

$$dA = dx \, dy \qquad \text{or} \qquad dA = dy \, dx.$$

Our first application of this summation process is the calculation of the mass of a plate of nonuniform density.

Example 1   Use the typical element to find

    a. the area of the region in Figure 9.36.
    b. the mass m, given that the density of the plate is xy.

Solution.

    a. The area of the typical element is $dA = dy \, dx$. Adding these elements, we obtain the integral

$$A = \int_0^1 \int_{x^2}^1 dy \, dx = \frac{2}{3}$$

evaluated before.

b. Since the density of the plate is xy, the mass of the typical element is $xy\,dy\,dx$. Adding these elements, we get

$$m = \int_0^1 \int_{x^2}^1 xy\,dy\,dx = \int_0^1 x \left(\frac{1}{2}y^2\right)\Big|_{x^2}^1 dx$$

$$= \frac{1}{2}\int_0^1 (x - x^5)\,dx = \frac{1}{6} \text{ unit of mass} \qquad \blacksquare$$

The new typical element in Figure 9.37 gives us a simple way to derive compact formulas for centroids and moments of inertia. From the definitions in Chapter 5 the moment of the typical element about the y-axis is

$$x \cdot \text{area} = x\,dy\,dx \tag{9.15}$$

and the moment of inertia about the y-axis is

$$x^2 \cdot \text{mass} = x^2 \cdot \rho\,dy\,dx = \rho x^2\,dy\,dx \tag{9.16}$$

Recalling that the x-coordinate of the centroid is $\bar{x} = M_y/A$, we now obtain directly

$$\bar{x} = \frac{\int_a^b \int_{g_1(x)}^{g_2(x)} x\,dy\,dx}{A} \tag{9.17}$$

For the moment of inertia with respect to the y-axis, we find

$$I_y = \rho \int_a^b \int_{g_1(x)}^{g_2(x)} x^2\,dy\,dx \tag{9.18}$$

To obtain the formulas for the moment and moment of inertia with respect to the x-axis, we had to do some extra work in Chapter 5. Now we simply note that the distance to the x-axis is y, so that the moment of the typical element is

$$y \cdot \text{area} = y\,dy\,dx \tag{9.19}$$

and the moment of inertia

$$y^2 \cdot \text{mass} = y^2 \cdot \rho\,dy\,dx = \rho y^2\,dy\,dx \tag{9.20}$$

Hence the y-coordinate of the centroid becomes

$$\bar{y} = \frac{\int_a^b \int_{g_1(x)}^{g_2(x)} y\,dy\,dx}{A} \qquad (\text{since } \bar{y} = M_x/A) \tag{9.21}$$

For the moment of inertia with respect to the x-axis, we get

$$I_x = \rho \int_a^b \int_{g_1(x)}^{g_2(x)} y^2\,dy\,dx \tag{9.22}$$

We also recall that the radius of gyration is defined by

$$R = \left(\frac{I}{m}\right)^{1/2} \tag{9.23}$$

where I is the moment of inertia and m is the mass.

The real advantage of the double-integral approach now becomes clear: in addition to resulting in simple derivations, the formulas themselves all have the same form! Furthermore, we can also obtain a new formula, the moment of inertia I_0, about an axis through the origin, perpendicular to the xy-plane. Since the distance to the origin is $\sqrt{x^2 + y^2}$,

$$I_0 = \rho \int_a^b \int_{g_1(x)}^{g_2(x)} (x^2 + y^2)\, dy\, dx \tag{9.24}$$

Example 2   Draw a typical element and find the centroid, I_y, and I_0 of the region in Figure 9.38. The region is bounded by $y = x^2, y = 2 - x, x = 0$.

Solution. From the figure,

$$M_y = \int_0^1 \int_{x^2}^{2-x} x\, dy\, dx = \int_0^1 xy \Big|_{x^2}^{2-x} dx \qquad \text{moment} = x \cdot \text{area}$$
$$= x \cdot dy\, dx$$

$$= \int_0^1 x(2 - x - x^2)\, dx$$

(Note that the last integral could have been obtained from the methods of Chapter 5.) Continuing, we get

$$M_y = \int_0^1 (2x - x^2 - x^3)\, dx = \frac{5}{12}$$

As long as the limits of integration have already been found, we may as well use them to find the area. Thus

$$A = \int_0^1 \int_{x^2}^{2-x} dy\, dx = \int_0^1 y \Big|_{x^2}^{2-x} dx = \int_0^1 (2 - x - x^2)\, dx = \frac{7}{6} \qquad \text{area} = dy\, dx$$

Hence

$$\bar{x} = \frac{M_y}{A} = \frac{5}{12} \cdot \frac{6}{7} = \frac{5}{14}$$

Next,

$$M_x = \int_0^1 \int_{x^2}^{2-x} y\, dy\, dx = \int_0^1 \frac{1}{2} y^2 \Big|_{x^2}^{2-x} dx \qquad \text{moment} = y \cdot \text{area} = y \cdot dy\, dx$$

$$= \frac{1}{2} \int_0^1 [(2 - x)^2 - x^4]\, dx = \frac{16}{15}$$

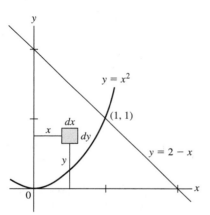

Figure 9.38

Therefore,

$$\bar{y} = \frac{M_x}{A} = \frac{16}{15} \cdot \frac{6}{7} = \frac{32}{35}$$

For the moments of inertia we have

$$I_y = \rho \int_0^1 \int_{x^2}^{2-x} x^2 \, dy \, dx = \rho \int_0^1 x^2 y \Big|_{x^2}^{2-x} dx \qquad \boldsymbol{x^2 \cdot \text{mass} = x^2 \cdot \rho \, dy \, dx}$$

$$= \rho \int_0^1 x^2 (2 - x - x^2) \, dx = \frac{13\rho}{60}$$

Finally,

$$I_0 = \rho \int_0^1 \int_{x^2}^{2-x} (x^2 + y^2) \, dy \, dx = \frac{149\rho}{105} \qquad \boldsymbol{\left(\sqrt{x^2 + y^2}\right)^2 \cdot \text{mass} = (x^2 + y^2) \cdot \rho \, dy \, dx}$$

after some calculations. ∎

Note that in Example 2 (since the mass $m = \rho A$)

$$R_y = \sqrt{\frac{I_y}{m}} = \sqrt{\frac{13\rho}{60} \cdot \frac{6}{7\rho}} = \sqrt{\frac{13}{70}} \approx 0.43$$

Triple Integrals*

The methods of this section can be extended to triple integrals, although we will not dwell at length on this topic. The three-dimensional analog of the typical element is the rectangular solid with dimensions $dx \cdot dy \cdot dz$ in Figure 9.39. The different moments of the typical element about the coordinate planes are readily seen to be the following:

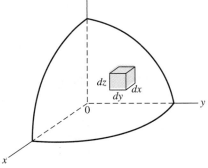

Figure 9.39

yz-plane: $x \, dz \, dy \, dx$

xz-plane: $y \, dz \, dy \, dx$

xy-plane: $z \, dz \, dy \, dx$

Example 3   Find $\bar{x}$ for the solid bounded by $z = x$, $x = 1$, and $y = 1$ and the coordinate planes (Figure 9.40).

Solution. The limits in the z-direction are always the lower surface and the upper surface, in this case, $z = 0$ and $z = x$. For the other limits we refer to the region R in the xy-plane, as usual. We now have

$$M_{yz} = \int_0^1 \int_0^1 \int_0^x x \, dz \, dy \, dx = \int_0^1 \int_0^1 xz \Big|_0^x dy \, dx$$

$$= \int_0^1 \int_0^1 x^2 \, dy \, dx = \frac{1}{3}$$

*The remainder of this section and Exercises 25–36 can be omitted without loss of continuity.

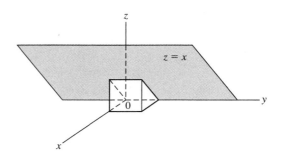

Figure 9.40

Also,

$$V = \int_0^1 \int_0^1 \int_0^x dz\,dy\,dx = \int_0^1 \int_0^1 x\,dy\,dx = \frac{1}{2}$$

so that

$$\bar{x} = \frac{M_{yz}}{V} = \frac{\frac{1}{3}}{\frac{1}{2}} = \frac{2}{3} \qquad\blacksquare$$

The different moments of inertia of the typical element are obtained similarly. About the

x-axis: $\rho(y^2 + z^2)\,dz\,dy\,dx$

y-axis: $\rho(x^2 + z^2)\,dz\,dy\,dx$

z-axis: $\rho(x^2 + y^2)\,dz\,dy\,dx$

Example 4   Find I_x, the moment of inertia about the x-axis, of the solid in Example 3.

Solution.

$$I_x = \int_0^1 \int_0^1 \int_0^x \rho(y^2 + z^2)\,dz\,dy\,dx = \rho \int_0^1 \int_0^1 \left(y^2 z + \frac{1}{3}z^3 \right) \Big|_0^x dy\,dx$$

$$= \rho \int_0^1 \int_0^1 \left(xy^2 + \frac{1}{3}x^3 \right) dy\,dx = \frac{\rho}{4} \qquad\blacksquare$$

Example 5   Set up the integral for finding I_z of the solid bounded by the paraboloid $z = x^2 + y^2 + 1$ and the planes $z = 1$, $x + y = 2$, $x = 0$, and $y = 0$.

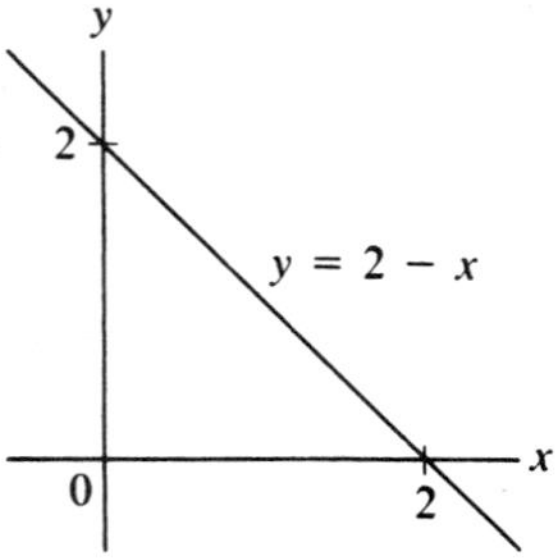

Solution. The solid lies above the region in the xy-plane bounded by the line $x + y = 2$ and the coordinate axes (Figure 9.41). The two-dimensional region is sufficient for finding the limits in the direction of x and y. In the z direction note that the lower surface is $z = 1$ and the upper surface is $z = x^2 + y^2 + 1$. So the integral is

$$I_z = \int_0^2 \int_0^{2-x} \int_1^{x^2+y^2+1} \rho(x^2 + y^2)\,dz\,dy\,dx \qquad\blacksquare$$

Figure 9.41

As in the case of plane regions, we can find the mass of a solid of nonuniform density. If in Example 5 the density of the solid is $3x^2y^2$, then the mass of the typical element of volume is $3x^2y^2 \, dz \, dy \, dx$. Hence the total mass is

$$m = \int_0^2 \int_0^{2-x} \int_1^{x^2+y^2+1} 3x^2y^2 \, dz \, dy \, dx$$

■ Exercises / Section 9.7

In Exercises 1–6, find the area and the mass of the region bounded by the given curves. (See Example 1.)

1. $y = (1/2)x$, $x = 4$, x-axis; density: $(1/4)x$

2. $y = x$, $y = 1$, y-axis; density: x^2

3. $y = \sqrt{x}$, $y = 2$, y-axis; density: xy

4. $y = \sqrt{x}$, $x = 4$, x-axis; density: $(1/6)y$

5. $y = 4 - x^2$, x-axis; density: x^2

6. $y = x$, $y = x^2$; density: $1 + y$

In Exercises 7–36, draw a typical element and set up the appropriate integral.

7. Find the coordinates of the centroid of the region bounded by $y = x^2$, $x = 2$, and the x-axis.

8. Find the coordinates of the centroid of the first-quadrant region bounded by $x^2 + y^2 = 9$ and the axes.

9. Find R_y of the region bounded by $y = x^3$, $x = 1$, and the x-axis.

10. Find R_x of the region bounded by $2x + y = 2$, $y = 2$, and $x = 1$.

11. Find I_0 of the region bounded by $y = x$ and $y = x^2$.

12. Find the coordinates of the centroid of the region bounded by $2y = x$ and $4y = x^2$.

13. Find R_y of the region bounded by $y = 1/x$, $x = 1$, $x = e$, and the x-axis.

14. Find R_0 of the region bounded by $x + y = 1$ and the coordinate axes.

15. Find $\bar{x}$ for the region in Exercise 13.

16. Find I_x and I_0 of the region bounded by $y = 1/x^2$, $x = 1$, $x = 2$, and the x-axis.

17. Find I_x of the region bounded by $y = 1/\sqrt[3]{x}$, $x = e$, $x = e^2$, and the x-axis.

18. Find the coordinates of the centroid of the region bounded by $y = \sin x$, $x = 0$, $x = \pi$, and the x-axis.

19. Find I_x of the region bounded by $x = 4y - 2y^2$ and the y-axis.

20. Find the coordinates of the centroid of the region bounded by $y = 4x$ and $y = x^3$ (first quadrant).

21. Find the coordinates of the centroid of the region bounded by $y = e^{-x}$, $y = 1$, and $x = 1$.

22. Find I_y of the region bounded by $y = \ln x$, $x = 1$, $x = e^2$, and the x-axis.

23. Find the mass of the region in Exercise 22 if the density is x.

24. Find I_x of the region bounded by $y = \sec^{2/3} x$, $x = 0$, $x = \pi/4$, and the x-axis.

25. Find the coordinates of the centroid of the solid in the first octant bounded by $y = 1 - x^2$ and $y + z = 1$.

26. Find I_x of the solid bounded by $y = x$, $x = 2$, $y = 0$, $z = 0$, and $z = y$.

27. Find I_z of the solid in the first octant bounded by $x + y = 1$ and $z = 2$.

28. Find the coordinates of the centroid of the solid in Exercise 26.

29. Find I_y of the solid bounded by $y = x$, $y = x^2$, $z = xy$, and $z = 0$.

30. Find I_z of the solid in the first octant bounded by $z = 4 + x^2$, $x = 2$, and $y = 3$.

31. Find I_x of the solid in the first octant bounded by $z = x^2$, $x = 1$, and $y = 2$.

32. Set up the integral for finding I_y of the solid bounded by $x^2 + z = 4$, $2x + z = 4$, $y = 0$, and $y = 3$.

33. Find the coordinates of the centroid of the solid in Exercise 27.

34. Find the coordinates of the centroid of the solid in Exercise 31.

35. Set up the integral for finding the mass of the solid in Exercise 32 if the density of the solid is $1 + x$.

36. Set up the integral for finding I_z of the solid in Exercise 32 if the density of the solid is $1 + y$.

9.8   Volumes in Cylindrical Coordinates

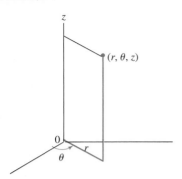

Figure 9.42

In some situations another coordinate system is particularly convenient. **Cylindrical coordinates** combine polar and rectangular coordinates in the sense that the x- and y-coordinates are replaced by polar coordinates, while the z-coordinates are kept the same. (See Figure 9.42.) For example, the point $(1, \sqrt{3})$ is $(2, 60°)$ in polar coordinates. So the point $(1, \sqrt{3}, 4)$ becomes $(2, 60°, 4)$ in cylindrical coordinates. (See Figure 9.43.) The conversion formulas from Chapter 8 are therefore the same, as long as we remember to leave z unchanged.

Conversion Formulas

$$x = r\cos\theta \qquad y = r\sin\theta \qquad z = z \tag{9.25}$$

$$r^2 = x^2 + y^2 \qquad \tan\theta = \frac{y}{x} \tag{9.26}$$

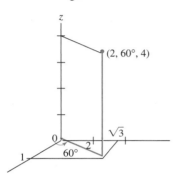

Figure 9.43

As another example, the point $(2, 5\pi/6, 5)$ in cylindrical coordinates becomes

$$x = 2\cos\frac{5\pi}{6} = 2\left(-\frac{\sqrt{3}}{2}\right) = -\sqrt{3}, \quad y = 2\sin\frac{5\pi}{6} = 2\left(\frac{1}{2}\right) = 1, \; z = 5$$

or $(-\sqrt{3}, 1, 5)$. Notice that z remains unchanged.

Since we are primarily interested in surfaces and volumes, consider the following example:

Example 1   The cylindrical surface $4x^2 + 4y^2 = 9$ becomes

$$4(x^2 + y^2) = 9 \qquad \text{or} \qquad 4r^2 = 9$$

Taking the positive square root of each side, the resulting equation is $r = 3/2$ in cylindrical coordinates.   ∎

Example 2   Write the equation of the paraboloid $z = 1 + 2x^2 + 2y^2$ in cylindrical coordinates.

Solution.

$$z = 1 + 2(x^2 + y^2) \qquad \text{or} \qquad z = 1 + 2r^2$$ ∎

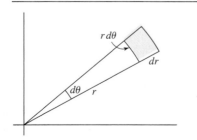

Figure 9.44

To calculate volumes in cylindrical coordinates, we recall from the previous section that the typical element of area for double integrals is $dA = dx\,dy$. In polar coordinates this becomes $dA = r\,dr\,d\theta$, as suggested by Figure 9.44. (For very small increments the region is approximately rectangular.)

As an example, let us set up the integral for finding the area of one leaf of the rose $r = \sin 3\theta$, which is traced out in the interval $\theta = 0$ to $\theta = \pi/3$. Using

$dA = r\, dr\, d\theta$, we could try

$$A = \int_0^{\pi/3} \int_0^{\sin 3\theta} r\, dr\, d\theta$$

As a check, observe that

$$A = \int_0^{\pi/3} \frac{1}{2} r^2 \bigg|_0^{\sin 3\theta} d\theta = \frac{1}{2} \int_0^{\pi/3} \sin^2 3\theta\, d\theta$$

which we know to be the correct form from Chapter 8.

It follows that the volume of a solid above the xy-plane bounded by $z = F(r, \theta)$, $r_1 = f_1(\theta), r_2 = f_2(\theta)$, and the rays $\theta = \alpha$ and $\theta = \beta$ is given by

$$V = \int_\alpha^\beta \int_{f_1(\theta)}^{f_2(\theta)} F(r, \theta)\, r\, dr\, d\theta \qquad (9.27)$$

Consider the next example.

Example 3

Find the first-octant volume of the solid bounded by the planes $z = x$ and $z = 0$ and by the cylinder $x^2 + y^2 = 4$.

Solution. The surface is shown in Figure 9.45. As usual, the limits of integration depend on the region in the xy-plane, shown in Figure 9.46. The circle $x^2 + y^2 = 4$ becomes $r = 2$, while θ ranges from $\theta = 0$ to $\theta = \pi/2$. The surface $z = x$ is $z = r \cos \theta$ in cylindrical coordinates. We therefore have

$$V = \int_0^{\pi/2} \int_0^2 (r \cos \theta)\, r\, dr\, d\theta = \int_0^{\pi/2} (\cos \theta) \left(\frac{1}{3} r^3\right) \bigg|_0^2 d\theta$$

$$= \frac{8}{3} \int_0^{\pi/2} \cos \theta\, d\theta = \frac{8}{3} \sin \theta \bigg|_0^{\pi/2} = \frac{8}{3}$$

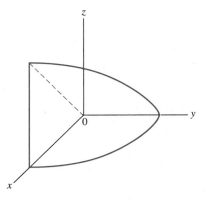

Figure 9.45

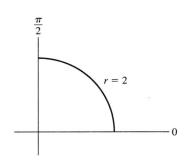

Figure 9.46 ■

■ Exercises / Section 9.8

In Exercises 1–12, find the volumes of the solids bounded by the given surfaces using cylindrical coordinates.

1. The cylinder $x^2 + y^2 = 9$ and the planes $z = 2$ and $z = 0$

2. First-octant-volume: the cylinder $x^2 + y^2 = 1$ and the planes $z = 4$ and $z = 0$

3. The paraboloid $z = 9 - x^2 - y^2$, above the xy-plane

4. Under the paraboloid $z = 9 + x^2 + y^2$, inside the cylinder $x^2 + y^2 = 4$, above the xy-plane

5. First-octant volume: the paraboloid $z = 4 - x^2 - y^2$, the cylinder $x^2 + y^2 = 1$, and $z = 0$

6. The planes $z = x$ and $z = 0$ and the cylinder $x^2 + y^2 = 1$, above the xy-plane

7. The planes $z = 3y$ and $z = 0$, and the cylinder $x^2 + y^2 = 4$, above the xy-plane

8. First-octant volume: the planes $y = 0$, $y = x$, the cylinder $x^2 + y^2 = 1$, and the paraboloid $z = 1 + x^2 + y^2$ (above the xy-plane)

9. First-octant volume: the planes $y = x$ and $x = 0$, the cylinder $x^2 + y^2 = 9$, and the paraboloid $z = x^2 + y^2$ (above the xy-plane)

10. The cylinders $x^2 + y^2 = 4$ and $x^2 + y^2 = 1$, and the planes $z = 4$ and $z = 0$

11. First-octant volume: the surface $z = xy$, the cylinder $x^2 + y^2 = 1$, and the plane $z = 0$

12. First-octant volume: the surface $z = \sqrt{x^2 + y^2}$, the cylinder $x^2 + y^2 = 9$, and the plane $z = 0$

Evaluate the following integrals by changing to cylindrical coordinates.

13. $\displaystyle\int_0^4 \int_0^{\sqrt{16-x^2}} x \, dy \, dx$

14. $\displaystyle\int_0^2 \int_0^{\sqrt{4-y^2}} y \, dx \, dy$

15. $\displaystyle\int_0^1 \int_0^{\sqrt{1-x^2}} (x^2 + y^2)^{3/2} \, dy \, dx$

16. $\displaystyle\int_0^3 \int_0^{\sqrt{9-y^2}} \sqrt{x^2 + y^2} \, dx \, dy$

17. $\displaystyle\int_{-2}^2 \int_0^{\sqrt{4-x^2}} x \, dy \, dx$

■ Review Exercises / Chapter 9

In Exercises 1–9, identify and sketch each of the given surfaces.

1. $x + y + 2z = 2$
2. $4z - 6x - 5y = 12$
3. $2y^2 + z^2 = 4$
4. $8x^2 + 4y^2 + z^2 = 8$
5. $y = 3x^2$
6. $z = 2x^2 + 2y^2$
7. $z^2 - 4x^2 - y^2 = 4$
8. $x^2 + y^2 + z^2 = 6$
9. $4x^2 + 2y^2 - z^2 = 9$

In Exercises 10 and 11, find:

a. $\dfrac{\partial z}{\partial x}$ b. $\dfrac{\partial z}{\partial y}$ c. $\dfrac{\partial^2 z}{\partial x^2}$

d. $\dfrac{\partial^2 z}{\partial y^2}$ e. $\dfrac{\partial^2 z}{\partial x \partial y}$

10. $z = y\sqrt{y^2 - x}$ 11. $z = \ln\sqrt{x^2 + y^2} + \sin xy$

12. The second-order partial differential equation

$$\frac{\partial^2 z}{\partial x^2} + \frac{\partial^2 z}{\partial y^2} = 0$$

is called *Laplace's equation*. Functions that satisfy this equation are called *harmonic*. Show that $z = e^x \cos y$ is harmonic.

13. The temperature (in °C) of a circular plate of radius 10 cm and center at the origin is $T = 2xy - x^2 - 2y^2 + 3x + 5$. Find the temperature at the warmest point.

14. A circular plate centered at the origin has a radius of 5 inches. If the temperature (in °C) is given by $T = 2x^2 + y^2 - 2xy + x + y$, find the temperature at the coldest point.

15. (Refer to Exercise 13.) Find the rate of change of the temperature with respect to the distance if we start at the point $(-3, 1)$ and move in the positive x-direction.

16. The range R (in feet) of a projectile hurled with velocity v at an angle θ with the ground is

$$R = \frac{v^2}{32} \sin 2\theta$$

Suppose v increases at the rate of 3.50 ft/s per second and θ increases at 0.100 rad/s. Determine the rate of change of R at the instant when $v = 60.0$ ft/s and $\theta = 52.0°$.

17. Find the volume of the solid inside the cylinder $x^2 + z^2 = 4$, above the xy-plane, and bounded by $y = x$ and $y = 0$ (first octant).

18. Find the volume of the solid bounded by $z = 2y$, $y = 4 - x^2$, and $z = 0$.

19. Find the volume of the solid bounded by $x + z = 1$, $y = 2$, $x = 0$, $y = 0$, $z = 0$.

20. Find the mass of the region bounded by $y = 2x$, $y = 2$, and the y-axis, given that the density is $\rho = (1/2)x$.

21. Use double integration to find I_y of the region bounded by $x = \sqrt{4 - y}$ and the coordinate axes.

22. Use double integration to find the centroid of the region in Exercise 21.

23. Use double integration to find the centroid of the region bounded by $y = x$ and $y = 2 - x^2$.

24. Find I_y and R_y of the region in Exercise 23.

25. Exchange the order of integration:

$$\int_0^4 \int_{\sqrt{y}}^2 F(x, y)\, dx\, dy$$

26. Exchange the order of integration:

$$\int_0^{16} \int_{-\sqrt{16-y}}^{\sqrt{16-y}} F(x, y)\, dx\, dy$$

27. The radius of a right circular cylinder is measured to be 10.00 cm with a possible error of ± 0.1 mm. The height is measured to be 15.000 cm with a possible error of ± 0.05 mm. Find the approximate resulting error and percentage error in the volume.

28. A certain box has a square base. If the base is measured to be 6.00 ft with a possible error of ± 0.01 ft and the height is measured to be 4.00 ft with a possible error of ± 0.02 ft, find the approximate error in the volume.

29. Find I_z of the solid bounded by $z = 9 - x^2$, $y = 0$, $y = 1$, and $z = 0$.

30. If the density of the solid in Exercise 29 is x^2, find its mass.

31. Set up the integral for finding I_x of the solid bounded by $z = x$, $z = 2x$, and $x = \sqrt{1 - y^2}$.

32. Find $\bar{x}$ for the solid bounded by $z = y^2$, $z = y$, $x = \sqrt{y}$, $y = 1$, and $x = 0$.

33. Use cylindrical coordinates to find the volume under the paraboloid $z = 16 - x^2 - y^2$, above the xy-plane

10

Infinite Series

10.1 Introduction to Infinite Series

In this section we will see how the idea of an ordinary sum can be extended to the sum of an infinite number of terms, called an *infinite series*.

Suppose that a man standing 1 m away from a wall steps $(1/2)$ m forward so that the distance to the wall is cut in half. Then he takes another step forward, again cutting the remaining distance in half (Figure 10.1). Imagine that this operation is continued indefinitely. Does it make sense to talk about the total distance covered? In one way it doesn't, for the total distance is given by

$$\frac{1}{2} + \frac{1}{4} + \frac{1}{8} + \frac{1}{16} + \cdots$$

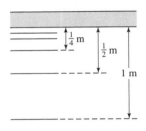

Figure 10.1

which involves an infinite number of terms. Even if you were immortal, you could never complete the addition. Yet according to Figure 10.1 the sum *ought* to be 1 m, even if you cannot physically (or even conceptually) add up infinitely many numbers.

The problem is one of definition: we first have to agree on the meaning of a sum of an infinite number of terms. To this end let a and r be two numbers and consider the expression

$$S = a + ar + ar^2 + ar^3 + \cdots \tag{10.1}$$

Infinite series

(The three dots indicate that the summation is to be continued indefinitely.) To distinguish this case from an ordinary sum, Equation (10.1) is called an **infinite series.** Suppose we consider only the first n terms:

$$S_n = a + ar + ar^2 + ar^3 + \cdots + ar^{n-1} \tag{10.2}$$

called the *nth partial sum.* The sum can be obtained by multiplying S_n by r and subtracting the result from S_n. Thus

$$
\begin{aligned}
S_n &= a + ar + ar^2 + ar^3 + \cdots + ar^{n-1} \\
r\,S_n &= ar + ar^2 + ar^3 + \cdots + ar^{n-1} + ar^n \\
\hline
S_n - r\,S_n &= a \phantom{+ ar + ar^2 + ar^3 + \cdots + ar^{n-1}} - ar^n \qquad \text{subtracting}
\end{aligned}
$$

or

$$S_n(1 - r) = a - ar^n \qquad \textbf{factoring } \pmb{S_n}$$

and

$$S_n = \frac{a - ar^n}{1 - r} = \frac{a(1 - r^n)}{1 - r} \qquad \textbf{dividing by 1} - \pmb{r}$$

The only legitimate way to pass from a finite to an infinite number of terms is through the limit process. We define the sum S as the limit of an infinite sequence of partial sums:

$$S = \lim_{n \to \infty} S_n = \lim_{n \to \infty} \frac{a(1 - r^n)}{1 - r} \qquad\qquad (10.3)$$

If $|r| < 1$, then $r^n \to 0$ as $n \to \infty$. Hence the limit is

$$S = \frac{a}{1 - r}$$

(If $|r| > 1$, the limit does not exist.)

The series in Equation (10.1) is called a *geometric series*.

Geometric Series

A geometric series has the form

$$S = a + ar + ar^2 + \cdots + ar^{n-1} + \cdots$$

If $|r| < 1$, the **sum** is given by

$$S = \frac{a}{1 - r} \qquad\qquad (10.4)$$

Example 1   Find the sum d of the individual distances in Figure 10.1:

$$\frac{1}{2} + \frac{1}{4} + \frac{1}{8} + \frac{1}{16} + \cdots$$

Solution. To find the sum we only need to observe that a is always the first term and ar the second. It follows that $a = 1/2$, $ar = (1/2)r = 1/4$. So $r = 1/2$. By Formula (10.4),

$$d = \frac{\frac{1}{2}}{1 - \frac{1}{2}} = 1$$

as expected.   ∎

Example 2   Find the sum of the following geometric series:

$$S = 4 - 3 + \frac{9}{4} - \frac{27}{16} + \frac{81}{64} - \cdots$$

Solution. Since $a = 4$ and $ar = -3$, it follows that $4r = -3$ and $r = -3/4$. Hence

$$S = \frac{a}{1-r} = \frac{4}{1 - \left(-\frac{3}{4}\right)} = \frac{4}{1} \cdot \frac{4}{7} = \frac{16}{7}$$

Observe that the given series can also be written in Form (10.1):

$$S = 4 + 4\left(-\frac{3}{4}\right) + 4\left(-\frac{3}{4}\right)^2 + 4\left(-\frac{3}{4}\right)^3 + 4\left(-\frac{3}{4}\right)^4 + \cdots \qquad \blacksquare$$

Formula (10.3) now provides the motivation for a general definition. Let

$$S = b_1 + b_2 + b_3 + \cdots + b_n + \cdots \qquad (10.5)$$

be an infinite series and let

$$S_n = b_1 + b_2 + b_3 + \cdots + b_n \qquad (10.6)$$

denote the nth partial sum. Then we define the sum S to be

$$S = \lim_{n \to \infty} S_n \qquad (10.7)$$

whenever the limit exists. This limit of partial sums leads to the definition of *convergence* and *divergence*.

Definition of Convergence and Divergence

If

$$S = b_1 + b_2 + \cdots + b_n + \cdots$$

is an **infinite series,** then

$$S_n = b_1 + b_2 + \cdots + b_n$$

is called the **nth partial sum.**

1. If $\lim_{n \to \infty} S_n$ exists, then the series is said to **converge** or to be convergent.
2. If $\lim_{n \to \infty} S_n$ does not exist, then the series is said to **diverge** or to be divergent.

The series

$$\frac{1}{2} + \frac{1}{4} + \frac{1}{8} + \frac{1}{16} + \cdots$$

considered earlier is an example of a convergent series. An example of a divergent series is

$$1 + 2 + 3 + \cdots + n + \cdots$$

since

$$S_n = 1 + 2 + 3 + \cdots + n = \frac{n(n+1)}{2}$$

(Section 4.2) and

$$\lim_{n \to \infty} S_n = \infty$$

On the other hand, the nth partial sum of the series

$$1 - 1 + 1 - 1 + 1 - \cdots$$

does not get very large, but since it oscillates between 0 and 1, it never settles down to a definite limit. As a consequence, the series has to be called divergent.

At first glance the definition of convergence appears to yield a method for finding the sum. Such is not the case: there does not exist a general method for finding the sum of an infinite series, and even (apparently) simple cases require considerable ingenuity. For example, it is known that

$$\frac{1}{1^2} + \frac{1}{2^2} + \frac{1}{3^2} + \frac{1}{4^2} + \cdots + \frac{1}{n^2} + \cdots = \frac{\pi^2}{6}$$

while the exact sum of

$$\frac{1}{1^3} + \frac{1}{2^3} + \frac{1}{3^3} + \frac{1}{4^3} + \cdots + \frac{1}{n^3} + \cdots$$

has never been found.

In the exercises, as well as in later sections, we are going to make regular use of the sigma notation in writing series. Thus

$$\sum_{n=1}^{\infty} b_n = b_1 + b_2 + b_3 + \cdots + b_n + \cdots$$

For example,

$$\sum_{n=1}^{\infty} \frac{1}{n} = \frac{1}{1} + \frac{1}{2} + \frac{1}{3} + \cdots + \frac{1}{n} + \cdots$$

■ Exercises / Section 10.1

In Exercises 1–14, find the sum of each of the geometric series.

1. $\displaystyle\sum_{n=0}^{\infty} \frac{1}{3^n} = 1 + \frac{1}{3} + \frac{1}{3^2} + \frac{1}{3^3} + \cdots + \frac{1}{3^n} + \cdots$

2. $\displaystyle\sum_{n=1}^{\infty} \left(\frac{2}{3}\right)^{n-1} = 1 + \frac{2}{3} + \left(\frac{2}{3}\right)^2 + \cdots + \left(\frac{2}{3}\right)^{n-1} + \cdots$

3. $\displaystyle\sum_{n=1}^{\infty} \frac{2}{3^n} = \frac{2}{3} + \frac{2}{3^2} + \frac{2}{3^3} + \cdots + \frac{2}{3^n} + \cdots$

4. $\displaystyle\sum_{n=1}^{\infty} \left(\frac{5}{6}\right)^n = \frac{5}{6} + \left(\frac{5}{6}\right)^2 + \left(\frac{5}{6}\right)^3 + \cdots + \left(\frac{5}{6}\right)^n + \cdots$

5. $\displaystyle\sum_{n=1}^{\infty} \left(\frac{3}{4}\right)^{n-1} = 1 + \frac{3}{4} + \frac{9}{16} + \cdots + \left(\frac{3}{4}\right)^{n-1} + \cdots$

6. $\displaystyle\sum_{n=1}^{\infty} \frac{3}{4^n} = \frac{3}{4} + \frac{3}{16} + \frac{3}{64} + \cdots + \frac{3}{4^n} + \cdots$

7. $\displaystyle\sum_{n=1}^{\infty} \frac{4}{9^n} = \frac{4}{9} + \frac{4}{9^2} + \frac{4}{9^3} + \cdots + \frac{4}{9^n} + \cdots$

8. $\displaystyle\sum_{n=1}^{\infty} \frac{(-1)^{n-1}}{2^{n-1}}$

$\qquad = 1 - \frac{1}{2} + \frac{1}{4} - \frac{1}{8} + \cdots + (-1)^{n-1}\frac{1}{2^{n-1}} + \cdots$

$\qquad$ (*Hint: r = -1/2.*)

9. $\displaystyle\sum_{n=1}^{\infty} (-1)^{n+1} \left(\frac{2}{3}\right)^n$

$\qquad = \frac{2}{3} - \frac{4}{9} + \frac{8}{27} - \cdots + (-1)^{n+1}\left(\frac{2}{3}\right)^n + \cdots$

10. $\displaystyle\sum_{n=0}^{\infty} (-1)^n \left(\frac{1}{10}\right)^n$

$\qquad = 1 - \frac{1}{10} + \left(\frac{1}{10}\right)^2 - \left(\frac{1}{10}\right)^3 + \left(\frac{1}{10}\right)^4 - \cdots$

11. $\displaystyle\sum_{n=0}^{\infty} (-1)^n \left(\frac{3}{4}\right)^n = 1 - \frac{3}{4} + \left(\frac{3}{4}\right)^2 - \left(\frac{3}{4}\right)^3 + \left(\frac{3}{4}\right)^4 - \cdots$

12. $\displaystyle\sum_{n=0}^{\infty} (-1)^n \left(\frac{3}{5}\right)^n = 1 - \frac{3}{5} + \left(\frac{3}{5}\right)^2 - \left(\frac{3}{5}\right)^3 + \left(\frac{3}{5}\right)^4 - \cdots$

13. $\displaystyle\sum_{n=0}^{\infty} (-1)^n \left(\frac{2}{7}\right)^n = 1 - \frac{2}{7} + \left(\frac{2}{7}\right)^2 - \left(\frac{2}{7}\right)^3 + \left(\frac{2}{7}\right)^4 - \cdots$

14. $\displaystyle\sum_{n=1}^{\infty} \left(\frac{4}{3}\right)^n = \frac{4}{3} + \left(\frac{4}{3}\right)^2 + \left(\frac{4}{3}\right)^3 + \cdots$

In Exercises 15–22, convert the repeating decimals to common fractions.

15. $0.212121\ldots$ $\left(Hint: 0.212121\ldots = \dfrac{21}{10^2} + \dfrac{21}{10^4} + \cdots\right)$

16. $0.050505\ldots$

17. $0.757575\ldots$

18. $0.55555\ldots$

19. $0.001001001\ldots$

20. $0.015015015\ldots$

21. $0.5070707\ldots$

22. $0.636363\ldots$

10.2 Tests for Convergence*

The purpose of this section is to discuss some of the standard tests for determining whether a given infinite series converges or diverges.

For many series it may be difficult to establish convergence. In some cases, however, it is possible to see at a glance that the series could not possibly converge. Let

$$S = a_1 + a_2 + a_3 + \cdots + a_n + \cdots$$

be an infinite series of positive terms and consider the sequence of partial sums

$$S_1 = a_1$$
$$S_2 = a_1 + a_2$$
$$S_3 = a_1 + a_2 + a_3$$
$$\vdots$$
$$S_n = a_1 + a_2 + a_3 + \cdots + a_n$$

If the infinite series converges, then $S = \lim_{n\to\infty} S_n$ exists by definition. Since n is just a dummy subscript, $\lim_{k\to\infty} S_k$ also exists and is equal to the first limit. In particular,

$$\lim_{n\to\infty} S_n = \lim_{n\to\infty} S_{n-1}$$

*This section may be omitted without loss of continuity.

Now, it can be seen from this sequence of partial sums that $a_n = S_n - S_{n-1}$. Consequently,

$$\lim_{n \to \infty} a_n = \lim_{n \to \infty} (S_n - S_{n-1}) = \lim_{n \to \infty} S_n - \lim_{n \to \infty} S_{n-1} = 0$$

We conclude that if $\sum_{n=1}^{\infty} a_n$ converges, then $\lim_{n \to \infty} a_n = 0$.

Necessary Condition for Convergence

If the series

$$S = a_1 + a_2 + \cdots + a_n + \cdots$$

is **convergent,** then

$$\lim_{n \to \infty} a_n = 0 \qquad\qquad\qquad (10.8)$$

Consequently, if $\lim_{n \to \infty} a_n$ is not equal to zero, then the series necessarily *diverges*. The converse is not true, however. If $\lim_{n \to \infty} a_n = 0$, then we may **not** conclude that the series converges—it may diverge anyway.

Example 1   For the series

$$\sum_{n=1}^{\infty} \frac{n}{n+1} = \frac{1}{2} + \frac{2}{3} + \frac{3}{4} + \cdots + \frac{n}{n+1} + \cdots$$

we have

$$\lim_{n \to \infty} a_n = \lim_{n \to \infty} \frac{n}{n+1} = 1 \qquad \textbf{by L'Hospital's rule}$$

so that the series diverges by Criterion (10.8).

On the other hand, for the series

$$\sum_{n=1}^{\infty} \frac{1}{n} = 1 + \frac{1}{2} + \frac{1}{3} + \cdots + \frac{1}{n} + \cdots$$

we have

$$\lim_{n \to \infty} a_n = \lim_{n \to \infty} \frac{1}{n} = 0$$

but the series *also diverges,* as we will see shortly.   ∎

Note that in Example 1, $a_n = n/(n+1)$ is a function of n. Using the notation $a_n = f(n)$, we can obtain the *integral test,* a simple and powerful test for convergence.

> **Integral Test**
>
> Suppose that $\sum_{n=1}^{\infty} a_n$ is a series of positive terms, with $a_n = f(n)$. Then $\sum_{n=1}^{\infty} a_n$ converges if the improper integral $\int_1^{\infty} f(x)\, dx$ exists and diverges if the integral does not exist. (The function $f(x)$ has to be continuous and decreasing.)

The validity of this test can be seen intuitively from a diagram (Figure 10.2). Note that the rectangles in each figure are constructed by plotting the ordinates $f(1), f(2), \ldots$, corresponding to $x = 1, 2, \ldots$.

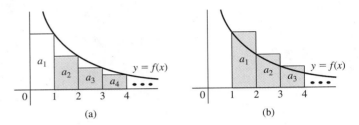

Figure 10.2

Suppose the integral exists. Then Figure 10.2(a) shows that the sum of the areas of the rectangles

$$f(2) + f(3) + \cdots + f(n) + \cdots = a_2 + a_3 + \cdots + a_n + \cdots$$

is less than $\int_1^{\infty} f(x)\, dx$, the area under the curve. Since the latter is finite, so is the sum of the rectangles, and the series converges. (The omission of the first term $a_1 = f(1)$ has no bearing on the result.) If $\int_1^{\infty} f(x)\, dx = \infty$, we see from Figure 10.2(b) that the sum of the rectangles

$$f(1) + f(2) + \cdots + f(n) + \cdots = a_1 + a_2 + \cdots + a_n + \cdots$$

cannot be finite either, so that the series diverges.

Example 2   Test the series

$$\sum_{n=2}^{\infty} \frac{1}{e^n} = \frac{1}{e^2} + \frac{1}{e^3} + \cdots + \frac{1}{e^n} + \cdots$$

for convergence by the integral test.

Solution. Since

$$\int_2^{\infty} \frac{1}{e^x}\, dx = \int_2^{\infty} e^{-x}\, dx = \lim_{b \to \infty} \int_2^{b} e^{-x}\, dx$$

$$= \lim_{b \to \infty} (-e^{-x}) \Big|_2^{b} = \lim_{b \to \infty} (-e^{-b} + e^{-2}) = \frac{1}{e^2}$$

the series converges. (Observe that the lower limit need not be $n = 1$.) ∎

Example 3   Find the values of p for which the series

$$\sum_{n=1}^{\infty} \frac{1}{n^p}$$

converges and diverges.

Solution.

$$\int_1^{\infty} \frac{dx}{x^p} = \lim_{b \to \infty} \int_1^b x^{-p} dx = \lim_{b \to \infty} \frac{x^{-p+1}}{-p+1} \Big|_1^b$$

$$= \lim_{b \to \infty} \left(\frac{b^{-p+1}}{-p+1} - \frac{1}{-p+1} \right) \quad (p \neq 1)$$

$$= \begin{cases} \dfrac{1}{p-1} & p > 1 \\ \infty & p < 1 \end{cases}$$

Consequently, the series converges for $p > 1$ and diverges for $p < 1$ by the integral test. If $p = 1$, we have

$$\int_1^{\infty} \frac{dx}{x} = \lim_{b \to \infty} \ln x \Big|_1^b = \lim_{b \to \infty} \ln b = \infty$$

so that the series

$$\sum_{n=1}^{\infty} \frac{1}{n}$$

is divergent.   ■

A series of the form

$$\sum_{n=1}^{\infty} \frac{1}{n^p} = \frac{1}{1^p} + \frac{1}{2^p} + \frac{1}{3^p} + \cdots + \frac{1}{n^p} + \cdots \tag{10.9}$$

p-series

is called a ***p*-series.** It was shown in Example 3 that a *p*-series converges if $p > 1$ and diverges if $p < 1$. The special case

$$\sum_{n=1}^{\infty} \frac{1}{n} = 1 + \frac{1}{2} + \frac{1}{3} + \cdots + \frac{1}{n} + \cdots \tag{10.10}$$

Harmonic series

($p = 1$) is called the **harmonic series.** The harmonic series is divergent.

In the process of discussing convergence and divergence we have encountered three special series: the geometric series, the *p*-series, and the harmonic series. There exist certain variations on the forms of these series, many of which can be tested by the *comparison test*.

Comparison Test

Let $\sum_{n=1}^{\infty} a_n$ be a series of positive terms.

1. If $\sum_{n=1}^{\infty} b_n$ is a convergent series, and if $a_n \leq b_n$ for all n, then $\sum_{n=1}^{\infty} a_n$ converges.

2. If $\sum_{n=1}^{\infty} c_n$ is a divergent series of positive terms, and if $a_n \geq c_n$, then $\sum_{n=1}^{\infty} a_n$ diverges.

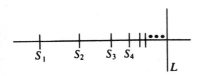

Figure 10.3

To check Part (1), let

$$S_n = a_1 + a_2 + \cdots + a_n \qquad \text{and} \qquad T_n = b_1 + b_2 + \cdots + b_n$$

Since $\sum_{n=1}^{\infty} b_n$ converges, $\lim_{n \to \infty} T_n = L$ exists. Now $S_n \leq T_n \leq L$; that is, the sequence $S_1, S_2, S_3, \ldots$ is increasing (since each a_n is positive) and bounded above by L. The diagram in Figure 10.3 suggests that $\lim_{n \to \infty} S_n$ also exists. (It is true in general that an increasing sequence bounded above is convergent.) Part (2) is proved similarly.

Example 4   Test the series

$$\sum_{n=1}^{\infty} \frac{1}{n(n+2)}$$

for convergence.

Solution. For each n,

$$\frac{1}{n(n+2)} = \frac{1}{n^2 + 2n} < \frac{1}{n^2}$$

Since $\sum_{n=1}^{\infty}(1/n^2)$ is a p-series with $p = 2$, it converges. Consequently, the given series also converges by the comparison test, Part (1). ∎

Example 5   Test the series

$$\sum_{n=2}^{\infty} \frac{1}{n-1}$$

for convergence.

Solution. For each $n \geq 2$

$$\frac{1}{n-1} > \frac{1}{n}$$

Since $\sum_{n=1}^{\infty}(1/n)$ is the Harmonic Series (10.10), which diverges, the given series also diverges by the comparison test, Part (2). ∎

For completeness we will state without proof the *ratio test*.

Ratio Test

Let $a_1 + a_2 + a_3 + \cdots + a_n + \cdots$ be a series of positive terms and

$$L = \lim_{n \to \infty} \frac{a_{n+1}}{a_n}$$

Then if:

1. $L < 1$, the series converges.
2. $L > 1$, the series diverges.
3. $L = 1$, the test fails. (Series may converge or diverge.)

Example 6   Test the series

$$\sum_{n=1}^{\infty} \frac{2^n}{n!}$$

for convergence. (Recall that $n! = n(n-1)(n-2)\cdots 2 \cdot 1$.)

Solution. Let us first note that the integral test cannot be applied to this series; for the comparison test we have no series with which to make a comparison. By the ratio test,

$$a_n = \frac{2^n}{n!} \quad \text{and} \quad a_{n+1} = \frac{2^{n+1}}{(n+1)!} = \frac{2^n \cdot 2}{(n+1)n!}$$

(To see that last step more clearly, observe that $10! = 10 \cdot 9!$ and $6! = 6 \cdot 5!$!) So we have

$$\lim_{n \to \infty} \frac{a_{n+1}}{a_n} = \lim_{n \to \infty} \frac{2^n \cdot 2}{(n+1)n!} \cdot \frac{n!}{2^n} = \lim_{n \to \infty} \frac{2}{n+1} = 0 = L$$

Since $L < 1$, the series converges. ∎

■ Exercises / Section 10.2

In Exercises 1–4, show that the series are divergent. (See Example 1.)

1. $\displaystyle\sum_{n=1}^{\infty} \frac{n}{2n+2}$

2. $\displaystyle\sum_{n=1}^{\infty} \frac{2n}{4n+1}$

3. $\displaystyle\sum_{n=2}^{\infty} \frac{5n^2}{2n^2-2}$

4. $\displaystyle\sum_{n=1}^{\infty} \frac{n!}{n^3+1}$

In Exercises 5–20, test the series for convergence by the integral test.

5. $\displaystyle\sum_{n=1}^{\infty} \frac{1}{(n+1)^2} = \frac{1}{2^2} + \frac{1}{3^2} + \frac{1}{4^2} + \cdots + \frac{1}{(n+1)^2} + \cdots$

6. $\displaystyle\sum_{n=2}^{\infty} \frac{1}{2n-2} = \frac{1}{2} + \frac{1}{4} + \frac{1}{6} + \cdots + \frac{1}{2n-2} + \cdots$

7. $\displaystyle\sum_{n=1}^{\infty} \frac{n}{n^2+1} = \frac{1}{2} + \frac{2}{5} + \frac{3}{10} + \cdots + \frac{n}{n^2+1} + \cdots$

8. $\displaystyle\sum_{n=1}^{\infty} \frac{1}{n^2+1} = \frac{1}{2} + \frac{1}{5} + \frac{1}{10} + \cdots + \frac{1}{n^2+1} + \cdots$

9. $\displaystyle\sum_{n=0}^{\infty} \frac{1}{(2n+2)^2}$

10. $\displaystyle\sum_{n=1}^{\infty} \frac{n}{2n^2-1}$

11. $\displaystyle\sum_{n=2}^{\infty} \frac{n^2}{n^3-2}$

12. $\displaystyle\sum_{n=3}^{\infty} \frac{1}{3n+2}$

13. $\displaystyle\sum_{n=1}^{\infty} \frac{n}{e^n}$

14. $\displaystyle\sum_{n=2}^{\infty} \frac{1}{n \ln n}$

15. $\displaystyle\sum_{n=1}^{\infty} \frac{n}{(n^2+1)^{3/2}}$

16. $\displaystyle\sum_{n=2}^{\infty} \frac{1}{n^2-2n+1}$

17. $\displaystyle\sum_{n=0}^{\infty} \frac{n}{(n^2+2)^2}$

18. $\displaystyle\sum_{n=3}^{\infty} \frac{1}{\sqrt{n-2}}$

19. $\displaystyle\sum_{n=2}^{\infty} \frac{n}{\sqrt{n^2+2}}$

20. $\displaystyle\sum_{n=0}^{\infty} \frac{n^2+1}{n^2+2}$

In Exercises 21–40, test the series for convergence or divergence by the comparison test.

21. $\displaystyle\sum_{n=1}^{\infty} \frac{1}{n^2+1}$

22. $\displaystyle\sum_{n=1}^{\infty} \frac{1}{n(n+4)}$

23. $\displaystyle\sum_{n=6}^{\infty} \frac{1}{n-5}$

24. $\displaystyle\sum_{n=1}^{\infty} \frac{1}{(n+2)(n+3)}$

25. $\displaystyle\sum_{n=0}^{\infty} \frac{1}{n^3+2}$

26. $\displaystyle\sum_{n=4}^{\infty} \frac{1}{n-3}$

27. $\displaystyle\sum_{n=2}^{\infty} \frac{1}{n^2+n}$

28. $\displaystyle\sum_{n=3}^{\infty} \frac{1}{n^3+4}$

29. $\displaystyle\sum_{n=0}^{\infty} \frac{1}{3^n+1}$

30. $\displaystyle\sum_{n=1}^{\infty} \frac{1}{4^n+2}$

31. $\displaystyle\sum_{n=2}^{\infty} \frac{1}{3^n-1}$

32. $\displaystyle\sum_{n=1}^{\infty} \frac{1}{4^n+n}$

33. $\displaystyle\sum_{n=4}^{\infty} \frac{1}{\sqrt{n-1}}$

34. $\displaystyle\sum_{n=2}^{\infty} \frac{1}{5^n-1}$

35. $\displaystyle\sum_{n=2}^{\infty} \frac{1}{n^3 - 1}$

36. $\displaystyle\sum_{n=4}^{\infty} \frac{1}{\sqrt[3]{n} - 1}$

37. $\displaystyle\sum_{n=1}^{\infty} \frac{1 + \sin n}{n^3}$

38. $\displaystyle\sum_{n=2}^{\infty} \frac{1}{n^4 - 1}$

39. $\displaystyle\sum_{n=2}^{\infty} \frac{1}{\ln n}$

40. $\displaystyle\sum_{n=1}^{\infty} \frac{1}{n^{5/4} + 1}$

In Exercises 41–60, test the series for convergence by the ratio test; if the test fails, use another test.

41. $\displaystyle\sum_{n=1}^{\infty} \frac{1}{3^n}$

42. $\displaystyle\sum_{n=1}^{\infty} \frac{2}{4^n}$

43. $\displaystyle\sum_{n=1}^{\infty} \frac{1}{n!}$

44. $\displaystyle\sum_{n=1}^{\infty} \frac{n}{n!}$

45. $\displaystyle\sum_{n=1}^{\infty} \frac{4^n}{n!}$

46. $\displaystyle\sum_{n=0}^{\infty} \frac{3^n}{(n+1)!}$

47. $\displaystyle\sum_{n=1}^{\infty} \frac{n^2}{2^n}$

48. $\displaystyle\sum_{n=1}^{\infty} \frac{n+1}{n!}$

49. $\displaystyle\sum_{n=0}^{\infty} n \left(\frac{2}{3}\right)^n$

50. $\displaystyle\sum_{n=1}^{\infty} \frac{7^{n+1}}{n!}$

51. $\displaystyle\sum_{n=1}^{\infty} n \left(\frac{3}{2}\right)^n$

52. $\displaystyle\sum_{n=1}^{\infty} \frac{n^3 + 1}{n!}$

53. $\displaystyle\sum_{n=1}^{\infty} \frac{n!}{7^n}$

54. $\displaystyle\sum_{n=1}^{\infty} \frac{e^n}{2n^2 + 1}$

55. $\displaystyle\sum_{n=1}^{\infty} \frac{n!}{1 \cdot 3 \cdot 5 \cdots (2n - 1)}$

56. $\displaystyle\sum_{n=1}^{\infty} \frac{1 \cdot 3 \cdot 5 \cdots (2n - 1)}{1 \cdot 4 \cdot 7 \cdots (3n - 2)}$

57. $\displaystyle\sum_{n=1}^{\infty} \frac{1 \cdot 4 \cdot 7 \cdots (3n - 2)}{2 \cdot 4 \cdot 6 \cdots (2n)}$

58. $\displaystyle\sum_{n=1}^{\infty} \frac{n - 1}{n^3}$

59. $\displaystyle\sum_{n=2}^{\infty} \frac{n^2 - 1}{n^3}$

60. $\displaystyle\sum_{n=2}^{\infty} \frac{1}{n \ln^2 n}$

10.3　Maclaurin Series

Power series

While the infinite series considered so far have contained only constant terms, many useful series consist of variable terms. The most important of these are series representing known functions. The main purpose of this section is to study a method by which a function $f(x)$ can be written as a **power series:**

$$f(x) = a_0 + a_1 x + a_2 x^2 + a_3 x^3 + \cdots + a_n x^n + \cdots \tag{10.11}$$

(Series expansions other than power series will be taken up in Section 10.6.)

To express a function as a power series, we need to determine the coefficients in Form (10.11). This can be done by means of a simple trick: we differentiate both sides of (10.11) repeatedly, as if it were a regular polynomial, and substitute zero for x. Hence $f(x)$ must be differentiable near $x = 0$ to start with. Moreover, it is shown in many books on advanced calculus that a power series may be differentiated term by term, provided that it converges for all x in some interval. We now get

$$f(x) = a_0 + a_1 x + a_2 x^2 + a_3 x^3 \qquad + a_4 x^4 \qquad + a_5 x^5 + \cdots$$

$$f'(x) = \qquad a_1 + 2a_2 x + 3a_3 x^2 \qquad + 4a_4 x^3 \qquad + 5a_5 x^4 + \cdots$$

$$f''(x) = \qquad 2 \cdot 1 a_2 + 3 \cdot 2 a_3 x \quad + 4 \cdot 3 a_4 x^2 \qquad + 5 \cdot 4 a_5 x^3 + \cdots$$

$$f'''(x) = \qquad\qquad 3 \cdot 2 \cdot 1 a_3 + 4 \cdot 3 \cdot 2 a_4 x \quad + 5 \cdot 4 \cdot 3 a_5 x^2 + \cdots$$

$$f^{(4)}(x) = \qquad\qquad\qquad 4 \cdot 3 \cdot 2 \cdot 1 a_4 + 5 \cdot 4 \cdot 3 \cdot 2 a_5 x + \cdots$$

$$f^{(5)}(x) = \qquad\qquad\qquad\qquad 5 \cdot 4 \cdot 3 \cdot 2 \cdot 1 a_5 + \cdots$$

If we let $x = 0$, all the terms on the right collapse to zero, except for the first in each row. Thus $f(0) = a_0$, $f'(0) = a_1$, $f''(0) = 2 \cdot 1 a_2$, $f'''(0) = 3 \cdot 2 \cdot 1 a_3$,

$f^{(4)}(0) = 4 \cdot 3 \cdot 2 \cdot 1 a_4$, and $f^{(5)}(0) = 5 \cdot 4 \cdot 3 \cdot 2 \cdot 1 a_5$. Solving for the constants and recalling that

$$n! = n(n-1)(n-2) \cdots 2 \cdot 1$$

we have

$$a_0 = f(0), \; a_1 = f'(0), \; a_2 = \frac{f''(0)}{2!}, \; \ldots, \; a_5 = \frac{f^{(5)}(0)}{5!}$$

The pattern is now clear:

$$a_n = \frac{f^{(n)}(0)}{n!}$$

Finally, after substituting in Series (10.11) we get the desired form of the *Maclaurin series* of $f(x)$.

Colin Maclaurin

> ### Maclaurin Series of *f(x)*
>
> $$f(x) = f(0) + f'(0)x + \frac{f''(0)}{2!}x^2 + \frac{f'''(0)}{3!}x^3$$
> $$+ \cdots + \frac{f^{(n)}(0)}{n!}x^n + \cdots \qquad (10.12)$$

The Maclaurin series is named after Colin Maclaurin (Scottish mathematician, 1698–1746). Maclaurin made many contributions to geometry, particularly to the development of higher algebraic curves. It is ironic that his name is now attached to a series which is only a special case of the *Taylor series* (Section 10.5). The latter series was published by Brook Taylor (English mathematician, 1685–1731) in 1715 (long before Maclaurin's work) but was known earlier to Johann Bernoulli.

The Maclaurin series can be written in particularly elegant form if we define $0! = 1$.

> ### Maclaurin Series (Sigma Form)
>
> $$f(x) = \sum_{n=0}^{\infty} \frac{f^{(n)}(0)}{n!} x^n \qquad (10.13)$$
> where $f^{(0)}(x) = f(x)$ and $0! = 1$.

(The Maclaurin series of a function is always unique.)

Example 1   Expand $f(x) = e^x$ in a Maclaurin series.

Solution. We differentiate first and let $x = 0$:

$$
\begin{array}{ll}
f(x) = e^x & f(0) = 1 \\
f'(x) = e^x & f'(0) = 1 \\
f''(x) = e^x & f''(0) = 1 \\
f'''(x) = e^x & f'''(0) = 1 \\
\quad \vdots & \quad \vdots \\
\text{and so on} & \text{and so on}
\end{array}
$$

Direct substitution in Series (10.12) yields

$$e^x = 1 + 1 \cdot x + \frac{1}{2!}x^2 + \frac{1}{3!}x^3 + \cdots + \frac{1}{n!}x^n + \cdots$$

or

$$e^x = 1 + x + \frac{x^2}{2!} + \frac{x^3}{3!} + \cdots + \frac{x^n}{n!} + \cdots$$

Suppose we take a peek ahead to Section 10.5 and replace x by 1; then

$$e = 1 + 1 + \frac{1}{2!} + \frac{1}{3!} + \cdots + \frac{1}{n!} + \cdots$$

Using the convention $0! = 1$, we now get the following elegant representation of the number e:

$$e = \sum_{n=0}^{\infty} \frac{1}{n!}$$

∎

Example 2 (Optional) Show that the series

$$e^x = 1 + x + \frac{x^2}{2!} + \frac{x^3}{3!} + \cdots + \frac{x^n}{n!} + \cdots$$

is convergent for all x.

Solution. Convergence may be proved by the ratio test. Since

$$a_n = \frac{x^n}{n!} \quad \text{and} \quad a_{n+1} = \frac{x^{n+1}}{(n+1)!} = \frac{x^n \cdot x}{(n+1)n!}$$

we have

$$\lim_{n \to \infty} \left| \frac{a_{n+1}}{a_n} \right| = \lim_{n \to \infty} \left| \frac{x^n \cdot x}{(n+1)n!} \cdot \frac{n!}{x^n} \right|$$

$$= |x| \lim_{n \to \infty} \frac{1}{n+1} = 0 = L < 1$$

Since $L < 1$ no matter what value we choose for x, the series is convergent for all x by the ratio test. ∎

Example 3 Find the Maclaurin expansion of the function $f(x) = \cos 2x$.

Solution. As before, we make a list of derivatives and let $x = 0$:

$$f(x) = \cos 2x \qquad f(0) = 1$$
$$f'(x) = -2 \sin 2x \qquad f'(0) = 0$$
$$f''(x) = -2^2 \cos 2x \qquad f''(0) = -2^2$$

$$f'''(x) = 2^3 \sin 2x \qquad f'''(0) = 0$$
$$f^{(4)}(x) = 2^4 \cos 2x \qquad f^4(0) = 2^4$$
$$f^{(5)}(x) = -2^5 \sin 2x \qquad f^5(0) = 0$$
$$f^{(6)}(x) = -2^6 \cos 2x \qquad f^6(0) = -2^6$$
$$\vdots \qquad\qquad \vdots$$

and so on   and so on

Substitution in (10.12) yields the desired series:

$$\cos 2x = 1 + 0x + \frac{-2^2}{2!}x^2 + \frac{0}{3!}x^3 + \frac{2^4}{4!}x^4 + \frac{0}{5!}x^5 + \frac{-2^6}{6!}x^6 + \cdots$$

or

$$\cos 2x = 1 - \frac{2^2}{2!}x^2 + \frac{2^4}{4!}x^4 - \frac{2^6}{6!}x^6 + \cdots$$

■

The behavior of a Maclaurin series can be explored with the help of a graphing utility. In particular, we can observe directly how a sequence of polynomials is able to approximate a transcendental function such as $y = \cos 2x$. Figure 10.4 shows the graph of $y = \cos 2x$ together with the sum of the first two terms, $1 - (2^2/2!)\,x^2$. Figure 10.5 shows the graph of the sum of the first three terms,

$$1 - \frac{2^2}{2!}x^2 + \frac{2^4}{4!}x^4$$

Try these explorations in the exercises.

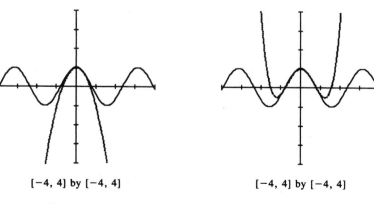

[−4, 4] by [−4, 4]   [−4, 4] by [−4, 4]

Figure 10.4   **Figure 10.5**

The following expansions are particularly important and are listed for later reference. (The first has already been obtained and the rest are left as exercises.)

$$e^x = 1 + x + \frac{x^2}{2!} + \frac{x^3}{3!} + \frac{x^4}{4!} + \cdots \qquad \text{(for all } x\text{)} \qquad (10.14)$$

$$\sin x = x - \frac{x^3}{3!} + \frac{x^5}{5!} - \cdots \qquad \text{(for all } x\text{)} \qquad (10.15)$$

$$\cos x = 1 - \frac{x^2}{2!} + \frac{x^4}{4!} - \cdots \qquad \text{(for all } x\text{)} \qquad (10.16)$$

$$\ln(1 + x) = x - \frac{x^2}{2} + \frac{x^3}{3} - \frac{x^4}{4} + \cdots \qquad (-1 < x \le 1) \qquad (10.17)$$

■ Exercises / Section 10.3

In Exercises 1–14, verify the Maclaurin series expansions. Confirm the results using your graphing utility.

1. $\sin x = x - \dfrac{x^3}{3!} + \dfrac{x^5}{5!} - \cdots$

2. $\cos x = 1 - \dfrac{x^2}{2!} + \dfrac{x^4}{4!} - \cdots$

3. $\sin 2x = 2x - \dfrac{2^3}{3!}x^3 + \dfrac{2^5}{5!}x^5 - \cdots$

4. $\cos 3x = 1 - \dfrac{3^2}{2!}x^2 + \dfrac{3^4}{4!}x^4 - \cdots$

5. $e^{-x} = 1 - x + \dfrac{x^2}{2!} - \dfrac{x^3}{3!} + \dfrac{x^4}{4!} - \cdots$

6. $e^{2x} = 1 + 2x + \dfrac{2^2}{2!}x^2 + \dfrac{2^3}{3!}x^3 + \dfrac{2^4}{4!}x^4 + \cdots$

7. $\ln(1 + x) = x - \dfrac{x^2}{2} + \dfrac{x^3}{3} - \dfrac{x^4}{4} + \cdots$

8. $\ln(1 - x) = -x - \dfrac{x^2}{2} - \dfrac{x^3}{3} - \dfrac{x^4}{4} - \cdots$

9. $\sinh x = \dfrac{1}{2}(e^x - e^{-x}) = x + \dfrac{x^3}{3!} + \dfrac{x^5}{5!} + \cdots$

10. $\cosh x = \dfrac{1}{2}(e^x + e^{-x}) = 1 + \dfrac{x^2}{2!} + \dfrac{x^4}{4!} + \cdots$

11. $\operatorname{Arctan} x = x - \dfrac{x^3}{3} + \dfrac{x^5}{5} - \cdots$

12. $\tan x = x + \dfrac{x^3}{3} + \dfrac{2x^5}{15} + \cdots$

13. (Optional) $\operatorname{Arcsin} x = x + \dfrac{1 \cdot x^3}{2 \cdot 3} + \dfrac{1 \cdot 3 \cdot x^5}{2 \cdot 4 \cdot 5} + \dfrac{1 \cdot 3 \cdot 5x^7}{2 \cdot 4 \cdot 6 \cdot 7}$
$$+ \dfrac{1 \cdot 3 \cdot 5 \cdot 7x^9}{2 \cdot 4 \cdot 6 \cdot 8 \cdot 9} + \cdots$$

14. $\dfrac{1}{1 - x} = 1 + x + x^2 + x^3 + \cdots \qquad |x| < 1$

15. (Optional) Show that the Series (10.15) and (10.16) converge for all x.

16. Verify the series expansion

$$(1 - x)^{-2} = \sum_{n=0}^{\infty} (n + 1) x^n$$

by **(a)** using the binomial theorem; **(b)** finding the Maclaurin series expansion; **(c)** dividing out $1/(1 - x)^2$.

10.4 Operations with Series

In this section we will study a number of operations that yield new series from series already known.

Example 1   Find the Maclaurin series for $\sin x^2$.

Solution. Consider the series

$$\sin x = x - \frac{x^3}{3!} + \frac{x^5}{5!} - \cdots$$

from the last section. If we replace x by x^2, we obtain

$$\sin x^2 = x^2 - \frac{(x^2)^3}{3!} + \frac{(x^2)^5}{5!} - \cdots$$

or

$$\sin x^2 = x^2 - \frac{x^6}{3!} + \frac{x^{10}}{5!} - \cdots$$

Since this series is a power series, it must be the Maclaurin series of $\sin x^2$, since such expansions are unique.   ∎

Example 2

Find the Maclaurin series for xe^x.

Solution. From

$$e^x = 1 + x + \frac{x^2}{2!} + \frac{x^3}{3!} + \cdots$$

we obtain by direct multiplication

$$xe^x = x\left(1 + x + \frac{x^2}{2!} + \frac{x^3}{3!} + \cdots\right)$$

$$= x + x^2 + \frac{x^3}{2!} + \frac{x^4}{3!} + \cdots$$ ∎

It has already been noted that convergent power series can be differentiated termwise; the same is true of integration.

Example 3

Show that $(d/dx)e^x = e^x$ by the use of Maclaurin series.

Solution.

$$\frac{d}{dx}e^x = \frac{d}{dx}\left(1 + x + \frac{x^2}{2!} + \frac{x^3}{3!} + \frac{x^4}{4!} + \cdots\right)$$

$$= 0 + 1 + \frac{2x}{2!} + \frac{3x^2}{3!} + \frac{4x^3}{4!} + \cdots$$

$$= 1 + x + \frac{x^2}{2!} + \frac{x^3}{3!} + \cdots = e^x$$ ∎

Example 4

Find the Maclaurin series of $\operatorname{Arctan} x$ by integrating $(d/dx)\operatorname{Arctan} x$ termwise.

Solution. We recall that

$$\frac{d}{dx}\operatorname{Arctan} x = \frac{1}{1 + x^2}$$

This expression can be written as a geometric series. Let $r = -x^2$ and $a = 1$; then

$$1 - x^2 + x^4 - x^6 + \cdots = \frac{1}{1 - (-x^2)} = \frac{1}{1 + x^2} \qquad s = \frac{a}{1 - r}$$

Consequently,

$$\text{Arctan}\, x = \int_0^x \frac{dx}{1+x^2} = \int_0^x (1 - x^2 + x^4 - x^6 + \cdots)\, dx$$

$$= x - \frac{x^3}{3} + \frac{x^5}{5} - \frac{x^7}{7} + \cdots \bigg|_0^x = x - \frac{x^3}{3} + \frac{x^5}{5} - \frac{x^7}{7} + \cdots$$

(It is actually poor practice to use x both for the variable of integration and for the upper limit, but the steps are much easier to see this way.)

Remark: Our main application of the integration of series will be discussed in the next section. ∎

The four fundamental operations—addition, subtraction, multiplication, and division—can theoretically be carried out with series. Two of these operations are demonstrated in the following examples.

Example 5

Find the power-series expansion of $e^x \sin x$ by multiplying the series for e^x and $\sin x$.

Solution. We first recall that

$$\sin x = x - \frac{x^3}{6} + \frac{x^5}{120} - \cdots$$

and

$$e^x = 1 + x + \frac{x^2}{2} + \frac{x^3}{6} + \frac{x^4}{24} + \frac{x^5}{120} + \cdots$$

We may now multiply each term in the second series by each term in the first series in exactly the same way that we multiply polynomials. If we decide to carry only powers up to the fifth power, we obtain

$$e^x = 1 + x + \frac{x^2}{2} + \frac{x^3}{6} + \frac{x^4}{24} + \frac{x^5}{120} + \text{higher powers}$$

$$\sin x = x - \frac{x^3}{6} + \frac{x^5}{120} + \text{higher powers}$$

$$\underline{\hspace{10cm}}$$

$$x + x^2 + \frac{x^3}{2} + \frac{x^4}{6} + \frac{x^5}{24} + \cdots \qquad \textbf{multiplying}$$

$$- \frac{x^3}{6} - \frac{x^4}{6} - \frac{x^5}{12} + \cdots$$

$$+ \frac{x^5}{120} + \cdots$$

$$\underline{\hspace{10cm}}$$

$$x + x^2 + \frac{x^3}{3} \qquad - \frac{x^5}{30} + \cdots$$

We now conclude that, up to the fifth power,

$$e^x \sin x = x + x^2 + \frac{1}{3}x^3 - \frac{1}{30}x^5 + \cdots$$

■

Example 6   Use the series in Exercises 7 and 8 of the last section to expand

$$\ln \frac{1+x}{1-x}$$

Solution. We have

$$\ln(1+x) = x - \frac{x^2}{2} + \frac{x^3}{3} - \frac{x^4}{4} + \frac{x^5}{5} - \cdots$$

and

$$\ln(1-x) = -x - \frac{x^2}{2} - \frac{x^3}{3} - \frac{x^4}{4} - \frac{x^5}{5} - \cdots$$

Hence

$$\ln \frac{1+x}{1-x} = \ln(1+x) - \ln(1-x)$$

$$= \left(x - \frac{x^2}{2} + \frac{x^3}{3} - \frac{x^4}{4} + \frac{x^5}{5} - \cdots \right)$$

$$- \left(-x - \frac{x^2}{2} - \frac{x^3}{3} - \frac{x^4}{4} - \frac{x^5}{5} - \cdots \right)$$

$$= 2 \left(x + \frac{x^3}{3} + \frac{x^5}{5} + \cdots \right)$$

■

As a final exercise we are going to uncover a relationship among three of our transcendental functions by making use of the basic imaginary unit $j = \sqrt{-1}$. As a starting point, notice that the expansion of the sine function has only odd powers and that of the cosine function only even powers. However, all the powers occur in the expansion of e^x; so e^x comes very close to being the sum of the other two—if only the signs matched! Now, by introducing j formally, we find that

$$e^{jx} = 1 + jx + \frac{j^2 x^2}{2!} + \frac{j^3 x^3}{3!} + \frac{j^4 x^4}{4!} + \frac{j^5 x^5}{5!} + \cdots$$

$$= 1 + jx - \frac{x^2}{2!} - \frac{jx^3}{3!} + \frac{x^4}{4!} + \frac{jx^5}{5!} - \cdots \qquad \begin{array}{l} \boldsymbol{j = j, j^2 = -1} \\ \boldsymbol{j^3 = j^2 j = -j, j^4 = j^3 j = 1} \end{array}$$

$$= 1 - \frac{x^2}{2!} + \frac{x^4}{4!} - \cdots + j \left(x - \frac{x^3}{3!} + \frac{x^5}{5!} - \cdots \right)$$

$$= \cos x + j \sin x$$

The resulting formula is known as *Euler's identity* after the Swiss mathematician Leonhard Euler (1707–1783).

Euler's Identity

$$e^{jx} = \cos x + j \sin x \qquad\qquad (10.18)$$

Euler's identity arises in the study of differential equations, as we will see in Chapter 12.

Although there is some room for opinion, it can be argued that the most interesting numbers in mathematics are $0, 1, j, e$, and π. By Euler's identity,

$$e^{j\pi} = \cos \pi + j \sin \pi = -1$$

or

$$e^{j\pi} + 1 = 0 \qquad\qquad (10.19)$$

which involves all five of these numbers. This astounding relationship has been called the eutectic point of mathematics, for no matter how you try to analyze it, it seems to retain an air of mystery not easily explained away.

■ Exercises / Section 10.4

In Exercises 1–14, use the method of Examples 1 and 2 to find the Maclaurin series of the functions.

1. $\sin 3x$
2. $\cos 2x$
3. e^{-x}
4. e^{-x^2}
5. $\cos \sqrt{x}$
6. $\dfrac{\sin x^2}{x}$
7. $x \cos x$
8. $x^2 e^x$
9. $\ln(1 + x^2)$
10. $\ln(1 - x)$
11. $\dfrac{\text{Arctan } x}{x}$
12. $x \ln(1 + x)$
13. $\dfrac{\ln(1 + x)}{x}$
14. $x^2 \sin 2x$

15. Show that $(d/dx) \sin x = \cos x$ by use of the Maclaurin series. (See Example 3.)

16. Show that $(d/dx) \cos x = -\sin x$ by use of the Maclaurin series.

17. Use the method of Example 4 to find the Maclaurin series of $\ln(1 + x)$.

18. Expand $(\sin x - x)/x^2$ in a Maclaurin series.

19. Use the method of Example 5 to find the Maclaurin series of $e^{-x} \cos x$.

20. Use the method of Example 6 to find the Maclaurin series of $\ln(1 + x)^2$.

21. Expand the function $\ln(1 + x^2)^3$ in a Maclaurin series.

22. Find the Maclaurin series of $(1/2)(e^x + e^{-x})$ by addition of series.

23. Find the Maclaurin series of $\ln(1 + x) + \text{Arctan } x$.

A complex number $a + bj$ can be written in polar form $r(\cos \theta + j \sin \theta)$, which by Euler's identity becomes $re^{j\theta}$, known as the *exponential form.* In Exercises 24–29, change the complex numbers to exponential form.

24. $1 + j$
25. $-\sqrt{3} + j$
26. $1 - \sqrt{3} j$
27. $3j$
28. $-4j$
29. $-2 + 2j$

10.5 Computations with Series; Applications

In this section we are going to do numerical computations by means of power series. By using a sufficiently large number of terms, we can obtain the values of some transcendental functions to any desired degree of accuracy. A particularly important application of these numerical techniques is the evaluation of certain definite integrals.

Before we consider computations involving series, we need to make a few additional observations about series of constants. Suppose that $a_1, a_2, a_3, \ldots, a_n, \ldots$ is a sequence of positive numbers such that each number is less than the preceding one, that is, $a_{n+1} < a_n$ for all n, and consider the series

$$\sum_{n=1}^{\infty} (-1)^{n+1} a_n = a_1 - a_2 + a_3 - a_4 + \cdots + (-1)^{n+1} a_n + \cdots \qquad (10.20)$$

Alternating series

called an **alternating series** since the signs alternate. If the series converges, then the sum may be obtained to any desired degree of accuracy by adding the first n terms and estimating the error from the $(n+1)$st term. To check this statement, suppose we add the first four terms of Series (10.20) and estimate the error by writing the series as follows:

$$(a_1 - a_2 + a_3 - a_4) + a_5 - (a_6 - a_7) - (a_8 - a_9) - \cdots$$

Since the numbers a_n are decreasing,

$$(a_6 - a_7) > 0, (a_8 - a_9) > 0, \text{ and so forth}$$

Hence

$$a_5 - (a_6 - a_7) - (a_8 - a_9) - \cdots < a_5$$

So, by adding $a_1 - a_2 + a_3 - a_4$, the error made is less than a_5.

If we wish to add the first five terms, then the error is estimated from

$$(a_1 - a_2 + a_3 - a_4 + a_5) - a_6 + (a_7 - a_8) + (a_9 - a_{10}) + \cdots$$

Again

$$(a_7 - a_8) > 0, (a_9 - a_{10}) > 0, \text{ and so on}$$

so that the error is no worse than $-a_6$.

The **error** made by adding the first n terms of a **convergent alternating series**

$$a_1 - a_2 + a_3 - a_4 + \cdots \qquad (a_n > 0, a_{n+1} < a_n)$$

is numerically less than the first term omitted.

(We state without proof that an alternating series converges if $\lim_{n \to \infty} a_n = 0$.)

Example 1

Compute the value of $e^{-0.1}$ by using the first four terms of the expansion of e^x. Find the maximum possible error and determine the accuracy of the result.

Solution. We let $x = -0.1$ in the series

$$e^x = 1 + x + \frac{x^2}{2!} + \frac{x^3}{3!} + \frac{x^4}{4!} + \frac{x^5}{5!} + \cdots$$

and find the sum of the first four terms:

$$e^{-0.1} = 1 - 0.1 + \frac{(0.1)^2}{2!} - \frac{(0.1)^3}{3!} = 0.904833$$

The error made is no worse than the fifth term:

$$+\frac{(0.1)^4}{4!} = +0.000004$$

$$\begin{array}{r} 0.904833 \\ + \, 0.000004 \\ \hline 0.904837 \end{array}$$

Based on these calculations, the correct value to six decimal places could be any one of the following: 0.904833, 0.904834, ..., or 0.904837. Consequently,

$$e^{-0.1} = 0.9048 \qquad \text{(correct to four decimal places)} \qquad \blacksquare$$

To see the relationship between e^x and its Maclaurin expansion, let us examine the graphs of the following approximations:

$$y = 1 + x \qquad y = 1 + x + \frac{x^2}{2} \qquad y = 1 + x + \frac{x^2}{2} + \frac{x^3}{6}$$

(See Figure 10.6.)

According to Figure 10.6, the approximation improves as we include more and more terms in the sum, producing a particularly good fit near the origin. As a result, for values near zero the Maclaurin series leads to a good approximation with very few terms, as Example 1 confirms. For values away from the origin, a good approximation can be obtained with few terms by means of the Taylor series discussed at the end of the section.

Example 2

Find an infinite-series representation of **(a)** e; **(b)** π.

Solution.

a. The representation of e was already obtained in Section 10.3.

$$e = \sum_{n=0}^{\infty} \frac{1}{n!}$$

b. Since

$$\text{Arctan}\, x = x - \frac{x^3}{3} + \frac{x^5}{5} - \cdots$$

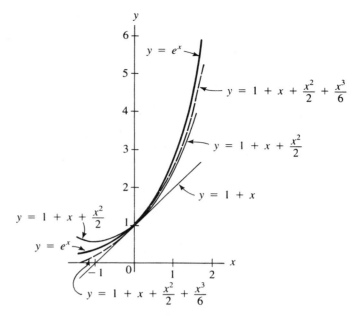

Figure 10.6

we let $x = 1$ and obtain

$$\text{Arctan } 1 = \frac{\pi}{4} = 1 - \frac{1}{3} + \frac{1}{5} - \frac{1}{7} + \frac{1}{9} - \cdots$$

or

$$\pi = 4\left(1 - \frac{1}{3} + \frac{1}{5} - \frac{1}{7} + \frac{1}{9} - \cdots\right)$$

Although it is a striking relationship, Series (b) does not provide us with a practical method of computing π, since the series converges too slowly. A better way is by means of the series for $f(x) = \text{Arcsin } x$ (Exercise 11). ∎

As noted in Chapter 7 (Section 7.10), many functions do not possess elementary antiderivatives. Since a power series can be integrated termwise, many such integrals can be worked out by means of Maclaurin series, leading to *nonelementary functions*. Study the next example and remark.

Example 3   Find the approximate value of

$$\int_0^1 \frac{\sin x^2\, dx}{x}$$

Solution. At $x = 0$ the integrand takes on the indeterminate form $0/0$. Now, by L'Hospital's rule,

$$\lim_{x \to 0} \frac{\sin x^2}{x} = \lim_{x \to 0} \frac{2x \cos x^2}{1} = 0$$

so that the function is bounded on (0, 1). (Otherwise we would be dealing with an improper integral.) From

$$\sin x = x - \frac{x^3}{3!} + \frac{x^5}{5!} - \frac{x^7}{7!} + \cdots$$

we have

$$\sin x^2 = x^2 - \frac{x^6}{3!} + \frac{x^{10}}{5!} - \frac{x^{14}}{7!} + \cdots \qquad \text{replacing } x \text{ by } x^2$$

and

$$\frac{\sin x^2}{x} = x - \frac{x^5}{3!} + \frac{x^9}{5!} - \frac{x^{13}}{7!} + \cdots \qquad \text{dividing by } x$$

Hence

$$\int_0^1 \frac{\sin x^2 \, dx}{x} = \int_0^1 \left(x - \frac{x^5}{3!} + \frac{x^9}{5!} - \frac{x^{13}}{7!} + \cdots \right) dx$$

$$= \left. \frac{x^2}{2} - \frac{x^6}{6 \cdot 3!} + \frac{x^{10}}{10 \cdot 5!} - \frac{x^{14}}{14 \cdot 7!} + \cdots \right|_0^1 \qquad \text{integrating}$$

$$= \frac{1}{2} - \frac{1}{6 \cdot 3!} + \frac{1}{10 \cdot 5!} - \frac{1}{14 \cdot 7!} + \cdots$$

Adding the first three terms, we get

$$\int_0^1 \frac{\sin x^2 \, dx}{x} = 0.473056$$

The error is numerically less than the fourth term:

$$-\frac{1}{14 \cdot 7!} = -0.000014 \qquad \begin{array}{r} \textbf{0.473056} \\ \textbf{- 0.000014} \\ \hline \textbf{0.473042} \end{array}$$

So the value to five decimal places is 0.47304, 0.47305, or 0.47306. Consequently,

$$\int_0^1 \frac{\sin x^2 \, dx}{x} = 0.473 \qquad \text{(correct to three decimal places)}$$

While the value of this integral could have been found by the methods of Section 4.9, the use of infinite series allows an easy determination of the error. ∎

Remark: The integral

$$\int \frac{\sin x^2}{x} \, dx$$

in Example 3 is nonelementary—that is, there does not exist an elementary function whose derivative is $(\sin x^2)/x$. The infinite-series form in Example 3,

$$\int \frac{\sin x^2\, dx}{x} = \frac{x^2}{2} - \frac{x^6}{6\cdot 3!} + \frac{x^{10}}{10\cdot 5!} - \frac{x^{14}}{14\cdot 7!} + \cdots$$

is an example of a *nonelementary function*. The extension of the function concept to include nonelementary functions is an important application of our study of infinite series.

Taylor Series

We observed in Example 1 that the Maclaurin series yields a good approximation with only a few terms, provided that x is close to zero. If x is large in absolute value, the number of terms needed for a good approximation may become prohibitive. We can get around this problem by generalizing the form of the series expansion as follows:

$$f(x) = a_0 + a_1(x - c) + a_2(x - c)^2 + \cdots + a_n(x - c)^n + \cdots$$

To compute the coefficients, we proceed as we did with the Maclaurin series: we find the derivatives of $f(x)$ and let $x = c$. Then the nth coefficient turns out to be

$$a_n = \frac{f^{(n)}(c)}{n!}$$

and the series becomes

$$f(x) = f(c) + f'(c)(x - c) + \frac{f''(c)}{2!}(x - c)^2 + \cdots + \frac{f^{(n)}(c)}{n!}(x - c)^n + \cdots$$

which is called the *Taylor series* of $f(x)$ in powers of $x - c$. We also say that $f(x)$ has been expanded about $x - c$.

Taylor Series of $f(x)$

$$f(x) = f(c) + f'(c)(x - c) + \frac{f''(c)}{2!}(x - c)^2 + \cdots$$
$$+ \frac{f^{(n)}(c)}{n!}(x - c)^n + \cdots \tag{10.21}$$

(If $c = 0$, then the Taylor series reduces to the Maclaurin series.)

Although c may have any value, in practice one selects a value that is particularly convenient, as we will see in the next example.

Example 4 Calculate $\sin 61°$ by means of the Taylor series.

Solution. The closest convenient value that we may select for c is $60° = \pi/3$:

$$f(x) = \sin x \qquad\qquad f(\pi/3) = \sqrt{3}/2$$

$$f'(x) = \cos x \qquad\qquad f'(\pi/3) = \frac{1}{2}$$

$$f''(x) = -\sin x \qquad\qquad f''(\pi/3) = -\sqrt{3}/2$$

$$f'''(x) = -\cos x \qquad\qquad f'''(\pi/3) = -\frac{1}{2}$$

$$f^{(4)}(x) = \sin x \qquad\qquad f^{(4)}(\pi/3) = \sqrt{3}/2$$

$$\vdots \qquad\qquad\qquad\qquad \vdots$$

$$\text{and so on} \qquad\qquad\qquad \text{and so on}$$

Substituting in Series (10.21), we get

$$\sin x = \frac{\sqrt{3}}{2} + \frac{1}{2}\left(x - \frac{\pi}{3}\right) - \frac{\sqrt{3}}{2}\frac{1}{2!}\left(x - \frac{\pi}{3}\right)^2 - \frac{1}{2}\frac{1}{3!}\left(x - \frac{\pi}{3}\right)^3$$

$$+ \frac{\sqrt{3}}{2}\frac{1}{4!}\left(x - \frac{\pi}{3}\right)^4 + \cdots$$

The reason for choosing $c = \pi/3$ now becomes clear: since $61° = 60° + 1° = \pi/3 + \pi/180$, we get

$$x - c = \left(\frac{\pi}{3} + \frac{\pi}{180}\right) - \frac{\pi}{3} = \frac{\pi}{180}$$

Consequently, the numerical value of $x - c$ is small and so the terms in the series become small as well. Using four terms, we get

$$\sin 61° = \frac{\sqrt{3}}{2} + \frac{1}{2}\left(\frac{\pi}{180}\right) - \frac{\sqrt{3}}{2}\frac{1}{2!}\left(\frac{\pi}{180}\right)^2 - \frac{1}{2}\frac{1}{3!}\left(\frac{\pi}{180}\right)^3 = 0.874620 \quad ■$$

■ Exercises / Section 10.5

In Exercises 1–10, find the indicated function values by means of Maclaurin series, using the number of terms indicated. Find the maximum possible error, and determine the accuracy of the result. (See Example 1.)

1. $\sin 0.7$ (3 terms)
2. $\ln 1.5$ (10 terms)
3. $\cos 10°$ (2 terms)
4. $\cos 20°$ (3 terms)
5. $e^{-0.2}$ (4 terms)
6. $\sin 35°$ (3 terms)
7. $\cos 1.2$ (4 terms)
8. $e^{-0.6}$ (6 terms)
9. $\ln 1.1$ (3 terms)
10. Arctan 0.2 (2 terms)

11. Use the expansion of Arcsin x (Exercise 13, Section 10.3) to compute π. (Let $x = 1/2$ and use 3 terms.)

12. Compute e by using 10 terms.

In Exercises 13–18, evaluate the integrals correct to five decimal places. (See Example 3.)

13. $\int_0^{1/2} \frac{1 - \cos x}{x} \, dx$
14. $\int_0^1 \frac{\sin x - x}{x^2} \, dx$

15. $\int_0^1 \cos \sqrt{x} \, dx$
16. $\int_0^{1/2} \frac{\text{Arctan } x}{x} \, dx$

17. $\int_0^{0.3} e^{-x^2} \, dx$
18. $\int_0^{0.6} \sin x^2 \, dx$

19. Evaluate $\sin 29°$ by expanding $f(x) = \sin x$ about $c = \pi/6$, using 3 terms. (See Example 4.)

20. Evaluate $\cos 50°$ by expanding $f(x) = \cos x$ about $c = \pi/4$, using 4 terms.

21. Evaluate $\cos 31°$ by expanding $f(x) = \cos x$ about $c = \pi/6$, using 4 terms.

22. Evaluate $\sin 64°$ by expanding $f(x) = \sin x$ about $c = \pi/3$, using 3 terms.

23. Evaluate $\cos 58°$ by expanding $f(x) = \cos x$ about $c = \pi/3$, using 3 terms.

24. Evaluate $\sin 44°$ by expanding $f(x) = \sin x$ about $c = \pi/4$, using 4 terms.

25. Show that

$$\ln x = (x - 1) - \frac{1}{2}(x - 1)^2 + \frac{1}{3}(x - 1)^3 - \cdots$$

26. Prove that for $x \geq 0$

$$e^x \geq 1 + x + \frac{1}{2}x^2$$

27. Show that the polynomial $5x^5 + 10x^4 - 2x^3 + x^2 + 5$ can be written

$$5(x - 1)^5 + 35(x - 1)^4 + 88(x - 1)^3 + 105(x - 1)^2$$
$$+ 61(x - 1) + 19$$

by using the Taylor series.

28. Find $\ln 3$ by using the series expansion of

$$\ln \frac{1 + x}{1 - x}$$

(Example 6, Section 10.4) with $x = 1/2$.

29. **a.** Use the Maclaurin expansion of $\sin \theta$ to show that $\sin \theta \approx \theta$ for small θ.

   b. A simple pendulum consists of a mass m hanging on a light string of length L. If θ is the angle made by the string with the vertical, then θ satisfies the relationship

$$mL \frac{d^2\theta}{dt^2} = -mg \sin \theta$$

   where g is the acceleration due to gravity. Use part **(a)** to show that if θ is sufficiently small, this equation represents a system oscillating with simple harmonic motion. (See Equation (6.29) in Section 6.10.)

   c. Show that for small oscillations the period P of a pendulum is

$$P = 2\pi \sqrt{\frac{L}{g}}$$

30. The following sum was first obtained by the Swiss mathematician Leonhard Euler:

$$\sum_{n=1}^{\infty} \frac{1}{n^2} = \frac{\pi^2}{6}$$

Use this sum and Exercise 25 to show that

$$\int_0^1 \frac{\ln x}{x - 1}\, dx = \frac{\pi^2}{6}$$

(This integral is nonelementary and improper.)

31. Find the value of the current $i = 2e^{-0.5t^2}$ when $t = 0.2$ s by using three terms of the Maclaurin series.

32. If $i = \sin t^2$ (in amperes), determine how many coulombs of charge pass a point from $t = 0.0$ s to $t = 0.8$ s, correct to five decimal places ($q = \int i\, dt$).

33. The top portion of the vertical loop on some roller coasters can be described by parametric equations of the form

$$x(t) = A \int_0^t \cos \frac{1}{2}\pi u^2\, du, \quad y(t) = B \int_0^t \sin \frac{1}{2}\pi u^2\, du$$

The integrals are known as Fresnel functions. Write the Fresnel function $x(t) = \int_0^t \cos u^2\, du$ as a Maclaurin series.

34. Integrating $\int dx/(1 + x^5)$ is extremely difficult. Even the finite answer obtained using a computer algebra system is so cumbersome that the infinite-series representation is actually easier to work with. Evaluate $\int_0^{0.5} dx/(1 + x^5)$ correct to four decimal places by using a geometric series.

35. The function

$$f(x) = \frac{1}{\sqrt{2\pi}} e^{(-1/2)x^2}$$

which is called the *standard normal distribution*, was first studied in the eighteenth century when scientists observed that the distribution of errors of measurement is closely approximated by this curve. Calculate to four decimal places the integral

$$P(0 \leq x \leq 1) = \frac{1}{\sqrt{2\pi}} \int_0^1 e^{(-1/2)x^2}\, dx$$

which is the probability that a measurement x having the standard normal distribution will take on a value between 0 and 1.

36. The charge q on a certain capacitor is

$$q = \int_0^{0.50} (1.00 - e^{-0.10t^2})\, dt$$

Evaluate this integral correct to four decimal places.

37. The charge on a capacitor decreases according to the equation

$$q(t) = \frac{\ln(1 + t^2)}{t^2}$$

Find the current at $t = 0.40$ s by using the first four terms of the Maclaurin series.

38. The study of blackbody radiation leads to a formula of the form

$$\rho(\omega) = \frac{A\omega^3}{e^{h\omega/a}} \frac{1}{1 - e^{-h\omega/a}}$$

Show that this formula can be written as an infinite series as follows:

$$\rho(\omega) = \frac{A\omega^3}{e^{h\omega/a}} \left(1 + e^{-h\omega/a} + e^{-2h\omega/a} + e^{-3h\omega/a} + \cdots\right)$$

39. (Refer to Exercise 29.) For an arbitrary displacement θ the period is given by

$$P_\theta = 2\pi \sqrt{\frac{L}{g}} \left(1 + \frac{1}{4}\sin^2\frac{\theta}{2} + \frac{9}{64}\sin^4\frac{\theta}{2} + \cdots\right)$$

Determine P_θ/P for $\theta = 5°$, correct to four decimal places.

40. The velocity v of a water wave is related to its length L and the depth h of the water by

$$v^2 = \frac{gL}{2\pi} \tanh \frac{2\pi h}{L}$$

a. Show that $\tanh x \approx x - (1/3)x^3$ if x is small.
b. Use the approximation $\tanh x \approx x$ to show that $v^2 \approx gh$ if h/L is small.

10.6 Fourier Series

In this section we will consider a different kind of series expansion, the **Fourier series,** which expresses a given function in terms of sines and cosines (instead of x^n as in the Maclaurin series). Such expansions are highly useful in the study of periodic phenomena. Moreover, the Fourier series is much less demanding: for example, it is not required that the function to be expanded be differentiable or even continuous.

Let f be the given periodic function with period $2p$ and consider the following expansion:

$$f(t) = \frac{1}{2}a_0 + a_1 \cos \frac{\pi t}{p} + a_2 \cos \frac{2\pi t}{p} + \cdots + a_n \cos \frac{n\pi t}{p} + \cdots$$
$$+ b_1 \sin \frac{\pi t}{p} + b_2 \sin \frac{2\pi t}{p} + \cdots + b_n \sin \frac{n\pi t}{p} + \cdots \qquad (10.22)$$

(Bowing to the usual convention, we will use t for the independent variable.) As in the case of the Maclaurin series, our main task is to find the coefficients in Expansion (10.22). To do so, we need the following definite integrals:

$$\int_{-p}^{p} \cos \frac{n\pi t}{p} \, dt = 0 \qquad (n \neq 0) \qquad (10.23)$$

$$\int_{-p}^{p} \sin \frac{n\pi t}{p} \, dt = 0 \qquad (10.24)$$

$$\int_{-p}^{p} \cos \frac{m\pi t}{p} \cos \frac{n\pi t}{p} \, dt = 0 \qquad (m \neq n) \qquad (10.25)$$

$$\int_{-p}^{p} \cos^2 \frac{n\pi t}{p}\, dt = p \qquad (n \neq 0) \tag{10.26}$$

$$\int_{-p}^{p} \cos \frac{m\pi t}{p} \sin \frac{n\pi t}{p}\, dt = 0 \tag{10.27}$$

$$\int_{-p}^{p} \sin \frac{m\pi t}{p} \sin \frac{n\pi t}{p}\, dt = 0 \qquad (m \neq n) \tag{10.28}$$

$$\int_{-p}^{p} \sin^2 \frac{n\pi t}{p}\, dt = p \qquad (n \neq 0) \tag{10.29}$$

Integrals (10.23), (10.24), (10.26), and (10.29) can be evaluated routinely by the methods of Chapter 7. The remaining integrals can be obtained by using the table of integrals in Appendix A (Table 2, Forms 61–63).

The coefficients in Expansion (10.22) can now be calculated by multiplying both sides by $\cos(n\pi t/p)$ and $\sin(n\pi t/p)$ and by integrating the result from $t = -p$ to $t = p$. Suppose we try this procedure for $\cos(n\pi t/p)$:

$$\begin{aligned}
\int_{-p}^{p} f(t) \cos \frac{n\pi t}{p}\, dt = &\int_{-p}^{p} \frac{a_0}{2} \cos \frac{n\pi t}{p}\, dt \\
&+ a_1 \int_{-p}^{p} \cos \frac{\pi t}{p} \cos \frac{n\pi t}{p}\, dt \\
&+ a_2 \int_{-p}^{p} \cos \frac{2\pi t}{p} \cos \frac{n\pi t}{p}\, dt + \cdots \\
&+ a_n \int_{-p}^{p} \cos \frac{n\pi t}{p} \cos \frac{n\pi t}{p}\, dt + \cdots \\
&+ b_1 \int_{-p}^{p} \sin \frac{\pi t}{p} \cos \frac{n\pi t}{p}\, dt \\
&+ b_2 \int_{-p}^{p} \sin \frac{2\pi t}{p} \cos \frac{n\pi t}{p}\, dt + \cdots \\
&+ b_n \int_{-p}^{p} \sin \frac{n\pi t}{p} \cos \frac{n\pi t}{p}\, dt + \cdots
\end{aligned}$$

Carefully inspecting all the terms in this series with an eye on the special integrals, we see that only one term survives (the nth one). Thus

$$\int_{-p}^{p} f(t) \cos \frac{n\pi t}{p}\, dt = a_n p$$

or

$$a_n = \frac{1}{p} \int_{-p}^{p} f(t) \cos \frac{n\pi t}{p}\, dt \qquad (n \neq 0) \tag{10.30}$$

By multiplying both sides of Expansion (10.22) by $\sin(n\pi t/p)$ and integrating, we obtain

$$b_n = \frac{1}{p}\int_{-p}^{p} f(t)\sin\frac{n\pi t}{p}\,dt \qquad (10.31)$$

To get the remaining term a_0, we simply integrate both sides of Expansion (10.22) to find that

$$a_0 = \frac{1}{p}\int_{-p}^{p} f(t)\,dt \qquad (10.32)$$

(The coefficient $1/2$ of a_0 in Expansion (10.22) was introduced so that Formulas (10.32) and (10.30) would have the same form.)

Series (10.22), together with Formulas (10.30) through (10.32), is called a **Fourier series,** after J. B. J. Fourier (1768–1830), a French mathematician and physicist who was a pioneer in the study of heat conduction. A confidant of Napoleon, Fourier participated in Napoleon's expedition to Egypt. (A child who met Fourier and viewed his collection of Egyptian antiquities—J. F. Champollion—was inspired by this experience to study Egyptian hieroglyphics and twenty years later succeeded in decoding them.)

Fourier

Fourier Series

The Fourier series for a function $f(t)$ is

$$f(t) = \frac{a_0}{2} + a_1\cos\frac{\pi t}{p} + a_2\cos\frac{2\pi t}{p} + \cdots + a_n\cos\frac{n\pi t}{p} + \cdots$$
$$+ b_1\sin\frac{\pi t}{p} + b_2\sin\frac{2\pi t}{p} + \cdots + b_n\sin\frac{n\pi t}{p} + \cdots \qquad (10.33)$$

where the coefficients are given by

$$a_0 = \frac{1}{p}\int_{-p}^{p} f(t)\,dt \qquad (10.34)$$

$$a_n = \frac{1}{p}\int_{-p}^{p} f(t)\cos\frac{n\pi t}{p}\,dt \qquad (n \neq 0) \qquad (10.35)$$

$$b_n = \frac{1}{p}\int_{-p}^{p} f(t)\sin\frac{n\pi t}{p}\,dt \qquad (10.36)$$

Since $\sin bx$ and $\cos bx$ are periodic functions with period $2\pi/b$, each term in the series is periodic with period

$$\frac{2\pi}{\frac{n\pi}{p}} = 2\pi \cdot \frac{p}{n\pi} = \frac{2p}{n} \qquad (n = 1, 2, \ldots)$$

Since a periodic function with period $2p/n$ is also periodic with period $2p$, **the entire series is periodic and represents a periodic function with period $2p$.** Formulas (10.34) through (10.36) yield the Fourier coefficients only for the interval $[-p, p]$. By the periodicity, however, the representation automatically extends over the entire real line.

Since currents and voltages are often periodic, Fourier series are particularly useful in electronics. As examples, we are going to calculate the series of two functions frequently encountered in this field.

Example 1   Obtain the Fourier series of the "square wave" in Figure 10.7.

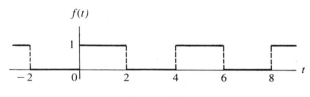

Figure 10.7

Solution.   When finding a Fourier series, we use only one period in the calculation. In this problem we use the period from $t = -2$ to $t = +2$, given by the function

$$f(t) = \begin{cases} 0 & -2 < t < 0 \\ 1 & 0 < t < 2 \end{cases}$$

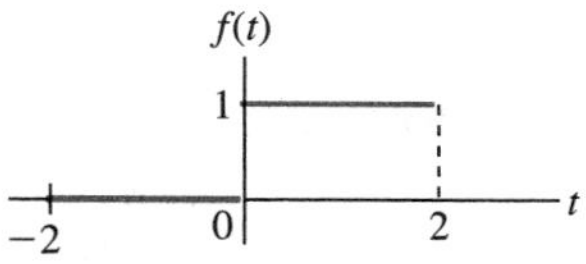

(See Figure 10.8.)

Since the period is 4, we have

$$2p = 4 \quad \text{and} \quad \textbf{\textit{p} = 2}$$

It is important to keep in mind that the first term of the series, $a_0/2$, is always computed separately. We now proceed by direct calculation:

$$a_0 = \frac{1}{p} \int_{-p}^{p} f(t)\, dt \qquad \textbf{Formula (10.34)}$$

Since the function is described differently over different intervals, we need to integrate over the intervals $[-2, 0]$ and $[0, 2]$ separately. Thus

$$a_0 = \frac{1}{2} \int_{-2}^{2} f(t)\, dt \qquad\qquad \textbf{\textit{p} = 2}$$

$$= \frac{1}{2} \int_{-2}^{0} 0 \, dt \qquad\qquad \textbf{\textit{f(t)} = 0 on (−2, 0)}$$

$$+ \frac{1}{2} \int_{0}^{2} 1 \, dt \qquad\qquad \textbf{\textit{f(t)} = 1 on (0, 2)}$$

$$= 0 + \frac{1}{2} \int_{0}^{2} dt = \frac{1}{2} t \Big|_{0}^{2} = 1$$

Figure 10.8 (margin):

$f(t)$

Figure 10.8

Since $a_0 = 1$, we get for our first term

$$\frac{a_0}{2} = \frac{1}{2}$$

To find a_n, let us evaluate a_1 and a_2 first and then generalize to a_n:

$$a_1 = \frac{1}{p} \int_{-p}^{p} f(t) \cos \frac{1\pi t}{p} \, dt \qquad \text{\textbf{\textit{n}} = 1 in Formula (10.35)}$$

$$= \frac{1}{2} \int_{-2}^{0} 0 \, dt + \frac{1}{2} \int_{0}^{2} 1 \cdot \cos \frac{\pi t}{2} \, dt \qquad \textbf{\textit{p} = 2}$$

$$= \frac{1}{2} \int_{0}^{2} \cos \frac{\pi t}{2} \, dt$$

$$= \frac{1}{2} \frac{2}{\pi} \sin \frac{\pi t}{2} \Big|_{0}^{2} \qquad \left[\begin{array}{l} \textbf{\textit{u}} = \dfrac{\pi t}{2} \\[2mm] \textbf{\textit{du}} = \dfrac{\pi}{2} \, \textbf{\textit{dt}} \end{array} \right]$$

$$= \frac{1}{\pi} \sin \pi = 0$$

Next we evaluate a_2:

$$a_2 = \frac{1}{p} \int_{-p}^{p} f(t) \cos \frac{2\pi t}{p} \, dt \qquad \text{\textbf{\textit{n}} = 2 in Formula (10.35)}$$

$$= \frac{1}{2} \int_{0}^{2} \cos \pi t \, dt \qquad \textbf{\textit{p} = 2}$$

$$= \frac{1}{2} \frac{1}{\pi} \sin \pi t \Big|_{0}^{2} \qquad \left[\begin{array}{l} \textbf{\textit{u}} = \pi t \\[1mm] \textbf{\textit{du}} = \pi \, \textbf{\textit{dt}} \end{array} \right]$$

$$= \frac{1}{2\pi} \sin 2\pi = 0$$

To find the general coefficient a_n, recall the following relationships:

$$\sin n\pi = 0 \qquad \text{(for all } n\text{)}$$
$$\cos n\pi = \begin{cases} 1 & \text{for } n \text{ even} \\ -1 & \text{for } n \text{ odd} \end{cases}$$
$$\cos(-n\pi) = \cos n\pi$$

Thus we have

$$a_n = \frac{1}{p} \int_{-p}^{p} f(t) \cos \frac{n\pi t}{p} \, dt \qquad \textbf{Formula (10.35)}$$

$$= \frac{1}{2} \int_{-2}^{0} 0 \, dt + \frac{1}{2} \int_{0}^{2} 1 \cdot \cos \frac{n\pi t}{2} \, dt \qquad \textbf{\textit{p} = 2}$$

$$= \frac{1}{2} \int_0^2 \cos \frac{n\pi t}{2} \, dt$$

$$= \frac{1}{2} \frac{2}{n\pi} \sin \frac{n\pi t}{2} \Big|_0^2 \qquad\qquad \left[\begin{array}{l} u = \dfrac{n\pi t}{2} \\[2mm] du = \dfrac{n\pi}{2} \, dt \end{array} \right]$$

$$= \frac{1}{n\pi} \sin n\pi = 0$$

since $\sin n\pi = 0$ for all n.

For our next calculation, we find b_1 and b_2 and then generalize to b_n:

$$b_1 = \frac{1}{p} \int_{-p}^p f(t) \sin \frac{1\pi t}{p} \, dt \qquad\qquad \textbf{n = 1 in Formula (10.36)}$$

$$= \frac{1}{2} \int_{-2}^0 0 \, dt + \frac{1}{2} \int_0^2 1 \cdot \sin \frac{\pi t}{2} \, dt \qquad \textbf{p = 2}$$

$$= \frac{1}{2} \int_0^2 \sin \frac{\pi t}{2} \, dt$$

$$= \frac{1}{2} \left(-\frac{2}{\pi} \right) \cos \frac{\pi t}{2} \Big|_0^2 \qquad\qquad \left[\begin{array}{l} u = \dfrac{\pi t}{2} \\[2mm] du = \dfrac{\pi}{2} \, dt \end{array} \right]$$

$$= -\frac{1}{\pi} (\cos \pi - \cos 0)$$

$$= -\frac{1}{\pi} (-1 - 1) = \frac{2}{\pi}$$

$$b_2 = \frac{1}{p} \int_{-p}^p f(t) \sin \frac{2\pi t}{p} \, dt \qquad\qquad \textbf{n = 2 in Formula (10.36)}$$

$$= \frac{1}{2} \int_0^2 \sin \pi t \, dt \qquad\qquad \textbf{p = 2}$$

$$= \frac{1}{2} \left(-\frac{1}{\pi} \right) \cos \pi t \Big|_0^2 \qquad\qquad \left[\begin{array}{l} u = \pi t \\[1mm] du = \pi \, dt \end{array} \right]$$

$$= -\frac{1}{2\pi} (\cos 2\pi - \cos 0)$$

$$= -\frac{1}{2\pi} (1 - 1) = 0$$

$$b_n = \frac{1}{p} \int_{-p}^p f(t) \sin \frac{n\pi t}{p} \, dt \qquad\qquad \textbf{Formula (10.36)}$$

$$= \frac{1}{2} \int_{-2}^0 0 \, dt + \frac{1}{2} \int_0^2 1 \cdot \sin \frac{n\pi t}{2} \, dt \qquad \textbf{p = 2}$$

$$= \frac{1}{2} \int_0^2 \sin \frac{n\pi t}{2} \, dt$$

$$= \frac{1}{2}\left(-\frac{2}{n\pi}\right)\cos\frac{n\pi t}{2}\Big|_0^2 \qquad \left[\begin{array}{c} u = \dfrac{n\pi t}{2} \\[2mm] du = \dfrac{n\pi}{2}\,dt \end{array}\right]$$

$$= -\frac{1}{n\pi}(\cos n\pi - \cos 0)$$

$$= -\frac{1}{n\pi}(\cos n\pi - 1)$$

$$= \begin{cases} 0 & \text{for } n \text{ even} \\[2mm] \dfrac{2}{n\pi} & \text{for } n \text{ odd} \end{cases}$$

since $\cos n\pi = 1$ for n even and $\cos n\pi = -1$ for n odd.

We have obtained the following coefficients:

$$\frac{a_0}{2} = \frac{1}{2}, \ a_n = 0, \ b_1 = \frac{2}{\pi}, \ b_2 = 0, \ b_3 = \frac{2}{3\pi}, \ b_4 = 0, \ b_5 = \frac{2}{5\pi}, \text{ and so on}$$

Substituting these values in Series (10.33), we get (since $p = 2$)

$$f(t) = \frac{1}{2} + \frac{2}{\pi}\sin\frac{1\pi t}{2} + 0 + \frac{2}{3\pi}\sin\frac{3\pi t}{2} + 0 + \frac{2}{5\pi}\sin\frac{5\pi t}{2} + \cdots$$

or

$$f(t) = \frac{1}{2} + \frac{2}{\pi}\left(\sin\frac{\pi t}{2} + \frac{1}{3}\sin\frac{3\pi t}{2} + \frac{1}{5}\sin\frac{5\pi t}{2} + \cdots\right)$$ ∎

It may seem strange that a series of sines and cosines may approximate a graph consisting of straight lines, especially if such a graph is not even continuous. Suppose we graph the sum of the first two terms of the series in Example 1:

$$\frac{1}{2} + \frac{2}{\pi}\sin\frac{\pi t}{2}$$

This graph (dashed curve in Figure 10.9) is only a crude approximation. If we graph the sum of the first three terms,

$$\frac{1}{2} + \frac{2}{\pi}\sin\frac{\pi t}{2} + \frac{2}{3\pi}\sin\frac{3\pi t}{2}$$

the approximation improves (solid curve in Figure 10.9). If we use more terms of the series, the graph will contain more small oscillations and approximate the given function more and more closely as the number of terms increases. (See Figure 10.10.)

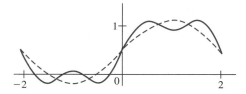

Figure 10.9

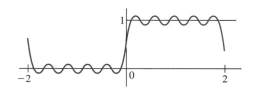

Figure 10.10

At a point of discontinuity, the graph passes through the point midway between the ordinates 0 and 1.

Example 2   Obtain the Fourier series of the "sawtooth function" in Figure 10.11, given over one period by

$$f(t) = t \qquad (-1 < t < 1)$$

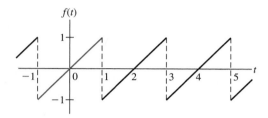

Figure 10.11

Solution.   In this problem we use the period $(-1, 1)$ to obtain the coefficients. Since the period is 2, we have

$$2p = 2 \qquad \text{or} \qquad p = 1$$

As always, we evaluate the first term separately:

$$a_0 = \frac{1}{p} \int_{-p}^{p} f(t)\, dt = \frac{1}{1} \int_{-1}^{1} t\, dt = \frac{1}{2} t^2 \Big|_{-1}^{1} = 0 \qquad \boldsymbol{p = 1}$$

Hence $a_0 = 0$.

Next we evaluate a_n for $n \geq 1$:

$$a_n = \frac{1}{p} \int_{-p}^{p} f(t) \cos \frac{n\pi t}{p}\, dt = \frac{1}{1} \int_{-1}^{1} t \cos \frac{n\pi t}{1}\, dt \qquad \boldsymbol{p = 1}$$

$$\begin{array}{c|c} u = t & dv = \cos n\pi t\, dt \\ \hline du = dt & v = \dfrac{1}{n\pi} \sin n\pi t \end{array} \qquad \textbf{integration by parts}$$

$$a_n = \frac{t}{n\pi} \sin n\pi t \Big|_{-1}^{1} - \frac{1}{n\pi} \int_{-1}^{1} \sin n\pi t\, dt$$

$$= \frac{t}{n\pi} \sin n\pi t \Big|_{-1}^{1} - \left(\frac{1}{n\pi} \right) \left(-\frac{1}{n\pi} \cos n\pi t \right) \Big|_{-1}^{1} \qquad \left[\begin{array}{l} \boldsymbol{u = n\pi t} \\ \boldsymbol{du = n\pi\, dt} \end{array} \right]$$

$$= \left[\frac{1}{n\pi} \sin n\pi - \frac{-1}{n\pi} \sin(-n\pi) \right] + \frac{1}{n^2\pi^2} [\cos n\pi - \cos(-n\pi)]$$

Since $\sin n\pi = 0$ and $\cos(-n\pi) = \cos n\pi$, we get

$$a_n = 0 + \frac{1}{n^2\pi^2}(\cos n\pi - \cos n\pi) = 0$$

Hence $a_n = 0$.

Finally, we evaluate b_n:

$$b_n = \frac{1}{p}\int_{-p}^{p} f(t)\sin\frac{n\pi t}{p}\,dt = \frac{1}{1}\int_{-1}^{1} t\sin\frac{n\pi t}{1}\,dt \qquad p=1$$

$u = t$	$dv = \sin n\pi t\,dt$	
$du = dt$	$v = -\dfrac{1}{n\pi}\cos n\pi t$	**integration by parts**

$$b_n = -\frac{t}{n\pi}\cos n\pi t\Big|_{-1}^{1} + \frac{1}{n\pi}\int_{-1}^{1}\cos n\pi t\,dt$$

$$= -\frac{t}{n\pi}\cos n\pi t\Big|_{-1}^{1} + \left(\frac{1}{n\pi}\right)\left(\frac{1}{n\pi}\sin n\pi t\right)\Big|_{-1}^{1} \qquad \begin{bmatrix} u = n\pi t \\ du = n\pi\,dt \end{bmatrix}$$

$$= -\frac{1}{n\pi}\cos n\pi + \frac{-1}{n\pi}\cos(-n\pi) + \frac{1}{n^2\pi^2}[\sin n\pi - \sin(-n\pi)]$$

Since $\sin n\pi = 0$ for all n and $\cos(-n\pi) = \cos n\pi$, we get

$$b_n = -\frac{1}{n\pi}\cos n\pi - \frac{1}{n\pi}\cos n\pi + 0$$

$$= -\frac{2}{n\pi}\cos n\pi$$

$$= \begin{cases} \dfrac{2}{n\pi} & \text{for } n \text{ odd} \qquad \boldsymbol{\cos n\pi = -1 \text{ for } n \text{ odd}} \\[2mm] -\dfrac{2}{n\pi} & \text{for } n \text{ even} \qquad \boldsymbol{\cos n\pi = 1 \text{ for } n \text{ even}} \end{cases}$$

We have obtained the following coefficients:

$$a_0 = 0,\ a_n = 0,\ b_1 = \frac{2}{1\pi},\ b_2 = -\frac{2}{2\pi},\ b_3 = \frac{2}{3\pi},\ b_4 = -\frac{2}{4\pi},\ \text{and so on}$$

Substituting in Series (10.33), we get (since $p = 1$)

$$f(t) = \frac{2}{1\pi}\sin\frac{1\pi t}{1} - \frac{2}{2\pi}\sin\frac{2\pi t}{1} + \frac{2}{3\pi}\sin\frac{3\pi t}{1} - \frac{2}{4\pi}\sin\frac{4\pi t}{1} + \cdots$$

or

$$f(t) = \frac{2}{\pi}\left(\sin \pi t - \frac{1}{2}\sin 2\pi t + \frac{1}{3}\sin 3\pi t - \frac{1}{4}\sin 4\pi t + \cdots\right) \qquad \blacksquare$$

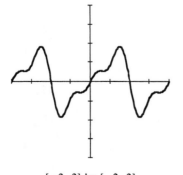

[−2, 2] by [−2, 2]

Figure 10.12

Figure 10.12 shows the graph of the sum of the first three terms, obtained with a graphing utility.

■ Exercises / Section 10.6

In Exercises 1–5, each function is given over one period only. Calculate the coefficients in parts **(a)** and **(b)**; then determine the Fourier series.

1. $f(t) = \begin{cases} 0 & -1 < t < 0 \\ 1 & 0 < t < 1 \end{cases}$

   a. $a_0, a_1,$ and a_2 **b.** $b_1, b_2,$ and b_3

2. $f(t) = \begin{cases} 0 & -\pi < t < 0 \\ 1 & 0 < t < \pi \end{cases}$

   a. $a_0, a_1,$ and a_2 **b.** $b_1, b_2,$ and b_3

3. $f(t) = \begin{cases} 0 & -5 < t < 0 \\ 1 & 0 < t < 5 \end{cases}$

   a. $a_0, a_1,$ and a_2 **b.** $b_1, b_2,$ and b_3

4. $f(t) = \begin{cases} -3 & -2 < t < 0 \\ 3 & 0 < t < 2 \end{cases}$

   a. $a_0, a_1,$ and a_2 **b.** $b_1, b_2,$ and b_3

5. $f(t) = t \qquad -a < t < a$
   a. $a_0, a_1,$ and a_2 **b.** $b_1, b_2,$ and b_3

In Exercises 6–8, determine the Fourier series for each periodic function.

6. $f(t) = \begin{cases} -t & -2 < t < 0 \\ t & 0 < t < 2 \end{cases}$

7. $f(t) = \begin{cases} 0 & -a < t < 0 \\ t & 0 < t < a \end{cases}$

8. $f(t) = \begin{cases} \pi + t & -\pi < t < 0 \\ \pi - t & 0 < t < \pi \end{cases}$

In Exercises 9 and 10, use the table of integrals in Appendix A (Table 2, Forms 61 and 63) to obtain the coefficients for the series.

9. Obtain the Fourier series of the "half-wave rectification" of the sine (Figure 10.13), whose definition in one period is given by

$$\begin{cases} 0 & -\pi < t < 0 \\ \sin t & 0 < t < \pi \end{cases}$$

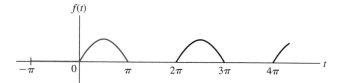

Figure 10.13

10. Show that the Fourier series of the "full-wave rectification" of the sine (Figure 10.14) is given by

$$\frac{2}{\pi} - \frac{4}{\pi}\left[\frac{\cos 2t}{2^2 - 1} + \frac{\cos 4t}{4^2 - 1} + \frac{\cos 6t}{6^2 - 1} + \cdots\right]$$

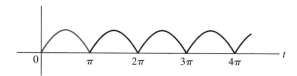

Figure 10.14

■ Review Exercises / Chapter 10

1. Find the sum of the geometric series

$$\sum_{n=1}^{\infty} \frac{(-1)^{n-1}}{3^{n-1}} = 1 - \frac{1}{3} + \frac{1}{9} - \frac{1}{27} + \cdots$$

$$+ (-1)^{n-1}\frac{1}{3^{n-1}} + \cdots$$

2. Convert $0.13232\ldots$ to a common fraction.

In Exercises 3–10 (optional), test the series for convergence or divergence.

3. $\displaystyle\sum_{n=1}^{\infty} \frac{2n}{4n+3}$ **4.** $\displaystyle\sum_{n=1}^{\infty} \frac{1}{n^2+4}$ **5.** $\displaystyle\sum_{n=2}^{\infty} \frac{1}{n \ln^2 n}$

6. $\displaystyle\sum_{n=1}^{\infty} \frac{1}{n(n+3)}$ **7.** $\displaystyle\sum_{n=2}^{\infty} \frac{1}{\ln n}$ **8.** $\displaystyle\sum_{n=0}^{\infty} \frac{n^3}{n!}$

9. $\displaystyle\sum_{n=1}^{\infty} \frac{6^n}{n!}$ **10.** $\displaystyle\sum_{n=2}^{\infty} \frac{n^2}{n^3-3}$

11. Use the formula for the Maclaurin series to verify that

$$e^{-x} = 1 - x + \frac{x^2}{2!} - \frac{x^3}{3!} + \cdots$$

12. Repeat Exercise 11 for

$$\ln(1+x) = x - (x^2/2) + (x^3/3) - \cdots$$

13. Use Expansion (10.15) to find the Maclaurin series of $\sin x^2$.

14. Use Expansion (10.16) to find the Maclaurin series of $\cos \sqrt{x}$.

15. Expand $f(x) = (1 - e^x)/x$ by using (10.14).

16. Expand $f(x) = \sinh x = (1/2)(e^x - e^{-x})$ by subtracting the appropriate series.

17. Evaluate $\cos(0.5)$ using three terms. Find the maximum possible error and determine the accuracy of the result.

18. Evaluate $\int_0^1 [(\sin x)/x]\,dx$ accurate to five decimal places.

19. Evaluate $\int_0^{0.9} \cos x^3\,dx$ accurate to four decimal places.

20. Evaluate $\int_0^{0.1} e^{-x^3}\,dx$ accurate to three decimal places.

21. The charge on a certain capacitor is given by

$$q = \int_0^{0.10} e^{-0.20t^2}\,dt$$

Find the value of q.

22. Evaluate $\cos 33°$ by expanding $f(x) = \cos x$ about $c = \pi/6$. (Use three terms.)

23. Evaluate $\sin 44°$ by expanding $f(x) = \sin x$ about $c = \pi/4$. (Use four terms.)

24. Show that $\lim_{x \to 0} (\sin x)/x = 1$ by means of Maclaurin series.

25. Show that $\lim_{x \to 0} (1 - \cos x)/x = 0$ by means of Maclaurin series.

26. Show that the Fourier series of the periodic function whose definition in one period is $f(t) = t$, $-2 < t < 2$, is given by

$$\frac{4}{\pi} \left(\sin \frac{\pi t}{2} - \frac{1}{2} \sin \frac{2\pi t}{2} + \frac{1}{3} \sin \frac{3\pi t}{2} - \cdots \right)$$

11

First-Order Differential Equations

Euler

Most of the material in this and the next chapter owes its development to the Swiss scientist and mathematician Leonhard Euler (1707–1783). Considered the greatest mathematician of the eighteenth century and easily the most prolific in the history of mathematics, Euler contributed greatly to the development of calculus, particularly to the study of infinite series. He was also actively engaged in analytic number theory (which he founded); physics, particularly celestial mechanics; the mathematics of finance (which he created), probability; geometry, particularly three-dimensional analytic geometry (which he practically created); parametric equations; and polar coordinates (which he did not create but developed fully for the first time). Perhaps most important of all, he systematized and unified much of the mathematics of his day. In the process he introduced many of the notations that are now standard; for example, the symbols $f(x)$, $\ln x$, $\sum$ (for sum), e, π, and i (for $\sqrt{-1}$) are all due to Euler. Even total blindness during the last seventeen years of his life could not retard the output of his research, and his collected works amount to seventy-five large volumes.

Euler was born in Basel but spent a large part of his life in St. Petersburg, Russia (1727–1741 and 1766–1783) at the invitation of the authorities at the St. Petersburg Academy of Science. Between his stays there he spent twenty-five years at the court of King Frederick the Great of Prussia. (Part of his time there coincided with Voltaire's shorter stay, from 1750 to 1753.) Euler died of a stroke on September 18, 1783.

11.1 What Is a Differential Equation?

Everyone who has studied algebra is familiar with the word *equation*. Equations always seem to involve an unknown, usually denoted by x, and your job is to find its value. In this chapter we are going to look at a completely different kind of equation, one for which the solution is not just a number, but a function. Such equations are therefore called **functional equations.** For example, can you guess the solution of the functional equation $f(x + y) = f(x)f(y)$? A little reflection shows that any exponential function $f(x) = a^x$ will do, since

$$a^{x+y} = a^x a^y$$

Now try $f'(x) - f(x) = 0$. The solution is easily found by inspection once the equation is written in the form

$$f'(x) = f(x)$$

Since $de^x/dx = e^x$, it follows that $f(x) = e^x$. We now define a **differential equation** to be a functional equation that contains derivatives or differentials. For example, the equation $f'(x) - f(x) = 0$ is now seen to be a differential equation; $f(x) = e^x$ is called a **solution.** We may also write $y' - y = 0$ or $dy/dx - y = 0$ for the equation and $y = e^x$ for the solution.

These ideas are summarized next.

A **differential equation** is a functional equation of two variables containing derivatives or differentials.

A **solution** of a differential equation is a relation between the variables that satisfies the equation.

One point should now be emphasized. In the equation under discussion, $y = e^x$ is *a* solution, not *the* solution, since $y = ce^x$, for any arbitrary constant c, satisfies the equation equally well.

It is not hard to see how differential equations arise, since we have already encountered a number of examples. For instance, if v is the velocity of an object moving with constant acceleration a, then $dv/dt = a$, which is indeed a differential equation. It can be readily solved by direct integration:

$$v = \int a\, dt = at + c$$

again suggesting why solutions are not unique.

Not all differential equations can be solved in this simple manner. Recall that the voltages across an inductor and resistor are $L(di/dt)$ and Ri, respectively. If connected in series with a generator of E volts, then

$$L\frac{di}{dt} + Ri = E$$

by Kirchhoff's voltage law. Here we have an example of a differential equation that cannot be solved by any technique considered so far.

We are going to consider many more of these practical uses of differential equations in Sections 11.4 and 12.4, but first we need to learn some methods for solving these equations.

Example 1
Find the function y if $dy/dx = 2x^3$ and the curve passes through the point $(1, 2)$. (See Section 4.8.)

Solution. The equation $dy/dx = 2x^3$ can be solved directly by integration—that is,

$$y = \frac{1}{2}x^4 + c \tag{11.1}$$

Because $(1, 2)$ lies on the curve, we let $x = 1$ and $y = 2$ and solve for c:

$$2 = \frac{1}{2}(1)^4 + c$$

so that $c = 3/2$. So the function is

$$y = \frac{1}{2}x^4 + \frac{3}{2} \tag{11.2}$$

∎

This example may serve to introduce some of our basic terms.

Solution (11.1) satisfies the equation no matter what value of c is chosen and is called the **general solution.** In Solution (11.2) the arbitrary constant was evaluated from the extra condition. Solution (11.2) is therefore called a **particular solution.**

Example 2
Show that the function

$$y = 3xe^{2x} + ce^{2x}$$

is the general solution of the differential equation

$$\frac{dy}{dx} - 2y = 3e^{2x}$$

Solution. To show that the given function solves the equation, we find dy/dx and substitute. By the product rule,

$$\frac{dy}{dx} = \frac{d}{dx}(3xe^{2x} + ce^{2x})$$
$$= 3xe^{2x}(2) + 3e^{2x} + 2ce^{2x}$$
$$= 6xe^{2x} + 3e^{2x} + 2ce^{2x}$$

Substituting in the left side of the given equation

$$\frac{dy}{dx} - 2y = 3e^{2x}$$

we get

$$(6xe^{2x} + 3e^{2x} + 2ce^{2x}) - 2(3xe^{2x} + ce^{2x})$$
$$= 6xe^{2x} + 3e^{2x} + 2ce^{2x} - 6xe^{2x} - 2ce^{2x}$$
$$= 3e^{2x} \quad \text{(the right side)}$$

The equation is therefore satisfied. ∎

Order

The **order** of a differential equation is the order of the highest derivative. The equation in Example 2 is of first order, while

$$\frac{d^3y}{dx^3} + x\frac{dy}{dx} + y = 0$$

is of third order. In this chapter we will study various types of first-order equations and their applications.

■ Exercises / Section 11.1

In Exercises 1–16, show that the given function is a solution of the differential equation. (See Example 2.)

1. $\dfrac{dy}{dx} - 3y = 0$, $y = 2e^{3x}$

2. $y' - 3y = e^{4x}$, $y = e^{4x}$

3. $\dfrac{dy}{dx} - y = e^x$, $y = xe^x + 2e^x$

4. $\dfrac{dy}{dx} + 2y = e^{-2x}$, $y = xe^{-2x} + 3e^{-2x}$

5. $y'' + 4y = 0$, $y = 2\cos 2x + 3\sin 2x$

6. $y'' + y = 0$, $y = c\cos x$

7. $y'' + y' - 6y = 0$, $y = c_1e^{-3x} + c_2e^{2x}$ (c_1 and c_2 are arbitrary constants)

8. $\dfrac{d^2y}{dx^2} - \dfrac{dy}{dx} - 2y = 0$, $y = c_1e^{2x} + c_2e^{-x}$

9. $x\dfrac{dy}{dx} - y = x^2 + 4$, $y = x^2 - 4$

10. $x\dfrac{dy}{dx} + y = x + 3$, $y = \dfrac{1}{2}x + \dfrac{1}{x} + 3$

11. $x^2\dfrac{dy}{dx} - xy = x^2 + x$, $y = x + x\ln x - 1$

12. $x^2\dfrac{dy}{dx} + xy = x + 1$, $y = 1 + \dfrac{\ln x}{x}$

13. $y'' + 9y = 6\sin 3x$, $y = c\cos 3x - x\cos 3x$

14. $\dfrac{d^2y}{dx^2} + 16y = 8\cos 4x$, $y = x\sin 4x + c\sin 4x$

15. $\dfrac{d^2y}{dx^2} + 5\dfrac{dy}{dx} - 6y = 7e^x$, $y = c_1e^x + c_2e^{-6x} + xe^x$

16. $y'' + 2y' - 8y = 6e^{2x}$, $y = c_1e^{-4x} + c_2e^{2x} + xe^{2x}$

In Exercises 17–22, solve the differential equations subject to the given conditions. (See Example 1.)

17. $\dfrac{dy}{dx} = 3x^2$, $y = 5$ when $x = 2$

18. $\dfrac{dy}{dx} = 2x^3$, $y = 6$ when $x = 2$

19. $\dfrac{dy}{dx} = \sec^2 x$, $y = 1$ when $x = \dfrac{\pi}{4}$

20. $\dfrac{dy}{dx} = \cos x$, $y = 2$ when $x = \dfrac{\pi}{2}$

21. $\dfrac{d^2y}{dx^2} = e^x$, $y = 0$ when $x = 0$ and $y = 1$ when $x = 1$ (The general solution has two arbitrary constants.)

22. $\dfrac{d^2y}{dx^2} = \sin x$, $y = 0$ when $x = 0$ and $\dfrac{dy}{dx} = 1$ when $x = 0$

11.2   Separation of Variables

In this section we discuss a method for solving a certain type of first-order equation. First recall that if $dy/dx = f'(x)$, then $dy = f'(x)\,dx$; the converse is also true. In other words, a derivative can always be expressed in differential form, and in this section it turns out to be convenient to do so. To see why, consider the following equation of first order:

$$\frac{dy}{dx} = F(x, y) \tag{11.3}$$

Using the differential notation, the equation can be written in the following form:

General First-Order Differential Equation

$$M(x, y)\,dx + N(x, y)\,dy = 0 \tag{11.4}$$

For some equations it is possible to *separate variables* by writing Equation (11.4) in the form $M(x)\,dx + N(y)\,dy = 0$. In other words, M is a function of x alone, and N is a function of y alone. In this form, each of the terms can be integrated separately. This procedure is known as the method of **separation of variables.**

Equation with Variables Separated

$$M(x)\,dx + N(y)\,dy = 0 \tag{11.5}$$

The next example illustrates the technique.

Example 1   Use separation of variables to solve the equation

$$\csc 2x \frac{dy}{dx} + e^{\cos 2x} = 0$$

Solution.   Writing the equation in Form (11.4), we get

$$e^{\cos 2x}\,dx + \csc 2x\,dy = 0 \qquad \textbf{formally multiplying by } \boldsymbol{dx}$$

Observe next that if we divide both sides of the equation by $\csc 2x$, then the equation becomes

$$\frac{e^{\cos 2x}\,dx}{\csc 2x} + dy = 0$$

$$e^{\cos 2x} \sin 2x\,dx + dy = 0 \qquad \textbf{1/csc } \boldsymbol{\theta} = \textbf{sin } \boldsymbol{\theta}$$

In this form all the terms are readily integrated since $M(x)$, the coefficient of dx, is a function of x alone and $N(y)$, the coefficient of dy, is a function of y alone. Thus

$$\int e^{\cos 2x} \sin 2x \, dx + \int dy = c_1$$

or

$$-\frac{1}{2} \int e^{\cos 2x} (-2 \sin 2x) \, dx + \int dy = c_1 \qquad \left[\begin{array}{l} \boldsymbol{u = \cos 2x} \\ \boldsymbol{du = -2\sin 2x\, dx} \end{array} \right]$$

and

$$-\frac{1}{2} e^{\cos 2x} + c_2 + y + c_3 = c_1 \qquad\qquad (11.6)$$

∎

The constants in the solution of the equation in Example 1 can be combined. Since Equation (11.6) can be written

$$-\frac{1}{2} e^{\cos 2x} + y = c_1 - c_2 - c_3$$

we may let $k = c_1 - c_2 - c_3$ to obtain the simpler form

$$-\frac{1}{2} e^{\cos 2x} + y = \boldsymbol{k}$$

Moreover, multiplying by -2, we get

$$e^{\cos 2x} - 2y = \boldsymbol{-2k}$$

which suggests letting $c = -2k$ to obtain

$$e^{\cos 2x} - 2y = c$$

(If k is arbitrary, so is $-2k$.)

From now on, whenever we integrate the terms in an equation, we simply *place a single arbitrary constant on the right side*.

Example 2   Find the general solution of the differential equation

$$2x \, dx + xy \, dx + \cos^2 x^2 \, dy = 0$$

Solution. We need to rewrite the equation in the form $M(x) \, dx + N(y) \, dy = 0$, where $M(x)$ is a function of x alone and $N(y)$ is a function of y alone:

$$2x \, dx + xy \, dx + \cos^2 x^2 \, dy = 0 \qquad \textbf{given equation}$$

$$(2 + y)x \, dx + \cos^2 x^2 \, dy = 0 \qquad \textbf{factoring } \boldsymbol{x\, dx}$$

$$\frac{x \, dx}{\cos^2 x^2} + \frac{dy}{2 + y} = 0 \qquad \textbf{dividing by } \boldsymbol{(2 + y)\cos^2 x^2}$$

$$(\sec^2 x^2)x \, dx + \frac{dy}{2 + y} = 0 \qquad \boldsymbol{\sec \theta = \dfrac{1}{\cos \theta}}$$

The equation now has the proper form, and we may integrate each term separately and introduce the arbitrary constant k on the right side:

$$\frac{1}{2} \int (\sec^2 x^2)(2x)\, dx + \int \frac{dy}{2+y} = k \qquad \left[\begin{array}{c} \textbf{\textit{u} = \textit{x}}^{\textbf{2}} \\ \textbf{\textit{du} = 2\textit{x dx}} \end{array} \right]$$

$$\frac{1}{2} \tan x^2 + \ln|2+y| = k \qquad \textbf{the second integral has the form } \int \frac{\textbf{\textit{du}}}{\textbf{\textit{u}}}$$

$$\tan x^2 + 2 \ln|2+y| = 2k \qquad \textbf{multiplying by 2}$$

$$\tan x^2 + 2 \ln|2+y| = c \qquad \textbf{letting \textit{c} = 2\textit{k}} \qquad \blacksquare$$

Example 3   Obtain the general solution of the equation $(y-2)\, dx + (1-x)\, dy = 0$.

Solution.   Separating variables, we get

$$\frac{dx}{1-x} + \frac{dy}{y-2} = 0 \qquad \textbf{dividing by (\textit{y} − 2)(1 − \textit{x})}$$

If we now integrate each term and introduce the arbitrary constant k on the right side, we get

$$-\int \frac{-dx}{1-x} + \int \frac{dy}{y-2} = k \qquad \left[\begin{array}{c} \textbf{\textit{u} = 1−\textit{x}} \\ \textbf{\textit{du} = −\textit{dx}} \end{array} \right]$$

$$-\ln|1-x| + \ln|y-2| = k$$

or

$$\ln \left| \frac{y-2}{1-x} \right| = k \qquad \textbf{ln \textit{A} − ln \textit{B} = ln } \frac{\textbf{\textit{A}}}{\textbf{\textit{B}}}$$

Using the properties of logarithms, we can write the solution much more compactly as follows: for every k, there exists a number $c' > 0$ such that $\ln c' = k$; hence

$$\ln \left| \frac{y-2}{1-x} \right| = \ln c'$$

and, since logarithms are unique, it follows that

$$\left| \frac{y-2}{1-x} \right| = c' \qquad (c' > 0)$$

By the definition of absolute value,

$$\frac{y-2}{1-x} = \pm c'$$

Finally, letting $c = \pm c'$, we get

$$y - 2 = c(1 - x) \qquad (c \neq 0)$$

This type of reduction is possible whenever the solution contains only logarithmic functions. ∎

If the equation satisfies an additional condition, then we get a unique particular solution.

Example 4   Find the particular solution of the equation $dx + y^3 \cot x \, dy = 0$, subject to the condition $y = 2$ when $x = 0$.

Solution. Separating variables, we get

$$\frac{dx}{\cot x} + y^3 \, dy = 0 \qquad \text{dividing by cot } x$$

$$\tan x \, dx + y^3 \, dy = 0 \qquad \frac{1}{\cot x} = \tan x$$

and, after performing the integration,

$$\ln|\sec x| + \frac{1}{4}y^4 = c$$

We now let $x = 0$ and $y = 2$, so that $0 + 4 = c$, and the solution becomes $\ln|\sec x| + (1/4)y^4 = 4$ or $4\ln|\sec x| + y^4 = 16$. ∎

Note that the solution (as well as the conditions) can always be checked. In Example 4 we see that the condition $y = 2$ when $x = 0$ is obviously satisfied. To check the solution, we differentiate implicitly to obtain

$$4\frac{\sec x \tan x}{\sec x} + 4y^3 \frac{dy}{dx} = 0$$

which reduces to the given equation.

Example 5   Show that $\tan x + e^y = c$ satisfies the equation

$$e^{-y} \, dx + \cos^2 x \, dy = 0$$

Solution. Differentiating $\tan x + e^y = c$ implicitly, we get

$$\sec^2 x + e^y \frac{dy}{dx} = 0 \qquad \text{since } \frac{d}{dx}e^u = e^u \frac{du}{dx}$$

$$\sec^2 x \, dx + e^y \, dy = 0$$

$$\frac{dx}{e^y} + \frac{dy}{\sec^2 x} = 0$$

and

$$e^{-y}\,dx + \cos^2 x\,dy = 0 \qquad \frac{1}{\sec\theta} = \cos\theta$$

which is the given equation.   ■

■ Exercises / Section 11.2

In Exercises 1–32, find the general solution of each of the equations.

1. $x^2\,dx + y\,dy = 0$

2. $x^5\,dx + y^2\,dy = 0$

3. $(1 + x^2)\,dx = 3y\,dy$

4. $(1 - x)\,dx + 4y^3\,dy = 0$

5. $2x + (1 + x^2)\dfrac{dy}{dx} = 0$

6. $x + 2 + 2(x + 1)y\dfrac{dy}{dx} = 0$

7. $(1 + x^2)\dfrac{dy}{dx} + y = 0$

8. $y\,dx + \sec x\,dy = 0$

9. $2y\,dx + 3x\,dy = 0$ (See Example 3.)

10. $3y\dfrac{dy}{dx} + \tan x = 0$

11. $1 + (x^2 y - x^2)\dfrac{dy}{dx} = 0$

12. $(1 + x)^2\dfrac{dy}{dx} = 1$

13. $dx - y\,dx + x\,dy = 0$

14. $xy\,dx + dy + x^2\,dy = 0$

15. $\dfrac{dV}{dP} = -\dfrac{V}{P}$

16. $\dfrac{di}{dR} = -\dfrac{i}{2R}$

17. $dx + (2\cos^2 x - y\cos^2 x)\,dy = 0$

18. $xe^{x^2+y}\,dx + dy = 0$

19. $\cos^2 t + y\csc t\dfrac{dy}{dt} = 0$

20. $(\tan t + y\tan t)\,dt + dy = 0$

21. $\sqrt{v^2 + 1}\,dt + vt^2\,dv = 0$

22. $r\dfrac{dr}{dt} = \dfrac{\sin t \sec t}{\ln r}$

23. $T_1\,dT_1 + (\csc T_1 + T_2\csc T_1)\,dT_2 = 0$

24. $2s_1\dfrac{ds_2}{ds_1} - s_2{}^3\ln s_1 = 0$

25. $(y^2 - 1)\cos x\,dx + 2y\sin x\,dy = 0$

26. $(x^3 - x^3 y^2)\,dx = y\,dy$

27. $xe^y\,dx + e^{-x}\,dy = 0$

28. $xy\,dx + \sqrt{1 - x^2}\,dy = 0$

29. $(e^x\tan y + \tan y)\dfrac{dy}{dx} + e^x = 0$

30. $\dfrac{dy}{dx} = \dfrac{y}{x^2 - 4x + 4}$

31. $y\dfrac{dy}{dx} + 2x\sec y = 0$

32. $3xy\,dx + (1 + x^2)\,dy = 0$

In Exercises 33–37, find the particular solution of the equations satisfying the given conditions.

33. $x\,dy - y\,dx = 0,\ y = 2$ when $x = 1$

34. $\sin x\,dx + \sec x\,dy = 0,\ y = -1/4$ when $x = \pi/4$

35. $(y + 2)\,dx + (x - 3)\,dy = 0,\ y = 5$ when $x = 2$

36. $(1 + x)\,dy - y^2\,dx = 0,\ y = -1$ when $x = 0$

37. $dx + x\tan y\,dy = 0,\ y = 0$ when $x = 1$

11.3   First-Order Linear Differential Equations

So far we have considered only differential equations that can be solved by separation of variables. In this section we are going to consider another type of equation called *linear*.

> **First-Order Linear Differential Equation**
>
> $$\frac{dy}{dx} + P(x)y = Q(x) \tag{11.7}$$

The method of solution of a first-order linear equation depends on a special device: we multiply both sides of the given equation by a certain function called an **integrating factor,** designed to facilitate the integration. The integrating factor is the function

$$e^{\int P(x)\,dx} \tag{11.8}$$

where $P(x)$ is the coefficient of y in Equation (11.7). After multiplying both sides of the equation by the integrating factor, we can write the left side in the form

$$\frac{d}{dx}\left(ye^{\int P(x)\,dx}\right)$$

as we will see. After removing the differentiation symbol (by integrating both sides), we can solve for y algebraically. Before we justify this procedure, let us consider an example.

Example 1   Solve the linear equation $dy/dx - 2xy = xe^{x^2}$.

Solution. By (11.8), the integrating factor (IF) is the exponential function

$$e^{\int(-2x)\,dx}$$

where $-2x$ is the coefficient of y. Thus

$$\text{IF} = e^{-x^2}$$

Multiplying both sides of the given equation by this function, we get

$$e^{-x^2}\left(\frac{dy}{dx} - 2xy\right) = xe^{x^2}e^{-x^2} = x$$

Now observe that the left side can be written

$$\frac{d}{dx}\left(ye^{-x^2}\right)$$

as can be readily checked by the product rule:

$$\frac{d}{dx}\left(ye^{-x^2}\right) = e^{-x^2}\frac{dy}{dx} + y\frac{d}{dx}\left(e^{-x^2}\right)$$

$$= e^{-x^2}\frac{dy}{dx} - 2xye^{-x^2} = e^{-x^2}\left(\frac{dy}{dx} - 2xy\right)$$

The resulting equation is therefore

$$\frac{d}{dx}\left(ye^{-x^2}\right) = x$$

In this form the equation can be readily solved by integrating both sides:

$$ye^{-x^2} = \frac{1}{2}x^2 + c \quad \text{or} \quad y = \frac{1}{2}x^2 e^{x^2} + ce^{x^2}$$ ∎

The method of Example 1 works with all linear equations. To see why, refer to General Form (11.7) and let

$$IF = e^{\int P(x)\,dx}$$

Multiplying both sides of the equation, we get

$$e^{\int P(x)\,dx}\left[\frac{dy}{dx} + P(x)y\right] = Q(x)e^{\int P(x)\,dx}$$

which can be written

$$\frac{d}{dx}(ye^{\int P(x)\,dx}) = Q(x)e^{\int P(x)\,dx} \tag{11.9}$$

In other words, the left side becomes the derivative of the product of y and IF. To check this statement, we use the product rule:

$$\frac{d}{dx}(ye^{\int P(x)\,dx}) = y\frac{d}{dx}e^{\int P(x)\,dx} + e^{\int P(x)\,dx}\frac{dy}{dx} \qquad \textbf{product rule}$$

$$= ye^{\int P(x)\,dx}\frac{d}{dx}\int P(x)\,dx + e^{\int P(x)\,dx}\frac{dy}{dx} \qquad \frac{d}{dx}e^u = e^u\frac{du}{dx}$$

$$= ye^{\int P(x)\,dx} \cdot P(x) + e^{\int P(x)\,dx}\frac{dy}{dx} \qquad \frac{d}{dx}\int P(x)\,dx = P(x)$$

$$= e^{\int P(x)\,dx}\left[\frac{dy}{dx} + P(x)y\right] \qquad \textbf{factoring}$$

If we now integrate both sides of Equation (11.9), namely,

$$\frac{d}{dx}(ye^{\int P(x)\,dx}) = Q(x)e^{\int P(x)\,dx}$$

we obtain the solution

$$ye^{\int P(x)\,dx} = \int Q(x)e^{\int P(x)\,dx}\,dx + c$$

Integrating Factor (IF)

An **integrating factor** of the linear equation

$$\frac{dy}{dx} + P(x)y = Q(x) \tag{11.10}$$

is given by

$$IF = e^{\int P(x)\,dx} \tag{11.11}$$

The procedure for solving first-order linear equations may be summarized as follows:

Procedure for Solving First-Order Linear Differential Equations

1. Write the equation in the *standard form,* Equation (11.10).
2. Determine the integrating factor (IF).
3. Multiply both sides of the equation by IF.
4. Write the left side as the derivative of the product of y and IF.
5. Integrate both sides.

Example 2   Solve the linear equation

$$\frac{dy}{dx} + 3y = 4$$

Solution. The equation is already in standard form (**Step 1**), so by (11.11)

$$\text{IF} = e^{\int 3\,dx} = e^{3x} \qquad \textbf{Step 2}$$

Multiplying both sides of the equation by IF, we get

$$e^{3x}\left(\frac{dy}{dx} + 3y\right) = 4e^{3x} \qquad \textbf{Step 3}$$

By (11.9), the left side can be written as the derivative of the product of y and IF:

$$\frac{d}{dx}(ye^{3x}) = 4e^{3x} \qquad \textbf{Step 4}$$

Integrating both sides, we get

$$ye^{3x} = \int 4e^{3x}\,dx = \frac{4}{3}e^{3x} + c \qquad \begin{bmatrix} u = 3x \\ du = 3\,dx \end{bmatrix} \qquad \textbf{Step 5}$$

or

$$y = \frac{4}{3} + ce^{-3x} \qquad\qquad \blacksquare$$

Before continuing with our next example, we need to examine a special form that frequently arises in the integrating factor. If we write $\log_e N$ for $\ln N$ and recall that

$$\log_e A = B \qquad \text{means} \qquad e^B = A$$

then it follows that

$$\log_e N = \log_e N$$

is equivalent to

$$e^{\log_e N} = N$$

or

$$e^{\ln N} = N \qquad\qquad (11.12)$$

Example 3   Solve the linear equation

$$x \, dy = (3y + 5x^5) \, dx$$

Solution. First we need to write the equation in Standard Form (11.10):

$$x \, dy = (3y + 5x^5) \, dx \qquad \text{given equation}$$

$$x \frac{dy}{dx} = 3y + 5x^5 \qquad \text{derivative form}$$

$$x \frac{dy}{dx} - 3y = 5x^5 \qquad \text{transposing } 3y$$

$$\frac{dy}{dx} - \frac{3}{x}y = 5x^4 \qquad \text{dividing by } x \qquad \textbf{Step 1}$$

The equation is now in Standard Form (11.10). The integrating factor is

$$\text{IF} = e^{\int(-3/x)\,dx} = e^{-3\ln x} \qquad \textbf{by (11.11)}$$

$$= e^{\ln x^{-3}} \qquad a \ln x = \ln x^a$$

$$= x^{-3} \qquad e^{\ln N} = N \qquad \textbf{Step 2}$$

Multiplying both sides of the equation

$$\frac{dy}{dx} - \frac{3}{x}y = 5x^4$$

by x^{-3}, we get

$$x^{-3}\left(\frac{dy}{dx} - \frac{3}{x}y\right) = 5x \qquad \textbf{Step 3}$$

or, by (11.9),

$$\frac{d}{dx}(x^{-3}y) = 5x \qquad \text{derivative of the product of } y \text{ and IF} \qquad \textbf{Step 4}$$

Integrating both sides, we have

$$x^{-3}y = 5\frac{x^2}{2} + c \qquad \textbf{Step 5}$$

or

$$y = \frac{5}{2}x^5 + cx^3$$

Example 4   Solve the linear equation

$$(\tan x)\frac{dy}{dx} + y = \sec^3 x$$

Solution. The equation can be written in Standard Form (11.10) by dividing both sides by $\tan x$:

$$\frac{dy}{dx} + y \cot x = \frac{\sec^3 x}{\tan x} = \frac{\sec^3 x}{\sin x / \cos x} = \frac{\sec^2 x}{\sin x}$$

Hence

$$\text{IF} = e^{\int \cot x \, dx} = e^{\ln \sin x} = \sin x \qquad e^{\ln N} = N$$

We now multiply both sides of the equation by IF to get

$$\sin x \left(\frac{dy}{dx} + y \cot x\right) = \sin x \left(\frac{\sec^2 x}{\sin x}\right) = \sec^2 x$$

or, by (11.9),

$$\frac{d}{dx}(y \sin x) = \sec^2 x \qquad \text{derivative of the product of } y \text{ and IF}$$

Integrating both sides, we get

$$y \sin x = \tan x + c$$

$$y = \frac{\tan x}{\sin x} + \frac{c}{\sin x} \qquad \text{or} \qquad y = \sec x + c \csc x \qquad \blacksquare$$

■ Exercises / Section 11.3

Solve the following linear equations.

1. $\dfrac{dy}{dx} + y = 1$

2. $\dfrac{dy}{dx} - y = 2$

3. $\dfrac{dy}{dx} - 2y = e^{3x}$

4. $\dfrac{dy}{dx} - 2xy = e^{x^2}$

5. $2\dfrac{dy}{dx} - 8xy = e^{2x^2}$

6. $3\dfrac{dy}{dx} + 6y = e^{-6x}$

7. $\dfrac{1}{2}\dfrac{dy}{dx} + y \cos x = \cos x$

8. $\dfrac{dz}{dr} + z \sin 2r = \sin 2r$

9. $x \, dy + (y - x) \, dx = 0$

10. $x \, dy + (y - 2x) \, dx = 0$

11. $y' = e^x - \dfrac{y}{x}$

12. $y' + \dfrac{2y}{x} = \dfrac{e^x}{x^2}$

13. $\dfrac{dy}{dx} - \dfrac{2y}{x} - x^2 \sec^2 x = 0$

14. $x\left(\dfrac{dy}{dx} - 1\right) + 2y = x^2$

15. $xy' = 3y + x^5 \sin x$

16. $xy' = y + x^3 e^x$

17. $xy' - 2y = x^3 e^x$

18. $x^3\dfrac{dy}{dx} + 3x^2 y = \csc^2 x$

19. $(y - 1)\sin x \, dx + dy = 0$

20. $y' - y \sec^2 x = \sec^2 x$

21. $y' - y \tan x - \cos x = 0$

22. $y' + y \cot x = x$

23. $(x + 1)y' + y = \dfrac{x + 1}{x - 1}$

24. $x^2\dfrac{dy}{dx} + xy = 1$

25. $y' - \dfrac{1}{x}y = x^2 \sin x^2$

26. $y' - \dfrac{1}{t}y = t^2 \cos t$

27. $y' + y \tan t = \sec t$

28. $v' + v \cot t = \cos t$

29. $t\dfrac{dr}{dt} + r = t \ln t$

30. $t\dfrac{ds}{dt} - 2s = t^4 \cos 2t$

31. $s\dfrac{dr}{ds} - r = s^3 e^{3s}$

In each of the remaining exercises, find the particular solution.

32. $\dfrac{dy}{dx} - 2y = 4$, $y = 3$ when $x = 0$

33. $\dfrac{dy}{dx} + y = 6e^{-x}$, $y = 2$ when $x = 0$

34. $x\,dy - (2y + x^3 \cos x)\,dx = 0$, $y = \dfrac{\pi^2}{4}$ when $x = \dfrac{\pi}{2}$

35. $dy + x^2 y\,dx = 2x^2\,dx$, $y = 3$ when $x = 0$

36. $\dfrac{dy}{dx} + \dfrac{y}{x} = \cos x + \dfrac{\sin x}{x}$, $y = 1$ when $x = \dfrac{\pi}{2}$

37. $L\dfrac{di}{dt} + Ri = E$, $i = 0$ when $t = 0$

38. $m\dfrac{dv}{dt} + kv = mg$, $v = 0$ when $t = 0$

11.4 Applications of First-Order Differential Equations

In this section we will study various applications of first-order equations. Our first application is concerned with **growth and decay**. Suppose $N = N(t)$ is an amount (or number) that changes at a rate proportional to the amount present. This statement may be expressed as $dN/dt = kN$, where k is the constant of proportionality.

Law of Growth and Decay

$$\frac{dN}{dt} = kN \tag{11.13}$$

where N, the amount present, is a function of time and k is a constant.

Example 1

A certain radioactive substance decays at a rate proportional to the amount present. Experimenters observe that after one year 10% of the original amount has decayed. Find the half-life of the substance.

Solution. By (11.13) the differential equation is $dN/dt = kN$. If we now let N_0 be the amount present when $t = 0$ (called the **initial condition**), we have the information needed to compute the arbitrary constant c. The equation is linear, but it may also be solved by separation of variables:

Initial condition

$$N = ce^{kt}$$

At $t = 0$, $N = N_0$; thus $N_0 = ce^0 = c$ and

$$N = N_0 e^{kt}$$

To find k, note that at $t = 1$ we have $N = 0.90N_0$ (90% of the original amount N_0). So

$$0.90N_0 = N_0 e^{k \cdot 1} \qquad \text{or} \qquad 0.90 = e^k \qquad \textbf{dividing by } N_0$$

We take the natural logarithm of both sides to get

$$\ln(0.90) = \ln e^k$$

$$\ln(0.90) = k \ln e = k \qquad \textbf{ln e = 1}$$

and from .9 $\boxed{\text{LN}}$ $\boxed{\text{STO}}$ (saving the value) we have

$$k = -0.1054$$

It follows that

$$N = N_0 e^{-0.1054t}$$

The last expression gives us the amount present as a function of time. In particular (to answer the question posed in the problem), we can find the so-called half life—that is, the time taken for half of the original amount to decay. So if t is the half-life, then $N = 0.5N_0$, so that

$$0.5N_0 = N_0 e^{-0.1054t} \qquad \text{or} \qquad 0.5 = e^{-0.1054t}$$

Taking logarithms again, we have

$$\ln 0.5 = \ln e^{-0.1054t} = -0.1054t \ln e = -0.1054t$$

and

$$t = \frac{\ln 0.5}{-0.1054} = 6.58 \text{ years, the half-life}$$

A possible sequence is .5 $\boxed{\text{LN}}$ $\boxed{\div}$ $\boxed{\text{RCL}}$ $\boxed{=}$ ∎

Corresponding problems involving growth lead to a positive constant k; otherwise the technique is the same.

A similar type of problem arises in **Newton's law of cooling:** *the time rate of change of the temperature of a body is proportional to the temperature difference between the body and its surrounding medium.* If T_m denotes the temperature of the medium, and T is the temperature of the body at any time, then Newton's law says that

$$\frac{dT}{dt} = -k(T - T_m) \qquad (k > 0)$$

Newton's Law of Cooling

$$\frac{dT}{dt} = -k(T - T_m) \qquad (k > 0) \qquad\qquad (11.14)$$

where T is the temperature of the body at any time, T_m is the temperature of the surrounding medium, and k is a constant.

Example 2   Suppose a chef baking a cake places the dough at temperature $21°C$ in an oven kept at a constant temperature of $175°C$. It is observed that the temperature of the dough has risen to $50°C$ after 10 min. How long will it take for the temperature of the dough to reach $100°C$?

Solution.   Since $T_m = 175$, by Equation (11.14) we find

$$\frac{dT}{dt} = -k(T - 175)$$

which is a linear equation with initial condition $T = 21$ when $t = 0$. From

$$\frac{dT}{dt} + kT = 175k$$

we see that IF $= e^{kt}$, so that the equation becomes

$$\frac{d}{dt}(Te^{kt}) = 175ke^{kt} \qquad \left[\begin{array}{l} \boldsymbol{u = e^{kt}} \\ \boldsymbol{du = ke^{kt}\ dt} \end{array} \right]$$

Integrating, we have $Te^{kt} = 175e^{kt} + c$ or $T = 175 + ce^{-kt}$. Substituting $t = 0$ and $T = 21$, we get $21 = 175 + c$, or $c = -154$. The solution is now written

$$T = 175 - 154e^{-kt}$$

To find k we make use of the fact that $T = 50$ when $t = 10$:

$$50 = 175 - 154e^{-10k}$$

Solving for k, we get $k = 0.0209$, so that

$$T = 175 - 154e^{-0.0209t}$$

Finally, letting $T = 100$ and solving for t, we get $t = 34.4$ min—that is, in 34.4 min the dough will be at $100°C$. The rising temperature is illustrated graphically in Figure 11.1. ∎

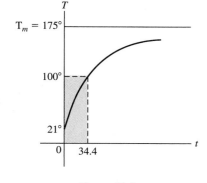

Figure 11.1

The force acting on a freely falling object is given by mg, where m is the mass of the object. If the object moves through a medium such as air, it is subjected to a resistance force that depends on the velocity and the size and shape of the object. At relatively low velocities the resistance force appears to be directly proportional to the velocity, or equal to kv, $k > 0$. Consequently, the net force F on the body is given by $F = mg - kv$. So, by Newton's second law, we have

$$m\frac{dv}{dt} = mg - kv \qquad \boldsymbol{mg = weight}$$

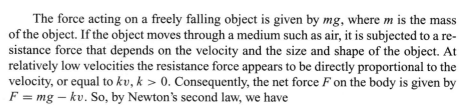

Motion Through a Resisting Medium

The velocity v of a body of mass m falling through a resisting medium is found from the equation

$$m\frac{dv}{dt} = mg - kv \qquad\qquad (11.15)$$

Since Equation (11.15) is linear and since k has to be given explicitly, the solution is quite straightforward. (However, for some objects the force due to air resistance is approximately kv^2. For this case see Exercise 19.)

Example 3   An object of mass 15 kg (weight = 150 N) is dropped from rest. The retarding force due to air resistance is numerically equal to 0.6 of the velocity (that is, $k = 0.6$). Find the velocity after 20 s, as well as the limiting velocity.

Solution. As noted earlier, the force due to air resistance is directly proportional to the velocity. In our problem, the constant of proportionality is $k = 0.6$, so that the force due to air resistance is $0.6v$. Also,

$$m = 15 \text{ kg} \quad \text{and} \quad mg = 150 \text{ N} \quad \textbf{(15 kg)(10 m/s}^2\textbf{) = 150 N}$$

So, by Equation (11.15),

$$15\frac{dv}{dt} = 150 - 0.6v$$

which is a linear equation with initial condition $v = 0$ when $t = 0$.

It is a straightforward exercise to show that

$$v = 250(1 - e^{-0.04t})$$

At $t = 20$ s,

$$v = 250(1 - e^{-0.04(20)}) = 140 \text{ m/s}$$

To find the limiting velocity, we let $t \to \infty$; thus

$$\lim_{t \to \infty} v = \lim_{t \to \infty} 250(1 - e^{-0.04t}) = 250$$

In other words, the velocity of the object approaches 250 m/s. (See Figure 11.2.)

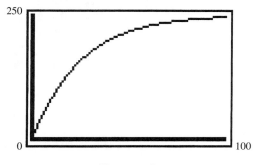

Figure 11.2

Our next application is concerned with **orthogonal trajectories**: *given a family of curves, find another family such that every member of the new family intersects every member of the old family at right angles.*

Example 4   Find the equation of the family of orthogonal trajectories of $y = ce^x$. (The curves corresponding to various values of the parameter c are shown in Figure 11.3.)

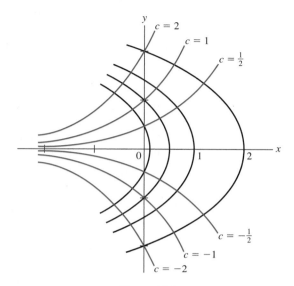

Figure 11.3

Solution. For any point (x, y) not on the x-axis, there exists a member of the given family passing through this point. (For example, the member passing through $(0, -2)$ is $y = -2e^x$.) The slope of the tangent line at any point is $dy/dx = ce^x$. Eliminating the parameter c between

$$\frac{dy}{dx} = ce^x \qquad \text{and the given equation} \qquad y = ce^x$$

yields $dy/dx = y$. (At $(0, -2)$ we get $dy/dx = y = -2$, which is the slope of $y = -2e^x$ at that point.) At any point the slope of the orthogonal trajectory is the negative reciprocal:

$$\frac{dy}{dx} = -\frac{1}{y}$$

(Recall that two lines are perpendicular if their slopes are negative reciprocals.) Hence this condition must be satisfied by the family of orthogonal trajectories. Separating variables, we get

$$y \, dy = -dx$$

and

$$\frac{y^2}{2} = -x + k \qquad \text{or} \qquad y^2 + 2x = k$$

which is a family of parabolas. (See Figure 11.3.)   ∎

Orthogonal trajectories play an important role in many technical areas. For example, in a gravitational, magnetic, or electrostatic field the lines of force are orthogonal to the lines of equal potential. On a map the curves of steepest descent are orthogonal to the contour lines. Similar properties hold for heat conduction and fluid flow. (See Exercises 29 and 30.)

For our final application we turn to certain mixing and flow problems.

Example 5

A 100-gallon tank of brine contains 30 lb of dissolved salt. Saltwater containing 1 lb of salt per gallon is poured in at the rate of 2 gal/min. Assume that the solution continues to be well stirred while the same amount of the mixture is running out each minute. Find an expression for the amount x of salt as a function of time. Determine the amount of salt after 1 h.

Solution.

$$\frac{dx}{dt} = \text{rate of gain} - \text{rate of loss}$$

$$\text{rate of gain} = \left(1\frac{\text{lb}}{\text{gal}}\right)\left(2\frac{\text{gal}}{\text{min}}\right) = 2\frac{\text{lb}}{\text{min}}$$

$$\text{rate of loss} = \frac{x \text{ lb}}{100 \text{ gal}}\left(2\frac{\text{gal}}{\text{min}}\right) = \frac{2x}{100} = \frac{x}{50}\frac{\text{lb}}{\text{min}}$$

Thus

$$\frac{dx}{dt} = 2 - \frac{x}{50} = \frac{100 - x}{50}$$

Separating variables,

$$\frac{dx}{100 - x} = \frac{dt}{50}$$

$$\frac{-dx}{100 - x} = -\frac{1}{50}dt \qquad \begin{bmatrix} u = 100 - x \\ du = -dx \end{bmatrix}$$

$$\ln(100 - x) = -\frac{1}{50}t + \ln c$$

$$\ln\left(\frac{100 - x}{c}\right) = -\frac{1}{50}t$$

$$\frac{100 - x}{c} = e^{-t/50}$$

$$100 - x = ce^{-t/50}$$

$$x = 100 - ce^{-t/50}$$

When $t = 0$, $x = 30$: $30 = 100 - c$, or $c = 70$. So

$$x(t) = 100 - 70e^{-t/50}$$

The solution gives us the amount x as a function of time.

Finally, when $t = 1$ h $= 60$ min, $x = 100 - 70e^{-60/50} = 78.9$ lb. ∎

■ Exercises / Section 11.4

1. An RL circuit with $R = 5 \Omega$ and $L = 0.2$ H has an applied voltage $e(t) = 5$ volts. Find the current as a function of time if $i = 0$ when $t = 0$. (Recall that $L \, di/dt + Ri = e(t)$.)

2. Repeat Exercise 1 if $e(t) = \sin t$.

3. A radioactive substance decays at a rate proportional to the amount present. If there are 100 g initially and 80 g after 10 days, find an expression for the mass at any time t.

4. A radioactive substance has a half-life of 1000 years. Find how long it takes for 5% of the substance to decay.

5. Radioactive cobalt has a half-life of 5.27 years. Find how long it takes for 80% of a given amount to decay.

6. It was determined experimentally that 5% of a quantity of polonium-210 decayed after 10.36 days. Calculate the half-life.

7. It is found that 1% of a certain quantity of some isotope of radium decays after 20 years. Determine the half-life of this isotope.

8. A bacteria culture is known to increase at a rate proportional to the number of bacteria present. It is observed that the size of the culture triples in 3 h. After how many hours should it be 10 times as large?

9. The age of a fossil can be determined by a procedure called *carbon dating*. The procedure is based on the fact that carbon-14 is found in all organisms in a fixed percentage. When the organisms die, the carbon-14 decays with a half-life of 5600 years. Suppose a fossil contains only 25% of the original amount. How long has the organism been dead?

10. A fossil contains 75% of the original amount of carbon-14. How long ago did the organism die? (Refer to Exercise 9.)

11. If S is an investment earning $100r\%$ annual (compound) interest and ΔS is the interest earned in the time interval Δt, then $\Delta S \approx Sr \Delta t$. (That is, ΔS is approximately equal to principal × rate × time, the simple interest earned in this time interval.) Thus $\Delta S / \Delta t \approx Sr$. As $\Delta t \to 0$, $dS/dt = Sr$. Show that $S = S_0 e^{rt}$, where S_0 is the amount invested initially. (The interest is said to be compounded continuously since the length Δt of each interest period approaches zero.)
   a. To what sum will $1000 accumulate in 10 years if invested at 7.75% per year compounded continuously?
   b. If it takes 10 years for $100 to double, what is the interest rate per year compounded continuously?
   c. If a sum is to be invested at 8% per year compounded continuously, how large would the investment have to be in order to accumulate to a sum of $500 at the end of 5 years? The amount sought is called the "present value."

12. A body whose temperature is 25°C is placed outside where the temperature is 0°C. One minute later the temperature of the body has dropped to 22°C. What will the temperature be after 10 min?

13. A thermometer reading 20°F is placed in a room kept at 70°F. Two minutes later the thermometer reads 35°F. Find the temperature as a function of t and the time it takes for the thermometer reading to rise to within one degree of the room temperature.

14. A body whose temperature is 20°C is placed in a freezer kept at -10°C. One minute later the temperature of the body has dropped to 16°C. How long will it take for the temperature to fall to 0°C?

15. A body taken out of a freezer kept at -15°F is placed in a room whose temperature is 60°F. After 3 min the temperature of the body has risen to -5°F. Find the time it takes for the temperature of the body to rise to 50°F.

16. The temperature of a room is 21°C. At 5 PM the thermometer is taken outside where the temperature is 10°C. At 5:01 PM the thermometer reads 15°C when it is taken back inside. What is the reading at 5:02 PM?

17. A body of mass 10 kg (weight = 100 N) is dropped from rest. If the retarding force due to air resistance is numerically equal to 0.2 of the velocity, find the velocity after 10 s. What is the limiting velocity, that is, $\lim_{t \to \infty} v$?

18. A body of mass 5 kg (weight = 50 N) is dropped with an initial velocity of 1 m/s. If the retarding force is directly proportional to the velocity ($k = 1$), find an expression for the velocity as a function of time, as well as its limiting velocity.

19. A 10-kg body (weight = 100 N) attached to a parachute is dropped from rest and encounters a force due to air resistance numerically equal to $4v^2$. Find the velocity as a function of time and the limiting velocity. (Use partial fractions or Table 2 in Appendix A.)

20. A body of weight 64 lb (2 slugs) is dropped from rest and encounters a retarding force numerically equal to 0.3 of the velocity. Determine the velocity as a function of time, as well as the limiting velocity.

21. A law similar to the law of growth and decay comes to us from chemistry: in certain chemical reactions in which a substance is converted into another substance, the time rate of change of the amount x of unconverted substance is proportional to x. If only a fourth of the substance has been converted (three-fourths unconverted) at the end of 10 s, find when nine-tenths of the substance will have been converted.

22. Find the orthogonal trajectories of the family of curves $y = cx^2$. Draw the graphs.

23. For a point mass at the origin, the curves of equal gravitational potential are $x^2 + y^2 = c^2$. Find the equations of the lines of force.

In Exercises 24–28, find the orthogonal trajectories of the families of curves.

24. $x^2 - y^2 = c$

25. $y^2 = 4px$

26. $y = cx^3$

27. $y = ce^{-2x}$

28. $x = ce^{y^2}$

29. Suppose the *streamlines* of a certain fluid motion, defined as curves such that the tangent at a point gives the direction of motion, are the hyperbolas $xy = c$. Find the *velocity equipotential curves,* which are the orthogonal trajectories.

30. Suppose in a problem in heat flow the family of isothermal curves, defined as curves joining points at the same temperature, are given by the ellipses $x^2 + (1/2)y^2 = c^2$. Find the curves along which the heat flows (that is, the orthogonal trajectories).

31. In 1845 the Belgian mathematician P. Verhulst proposed that the constant k in Equation (11.13), applied to population growth, be replaced by $k(M - N)$, where M is the ultimate maximum population. The resulting equation, $dN/dt = kN(M - N)$, is called the *logistic equation.* Suppose that a bacterial population has an initial population of 1000, a population of 2000 after 1 h, and a maximum of 10,000. Use partial fractions or Table 2 in Appendix A to determine the population N as a function of time.

32. Refer to Exercise 31. Solve the general logistic equation with initial condition $N = N_0$ when $t = 0$.

33. A 200-gallon tank is filled with a salt solution containing 20 lb of dissolved salt. Saltwater containing 0.2 lb of salt per gallon is poured into the tank at the rate of 4 gal/min. Assume that the solution is well stirred while the same amount of the mixture is running out each minute. Find an expression for the amount x of salt as a function of time. Determine the amount of salt after 30 min.

34. Fifty gallons of brine originally contained 10 lb of dissolved salt. Saltwater containing 0.5 lb of salt per gallon is poured in at the rate of 3 gal/min. As before, assume that the solution continues to be well stirred and that the same amount of the mixture is running out each minute. Find the amount x of salt as a function of time and determine the amount of salt after 20 min.

35. Repeat Exercise 33, assuming now that pure water is poured into the tank. (*Hint:* the rate of gain is zero.)

36. Repeat Exercise 34, assuming now that pure water is poured into the tank.

11.5 Numerical Solutions

Many differential equations do not have an exact (closed-form) solution. For example, we cannot find an explicit function that solves the equation $dy/dx = y^2 - x^2$. We may instead use a numerical approximation technique that gives us a table of x- and y-values.

Our first method, called **Euler's method,** involves nothing more complicated than the *differential* from Chapter 3. Recall that the differential $dy = f'(x)\,dx$ is approximately equal to the increment Δy, provided that we identify dx with Δx, assumed to be small. Suppose that the differential equation $dy/dx = f(x, y)$ is written in the form $dy = f(x, y)\,dx$. Suppose further that the initial conditions are $x = x_0$ and $y = y_0$, so that the solution curve passes through the point (x_0, y_0). Let's call the next point (x_1, y_1), where $x_1 = x_0 + dx$. (As before, $dx = \Delta x$.) Then

$$y_1 = y_0 + f(x, y)\,dx \qquad \text{(approximately)}$$

So (x_1, y_1) is a point on the curve or nearly on the curve. We now repeat the process with (x_1, y_1) to obtain (x_2, y_2).

Let us try this procedure with an equation whose solution is known. This enables us to see how well our approximation scheme works.

Example 1

Use Euler's method to solve the differential equation $dy/dx = -y + e^{-x}\cos x$, that is, find the y-values of the solution from $x = 0$ to $x = 1$, given that the curve passes through the point (0, 0). Use $dx = \Delta x = 0.1$.

Solution. Write the equation in the form

$$dy = (-y + e^{-x}\cos x)\,dx$$

and substitute the initial values $x = 0$ and $y = 0$:

$$dy = (0 + 1)\,dx = 1(0.1) = 0.1$$

Thus

$$x_1 = 0 + 0.1 \qquad \text{and} \qquad y_1 = y_0 + dy = 0 + 0.1 = 0.1$$

So the curve passes or nearly passes through the point (0.1, 0.1).

Using (0.1, 0.1) as our new point, we get

$$dy = [-0.1 + e^{-0.1}\cos(0.1)]\,dx = 0.0800317$$

It follows that the next point has coordinates

$$x_2 = 0.1 + 0.1 = 0.2$$

and

$$y_2 = y_1 + dy = 0.1 + 0.0800317 = 0.180032$$

that is, $(x_2, y_2) = (0.2, 0.180032)$.

These calculations can be done very quickly with a spreadsheet. For convenience of description, let us place the initial x-value ($x = 0$) into cell A1 and the initial y-value ($y = 0$) into cell B1. To generate the x-values, enter the expression +A1 + 0.1 in cell A2. Now copy the contents of this cell, mark the block extending from A3 through A11, and paste.

Shortcut: Some spreadsheets allow you to generate these values more quickly: drag the "fill handle" in the lower right-hand corner of cell A2 to cell A11.

To obtain the corresponding y-values, enter the following expression in cell B2:

$$+ (-B1 + \exp(-A1) * \cos(A1)) * 0.1 + B1$$

Now copy the contents of this cell, mark the block B3 − B11, and paste. The tabulated results, including the correct values, are given in the next table. (The correct values were found using the exact solution $y = e^{-x}\sin x$, found by the method of Section 11.3.)

	A	B	
Cell	x	y	Correct y-value
1	0	0	0
2	0.1	0.1	0.0903330
3	0.2	0.180032	0.1626567
4	0.3	0.24227	0.2189268
5	0.4	0.288816	0.2610349
6	0.5	0.321675	0.2907863
7	0.6	0.342735	0.3098824
8	0.7	0.353757	0.3199090
9	0.8	0.356362	0.3223289
10	0.9	0.352031	0.3184770
11	1.0	0.342101	0.3095599

Using the graphing capability of the spreadsheet, we get the graph in Figure 11.4. (The horizontal axis gives the cell number.)

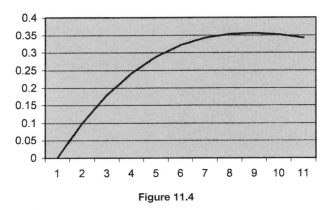

Figure 11.4

Since each calculated value depends on the previous value, the error increases as we move down the table. We can improve the accuracy by using a smaller dx. Using a spreadsheet, this approach is feasible: in Example 1, if $dx = 0.01$ (instead of the previous value 0.1), we get $y(1.0) = 0.312671$.

The fact remains, however, that for a given value of dx, Euler's method is not as accurate as other, more sophisticated, methods, an example of which is the **Runge-Kutta method.** This method calculates slopes at $x = x_0$, $x = x_0 + (1/2)\,dx$, and $x = x_0 + dx$ and may be viewed as a generalization of Simpson's rule in Section 4.9.

Once again the differential equation is written $dy = f(x, y)\,dx$. Let (x_0, y_0) be the initial point and, for convenience, let $h = dx$. The formulas for x_1 and y_1 are

$$x_1 = x_0 + h \quad \text{and} \quad y_1 = y_0 + \frac{1}{6}h(K_1 + 2K_2 + 2K_3 + K_4),$$

where

$$K_1 = f(x_0, y_0)$$

$$K_2 = f\left(x_0 + \frac{1}{2}h, y_0 + \frac{1}{2}hK_1\right)$$

$$K_3 = f\left(x_0 + \frac{1}{2}h, y_0 + \frac{1}{2}hK_2\right)$$

$$K_4 = f(x_0 + h, y_0 + hK_3)$$

Example 2   Solve the equation in Example 1 by the Runge-Kutta method.

Solution.   Restating the equation and conditions,

$$\frac{dy}{dx} = -y + e^{-x}\cos x, \ x_0 = 0, \ y_0 = 0, \ h = 0.1$$

From the above formulas we get

$$x_0 = 0 \qquad y_0 = 0 \qquad h = 0.1$$

$$K_1 = -y_0 + e^{-x_0}\cos x_0 = -0 + 1 = 1$$

$$K_2 = -\left[0 + \frac{1}{2}(0.1)(1)\right] + e^{-(1/2)(0.1)}\cos\left[\frac{1}{2}(0.1)\right] = 0.9000406$$

$$K_3 = -\left[0 + \frac{1}{2}(0.1)(0.9000406)\right] + e^{-(1/2)(0.1)}\cos\left[\frac{1}{2}(0.1)\right] = 0.905038604$$

$$K_4 = -[0 + (0.1)(0.905038604)] + e^{-0.1}\cos(0.1) = 0.809813139$$

It follows that $x_1 = 0 + 0.1 = 0.1$ and

$$y_1 = 0 + \frac{1}{6}(0.1)(1 + 2 \times 0.9000406 + 2 \times 0.905038604$$

$$+ 0.809813139) = 0.0903329$$

The values are given in the following table:

x	y
0	0
0.1	0.0903329
0.2	0.1626564
0.3	0.2189264
0.4	0.2610345
0.5	0.2907858
0.6	0.3098819
0.7	0.3199085
0.8	0.3223283
0.9	0.3184764
1.0	0.3095594

Observe that the values are much closer to the correct values than those obtained in Example 1. ∎

■ Exercises / Section 11.5

In Exercises 1–8, use Euler's method to find the approximate solution (y-values) for the given values of x and $\Delta x = dx$, using the given initial conditions. Graph the solution.

1. $\dfrac{dy}{dx} = 1 - y$, $x = 0$ to $x = 2$, $(0, 2)$
   a. $dx = 0.05$ **b.** $dx = 0.025$

2. $\dfrac{dy}{dx} = -xy^2$, $x = 0$ to $x = 2$, $(0, 1)$
   a. $dx = 0.1$ **b.** $dx = 0.05$

3. $\dfrac{dy}{dx} = x - y$, $x = 0$ to $x = 0.4$, $(0, 3)$
   a. $dx = 0.01$ **b.** $dx = 0.002$

4. $\dfrac{dy}{dx} = \dfrac{y}{x} + x$, $x = 1$ to $x = 3$, $(1, 2)$
   a. $dx = 0.1$ **b.** $dx = 0.05$

5. $\dfrac{dy}{dx} = y^2 - x^2$, $x = 0$ to $x = 0.5$, $(0, 1)$, $dx = 0.01$

6. $\dfrac{dy}{dx} = 1 + y^2$, $x = 0$ to $x = 0.5$, $(0, 0)$, $dx = 0.01$

7. $\dfrac{dy}{dx} = e^{-xy}$, $x = 0$ to $x = 1$, $(0, 0)$
   a. $dx = 0.05$ **b.** $dx = 0.01$

8. $\dfrac{dy}{dx} = \sqrt{\cos x + \sin y}$, $x = 0$ to $x = 1$, $(0, 0)$
   a. $dx = 0.05$ **b.** $dx = 0.02$

In Exercises 9–14, use the Runge-Kutta method to find the approximate solutions.

9. Exercise 1 with $dx = 0.05$
10. Exercise 2 with $dx = 0.05$
11. Exercise 3 with $dx = 0.01$
12. Exercise 4 with $dx = 0.05$
13. Exercise 5 with $dx = 0.05$
14. Exercise 6 with $dx = 0.01$

■ Review Exercises / Chapter 11

In Exercises 1–14, solve the differential equations.

1. $y' = x - y$
2. $(x^2 + 1)\,dx + x^2 y^2\,dy = 0$
3. $y' = x - 2xy$ (Solve by two methods.)
4. $dy - (2x + 2xy)\,dx = 0$ (Solve by two methods.)
5. $(1 + y^2)\,dx + (x^2 y + y)\,dy = 0$
6. $\sin x \dfrac{dy}{dx} = 1 - 2y \cos x$
7. $2y\,dx + x\,dy = 0$
8. $(3x^2 y - 3x^2 \tan y)\,dx - \tan^2 y\,dy = 0$
9. $(x^4 + 2y)\,dx - x\,dy = 0$
10. $(y + \cos^2 x)\,dx + \cos x\,dy = 0$
11. $x \sin^2 y\,dx - \cot y\,dy = 0$
12. $e^{2x+y}\,dx + y\,dy = 0$
13. $dy/dx + y \sec x = 0$, $y = 2$ when $x = \pi/4$ (Solve by two methods.)
14. $x(\ln x) \ln y\,dy + dx = 0$
15. A bacteria culture grows at a rate proportional to the number of bacteria present. If the size of the culture doubles in 2 h, how long will it take for the size to triple?
16. Radium decays at a rate proportional to the amount present. Experimenters have determined that 1.3% of a certain quantity has decayed after 30 years. Determine the half-life of radium.
17. A body is taken out of a freezer of unknown temperature and placed in a room kept at 65°F. After 15 min the temperature of the body is 0°F and after 30 min the temperature of the body is 20°F. Determine the temperature of the freezer.
18. A thermometer reading 65°F is taken outdoors where the temperature is 40°F. One minute later the thermometer reads 60°F. How long will it take for the reading to drop within one degree of the temperature outside?
19. A 6-kg object (weight = 60 N) dropped from rest experiences a retarding force directly proportional to the velocity ($k = 1$). Find its velocity as a function of time.
20. For a point mass at $(-3, 4)$ the curves of equal gravitational potential are given by $(x + 3)^2 + (y - 4)^2 = c^2$. Find the equations of the lines of force.
21. The isothermal curves of a metal plate are given by $x^2 - 2y^2 = c$. Find the curves along which the heat flows (the orthogonal trajectories).
22. Find the approximate solution of the differential equation
$$\frac{dy}{dx} = y + 2\sin x$$
from $x = 0$ to $x = 0.2$ with $dx = 0.02$, given that the solution passes through the point $(0, 0)$ using **(a)** Euler's method; **(b)** the Runge-Kutta method.

12

Higher-Order Linear Differential Equations

12.1 Higher-Order Homogeneous Differential Equations

In the last chapter we considered only differential equations of first order. In this chapter we will solve linear equations of second and higher order, defined next.

An **nth-order linear differential equation** has the form

$$b_0(x)\frac{d^n y}{dx^n} + b_1(x)\frac{d^{n-1}y}{dx^{n-1}} + \cdots + b_{n-1}(x)\frac{dy}{dx} + b_n(x)y = f(x) \quad (12.1)$$

If $f(x) = 0$ for all x, the equation is called **homogeneous.** If $f(x) \neq 0$, the equation is called **nonhomogeneous.**

In this section we will restrict our attention to the solution of homogeneous equations.

Linear Combinations

Linear combination

The method of solving linear equations depends on the following property: if y_1 and y_2 are two distinct solutions of a homogeneous linear equation, then the **linear combination**

$$y = c_1 y_1 + c_2 y_2$$

is also a solution. To see why, consider the second-order linear equation

$$\frac{d^2 y}{dx^2} - \frac{dy}{dx} - 6y = 0 \quad (12.2)$$

If y_1 and y_2 are solutions, then

$$\frac{d^2 y_1}{dx^2} - \frac{dy_1}{dx} - 6y_1 = 0 \quad \text{and} \quad \frac{d^2 y_2}{dx^2} - \frac{dy_2}{dx} - 6y_2 = 0 \quad (12.3)$$

Consider next the linear combination $y = c_1 y_1 + c_2 y_2$. Substituting into the left side of Equation (12.2), we get

$$\frac{d^2}{dx^2}(c_1 y_1 + c_2 y_2) - \frac{d}{dx}(c_1 y_1 + c_2 y_2) - 6(c_1 y_1 + c_2 y_2) \quad (12.4)$$

Now observe that

$$\frac{d}{dx}(c_1 y_1 + c_2 y_2) = c_1 \frac{dy_1}{dx} + c_2 \frac{dy_2}{dx}$$

and

$$\frac{d^2}{dx^2}(c_1 y_1 + c_2 y_2) = c_1 \frac{d^2 y_1}{dx^2} + c_2 \frac{d^2 y_2}{dx^2}$$

(Because of this property, the derivative is said to be *linear*.) As a result, Expression (12.4) can be rewritten

$$c_1 \frac{d^2 y_1}{dx^2} - c_1 \frac{dy_1}{dx} - 6c_1 y_1 + c_2 \frac{d^2 y_2}{dx^2} - c_2 \frac{dy_2}{dx} - 6c_2 y_2$$

$$= c_1 \left(\frac{d^2 y_1}{dx^2} - \frac{dy_1}{dx} - 6y_1 \right) + c_2 \left(\frac{d^2 y_2}{dx^2} - \frac{dy_2}{dx} - 6y_2 \right)$$

$$= c_1(0) + c_2(0) = 0 \qquad \textbf{by (12.3)}$$

We have therefore shown that the linear combination $y = c_1 y_1 + c_2 y_2$ is also a solution of (12.2). (We will see shortly how y_1 and y_2 can actually be obtained.)

Solution of Linear Equations

Before we turn to the solution of linear equations, we need to make some observations about the left side of (12.1). If the coefficients $b_0, b_1, b_2, \ldots, b_n$ are constants and if we use the symbol D^n for the nth derivative with respect to x, then the left side of (12.1) has the form of a polynomial. For example, a second-order linear equation is now written

$$b_0 D^2 y + b_1 Dy + b_2 y = 0$$

or, more commonly,

$$(b_0 D^2 + b_1 D + b_2)y = 0$$

Operator

    Now the "coefficient" of y, called an **operator,** not only looks like a polynomial but turns out to have similar algebraic properties. Consider, for example, the equation

$$(D^2 - D - 6)y = 0 \qquad (12.5)$$

If we "factor" the polynomial, we get

$$(D - 3)(D + 2)y = 0$$

to be understood in the following sense. First apply the operator $D + 2$ to y and then apply $D - 3$ to the result. As a check, note that

$$(D + 2)y = Dy + 2y$$

Then

$$(D-3)(D+2)y = (D-3)(Dy+2y)$$
$$= D(Dy+2y) - 3(Dy+2y)$$
$$= D^2y + 2Dy - 3Dy - 6y$$
$$= D^2y - Dy - 6y = (D^2 - D - 6)y$$

in agreement with Equation (12.5). Moreover,

$$(D-3)(D+2)y = (D+2)(D-3)y$$

as can be readily shown.

As a consequence, solving a homogeneous equation with constant coefficients can be reduced essentially to finding roots of polynomial equations. Suppose we split Equation (12.5) into two equations,

$$(D-3)y_1 = 0 \quad \text{and} \quad (D+2)y_2 = 0$$

each a first-order linear equation. We proceed to solve each one separately:

$$Dy_1 - 3y_1 = 0 \qquad Dy_2 + 2y_2 = 0$$
$$\text{IF} = e^{-3x} \qquad \text{IF} = e^{2x}$$
$$\frac{d}{dx}(y_1 e^{-3x}) = 0 \qquad \frac{d}{dx}(y_2 e^{2x}) = 0$$
$$y_1 e^{-3x} = c_1 \qquad y_2 e^{2x} = c_2$$
$$y_1 = c_1 e^{3x} \qquad y_2 = c_2 e^{-2x}$$

Since Equation (12.5) can be written in factored form, it is easily seen that both y_1 and y_2 satisfy the equation. In particular, both e^{3x} and e^{-2x} are solutions. Consequently, the *linear combination*

$$y = c_1 e^{3x} + c_2 e^{-2x}$$

is the general solution of the equation, as we saw earlier. In retrospect the solution could have been obtained directly from

$$(D-3)(D+2)y = 0$$

since the coefficients of x are the roots of the polynomial equation

$$(m-3)(m+2) = 0 \tag{12.6}$$

Auxiliary equation

so that $m = 3$ and $m = -2$. Equation (12.6) is called the **auxiliary equation.** We conclude, then, that the solution of the auxiliary equation may be used to find the general solution of the corresponding differential equation.

Definition of Auxiliary Equation

Let

$$(b_0 D^2 + b_1 D + b_2)y = 0$$

be a homogeneous second-order linear differential equation with constant coefficients. Then

$$b_0 m^2 + b_1 m + b_2 = 0$$

is called the **auxiliary equation.**

The solution of the auxiliary equation determines the solution of the differential equation.

General Solution of a Homogeneous Second-Order Linear Differential Equation

If m_1 and m_2 are distinct real roots of the auxiliary equation, then

$$y = c_1 e^{m_1 x} + c_2 e^{m_2 x}$$

is the **general solution** of the differential equation.

Example 1 Find the general solution of the equation

$$2\frac{d^2 y}{dx^2} + 5\frac{dy}{dx} - 12y = 0$$

Solution. In terms of the D-operator the equation can be written

$$(2D^2 + 5D - 12)y = 0$$

Hence the auxiliary equation is

$$2m^2 + 5m - 12 = 0 \qquad \text{or} \qquad (2m - 3)(m + 4) = 0$$

Since the roots are $m = 3/2$ and $m = -4$, the solution of the differential equation is given by

$$y = c_1 e^{(3/2)x} + c_2 e^{-4x}$$

or

$$y = c_1 e^{3x/2} + c_2 e^{-4x}$$

Example 2   Solve the equation

$$\frac{d^2y}{dx^2} + 2\frac{dy}{dx} - 2y = 0$$

Solution. Even without the D-operator the auxiliary equation is seen to be

$$m^2 + 2m - 2 = 0$$

By the quadratic formula we get

$$m = \frac{-2 \pm \sqrt{2^2 - 4(-2)}}{2} \qquad \text{See inside front cover.}$$

$$= \frac{-2 \pm \sqrt{12}}{2} = \frac{-2 \pm 2\sqrt{3}}{2} = \frac{2(-1 \pm \sqrt{3})}{2} = -1 \pm \sqrt{3}$$

Hence

$$y = c_1 e^{(-1+\sqrt{3})x} + c_2 e^{(-1-\sqrt{3})x}$$

$$= c_1 e^{-x} e^{\sqrt{3}x} + c_2 e^{-x} e^{-\sqrt{3}x}$$

$$= e^{-x}\left(c_1 e^{\sqrt{3}x} + c_2 e^{-\sqrt{3}x}\right) \qquad \textbf{factoring } e^{-x} \qquad ■$$

Example 3   Solve the differential equation $(D^2 - 4)y = 0$, subject to the following conditions:

$$y = 1 \qquad \text{and} \qquad Dy = 2 \qquad \text{when } x = 0$$

Solution.

Auxiliary equation: $m^2 - 4 = 0$

Roots of auxiliary equation: $m = \pm 2$

1.   General solution: $y = c_1 e^{2x} + c_2 e^{-2x}$
    To evaluate the constants c_1 and c_2, we substitute the given conditions and solve the resulting system of equations for c_1 and c_2:
2. $Dy = y' = 2c_1 e^{2x} - 2c_2 e^{-2x}$ **derivative of y**

$$\begin{array}{ll} 1 = c_1 + c_2 & \textbf{equation 1:} \quad \textbf{\textit{x} = 0, \textit{y} = 1} \\ 2 = 2c_1 - 2c_2 & \textbf{equation 2:} \quad \textbf{\textit{x} = 0, \textit{Dy} = 2} \\ \end{array}$$

$$\begin{array}{ll} 1 = c_1 + c_2 & \\ 1 = c_1 - c_2 & \textbf{dividing equation 2 by 2} \\ 2 = 2c_1 + 0 & \textbf{adding} \\ \end{array}$$

$$c_1 = 1, c_2 = 0$$

Substituting the values of c_1 and c_2 in the general solution (1), we get the particular solution

$$y = e^{2x} \qquad \textbf{\textit{c}}_1 = 1, \textbf{\textit{c}}_2 = 0 \qquad ■$$

The technique for solving second-order linear equations can be extended to linear equations of higher order. If

$$b_0 \frac{d^n y}{dx^n} + b_1 \frac{d^{n-1} y}{dx^{n-1}} + \cdots + b_{n-1} \frac{dy}{dx} + b_n y = 0$$

is a homogeneous linear equation with constant coefficients, then

$$b_0 m^n + b_1 m^{n-1} + \cdots + b_{n-1} m + b_n = 0$$

is the **auxiliary equation.**

General Solution of a Homogeneous Linear Equation

If $m_1, m_2, \ldots m_n$ are distinct real roots of the auxiliary equation, then

$$y = c_1 e^{m_1 x} + c_2 e^{m_2 x} + \cdots + c_n e^{m_n x}$$

is the **general solution** of the differential equation.

Example 4   Solve the equation $(D^3 - 3D^2 - D + 3)y = 0$.

Solution.   The auxiliary equation

$$m^3 - 3m^2 - m + 3 = 0$$

is a cubic equation. We can use the root-finding capability of a graphing utility to obtain the roots $m = -1$, 1, and 3.

Alternatively, we try to find one root by inspection and apply the factor theorem from algebra. Note first that the only possible rational roots are ± 1 and ± 3. It is easily checked that $m = 1$ is a root, so $m - 1$ must be a factor. To divide $m - 1$ into the polynomial, we may use synthetic division:

$$
\begin{array}{r}
1 - 3 - 1 + 3 \underline{)\,1} \\
1 - 2 - 3 \\
\hline
1 - 2 - 3 + 0
\end{array}
$$

We now have

$$(m - 1)(m^2 - 2m - 3) = (m - 1)(m + 1)(m - 3) = 0$$

or $m = 1$, -1, and 3. Thus

$$y = c_1 e^x + c_2 e^{-x} + c_3 e^{3x}$$ ∎

Note that the solution of the equation in Example 4 has three arbitrary constants. In general, an nth-order equation has n arbitrary constants in its general solution.

Remark: We saw in this section that forming linear combinations of solutions to obtain new solutions is essential to the technique for solving linear equations. This technique does not work with nonlinear equations, however, and as a consequence, *there does not exist a general method for solving nonlinear equations.* Some nonlinear equations can be solved by special techniques, as we have seen in the case of separable equations.

■ Exercises / Section 12.1

Solve each differential equation.

1. $(D^2 - 13D + 42)y = 0$ **2.** $\dfrac{d^2y}{dx^2} + \dfrac{dy}{dx} - 20y = 0$

3. $6\dfrac{d^2y}{dx^2} - \dfrac{dy}{dx} - 2y = 0$ **4.** $(D^2 - 4)y = 0$

5. $4D^2y + 7Dy - 2y = 0$ **6.** $(D^2 - 3)y = 0$

7. $\dfrac{d^2y}{dx^2} - \dfrac{dy}{dx} - y = 0$ **8.** $(D^2 + D - 3)y = 0$

9. $2D^2y - 3Dy + y = 0$ **10.** $(D^2 + 3D - 3)y = 0$

11. $(D^2 - 9)y = 0$; $y = 0$ and $Dy = 6$ when $x = 0$

12. $(D^2 - 3D + 2)y = 0$; if $x = 0$, then $y = 0$ and $Dy = 1$

13. $(D^2 - D - 2)y = 0$; $y = 0$ when $x = 0$, and $y = 1$ when $x = 1$

14. $(D^2 + 4D - 5)y = 0$; $y = 1$ and $Dy = 3$ when $x = 0$

15. $(D^3 - 7D + 6)y = 0$

16. $(D^3 - 2D^2 - 2D + 3)y = 0$

17. $(D^3 - D^2 - 4D - 2)y = 0$

18. $(4D^3 - 4D^2 - 11D + 6)y = 0$

19. $(D^2 + D - 1)y = 0$ **20.** $(D^2 + 2D - 4)y = 0$

21. $(D^2 - 2D - 2)y = 0$ **22.** $(D^2 + 2D - 1)y = 0$

23. $(D^2 - 4D - 2)y = 0$ **24.** $(D^2 - 6D - 2)y = 0$

25. $(D^2 + 6D - 6)y = 0$ **26.** $(2D^2 - 2D - 1)y = 0$

27. $(2D^2 + 4D + 1)y = 0$ **28.** $(3D^2 - 7D + 2)y = 0$

29. $(3D^2 - D - 2)y = 0$

30. The equation in Exercise 29, subject to the following conditions: $y = 1$ and $Dy = 0$ when $x = 0$.

12.2   Auxiliary Equations with Repeating or Complex Roots

In the preceding section all the roots of the auxiliary equations were distinct and real. In this section we will consider the cases of **repeating** and **complex roots.**

Real Repeating Roots

Consider the linear differential equation $(D^2 - 2D + 1)y = 0$. The auxiliary equation is

$$m^2 - 2m + 1 = (m - 1)^2 = 0$$

leading to the repeating root $m = 1, 1$. The method of the previous section now gives

$$y = c_1 e^x + c_2 e^x = (c_1 + c_2)e^x = ce^x$$

containing only one arbitrary constant. It is easily checked, however, that $y = xe^x$ is also a solution, so that the general solution is now given by the linear combination

$$y = c_1 e^x + c_2 x e^x$$

> **Real Repeating Roots**
>
> If the auxiliary equation has real repeating roots
>
> $$m = a, a$$
>
> then the **general solution** of the differential equation is
>
> $$y = c_1 e^{ax} + c_2 x e^{ax}$$

Example 1   The auxiliary equation of $(4D^2 + 12D + 9)y = 0$ is

$$4m^2 + 12m + 9 = (2m + 3)^2 = 0$$

whence

$$m = -\frac{3}{2}, -\frac{3}{2}$$

The general solution of the differential equation is therefore given by

$$y = c_1 e^{-(3/2)x} + c_2 x e^{-(3/2)x}$$

It is true in general that

$$(D - a)^n (x^k e^{ax}) = 0 \qquad \text{(for } k = 0, 1, 2, \ldots, n - 1)$$

Thus if

$$m = a, a, \ldots, a \qquad (n \text{ times})$$

then the general solution of the differential equation is

$$y = c_1 e^{ax} + c_2 x e^{ax} + c_3 x^2 e^{ax} + \cdots + c_n x^{n-1} e^{ax}$$

Combinations of distinct and repeating roots may also occur. Study the next example.

Example 2   Solve the equation $D^3(D - 2)y = 0$.

Solution. From the auxiliary equation

$$m^3(m - 2) = 0$$

we get $m = 0, 0, 0, 2$. Since zero is a repeating root but 2 is a single root, we get

$$y = c_1 e^{0x} + c_2 x e^{0x} + c_3 x^2 e^{0x} + c_4 e^{2x}$$
$$= c_1 + c_2 x + c_3 x^2 + c_4 e^{2x}$$

Complex Roots

If the auxiliary equation with real coefficients

$$b_0 m^2 + b_1 m + b_2 = 0$$

has complex roots, then we know from the quadratic formula that the roots have the form

$$m = a \pm bj$$

a pair of complex conjugates. (If the coefficients b_0, b_1, and b_2 are not real, this is not true.) The solution of the corresponding differential equation then becomes

$$y = c_3 e^{(a+bj)x} + c_4 e^{(a-bj)x}$$

which can be put into a more convenient form by applying Euler's identity from Section 10.4:

$$\begin{aligned} y &= e^{ax}(c_3 e^{bxj} + c_4 e^{-bxj}) \\ &= e^{ax}\{c_3(\cos bx + j \sin bx) + c_4[\cos(-bx) + j \sin(-bx)]\} \\ &= e^{ax}[c_3(\cos bx + j \sin bx) + c_4(\cos bx - j \sin bx)] \\ &= e^{ax}[(c_3 + c_4)\cos bx + j(c_3 - c_4)\sin bx] \\ &= e^{ax}(c_1 \cos bx + c_2 \sin bx) \end{aligned}$$

where c_1 and c_2 are new arbitrary (complex) constants.

Complex Roots

If the auxiliary equation has complex roots

$$m = a \pm bj$$

then the **general solution** of the differential equation is

$$y = e^{ax}(c_1 \cos bx + c_2 \sin bx) \tag{12.7}$$

Example 3   Find the general solution of the equation $(D^2 - 4D + 6)y = 0$.

Solution. Solving the auxiliary equation

$$m^2 - 4m + 6 = 0$$

we get by the quadratic formula

$$m = \frac{4 \pm \sqrt{16 - 24}}{2} = \frac{4 \pm \sqrt{-8}}{2} = \frac{4 \pm 2\sqrt{2}\,j}{2}$$

$$m = 2 \pm \sqrt{2}\,j$$

By Solution (12.7),

$$y = e^{2x}(c_1 \cos \sqrt{2}x + c_2 \sin \sqrt{2}x)$$

■

Combinations of real and complex roots may also occur. For example, if

$$m = 2, 3 \pm 4j$$

then

$$y = c_1 e^{2x} + e^{3x}(c_2 \cos 4x + c_3 \sin 4x)$$

(For the case of repeating complex roots, see Exercise 49.)

■ Exercises / Section 12.2
_____

In Exercises 1–47, solve the differential equations.

1. $(D^2 + 6D + 9)y = 0$ **2.** $(D^2 - 8D + 16)y = 0$

3. $(4D^2 - 4D + 1)y = 0$ **4.** $(4D^2 + 4D + 1)y = 0$

5. $9\dfrac{d^2y}{dx^2} + 12\dfrac{dy}{dx} + 4y = 0$ **6.** $9\dfrac{d^2y}{dx^2} + 30\dfrac{dy}{dx} + 25y = 0$

7. $(4D^2 - 20D + 25)y = 0$ **8.** $(16D^2 - 8D + 1)y = 0$

9. $(D^2 - 4D + 5)y = 0$ **10.** $(D^2 - D + 1)y = 0$

11. $\dfrac{d^2y}{dx^2} + 4\dfrac{dy}{dx} + 8y = 0$ **12.** $\dfrac{d^2y}{dx^2} - 2\dfrac{dy}{dx} + 5y = 0$

13. $2\dfrac{d^2y}{dx^2} - 2\dfrac{dy}{dx} + y = 0$ **14.** $\dfrac{d^2y}{dx^2} - 2\dfrac{dy}{dx} + 6y = 0$

15. $(D^2 + 25)y = 0$ **16.** $(4D^2 + 9)y = 0$

17. $(D^2 - 6D + 9)y = 0$ **18.** $\dfrac{d^2y}{dx^2} + 8\dfrac{dy}{dx} + 16y = 0$

19. $(D^4 + 2D^3)y = 0$ **20.** $D^5 y = 0$

21. $\dfrac{d^2y}{dx^2} + \dfrac{dy}{dx} + 2y = 0$ **22.** $\dfrac{d^2y}{dx^2} - 3\dfrac{dy}{dx} + 2y = 0$

23. $(D^2 - 3D + 5)y = 0$ **24.** $(2D^2 - 2D + 1)y = 0$

25. $(2D^2 - 4D + 5)y = 0$ **26.** $(3D^2 - 2D + 2)y = 0$

27. $(2D^2 + 4D - 1)y = 0$ **28.** $3\dfrac{d^2y}{dx^2} + 6\dfrac{dy}{dx} - 2y = 0$

29. $\dfrac{d^2y}{dx^2} - 100y = 0$ **30.** $(D^2 - 256)y = 0$

31. $\dfrac{d^2y}{dx^2} + 100y = 0$ **32.** $(D^2 + 256)y = 0$

33. $(3D^3 - 2D^2 + D)y = 0$ **34.** $(D^2 - 5D + 7)y = 0$

35. $(D^2 - 4D + 2)y = 0$ **36.** $(2D^2 - 4D + 3)y = 0$

37. $(D^2 + 4)y = 0$ **38.** $(D^3 + 2D)y = 0$

39. $(D^3 - 4D^2 + 4D)y = 0$

40. $(D^3 + 9D)y = 0$

41. $(D^3 - 9D)y = 0$

42. $(D^3 + 3D^2 + 3D + 1)y = 0$

43. $(D - 2)^4 y = 0$

44. $(D - 1)^2(D + 2)^3 y = 0$

45. $(D^2 + 1)y = 0$; $y = 0$ when $x = 0$, and $y = 1$ when $x = \pi/2$

46. $(D^2 + 4D + 4)y = 0$; if $x = 0$, $y = 0$ and if $x = -1$, $y = 2$

47. $(D^2 - 2D + 2)y = 0$; if $x = 0$, then $y = 0$ and $Dy = -1$

48. Repeat Exercise 47 for the following conditions: $y = 0$ when $x = 0$ and $y = 2$ when $x = \pi/4$.

49. Solve the equation $(D^4 + 18D^2 + 81)y = 0$ by writing the solution in the form $y = c_1 e^{Ax} + c_2 x e^{Ax} + c_3 e^{Bx} + c_4 x e^{Bx}$ and using Euler's identity.

In Exercises 50–56, use the root-finding capability of your graphing utility to find the roots of the auxiliary equation.

50. $(D^3 - 2D^2 + 2D - 1)y = 0$

51. $(4D^3 - D + 3)y = 0$

52. $(D^3 - 11D^2 + 45D - 63)y = 0$

53. $(D^3 + 2D^2 - 32D - 96)y = 0$

54. $(D^3 - 11D^2 + 39D - 45)y = 0$

55. $(4D^4 - 3D^2 + 10D + 9)y = 0$

56. $(D^4 + 2D^3 - 8D^2 - 24D + 48)y = 0$

12.3   Nonhomogeneous Equations

So far in this chapter all the equations have been homogeneous. We are now going to turn our attention to **nonhomogeneous equations,** equations for which the right side is not zero.

To get an overview of the problem, let us examine a typical nonhomogeneous equation of second order:

$$b_0 D^2 y + b_1 Dy + b_2 y = f(x) \tag{12.8}$$

If the equation *were* homogeneous, then we would know how to solve it. Let us denote by y_c the solution of the corresponding homogeneous equation, so that

$$b_0 D^2 y_c + b_1 Dy_c + b_2 y_c = 0 \tag{12.9}$$

However, being able to find y_c does not seem to accomplish much. So let us pretend that we have somehow managed to get some particular solution y_p (no arbitrary constants) to Equation (12.8). In other words, y_p is such that

$$b_0 D^2 y_p + b_1 Dy_p + b_2 y_p = f(x) \tag{12.10}$$

Now we draw a surprising conclusion: the function

$$y = y_c + y_p$$

is the general solution of Equation (12.8). This statement can be checked by direct substitution:

$$b_0 D^2 (y_c + y_p) + b_1 D(y_c + y_p) + b_2 (y_c + y_p)$$
$$= (b_0 D^2 y_c + b_1 Dy_c + b_2 y_c) + (b_0 D^2 y_p + b_1 Dy_p + b_2 y_p)$$
$$= 0 + f(x) = f(x)$$

by (12.9) and (12.10). Since $y = y_c + y_p$ contains the required number of arbitrary constants, it must be the general solution. The physical significance of y_c and y_p will be considered in the next section.

> ### General Solution of a Nonhomogeneous Equation
> The **general solution of a nonhomogeneous equation** is
> $$y = y_c + y_p$$
> where y_p is some particular solution of the equation and y_c is the general solution of the corresponding homogeneous equation.

To solve a nonhomogeneous equation, we find y_c, the solution of the corresponding homogeneous equation, by the methods studied in the last two sections; y_c is called the **complementary solution.** The real task is to find a particular solution y_p. Several techniques for finding y_p are available. We will confine ourselves to the **method of undetermined coefficients,** which is entirely adequate for most physical problems. This method has two subheadings, the **method of inspection,** which we try to use whenever possible, and the **annihilator method,** to be used whenever mere inspection fails. We will illustrate the different cases by several examples.

Inspection Method

Example 1   Solve the equation $(D^2 + 2)y = 2e^x$.

Solution. The first step is to find y_c, the solution of

$$(D^2 + 2)y = 0$$

From the auxiliary equation $m^2 + 2 = 0$, we find that $m = \pm\sqrt{2}\,j$, so that

$$y_c = c_1 \cos \sqrt{2}x + c_2 \sin \sqrt{2}x$$

That was easy. Since we really have no idea how to find y_p, we first ask ourselves: what would a particular solution look like? Could it have the form $y_p = Ax$ or $y_p = A \sin x$? Not likely. A reasonable assumption is $y_p = Ae^x$ for some constant A, as we can readily see from the equation itself. This is what is meant by "inspection." To find the constant A, we simply substitute $y_p = Ae^x$ in the given equation, noting that $y_p'' = Ae^x$. We get

$$(D^2 + 2)y = 2e^x \qquad \textbf{given equation}$$

$$y'' + 2y = 2e^x \qquad \textbf{same equation}$$

$$Ae^x + 2Ae^x = 2e^x \qquad \textbf{\textit{y}}_\textbf{\textit{p}} = \textbf{\textit{A}}\textbf{\textit{e}}^\textbf{\textit{x}}, \textbf{\textit{y}}_\textbf{\textit{p}}'' = \textbf{\textit{A}}\textbf{\textit{e}}^\textbf{\textit{x}}$$

or

$$3Ae^x = 2e^x$$

Equality holds only if $3A = 2$ or $A = 2/3$. (Now the "undetermined coefficient" has been determined.) So $y_p = (2/3)e^x$, and the solution of the equation is therefore given by

$$y = y_c + y_p = c_1 \cos \sqrt{2}x + c_2 \sin \sqrt{2}x + \frac{2}{3}e^x$$

The same method can be applied to equations of higher order.   ∎

Inspection Method

To see how the form y_p can be obtained by inspection, observe that the function $f(x)$ on the right side must be a linear combination of y_p and its derivatives. Consequently, y_p has to contain $f(x)$ as well as its derivatives. For example, if $f(x) = \sin x$ or $f(x) = \cos x$, we would choose $y_p = A \cos x + B \sin x$. If $f(x) = 2x^2$, then $y_p = Ax^2 + Bx + C$. If $f(x) = 2 \sin 3x - 3e^{-x}$, we let $y_p = A \sin 3x + B \cos 3x + Ce^{-x}$, and so on.

Example 2   **a.** If the right side of the equation is $f(x) = 5x^2$, then y_p has the form $y_p = Ax^2 + Bx + C$. Here Ax^2 is included to account for the term $5x^2$, Bx to account for any derivative of the Ax^2 term, and C to account for any second derivative of Ax^2.

b. If the right side is $2x - e^{3x}$, then y_p has the form $y_p = Ax + B + Ce^{3x}$. Here Ax is included to account for the term $2x$ and B to account for any derivative of the Ax term. To account for $-e^{3x}$, we include Ce^{3x}. No other terms are needed, however, since the derivatives of Ce^{3x} all have this same form.

c. If the right side is $2e^{-x} + 4\cos 2x$, then $y_p = Ae^{-x} + B\cos 2x + C\sin 2x$. As in part (**b**), Ae^{-x} is included to account for the term $2e^{-x}$, but no other terms are required. To account for $4\cos 2x$, we include $B\cos 2x$; $C\sin 2x$ is needed to account for any derivative of $B\cos 2x$. Since the second derivative of $B\cos 2x$ is another cosine function, no other terms are required.

d. If $f(x) = 3 + 6\cos 4x - 7\sin 4x$, then $y_p = A + B\cos 4x + C\sin 4x$.

e. If $f(x) = 6x^2 - 4$, then $y_p = Ax^2 + Bx + C$.   ∎

Example 3   Solve the differential equation $(D^2 - 3D + 2)y = 3x$.

Solution. To find y_c we note that the auxiliary equation $m^2 - 3m + 2$ has roots $m = 1$ and 2. Hence

$$y_c = c_1 e^x + c_2 e^{2x}$$

The right side, $3x$, must be a linear combination of y_p and its derivatives. Since $d(3x)/dx = 3$, a constant, we choose

$$y_p = Ax + B$$

Now we compute $y_p' = A$ and $y_p'' = 0$ and substitute into the given equation:

$$(D^2 - 3D + 2)y = 3x$$
$$D^2 y - 3Dy + 2y = 3x$$
$$0 - 3A + 2(Ax + B) = 3x \qquad y_p'' = 0,\ y_p' = A,\ y_p = Ax + B$$

or

$$2Ax + (-3A + 2B) = 3x + 0$$

Comparing the corresponding coefficients, we get the following system of equations:

$$\left.\begin{array}{r} 2A = 3 \\ -3A + 2B = 0 \end{array}\right\} \qquad \begin{array}{l} \textbf{coefficients of } x \\ \textbf{constants} \end{array}$$

The solution is $A = 3/2$ and $B = 9/4$. Hence

$$y_p = \frac{3}{2}x + \frac{9}{4}$$

Since $y = y_c + y_p$, the general solution is

$$y = c_1 e^x + c_2 e^{2x} + \frac{3}{2}x + \frac{9}{4}$$ ∎

> **Caution:**
>
> Lack of experience often results in the wrong choice for y_p. Suppose in Example 3 we had used $y_p = Ax$, ignoring the derivative of $3x$; then $y'_p = A$ and, after substituting,
>
> $$-3A + 2Ax = 3x$$
>
> This implies that $A = 0$ and $A = 3/2$, which is impossible. We have simply "overworked" the A.

Example 4

Solve the equation $(D^2 + 1)y = 4e^x - \sin 2x$.

Solution. The complementary solution is

$$y_c = c_1 \cos x + c_2 \sin x$$

Since $d(4e^x)/dx = 4e^x$ and $d(-\sin 2x)/dx = -2 \cos 2x$, we use

$$y_p = Ae^x + B \cos 2x + C \sin 2x$$

Also,

$$y'_p = Ae^x - 2B \sin 2x + 2C \cos 2x$$

and

$$y''_p = Ae^x - 4B \cos 2x - 4C \sin 2x$$

Substituting in $(D^2 + 1)y = 4e^x - \sin 2x$, we get

$$D^2 y_p + y_p = (Ae^x - 4B \cos 2x - 4C \sin 2x) + (Ae^x + B \cos 2x + C \sin 2x)$$
$$= 4e^x - \sin 2x$$

or

$$2Ae^x - 3B \cos 2x - 3C \sin 2x = 4e^x + 0 \cos 2x - \sin 2x$$

Comparing coefficients,

$$2A = 4 \qquad -3B = 0 \qquad -3C = -1$$

so that $A = 2$, $B = 0$, and $C = 1/3$. So $y_p = 2e^x + (1/3) \sin 2x$, and the general solution is

$$y = c_1 \cos x + c_2 \sin x + 2e^x + \frac{1}{3} \sin 2x$$

∎

Annihilator Method (Optional)

Remark: The remainder of this section is devoted to a brief discussion of the "annihilator method." If desired, this topic may be omitted. The exercises in which this method is needed are designated.

Example 5   Solve the equation

$$(D^2 - 4D + 3)y = e^x \qquad (12.11)$$

Solution. From the auxiliary equation

$$m^2 - 4m + 3 = (m - 1)(m - 3) = 0$$

we obtain $m = 1$ and $m = 3$. Hence

$$y_c = c_1 e^{3x} + c_2 e^x$$

Proceeding exactly as we did before, we choose

$$y_p = Ae^x$$

Substituting, we get

$$Ae^x - 4Ae^x + 3Ae^x = 0 \neq e^x$$

The failure of the inspection method could have been predicted from the outset since Ae^x is one of the terms in y_c. Consequently, Ae^x satisfies the corresponding homogeneous equation automatically and so cannot be a solution to the given equation.

To find the correct form of y_p, we start with the right side e^x and work backward: if e^x is a solution to an equation, this equation can be constructed by noting that the root of its auxiliary equation is $m' = 1$ (to yield ce^x). Hence $m' - 1 = 0$ is the auxiliary equation and

$$(D - 1)y = 0$$

is the corresponding differential equation. Hence

$$(D - 1)e^x = 0$$

Annihilator

The operator $D - 1$ is called the **annihilator** since applying $D - 1$ to both sides of the given equation "annihilates" the right side:

$$(D - 1)(D^2 - 4D + 3)y = (D - 1)e^x = 0 \qquad (12.12)$$

Moreover, any solution of Equation (12.11) must be a solution of Equation (12.12). Solving the Homogeneous Equation (12.12), from

$$(D - 1)^2(D - 3)y = 0$$

we get the solution

$$y = c_1 e^{3x} + c_2 e^x + c_3 x e^x$$

Since the first two terms on the right side coincide with y_c, the last term must be the form of y_p. Hence

$$y_p = Axe^x$$

The constant A is found in the usual way and turns out to be $-1/2$. The solution is therefore given by

$$y = c_1e^{3x} + c_2e^x - \frac{1}{2}xe^x$$ ∎

The annihilator method can be employed to solve any nonhomogeneous equation in this section but is needlessly complicated if y_p can be determined directly by inspection.

Example 6   Determine the form of y_p in the solution of the equation

$$(D^2 - 4D + 3)y = 4xe^x$$

Solution. The left side is identical to that of the equation in Example 5. To construct the annihilator, note that if $4xe^x$ is the solution of an equation, then $m' = 1$ and 1, a repeating root. Hence the auxiliary equation is $(m' - 1)^2 = 0$ and the annihilator $(D - 1)^2$. Applying this operator to both sides of the given equation, we get

$$(D - 1)^2(D^2 - 4D + 3)y = (D - 1)^2(4xe^x) = 0$$

or

$$(D - 1)^3(D - 3)y = 0$$

Hence

$$y = c_1e^{3x} + c_2e^x + c_3xe^x + c_4x^2e^x$$

Since

$$y_c = c_1e^{3x} + c_2e^x$$

we conclude that

$$y_p = Axe^x + Bx^2e^x$$

(Note that $m = 1$ is also one of the roots of the auxiliary equation $m^2 - 4m + 3 = 0$. In fact, *the annihilator method is ordinarily used whenever a root of the auxiliary equation coincides with a root associated with the annihilator.*) ∎

■ Exercises / Section 12.3

In Exercises 1–30, solve the differential equations.

1. $(D^2 - 6D + 9)y = e^x$

2. $(D^2 - 6D + 9)y = 2$

3. $(D^2 - 6D + 9)y = 9x$

4. $(D^2 - 6D + 9)y = 18x^2$

5. $(D^2 - D - 2)y = 2x^2$

6. $(D^2 - 3D - 4)y = -4x$

7. $(D^2 - D + 1)y = 1 - x^2$

8. $(D^2 + 4)y = 2x^2 - x$

9. $(D^2 - D + 2)y = 4e^{3x}$

10. $(D^2 + D + 2)y = 12e^{2x}$

11. $(D^2 - 6D + 9)y = 9\cos 3x$

12. $(D^2 + 4D - 5)y = 27e^{4x}$

13. $(D^2 + 4D + 3)y = 6 + e^x$

14. $(D^2 - 4)y = -5\cos x$

15. $(D^2 + 1)y = 6\sin 2x$ **16.** $(D^2 - 3D - 4)y = \cos x$

17. $(D^2 + 5D + 6)y = 4\cos x + 6\sin x$

18. $(D^2 - 4)y = 8x + 12$ **19.** $(D^2 - 2D + 1)y = 3e^{2x}$

20. $(D^2 - 1)y = e^{2x} + 4$

21. $(D^3 - 2D^2 - D + 2)y = 8e^{3x}$

22. $(D^2 - 2D + 1)y = x^2 - 1$

23. $(D^2 - 2D + 5)y = 4xe^x$

   (Since $(d/dx)(4xe^x) = 4xe^x + 4e^x$, $y_p = Axe^x + Be^x$.)

24. $(D^2 + 3)y = 8xe^x + 3$

25. $(D^2 + 9)y = 9e^{3x}$; if $x = 0$, then $y = 1$ and $Dy = 3/2$

26. $(D^2 + 1)y = x$; if $x = 0$, then $y = 1$ and $Dy = 1$

27. $(D^2 + 1)y = 6\cos 2x$; if $x = 0$, then $y = 3$ and $Dy = 1$

28. $(D^2 - 1)y = -16\sin x$; if $x = 0$, then $y = 0$ and $Dy = 1$

29. $(D^2 + 2D + 5)y = 10\cos x$; if $x = 0$, then $y = 5$ and $Dy = 6$

30. $(D^2 - 2D + 5)y = 5\sin x$; if $x = 0$, then $y = -2$ and $Dy = 0$

In Exercises 31–44, find y_p.

31. $(D^2 + 4)y = 10\sin 3x$ **32.** $(D^2 - 6)y = 5\cos 2x$

33. $(D^2 - D + 2)y = \cos x$

34. $(D^2 - 2D + 2)y = 5\sin 2x$

35. $(D^2 + D - 2)y = 2x^2 + 1$

36. $(D^2 + 5D + 1)y = x$ **37.** $(D^2 + 2)y = 2e^{2x} + 2$

38. $(D^2 + 5)y = 3xe^x$

39. $(D^2 - D - 4)y = x + 2e^{3x}$

40. $(D^2 - 6)y = e^x - \sin x$ **41.** $(D^2 + 3)y = 2x + \cos x$

42. $(D^2 - 2)y = x^2 + 2e^{-x}$ **43.** $(D^2 - D - 3)y = 6xe^x$

44. $(D^2 - 2D + 3)y = 4xe^x$

In Exercises 45–55, use the annihilator method.

45. $(D^2 - 4)y = 8e^{2x}$ **46.** $(D^2 - 9)y = 10e^{-3x}$

47. $(D^2 - D - 6)y = 10e^{-2x}$ **48.** $(D^2 + 2D - 15)y = 12e^{3x}$

49. $(D^2 + 3D - 28)y = 11e^{4x}$ **50.** $(D^2 + 4D - 12)y = 4e^{2x}$

51. $(D^2 - 1)y = 2e^x$ **52.** $(D^2 + D - 2)y = 3e^x$

53. $(D^2 - 4)y = 4 + e^{2x}$

54. $(D^3 - 3D^2 + 3D - 1)y = e^x - 3$

55. $(D^2 + 1)y = 2\sin x$ (*Hint:* Since $m' = \pm j$, the annihilator is $D^2 + 1$.)

12.4 Applications of Second-Order Equations

The main applications of second-order linear equations involve a weight oscillating on a spring and the electrical analog.

   Recall that a spring stretched x units beyond its natural length pulls back with a force kx by Hooke's law; k is called the **spring constant.** If the spring is compressed, then it pushes back with the same force kx. Suppose a mass hanging on a spring is allowed to come to rest (Figure 12.1). Let $x = 0$ be this **equilibrium position;** the x-coordinate in the downward direction is considered positive and the upward,

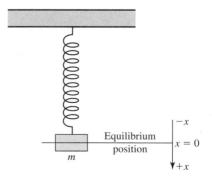

Figure 12.1

negative. Then by Newton's second law, the force exerted by the mass on the spring is given by

$$m\frac{d^2x}{dt^2} \qquad \textbf{mass × acceleration}$$

Since the spring exerts a force kx by Hooke's law, it follows that

$$m\frac{d^2x}{dt^2} = -kx \qquad\qquad\qquad (12.13)$$

The negative sign indicates that the forces act in opposite directions. If, as in the case of a falling body, we assume that the damping force due to the resistance of the surrounding medium is proportional to the velocity dx/dt, then Formula (12.13) must be modified as follows:

$$m\frac{d^2x}{dt^2} = -kx - b\frac{dx}{dt} \qquad \text{or} \qquad m\frac{d^2x}{dt^2} + b\frac{dx}{dt} + kx = 0$$

Fundamental Equation of Damped Oscillatory Motion

$$m\frac{d^2x}{dt^2} + b\frac{dx}{dt} + kx = 0 \qquad\qquad\qquad (12.14)$$

Example 1

A 2-lb weight stretches a spring 6 in. The weight is attached to the spring and allowed to come to rest (equilibrium position). The weight is then pulled 4 in. below the equilibrium position and released. Determine the motion of the weight as a function of time, assuming that the damping force is negligible.

Solution. To determine the spring constant, we use the information in the first sentence: 2 lb stretches the spring 6 in. $= 1/2$ ft. So by Hooke's law,

$$F = kx$$

$$2 = k\left(\frac{1}{2}\right) \qquad \text{or} \qquad k = 4$$

Since $g = 32$ ft/s^2, we have 32 lb $= 1$ slug, so that the mass of the 2-lb weight is

$$\frac{2}{32} = \frac{1}{16} \text{ slug}$$

Finally, since the damping force is negligible, $b = 0$. So by Equation (12.14),

$$m\frac{d^2x}{dt^2} + b\frac{dx}{dt} + kx = 0$$

$$\frac{1}{16}\frac{d^2x}{dt^2} + 4x = 0 \qquad \textbf{\textit{m} = $\frac{1}{16}$, \textit{b} = 0, \textit{k} = 4}$$

or

$$\frac{d^2x}{dt^2} + 64x = 0 \qquad\qquad \textbf{multiplying by 16}$$

Since the weight is initially at rest 4 inches below the equilibrium position, we have the following initial conditions:

1. When $t = 0$, $x = 4$ in. $= 1/3$ ft.   **initial position**
2. When $t = 0$, $dx/dt = 0$.   **initial velocity**

We now solve the equation, making use of the initial conditions:

$$\frac{d^2x}{dt^2} + 64x = 0$$

$$m^2 + 64 = 0 \qquad \textbf{auxiliary equation}$$

$$m = \pm\sqrt{-64} = \pm 8j$$

It follows that

$$x(t) = c_1 \cos 8t + c_2 \sin 8t$$

Substituting $t = 0$ and $x = 1/3$, we get

$$\frac{1}{3} = c_1 \cos 0 + c_2 \sin 0 \qquad \text{or} \qquad c_1 = \frac{1}{3}$$

Thus

$$x(t) = \frac{1}{3} \cos 8t + c_2 \sin 8t$$

Now

$$\frac{dx}{dt} = x'(t) = -\frac{8}{3} \sin 8t + 8c_2 \cos 8t$$

From the condition $dx/dt = 0$ when $t = 0$, we have

$$0 = 0 + 8c_2 \qquad \text{or} \qquad c_2 = 0$$

Hence from

$$x(t) = \frac{1}{3} \cos 8t + c_2 \sin 8t$$

the final solution is (since $c_2 = 0$)

$$x(t) = \frac{1}{3} \cos 8t \qquad\qquad\qquad \blacksquare$$

So far we have assumed that $g = 10 \text{ m/s}^2$ for the acceleration due to gravity. In this section we will use the more accurate value $g = 9.8 \text{ m/s}^2$.

Example 2
_____

A weight of mass 0.50 kg (weight $= 4.9$ N) stretches a spring 0.70 m. The weight is pushed 0.40 m above the equilibrium position and released. Find the position of the weight as a function of time, if a damping force numerically equal to twice the velocity is present.

Solution. First we determine the spring constant. Since $F = 4.9$ N when $x = 0.70$ m, we get by Hooke's law

$$4.9 = k(0.70) \qquad \text{or} \qquad k = 7.0$$

The initial conditions are:

1. When $t = 0$, $x = -0.40$.   **upward negative**
2. When $t = 0$, $dx/dt = 0$.   **initial velocity**

Since the damping force is numerically equal to twice the velocity, this force must be $2\,dx/dt$, so that $b = 2$. The resulting equation is

$$m\frac{d^2x}{dt^2} + b\frac{dx}{dt} + kx = 0$$

$$0.50\frac{d^2x}{dt^2} + 2\frac{dx}{dt} + 7.0x = 0 \qquad \textbf{\textit{m} = 0.50, \textit{b} = 2, \textit{k} = 7.0}$$

or

$$\frac{d^2x}{dt^2} + 4\frac{dx}{dt} + 14x = 0 \qquad \textbf{multiplying by 2}$$

$$m^2 + 4m + 14 = 0 \qquad \textbf{auxiliary equation}$$

$$m = \frac{-4 \pm \sqrt{16 - 4(14)}}{2} = \frac{-4 \pm \sqrt{-40}}{2}$$

$$= -2 \pm \sqrt{10}\,j$$

The general solution is

$$x(t) = e^{-2t}\left(c_1 \cos\sqrt{10}t + c_2 \sin\sqrt{10}t\right)$$

From the condition $x = -0.40$ when $t = 0$, we get

$$-0.40 = e^0(c_1 \cos 0 + c_2 \sin 0)$$

or

$$c_1 = -0.40$$

So

$$x(t) = e^{-2t}\left(-0.40 \cos\sqrt{10}t + c_2 \sin\sqrt{10}t\right)$$

Now

$$\frac{dx}{dt} = e^{-2t}\left(0.40\sqrt{10} \sin\sqrt{10}t + c_2\sqrt{10}\cos\sqrt{10}t\right)$$

$$+ (-2e^{-2t})\left(-0.40\cos\sqrt{10}t + c_2\sin\sqrt{10}t\right)$$

Since $dx/dt = 0$ when $t = 0$, we have

$$0 = \left(0 + c_2\sqrt{10}\right) - 2(-0.40 + 0)$$

$$0 = c_2\sqrt{10} + 0.80$$

$$c_2 = -\frac{0.80}{\sqrt{10}}$$

The solution is therefore given by

$$x(t) = e^{-2t}\left(-0.40\cos\sqrt{10}t - \frac{0.80}{\sqrt{10}}\sin\sqrt{10}t\right)$$

Using two significant digits, we get

$$x(t) = e^{-2.0t}(-0.40\cos 3.2t - 0.25\sin 3.2t)$$

The graph of the solution is shown in Figure 12.2. (In the figure, the downward direction is negative.)

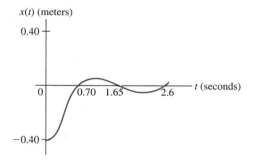

Figure 12.2   ∎

Note that the motion of the weight in Example 2 is oscillatory, but the oscillations gradually damp out due to the factor $e^{-2.0t}$. (See Figure 12.2.) If the damping force is neglected (as in Example 1), the equation becomes

$$0.50\frac{d^2x}{dt^2} + 7.0x = 0$$

$$\frac{d^2x}{dt^2} + 14x = 0$$

and

$$x(t) = c_1\cos\sqrt{14}t + c_2\sin\sqrt{14}t$$

or

$$x(t) = c_1\cos 3.7t + c_2\sin 3.7t$$

From the initial conditions we now obtain

$$x(t) = -0.40\cos 3.7t$$

This last equation describes **simple harmonic motion.**

In the General Case (12.14) the roots of the auxiliary equation are given by

$$\frac{-b \pm \sqrt{b^2 - 4mk}}{2m}$$

Consequently, the solution is sinusoidal (containing sines and cosines) and the motion oscillatory whenever these roots are complex numbers—that is, whenever

$$b^2 - 4mk < 0 \qquad \text{or} \qquad b < 2\sqrt{mk}$$

Such a system is said to be **underdamped.** If

$$b > 2\sqrt{mk}$$

then the motion is not oscillatory and the system is said to be **overdamped.** If

$$b = 2\sqrt{mk}$$

the system is said to be **critically damped.**

An additional vertical force $f(t)$ may act on the system. (For example, $f(t)$ may be due to the motion of the support or the presence of a magnetic field.) Then Equation (12.14) becomes

$$m\frac{d^2x}{dt^2} + b\frac{dx}{dt} + kx = f(t) \tag{12.15}$$

Such cases are called **forced oscillations.**

Example 3

Suppose the system in Example 1 is acted on by an external force $f(t) = (1/8)\sin 4t$. Determine the motion of the weight.

Solution. The equation in Example 1,

$$\frac{1}{16}\frac{d^2x}{dt^2} + 4x = 0$$

must be modified to take into account the external force:

$$\frac{1}{16}\frac{d^2x}{dt^2} + 4x = \frac{1}{8}\sin 4t$$

or

$$\frac{d^2x}{dt^2} + 64x = 2\sin 4t$$

The initial conditions are still the same:

1. When $t = 0$, $x = 1/3$ ft.
2. When $t = 0$, $dx/dt = 0$.

The solution obtained in Example 1 now serves as the complementary solution:

$$x_c = c_1 \cos 8t + c_2 \sin 8t$$

By the method of Section 12.3,

$$x_p = \frac{1}{24} \sin 4t$$

so that the general solution is

$$x(t) = c_1 \cos 8t + c_2 \sin 8t + \frac{1}{24} \sin 4t$$

The constants are now evaluated as in Examples 1 and 2. The final solution is

$$x(t) = \frac{1}{3} \cos 8t - \frac{1}{48} \sin 8t + \frac{1}{24} \sin 4t$$

The graph is shown in Figure 12.3.

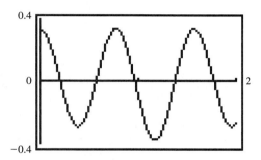

Figure 12.3

The physical situation we have dealt with so far has an electrical analog (Figure 12.4). Recall that the voltage across an inductor, resistor, and capacitor is given by

$$L \frac{di}{dt}, \quad Ri, \quad \text{and} \quad \frac{q}{C}$$

respectively. Since $i = dq/dt$, these expressions can be written

$$L \frac{d^2 q}{dt^2}, \quad R \frac{dq}{dt}, \quad \text{and} \quad \frac{q}{C}$$

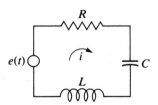

Figure 12.4

If the components are connected in series with a generator, then the impressed voltage $e(t)$ is equal to the sum of the voltages across the components, known as **Kirchhoff's voltage law.**

Differential Equation of the Circuit in Figure 12.4

$$L \frac{d^2 q}{dt^2} + R \frac{dq}{dt} + \frac{q}{C} = e(t) \tag{12.16}$$

As before, if $e(t) = 0$, then the roots of the auxiliary equation tell us whether or not the solution is sinusoidal (Exercise 24). In the oscillatory case, the damping factor will cause the current to die out quickly unless a voltage source is present.

Example 4   Find an expression for the charge on the capacitor in Figure 12.4 as a function of time if $L = 1$ H, $R = 15\,\Omega$, $C = 10^{-2}$ F, and $e(t) = 100 \sin 60t$.

Solution. By Equation (12.16)

$$\frac{d^2q}{dt^2} + 15\frac{dq}{dt} + 100q = 100 \sin 60t$$

To obtain the complementary solution q_c, we note that

$$m^2 + 15m + 100 = 0$$

and

$$m = \frac{-15 \pm \sqrt{225 - 400}}{2} = -\frac{15}{2} \pm \frac{5}{2}\sqrt{7}j$$

Hence

$$q_c = e^{-15t/2}\left(c_1 \cos \frac{5}{2}\sqrt{7}t + c_2 \sin \frac{5}{2}\sqrt{7}t\right)$$

For the particular solution we choose

$$q_p = A \cos 60t + B \sin 60t$$

so that

$$\frac{dq_p}{dt} = -60A \sin 60t + 60B \cos 60t$$

$$\frac{d^2q_p}{dt^2} = -3600A \cos 60t - 3600B \sin 60t$$

Substituting, we get

$$-3600A \cos 60t - 3600B \sin 60t - 900A \sin 60t$$

$$+900B \cos 60t + 100A \cos 60t + 100B \sin 60t = 100 \sin 60t$$

Collecting terms,

$$(-3500A + 900B) \cos 60t + (-900A - 3500B) \sin 60t = 100 \sin 60t$$

Comparing coefficients, we get the system of equations

$$\left.\begin{array}{l} -3500A + 900B = 0 \\ -900A - 3500B = 100 \end{array}\right\}$$

whose solution set is $A = -0.0069$ and $B = -0.027$. Finally,

$$q(t) = q_c + q_p = e^{-15t/2}\left(c_1 \cos \frac{5}{2}\sqrt{7}t + c_2 \sin \frac{5}{2}\sqrt{7}t\right)$$

$$- 0.0069 \cos 60t - 0.027 \sin 60t \qquad \blacksquare$$

Suppose we look at the solution in Example 4 more closely. The complementary solution q_c contains the exponential decaying factor $e^{-15t/2}$, but q_p does not. Consequently, q_c will die out quickly, leaving only q_p. For this reason q_c is called the **transient** part of the solution, while q_p is called the **steady-state solution.** In other words, after a certain time period the solution is essentially given by q_p, hence the name *steady state*. Moreover, in many problems only the steady-state solution may actually be of interest.

Transient
Steady state

Example 5 Find the steady-state current in the circuit in Example 4.

Solution. As noted above, the steady-state solution is

$$q_p = -0.0069 \cos 60t - 0.027 \sin 60t$$

Consequently, the steady-state current is

$$i = \frac{dq_p}{dt} = 0.41 \sin 60t - 1.62 \cos 60t \qquad\blacksquare$$

■ Exercises / Section 12.4

Solve the given problems. In Exercises 1–16, use your graphing utility when applicable to graph the solutions.

1. A spring is such that a 4-lb weight stretches it 6 in. The weight is attached to the spring and allowed to reach the equilibrium position. The weight is then pulled 3 in. below the equilibrium position and released. Find the motion of the weight as a function of time, assuming no damping.

2. A 2-lb weight stretches a spring 6 in. The weight is pushed 7 in. above the equilibrium position and released. Find the motion of the weight as a function of time, assuming no damping.

3. Find the motion of the weight in Exercise 1 if an external force $f(t) = (1/4)\cos 6t$ acts on the system.

4. Find the motion of the weight in Exercise 2 if the external force is $(1/4)\sin 4t$.

5. A 12-lb weight stretches a spring 2 ft. The weight is pulled 8 in. below the equilibrium position and given an initial downward velocity of 3 ft/s. (Recall that the downward direction is positive.) Find the motion of the weight as a function of time, assuming that the damping force may be neglected.

6. A 5-lb weight stretches a spring 6 in. The weight is pulled 4 in. below the equilibrium position and given an initial downward velocity of 4 ft/s. Find the motion of the weight as a function of time. (Assume that the damping force may be neglected.)

7. Find the motion of the weight in Exercise 5 if the initial velocity is 4 ft/s in the upward direction.

8. Find the motion of the weight in Exercise 6 if the initial velocity is 6 ft/s in the upward direction.

9. A 4-lb weight stretches a spring 2 ft. The weight is pulled 6 in. below the equilibrium position and released. Find the motion of the weight as a function of time, given that a damping force numerically equal to 1/8 of the velocity is present.

10. A 2-lb weight stretches a spring 6 in. The weight is pushed 4 in. above the equilibrium position and released. Find the motion of the weight as a function of time, given that a damping force numerically equal to 1/16 of the velocity is present.

11. A spring having a spring constant $k = 3$ lb/ft is subject to a retarding force equal to 1/4 of the velocity. An 8-lb weight is attached to the spring, pushed 3 in. above the equilibrium position, and released. Find $x(t)$.

12. In Exercise 11, show that the motion is oscillatory whenever $b < \sqrt{3}$.

13. A spring having a spring constant $k = 10$ lb/ft is subject to a retarding force equal to 5/8 of the velocity. A 10-lb weight is attached to the spring, pulled 6 in. below the equilibrium position and given an initial upward velocity of 3 ft/s. Find $x(t)$.

14. A 2-lb weight stretches a spring 6 in. The weight is pulled 3 in. below the equilibrium position and released. Find $x(t)$, given that a damping force numerically equal to the velocity is present.

15. Repeat Exercise 14, given that an external force $2\sin 8t$ acts on the system.

16. Repeat Exercise 14, given that the system is acted on by an external force $(1/4) \cos 8t$.

17. A spring is such that a weight of mass 2.0 kg (19.6 N) stretches it 0.098 m. If this weight is pulled 0.25 m below the equilibrium position and released, find the motion of the weight as a function of time. (We are assuming that the damping force may be neglected.

18. Find the motion of the weight in Exercise 17 if the system is acted on by an external force $20 \sin 5t$.

19. Find the motion of the weight in Exercise 17 if a damping force equal to four times the velocity is present.

20. In Exercise 19, for what values of b will the motion be oscillatory?

21. A mass of 2.0 kg (19.6 N) stretches a certain spring 0.098 m. Suppose that the damping force is numerically equal to 4 times the velocity and that the external force is $20 \sin 5t$. If the weight is attached to the spring and pulled 0.25 m below the equilibrium position and released, find the transient and steady-state solutions. (Refer to Exercise 19.)

22. A weight of 0.50 kg (4.9 N) stretches a spring 0.49 m. If a damping force equal to 2.8 times the velocity and an external force of $3.0 \cos 2t$ are present, find the steady-state motion of the attached weight.

23. For a given LC circuit, $L = 1$ H and $C = 1.0 \times 10^{-4}$ F. Find the charge and current as functions of time if $i = 10$ and $q = 0$ when $t = 0$. (Assume that $e(t) \equiv 0$.)

24. If $L = 1$ H and $C = 1.0 \times 10^{-4}$ F in an LRC circuit, find the range on R for which the current is oscillatory. (Assume that $e(t) \equiv 0$.)

25. For a given LC circuit, $L = 0.5$ H, $C = 8 \times 10^{-4}$ F, and $e(t) = 50 \sin 100t$. Find the charge as a function of time if $i(0) = q(0) = 0$.

26. Find the steady-state current of the following LRC circuit: $L = 1$ H, $C = 10^{-2}$ F, $R = 50\,\Omega$, and $e(t) = 100 \sin 50t$.

27. A weight of mass 1.2 kg is attached to a spring with spring constant $k = 80$ N/m. If the damping force is numerically equal to $1.5\,dx/dt$ and the external force is $10 \sin 5t$, find the steady-state motion of the weight.

28. A weight of mass 2.4 kg is attached to a spring with spring constant $k = 150$ N/m. If the damping force is numerically equal to $1.25\,dx/dt$ and an external force of $20 \sin 3t$ is present, find the steady-state motion of the attached weight.

29. Find the steady-state current of the following LRC circuit: $L = 1$ H, $R = 10\,\Omega$, $C = 1/100$ F, and $e(t) = 50 \cos 10t$.

■ Review Exercises / Chapter 12

In Exercises 1–22, solve the differential equations.

1. $D^4 y = 0$

2. $(D^2 - 4)y = 0$

3. $(D^2 + 4)y = 0$

4. $(D - 4)^2 y = 0$

5. $(D^2 - 2D - 2)y = 0$

6. $(D^2 - 2D + 2)y = 0$

7. $(3D^2 - D + 1)y = 0$

8. $(2D^2 + D + 2)y = 0$

9. $(D - 2)^2(D^2 + 1)y = 0$

10. $(D^5 + 9D^3)y = 0$

11. $2\dfrac{d^2y}{dx^2} - \dfrac{dy}{dx} + y = 0$

12. $2\dfrac{d^2y}{dx^2} - 2\dfrac{dy}{dx} - y = 0$

13. $3\dfrac{d^2y}{dx^2} - 2\dfrac{dy}{dx} - 2y = 0$

14. $\dfrac{d^2y}{dx^2} - 2\dfrac{dy}{dx} + 2y = 0$

15. $(D^2 - 3D - 4)y = 6e^x$

16. $(D^2 - 3D - 4)y = 8x$

17. $(D^2 - 3D - 4)y = 2 \sin x$

18. $(D^2 - 3D - 4)y = 20e^{-x}$ (annihilator)

19. $(D^2 - 3D - 4)y = 10xe^{-x}$ (annihilator)

20. $(D^2 - 3D - 4)y = e^{4x}$ (annihilator)

21. $(D^2 - 2D - 3)y = 0$; if $x = 0$, then $y = 0$ and $Dy = -4$

22. $(D^2 + 1)y = 0$; if $x = 0$, then $y = 2$ and $Dy = 0$

23. Find the steady-state current of the following LRC circuit: $L = 0.100$ H, $C = 2.00 \times 10^{-3}$ F, $R = 40.0\,\Omega$, and $e(t) = 100.0 \cos 20.0t$.

24. A weight of 2.45 N (0.25 kg) stretches a spring 4.9 cm. If the weight is attached to the spring, show that the motion is oscillatory if $b < 5\sqrt{2} \approx 7.1$.

25. A 5-lb weight stretches a spring 6 in. The weight is pulled 3 in. below the equilibrium position and given an initial upward velocity of 5 ft/s. Assume that an external force $(1/8) \cos 4t$ acts on the system and that the damping force may be neglected. Find the motion of the weight as a function of time.

13

The Laplace Transform

Introduction and Basic Properties

With the **Laplace transform** we are finally leaving the age of Euler for a brief glimpse into the twentieth century. Because of our sudden jump in time, it is difficult to motivate the definition of Laplace transform. Also, this concept was developed gradually over a period spanning several decades. But it is safe to say that in its modern form the Laplace transform can be traced to the English electrical engineer Oliver Heaviside (1850–1925), who discovered a unique method for solving differential equations arising in electrical circuit theory. Later attempts to justify Heaviside's methods led to the following definition:

Definition of the Laplace Transform

If $f(t)$ is defined for $0 \leq t < \infty$, then the **Laplace transform** of f is defined to be the integral

$$F(s) = L\{f(t)\} = \int_0^\infty e^{-st} f(t)\, dt \qquad (13.1)$$

Similar transforms had been studied earlier by the French mathematician Pierre Simon de Laplace (1749–1827) and even by Euler.

It is not difficult to see how Laplace transforms may be employed to solve differential equations. Since that is the topic of Section 13.4, we will first try to become acquainted with the transform and its basic properties.

Our main task in this section is to find the Laplace transforms of some special functions. Consider the following examples.

Example 1   Find the Laplace transform of the function $f(t) = t$, $t \geq 0$.

Solution. By Definition (13.1)

$$L\{t\} = \int_0^\infty e^{-st} \cdot t\, dt = \lim_{b \to \infty} \int_0^b t e^{-st}\, dt \qquad \textbf{improper integral}$$

We integrate by parts, letting $u = t$ and $dv = e^{-st}dt$. Since the variable s is a constant as far as the integration is concerned, we obtain $du = dt$ and $v = (-1/s)e^{-st}$. Hence

$$L\{t\} = \lim_{b \to \infty} \left(-\frac{t}{s}e^{-st} \Big|_0^b + \frac{1}{s} \int_0^b e^{-st} dt \right)$$

$$= \lim_{b \to \infty} \left[-\frac{t}{s}e^{-st} \Big|_0^b + \frac{1}{s}\left(-\frac{1}{s} \right)e^{-st} \Big|_0^b \right] \qquad \int e^{-st}\, dt = -\frac{1}{s}e^{-st}$$

$$= \lim_{b \to \infty} \left(-\frac{b}{s}e^{-sb} - \frac{1}{s^2}e^{-sb} + \frac{1}{s^2} \right)$$

$$= \lim_{b \to \infty} \left(-\frac{b}{se^{sb}} \right) - \frac{1}{s^2} \lim_{b \to \infty} \frac{1}{e^{sb}} + \lim_{b \to \infty} \frac{1}{s^2}$$

$$= \lim_{b \to \infty} \left(-\frac{b}{se^{sb}} \right) - 0 + \frac{1}{s^2}$$

provided that $s > 0$. The remaining limit may be evaluated using L'Hospital's rule:

$$L\{t\} = \lim_{b \to \infty} \left(-\frac{1}{s^2 e^{sb}} \right) + \frac{1}{s^2} = \frac{1}{s^2}$$

We conclude that

$$L\{t\} = \frac{1}{s^2} \qquad (s > 0) \tag{13.2}$$

(Observe that s has to be positive to ensure the existence of the improper integral. Otherwise the variable s will play no role in our work, as we will see.) ∎

Example 2   Find the Laplace transform of the function $f(t) = e^{at}$.

Solution. Again, by Formula (13.1),

$$L\{e^{at}\} = \int_0^{\infty} e^{-st} e^{at}\, dt$$

$$= \lim_{b \to \infty} \int_0^b e^{-(s-a)t}\, dt \qquad \begin{bmatrix} u = -(s - a)t \\ du = -(s - a)\,dt \end{bmatrix}$$

$$= \lim_{b \to \infty} \left(\frac{1}{-(s - a)} \right) \int_0^b e^{-(s-a)t}\, [-(s - a)]\, dt$$

$$= \lim_{b \to \infty} \frac{1}{-(s - a)} e^{-(s-a)t} \Big|_0^b$$

$$= \lim_{b \to \infty} \left(\frac{1}{-(s - a)} e^{-(s-a)b} + \frac{1}{s - a} \right) = \frac{1}{s - a}$$

provided that $s > a$. Hence

$$L\{e^{at}\} = \frac{1}{s - a} \qquad (s > a) \tag{13.3}$$

∎

We can see from these examples that a function $f(t)$ has a Laplace transform whenever the improper integral

$$\int_0^\infty e^{-st} f(t)\, dt$$

exists. For example, the functions $f(t) = \tan t$ and $f(t) = e^{t^2}$ do not possess transforms.

Table of Transforms

Rather than continuing with these calculations, we refer you to the accompanying table of common transforms, leaving a few additional cases as exercises.

Table 13.1 *Short table of Laplace transforms*

$f(t)$	$F(s)$	$f(t)$	$F(s)$
1. 1	$\dfrac{1}{s}$	**9.** $t^n e^{at}$	$\dfrac{n!}{(s-a)^{n+1}}$
2. t	$\dfrac{1}{s^2}$	**10.** $1 - \cos at$	$\dfrac{a^2}{s(s^2 + a^2)}$
3. t^n	$\dfrac{n!}{s^{n+1}}$	**11.** $at - \sin at$	$\dfrac{a^3}{s^2(s^2 + a^2)}$
4. e^{at}	$\dfrac{1}{s-a}$	**12.** $\sin at - at \cos at$	$\dfrac{2a^3}{(s^2 + a^2)^2}$
5. $\sin at$	$\dfrac{a}{s^2 + a^2}$	**13.** $t \sin at$	$\dfrac{2as}{(s^2 + a^2)^2}$
6. $\cos at$	$\dfrac{s}{s^2 + a^2}$	**14.** $\sin at + at \cos at$	$\dfrac{2as^2}{(s^2 + a^2)^2}$
7. $e^{at} \sin bt$	$\dfrac{b}{(s-a)^2 + b^2}$	**15.** $t \cos at$	$\dfrac{s^2 - a^2}{(s^2 + a^2)^2}$
8. $e^{at} \cos bt$	$\dfrac{s-a}{(s-a)^2 + b^2}$		

Observe next that the definition of the Laplace transform implies that

$$L\{af(t) + bg(t)\} = aL\{f(t)\} + bL\{g(t)\} \qquad (13.4)$$

Since it possesses Property (13.4), the Laplace transform is said to be *linear*.

Example 3   Use the table and the linearity property to obtain

$$L\{4t^3 + 2 \sin 3t\}$$

Solution. By Property (13.4) and Transforms 3 and 5 in the table,

$$L\{4t^3 + 2\sin 3t\} = 4L\{t^3\} + 2L\{\sin 3t\}$$

$$= 4\frac{3!}{s^4} + 2\frac{3}{s^2 + 9} = \frac{24}{s^4} + \frac{6}{s^2 + 9} \qquad \blacksquare$$

Example 4   Find $L\{e^{-3t} + e^t \cos 4t\}$.

Solution. By Transforms 4 and 8 in the table we get

$$\frac{1}{s+3} + \frac{s-1}{(s-1)^2 + 16}$$

◼

13.2   Inverse Laplace Transforms

The procedure in the last section can be reversed: we can look up $f(t)$, given $F(s)$. For example, if $F(s) = 1/(s+4)$, then $f(t) = e^{-4t}$ by Transform 4. Here $f(t)$ is called the **inverse transform,** denoted by L^{-1}. Thus we may write

$$L^{-1}\left\{\frac{1}{s+4}\right\} = e^{-4t}$$

Unfortunately, the forms in the table do not always fit, in which case an adjustment is required. For example,

$$L^{-1}\left\{\frac{3}{s+4}\right\} = 3L^{-1}\left\{\frac{1}{s+4}\right\} = 3e^{-4t}$$

which can be readily checked by reversing the steps. In general, then,

$$L^{-1}\{aF(s) + bG(s)\} = aL^{-1}\{F(s)\} + bL^{-1}\{G(s)\}$$

So the inverse Laplace transform is also a linear operator. For example,

$$L^{-1}\left\{\frac{\sqrt{3}}{s-6} - \frac{4s}{s^2+5}\right\} = \sqrt{3}e^{6t} - 4\cos\sqrt{5}t$$

by Transforms 4 and 6, respectively.

Example 1   Find

$$L^{-1}\left\{\frac{1}{s(s^2+4)}\right\}$$

Solution. By Transform 10 for $a = 2$, we get

$$L^{-1}\left\{\frac{1}{s(s^2+4)}\right\} = L^{-1}\left\{\frac{\frac{1}{4}\cdot 4}{s(s^2+4)}\right\}$$

$$= \frac{1}{4}L^{-1}\left\{\frac{4}{s(s^2+4)}\right\} = \frac{1}{4}(1 - \cos 2t)$$

◼

The next example is similar.

Example 2

$$L^{-1}\left\{\frac{1}{s^2+7}\right\} = \frac{1}{\sqrt{7}}L^{-1}\left\{\frac{\sqrt{7}}{s^2+7}\right\} = \frac{1}{\sqrt{7}}\sin\sqrt{7}t$$

by Transform 5.

◼

Example 3   Find

$$L^{-1}\left\{\frac{s+1}{s^2+4s+8}\right\}$$

Solution. For trinomial denominators that are not factorable we may use Transforms 7 and 8 after completing the square:

$$L^{-1}\left\{\frac{s+1}{s^2+4s+8}\right\}=L^{-1}\left\{\frac{s+1}{(s+2)^2+4}\right\}$$

Even now the form does not quite fit Transform 8, but noting that $a=-2$, we may proceed as follows:

$$L^{-1}\left\{\frac{s+1}{(s+2)^2+4}\right\}=L^{-1}\left\{\frac{s+2-2+1}{(s+2)^2+4}\right\}$$

$$=L^{-1}\left\{\frac{(s+2)-1}{(s+2)^2+4}\right\}$$

$$=L^{-1}\left\{\frac{s+2}{(s+2)^2+4}\right\}-L^{-1}\left\{\frac{1}{(s+2)^2+4}\right\}$$

$$=L^{-1}\left\{\frac{s+2}{(s+2)^2+4}\right\}-\frac{1}{2}L^{-1}\left\{\frac{2}{(s+2)^2+4}\right\}$$

$$=e^{-2t}\cos 2t-\frac{1}{2}e^{-2t}\sin 2t \qquad \boldsymbol{a=-2,b=2}$$

by Transforms 8 and 7, respectively.   ■

13.3   Partial Fractions

Transforms more complex than those considered in the previous section can often be broken up to fit the forms in the table. For example,

$$L^{-1}\left\{\frac{5}{(s-1)(s+4)}\right\}=L^{-1}\left\{\frac{1}{s-1}-\frac{1}{s+4}\right\} \tag{13.5}$$

$$=e^t-e^{-4t}$$

by Transform 4. The fractions on the right in Equation (13.5) are called **partial fractions.** Certain proper fractions (degree of the numerator less than the degree of the denominator) can be written as a sum of partial fractions according to the following rules:

I.   If a *linear* factor $as+b$ occurs n times in the denominator, then there exist n partial fractions

$$\frac{A_1}{as+b}+\frac{A_2}{(as+b)^2}+\cdots+\frac{A_n}{(as+b)^n}$$

where $A_1, A_2, \ldots, A_n$ are constants.

II. If a *quadratic* factor $as^2 + bs + c$ occurs n times in the denominator, then there exist n partial fractions

$$\frac{A_1 s + B_1}{as^2 + bs + c} + \frac{A_2 s + B_2}{(as^2 + bs + c)^2} + \cdots + \frac{A_n s + B_n}{(as^2 + bs + c)^n}$$

where the A's and B's are constants. In all cases n may be equal to 1.

Rules I and II are only intended to be a general guide. How they are put to use will be illustrated in the following examples.

Example 1   (*Distinct linear factors.*) Find

$$L^{-1} \left\{ \frac{6s^2 + 12s - 6}{(s - 1)(s + 2)(s + 3)} \right\}$$

Solution. Since the factors are all distinct, by Rule I we get

$$\frac{6s^2 + 12s - 6}{(s - 1)(s + 2)(s + 3)} = \frac{A}{s - 1} + \frac{B}{s + 2} + \frac{C}{s + 3} \tag{13.6}$$

(Since there are only three constants, it is more convenient to use A, B, and C, rather than subscripts.) The main task is to determine the constants. To this end we add the fractions on the right of Equation (13.6) to obtain

$$\frac{A(s + 2)(s + 3) + B(s - 1)(s + 3) + C(s - 1)(s + 2)}{(s - 1)(s + 2)(s + 3)}$$

The numerator of this fraction must be equal to the numerator of the left side of Equation (13.6). Thus

$$A(s + 2)(s + 3) + B(s - 1)(s + 3) + C(s - 1)(s + 2) = 6s^2 + 12s - 6 \tag{13.7}$$

To find the constants, we let s be equal to certain convenient values. For example, if $s = -2$, then Equation (13.7) collapses to

$$0 + B(-3)(1) + 0 = -6$$

so that $B = 2$. Similarly, if $s = -3$, we get

$$0 + 0 + C(-4)(-1) = 12$$

or $C = 3$. Finally, if $s = 1$, then

$$A(3)(4) + 0 + 0 = 12$$

or $A = 1$. We now have (since $A = 1$, $B = 2$, and $C = 3$)

$$L^{-1} \left\{ \frac{6s^2 + 12s - 6}{(s - 1)(s + 2)(s + 3)} \right\} = L^{-1} \left\{ \frac{1}{s - 1} + \frac{2}{s + 2} + \frac{3}{s + 3} \right\}$$

$$= e^t + 2e^{-2t} + 3e^{-3t}$$

by Transform 4.   ■

Example 2   (*Repeating linear factors.*) Find

$$L^{-1}\left\{\frac{s}{(s-2)^2(s+1)}\right\}$$

Solution. This form contains only linear factors, but one of them is repeating. Even with this repetition Rule I applies:

$$\frac{s}{(s-2)^2(s+1)} = \frac{A}{s-2} + \frac{B}{(s-2)^2} + \frac{C}{s+1}$$

(Note that the factor $s+1$ does not repeat and hence occurs only once on the right side.) As before, we combine the fractions on the right side to obtain

$$\frac{s}{(s-2)^2(s+1)} = \frac{A(s-2)(s+1) + B(s+1) + C(s-2)^2}{(s-2)^2(s+1)}$$

Equating numerators, we get

$$A(s-2)(s+1) + B(s+1) + C(s-2)^2 = s.$$

To determine the constants, we make the appropriate substitutions:

$$s = 2: \qquad 0 + B(3) + 0 = 2 \qquad \text{or} \qquad B = \frac{2}{3}$$

$$s = -1: \qquad 0 + 0 + C(-3)^2 = -1 \qquad \text{or} \qquad C = -\frac{1}{9}$$

At this point we seem to have run out of values to substitute. However, if we **use the values already obtained for B and C,** we can let s be equal to any number, say $s = 0$.

$$A(s-2)(s+1) + \frac{2}{3}(s+1) + \left(-\frac{1}{9}\right)(s-2)^2 = s \qquad \boldsymbol{B = \frac{2}{3}, C = -\frac{1}{9}}$$

$$s = 0: \qquad A(-2)(1) + \frac{2}{3}(1) + \left(-\frac{1}{9}\right)(-2)^2 = 0 \qquad \text{or} \qquad A = \frac{1}{9}$$

It follows that

$$L^{-1}\left\{\frac{s}{(s-2)^2(s+1)}\right\} = L^{-1}\left\{\frac{1}{9}\frac{1}{s-2} + \frac{2}{3}\frac{1}{(s-2)^2} - \frac{1}{9}\frac{1}{s+1}\right\}$$

$$= \frac{1}{9}e^{2t} + \frac{2}{3}te^{2t} - \frac{1}{9}e^{-t}$$

by Transforms 4 and 9, respectively.   ∎

Example 3   (*Distinct quadratic factors.*) Find

$$L^{-1}\left\{\frac{4}{(s-1)(s+1)(s^2+1)}\right\}$$

Solution. Since one of the factors is quadratic, Rule II applies. (The linear factors lead to the usual form by Rule I.) Thus

$$\frac{4}{(s-1)(s+1)(s^2+1)} = \frac{A}{s-1} + \frac{B}{s+1} + \frac{Cs+D}{s^2+1}$$

$$= \frac{A(s+1)(s^2+1) + B(s-1)(s^2+1) + (Cs+D)(s-1)(s+1)}{(s-1)(s+1)(s^2+1)}$$

Equating numerators, we get

$$A(s+1)(s^2+1) + B(s-1)(s^2+1) + (Cs+D)(s-1)(s+1) = 4$$

Once again we substitute convenient values for s:

$s = 1$: $A(2)(2) + 0 + 0 = 4$ $A = 1$

$s = -1$: $0 + B(-2)(2) + 0 = 4$ $B = -1$

To get the remaining coefficients, we use the values already found for A and B:

$$1(s+1)(s^2+1) + (-1)(s-1)(s^2+1) + (Cs+D)(s-1)(s+1) = 4$$

Now, because of the factor $Cs + D$, we let $s = 0$ and solve for D:

$s = 0$: $1(1)(1) + (-1)(-1)(1) + D(-1)(1) = 4$

whence $D = -2$:

$$1(s+1)(s^2+1) + (-1)(s-1)(s^2+1) + (Cs-2)(s-1)(s+1) = 4$$

Finally, to get C, we let $s =$ any value, say $s = 2$:

$s = 2$: $1(3)(5) + (-1)(1)(5) + (2C-2)(1)(3) = 4$ **A = 1, B = −1, D = −2**

yielding $C = 0$. Thus

$$L^{-1}\left\{\frac{4}{(s-1)(s+1)(s^2+1)}\right\} = L^{-1}\left\{\frac{1}{s-1} - \frac{1}{s+1} + \frac{-2}{s^2+1}\right\}$$

$$= e^t - e^{-t} - 2\sin t$$

by Transforms 4 and 5, respectively.   ∎

Example 4   (*Trinomial factor.*) Find

$$L^{-1}\left\{\frac{3s-7}{(s^2-2s+5)(s+2)}\right\}$$

Solution. Since one factor is quadratic and one linear, we get

$$\frac{3s-7}{(s^2-2s+5)(s+2)} = \frac{As+B}{s^2-2s+5} + \frac{C}{s+2}$$

$$= \frac{(As+B)(s+2) + C(s^2-2s+5)}{(s^2-2s+5)(s+2)}$$

Equating numerators, we get

$$(As+B)(s+2) + C(s^2-2s+5) = 3s-7$$

$s = -2$: $C(4+4+5) = -6-7$ or $C = -1$

Next, we use the value for C already found and let $s = 0$ to find B.

$$(As+B)(s+2) - 1(s^2-2s+5) = 3s-7$$

$s = 0$: $B(2) - 1(5) = -7$ or $B = -1$

$$(As-1)(s+2) - 1(s^2-2s+5) = 3s-7$$

Finally, let $s =$ any value, say $s = 1$:

$$s = 1: \quad (A - 1)(3) - 1(4) = -4 \quad \text{or} \quad A = 1 \quad \textbf{C = -1, B = -1}$$

Hence

$$L^{-1}\left\{\frac{3s - 7}{(s^2 - 2s + 5)(s + 2)}\right\} = L^{-1}\left\{\frac{s - 1}{s^2 - 2s + 5} - \frac{1}{s + 2}\right\}$$

$$= L^{-1}\left\{\frac{s - 1}{(s - 1)^2 + 4} - \frac{1}{s + 2}\right\}$$

$$= e^t \cos 2t - e^{-2t}$$

by Transforms 8 and 4, respectively. ∎

■ Exercises / Sections 13.1–13.3

1. Use Definition (13.1) to verify Transforms 1, 5, 3 (for $n = 2$), and 9 (for $n = 1$).

In Exercises 2–13, use the table to find the transforms of the functions.

2. $f(t) = 5e^{2t}$

3. $f(t) = 2 + 3e^{-t}$

4. $f(t) = 1 - \sin t$

5. $f(t) = t + \cos 2t$

6. $f(t) = 2t^2 - 3\cos t$

7. $f(t) = e^{2t} \sin 5t$

8. $f(t) = e^{-2t} \cos 3t$

9. $f(t) = t^3 e^{-4t}$

10. $f(t) = 2t^2 e^{3t}$

11. $f(t) = 2t^4 e^{-t}$

12. $f(t) = 5t + 3e^{2t}$

13. $f(t) = 4 - 5\sin 2t$

In Exercises 14–29, use the table to find the inverse transforms of the functions.

14. $F(s) = \dfrac{3}{s - 5}$

15. $F(s) = \dfrac{10}{s^2 + 4}$

16. $F(s) = \dfrac{s}{s^2 + 7}$

17. $F(s) = \dfrac{1}{(s + 2)^3}$

18. $F(s) = \dfrac{1}{s^2(s^2 + 9)}$

19. $F(s) = \dfrac{4}{(s^2 + 4)^2}$

20. $F(s) = \dfrac{s^2 - 4}{(s^2 + 4)^2}$

21. $F(s) = \dfrac{2s}{(s^2 + 9)^2}$

22. $F(s) = \dfrac{1}{(s - 4)^3}$

23. $F(s) = \dfrac{5s}{s^2 + 6}$

24. $F(s) = \dfrac{5}{s^2 + 9}$

25. $F(s) = \dfrac{2}{s^2(s^2 + 4)}$

26. $F(s) = \dfrac{s}{(s^2 + 16)^2}$

27. $F(s) = \dfrac{2s + 4}{(s + 2)^2 + 4}$

28. $F(s) = \dfrac{3s - 3}{(s - 1)^2 + 16}$

29. $F(s) = \dfrac{1}{(s + 3)^2 + 5}$

In Exercises 30–36, find the inverse transforms of the functions by the method of Example 3 (Section 13.2).

30. $\dfrac{s + 1}{s^2 + 2s + 5}$

31. $\dfrac{\sqrt{10}}{s^2 - 2s + 11}$

32. $\dfrac{6}{s^2 + 2s + 5}$

33. $\dfrac{s}{s^2 - 6s + 10}$

34. $\dfrac{1}{s^2 + 6s + 12}$

35. $\dfrac{s}{s^2 - 2s + 6}$

36. $\dfrac{2s + 1}{s^2 + 4s + 9}$

In Exercises 37–55, use the method of partial fractions to find the inverse transforms of the functions.

37. $\dfrac{1}{s(s + 1)}$

38. $\dfrac{2}{s(s - 1)(s + 1)}$

39. $\dfrac{2s + 1}{(s - 2)(s + 3)}$

40. $\dfrac{s}{(s - 1)(s + 3)}$

41. $\dfrac{s^2}{(s - 2)(s + 2)(s - 4)}$

42. $\dfrac{1}{s^2(s - 2)}$

43. $\dfrac{3s^2}{(s + 2)^2(s - 1)}$

44. $\dfrac{1}{(s + 2)(s^2 + 4)}$

45. $\dfrac{1}{(s + 1)(s^2 + 1)}$

46. $\dfrac{5}{(s - 1)(s^2 + 4)}$

47. $\dfrac{s}{(s + 1)(s^2 + 1)}$

48. $\dfrac{1}{(s + 1)^2(s + 2)}$

49. $\dfrac{9s}{(s + 2)^2(s - 1)}$

50. $\dfrac{2s}{s^4 - 1}$

51. $\dfrac{s}{(s + 1)^2}$

52. $\dfrac{s^2}{(s - 1)^3}$

53. $\dfrac{2s^2 + 2s + 1}{(s^2 + 2s + 2)(s - 1)}$

54. $\dfrac{5}{(s^2 + 2s + 5)(s + 2)}$

55. $\dfrac{1}{(s^2 + 4s + 7)(s + 4)}$

13.4 Solution of Linear Equations by Laplace Transforms

In this section we are finally going to return to differential equations. To see how differential equations may be solved by using Laplace transforms, let us find the transform of $f'(t)$ in terms of the transform of $f(t)$. Assume that $F(s) = L\{f(t)\}$ exists. Then

$$L\{f'(t)\} = \int_0^\infty e^{-st} f'(t) \, dt$$

$u = e^{-st}$	$dv = f'(t) \, dt$	**integration by parts**
$du = -se^{-st} dt$	$v = f(t)$	

It follows that

$$L\{f'(t)\} = \lim_{b \to \infty} \left[e^{-st} f(t) \Big|_0^b + s \int_0^b e^{-st} f(t) \, dt \right]$$

$$= \lim_{b \to \infty} \left[e^{-sb} f(b) - f(0) \right] + s \int_0^\infty e^{-st} f(t) \, dt$$

So, if

$$\lim_{t \to \infty} e^{-st} f(t) = 0$$

then

$$L\{f'(t)\} = -f(0) + s \int_0^\infty e^{-st} f(t) \, dt$$

or

$$L\{f'(t)\} = sL\{f(t)\} - f(0) \tag{13.8}$$

We can see, then, that the transform of $f'(t)$ may be expressed in terms of the transform of $f(t)$ itself. Since we are now interested in differential equations, let us adopt the following notation: if $y = f(t)$, denote the transform of y by $Y(s)$ and $f(0)$ by $y(0)$. Formula (13.8) then becomes

$$L\{y'\} = sY(s) - y(0) \tag{13.9}$$

Now consider the differential equation

$$y' - 2y = e^t \qquad y(0) = 0$$

To solve this equation, we take the Laplace transform of both sides, making use of Formula (13.9):

$$L\{y'\} - 2L\{y\} = L\{e^t\}$$

or

$$sY(s) - y(0) - 2Y(s) = \frac{1}{s - 1} \qquad \textbf{by (13.9)}$$

Since $y(0) = 0$, the initial condition, we have

$$sY(s) - 2Y(s) = \frac{1}{s-1} \qquad \textbf{\textit{y}(0) = 0}$$

Note that Formula (13.9) has "destroyed" the derivative, so that we are left with a simple algebraic equation. Solving this equation for $Y(s)$, we get

$$Y(s) = \frac{1}{(s-2)(s-1)} = \frac{1}{s-2} - \frac{1}{s-1}$$

Hence the inverse transform—namely,

$$y = e^{2t} - e^t$$

must be the solution. This example points out another critical feature of the transform method: since the condition $y(0) = 0$ was used in the third step, no arbitrary constants appear in the solution.

The method for solving equations by Laplace transforms will now be summarized.

Solution of Differential Equations by Laplace Transforms

1. Find the Laplace transform of both sides of the differential equation.
2. Substitute the initial value(s).
3. Solve the resulting algebraic equation for $Y(s)$.
4. Find the inverse transform $y = L^{-1}\{Y(s)\}$.

Repeated use of Formula (13.9) yields derivative formulas for higher derivatives:

$$L\{y''\} = L\{(y')'\} = sL\{y'\} - y'(0) = s[sY(s) - y(0)] - y'(0)$$

or

$$L\{y''\} = s^2Y(s) - sy(0) - y'(0)$$

Similarly,

$$L\{y'''\} = s^3Y(s) - s^2y(0) - sy'(0) - y''(0) \qquad (13.10)$$

and so on.

The first two formulas will now be restated for easy reference.

Laplace Transforms of Derivatives

$$L\{y'\} = sY(s) - y(0) \qquad\qquad (13.11)$$

$$L\{y''\} = s^2Y(s) - sy(0) - y'(0) \qquad\qquad (13.12)$$

Example 1   Solve the equation $y'' - 4y' + 8y = 0$, with $y(0) = 1$, $y'(0) = 0$.

Solution. Taking the transform of both sides, we get

$$L\{y''\} - 4L\{y'\} + 8L\{y\} = L\{0\} = 0$$

or, using Formulas (13.11) and (13.12),

$$s^2 Y(s) - sy(0) - y'(0) - 4[sY(s) - y(0)] + 8Y(s) = 0 \qquad \text{Step 1}$$

Making use of the initial conditions, the last equation reduces to

$$s^2 Y(s) - s - 4s\, Y(s) + 4 + 8Y(s) = 0 \qquad y(0) = 1, y'(0) = 0 \qquad \text{Step 2}$$

We now solve for $Y(s)$:

$$s^2 Y(s) - 4s Y(s) + 8Y(s) = s - 4 \qquad \text{Step 3}$$

Thus

$$(s^2 - 4s + 8)Y(s) = s - 4 \qquad \text{factoring } Y(s)$$

or

$$Y(s) = \frac{s - 4}{s^2 - 4s + 8} \qquad \text{dividing by } (s^2 - 4s + 8)$$

To find the inverse transform, we complete the square in the denominator. Thus

$$Y(s) = \frac{s - 4}{(s - 2)^2 + 4} = \frac{s - 2 - 2}{(s - 2)^2 + 4} \qquad \text{Step 4}$$
$$= \frac{s - 2}{(s - 2)^2 + 4} - \frac{2}{(s - 2)^2 + 4}$$

From the table

$$y = e^{2t} \cos 2t - e^{2t} \sin 2t \qquad\blacksquare$$

Example 2   Solve the differential equation $y'' + 2y' + y = te^{-t}$, with $y(0) = 0$, $y'(0) = -2$.

Solution. Transforming, by Transform 9 we get

$$s^2 Y(s) - sy(0) - y'(0) + 2\,[sY(s) - y(0)] + Y(s) = \frac{1}{(s + 1)^2} \qquad \text{Step 1}$$

Using the initial conditions, we have

$$s^2 Y(s) - (-2) + 2s Y(s) + Y(s) = \frac{1}{(s+1)^2} \qquad y(0) = 0,\, y'(0) = -2 \qquad \textbf{Step 2}$$

$$s^2 Y(s) + 2s Y(s) + Y(s) = \frac{1}{(s+1)^2} - 2 \qquad \textbf{Step 3}$$

$$(s^2 + 2s + 1)Y(s) = \frac{1}{(s+1)^2} - 2$$

$$(s+1)^2 Y(s) = \frac{1}{(s+1)^2} - 2$$

$$Y(s) = \frac{1}{(s+1)^4} - \frac{2}{(s+1)^2}$$

$$= \frac{1}{3!}\frac{3!}{(s+1)^4} - 2 \cdot \frac{1}{(s+1)^2}$$

Finally,

$$y = \frac{1}{6} t^3 e^{-t} - 2t e^{-t} \qquad \textbf{Step 4}$$

by Transform 9 in the table.   ∎

An application that is beyond the scope of the methods of Chapter 12 concerns the phenomenon of resonance. Study the next example.

Example 3   A weight having a mass of 1 slug is attached to a spring whose spring constant is 4 lb/ft. The weight is initially at rest. Find its motion as a function of time, assuming no damping but an external force $f(t) = \sin 2t$.

Solution. The equation is

$$\frac{d^2 x}{dt^2} + 4x = \sin 2t; \qquad x(0) = x'(0) = 0$$

Transforming and solving for $X(s)$, we get

$$s^2 X(s) - sx(0) - x'(0) + 4X(s) = \frac{2}{s^2 + 4}$$

$$s^2 X(s) + 4X(s) = \frac{2}{s^2 + 4}$$

$$X(s)(s^2 + 4) = \frac{2}{s^2 + 4}$$

$$X(s) = \frac{2}{(s^2 + 4)^2}$$

$$= \frac{1}{2^3}\frac{2 \cdot 2^3}{(s^2 + 4)^2} \qquad \textbf{Transform 12}$$

$$x(t) = \frac{1}{2^3}(\sin 2t - 2t \cos 2t)$$

$$= \frac{1}{8}\sin 2t - \frac{1}{4}t \cos 2t$$

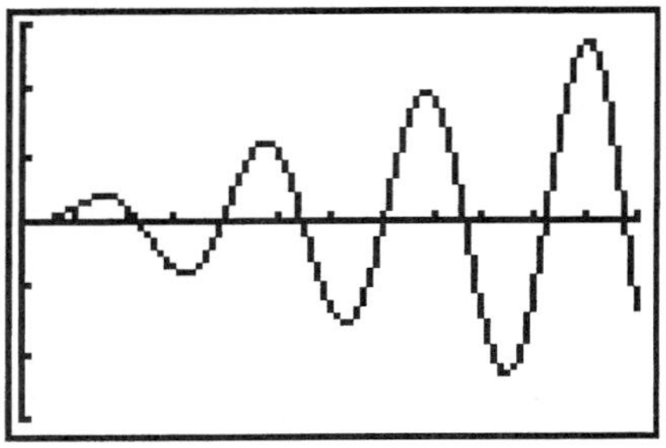

Figure 13.1

Because of the t-factor, the second term in the solution gets ever larger in absolute value. This building up of large amplitudes in the vibration is called *resonance*. (See Figure 13.1.) ∎

■ Exercises / Section 13.4

In Exercises 1–28, solve the differential equations by the method of Laplace transforms.

1. $y' - y = 0,\ y(0) = 1$
2. $y' + 3y = 0,\ y(0) = 2$
3. $y' - 2y = 4,\ y(0) = 0$
4. $y' + 2y = 1,\ y(0) = 1$
5. $y' - 2y = e^{2t},\ y(0) = 0$
6. $y' - 3y = e^{3t},\ y(0) = -2$
7. $y' + 4y = te^{-4t},\ y(0) = 3$
8. $y'' + 4y = 0,\ y(0) = 1,\ y'(0) = 0$
9. $y'' + 9y = 0,\ y(0) = 1,\ y'(0) = -2$
10. $y' - y = 4e^{-3t},\ y(0) = 0$
11. $y'' + y = 2\sin t,\ y(0) = y'(0) = 0$
12. $y'' + 4y = \sin 2t,\ y(0) = 0,\ y'(0) = 1$
13. $y'' + 4y = 2\cos 2t,\ y(0) = -2,\ y'(0) = 0$
14. $y'' + y = \sin t,\ y(0) = 0,\ y'(0) = 1$
15. $y' - y = \cos 2t,\ y(0) = 0$
16. $y' + 6y = 5\sin 3t,\ y(0) = 0$
17. $y'' - 4y' + 4y = e^{3t},\ y(0) = 0,\ y'(0) = -2$
18. $y'' + 6y' + 9y = 25\,e^{2t},\ y(0) = 1,\ y'(0) = 0$
19. $y'' - 6y' + 9y = 12t^2 e^{3t},\ y(0) = y'(0) = 0$
20. $y'' + 6y' + 13y = 0,\ y(0) = 1,\ y'(0) = -2$
21. $y'' - 4y' + 10y = 0,\ y(0) = -3,\ y'(0) = 0$
22. $y'' - 4y' + 6y = 0,\ y(0) = 2,\ y'(0) = 0$
23. $y'' + y = 4e^t,\ y(0) = y'(0) = 0$

24. $y'' - 4y = 4e^{3t},\ y(0) = y'(0) = 0$
25. $y'' - 4y = 3\cos t,\ y(0) = y'(0) = 0$
26. $y'' - y' - 6y = 50\sin t,\ y(0) = y'(0) = 0$
27. $y'' + 2y' + 5y = 8e^t,\ y(0) = y'(0) = 0$
28. $y'' - 4y' + 5y = 4e^t,\ y(0) = 1,\ y'(0) = 0$

29. In an LRC circuit, $L = 0.1$ H, $R = 6.0\,\Omega$, $C = 0.02$ F, and $e(t) = 6.0$ V. Find q as a function of time if $q(0) = 0$ and $i(0) = q'(0) = 0$.

30. A 64-lb weight (2 slugs) stretches a certain spring 2 ft. With this weight attached, the spring is stretched 2 ft below its equilibrium position and released. Find the resulting motion, assuming no damping but an external force $f(t) = 2\sin 4t$.

31. Rework Exercises 1 and 9, Section 12.4, using the method of Laplace transforms.

In the remaining exercises, use a computer algebra system to find the partial fraction expansions.

32. $y'' + 9y = 2\sin 2t,\ y(0) = y'(0) = 0$
33. $y'' + 4y = 5\cos 3t,\ y(0) = y'(0) = 0$
34. $y'' + 2y' + 4y = \cos 2t,\ y(0) = y'(0) = 0$
35. $y'' + 9y = 2\sin 2t,\ y(0) = 0,\ y'(0) = 2$
36. $y'' + 4y' + 9y = 2\cos 3t,\ y(0) = 0,\ y'(0) = -1$
37. $y'' + 4y' + 6y = 3\cos\sqrt{6}t,\ y(0) = -2,\ y'(0) = 0$
38. $y'' + 2y' + 6y = \sin 2t,\ y(0) = y'(0) = 0$

39. A 12-lb weight stretches a spring 2 ft. The weight is pulled 1 ft below the equilibrium position and released. Find the motion of the weight as a function of time assuming no damping but an external force $f(t) = 12\sin t$.

40. A 12-lb weight stretches a spring 2 ft. The weight is pushed 1/5 ft above the equilibrium position and released. Find $x(t)$, given that a damping force numerically equal to 3/4 of the velocity is present, while an external force $f(t) = 12\cos 2t$ acts on the weight.

41. In an LRC-circuit, $L = 0.100$ H, $R = 10.0\,\Omega$, $C = 1.00 \times 10^{-3}$ F, and $e(t) = 10.0\sin 20.0t$. Find q as a function of time, if $q(0) = q'(0) = 0$.

■ Review Exercises / Chapter 13

In Exercises 1–6, use the table to find the transform of each of the given functions.

1. $f(t) = 2e^{-3t}$

2. $f(t) = 1 - 2\cos 2t$

3. $f(t) = 2t^3 + \sin 3t$

4. $f(t) = e^{-3t}\cos 6t$

5. $f(t) = 2t - \sin 2t$

6. $f(t) = 5t^4 e^{3t}$

In Exercises 7–14, find the inverse transform in each case.

7. $F(s) = \dfrac{s-2}{(s-2)^2+5}$

8. $F(s) = \dfrac{1}{(s+3)^4}$

9. $F(s) = \dfrac{s}{s^2-2s+5}$

10. $F(s) = \dfrac{1}{(s^2+4)^2}$

11. $F(s) = \dfrac{s}{(s+1)(s-2)}$

12. $F(s) = \dfrac{s}{(s+1)^2(s+4)}$

13. $F(s) = \dfrac{1}{(s+2)(s-3)(s-4)}$

14. $F(s) = \dfrac{5}{(s+2)(s^2+1)}$

In Exercises 15–24, use the method of Laplace transforms to solve the differential equations.

15. $y' + 2y = 0,\ y(0) = 1$

16. $y' - y = e^t,\ y(0) = 0$

17. $y' + 2y = te^{-2t},\ y(0) = -1$

18. $y'' + 2y = 0,\ y(0) = 0,\ y'(0) = 2$

19. $y'' - 2y' - 3y = 0,\ y(0) = 0,\ y'(0) = -4$

20. $y'' + y = e^{-t},\ y(0) = y'(0) = 0$

21. $y'' + 2y' + 5y = 0,\ y(0) = 1,\ y'(0) = 0$

22. $y'' - 2y' + 5y = 0,\ y(0) = 0,\ y'(0) = 1$

23. $y'' + 2y' + 5y = 3e^{-2t},\ y(0) = 1,\ y'(0) = 1$

24. $y'' + 4y' + 4y = 2\sin t,\ y(0) = y'(0) = 0$

25. A 4-lb weight stretches a spring 6 in. The weight is attached to the spring and allowed to reach the equilibrium position. It is then given an initial upward velocity of 4 ft/s. Find the motion of the weight as a function of time if a damping force numerically equal to twice the velocity is present.

26. A 12-lb weight stretches a certain spring 2 ft. The weight is pushed 6 in. above the equilibrium position and released. Find the motion of the weight as a function of time if an external force $f(t) = (3/4)\cos 4t$ is present and the damping force is negligible.

Tables

Table 1 *Common units of measurement*

	British System			International System (SI)		
Quantity	Name	Symbol		Name	Symbol	In terms of other units
Length	foot	ft		meter	m	
Mass	slug			kilogram	kg	
Force	pound	lb		newton	N	$kg \cdot m/s^2$
Capacity	gallon	gal		liter	L or ℓ	$1L = 1000 \ cm^3$
Pressure		lb/ft^2		pascal	Pa	N/m^2
Work, energy	foot-pound	$ft \cdot lb$		joule	J	$N \cdot m$
Power	horsepower	hp		watt	W	J/s
Current	ampere	A		ampere	A	
Charge	coulomb	C		coulomb	C	
Electric potential	volt	V		volt	V	
Capacitance	farad	F		farad	F	
Inductance	henry	H		henry	H	
Resistance	ohm	Ω		ohm	Ω	
Quantity of heat	British thermal unit	Btu		joule	J	
Temperature	degree Fahrenheit	°F		degree Celsius	°C	
Absolute temperature				kelvin	K	
Time	second	s		second	s	
	minute	min		minute	min	
	hour	h		hour	h	

Table 2 *A short table of integrals*

Forms Containing $a + bu$

---

1. $\displaystyle\int (a + bu)^n\, du = \frac{(a + bu)^{n+1}}{b(n + 1)} + C, n \neq -1$

2. $\displaystyle\int \frac{du}{a + bu} = \frac{1}{b}\ln|a + bu| + C$

3. $\displaystyle\int \frac{u\, du}{a + bu} = \frac{1}{b^2}[(a + bu) - a\ln|a + bu|] + C$

4. $\displaystyle\int \frac{u^2\, du}{a + bu} = \frac{1}{b^3}\left[\frac{1}{2}(a + bu)^2 - 2a(a + bu) + a^2\ln|a + bu|\right] + C$

5. $\displaystyle\int \frac{du}{u(a + bu)} = \frac{1}{a}\ln\left|\frac{u}{a + bu}\right| + C$

6. $\displaystyle\int \frac{du}{u^2(a + bu)} = -\frac{1}{au} + \frac{b}{a^2}\ln\left|\frac{a + bu}{u}\right| + C$

7. $\displaystyle\int \frac{u\, du}{(a + bu)^2} = \frac{1}{b^2}\left(\ln|a + bu| + \frac{a}{a + bu}\right) + C$

Forms Containing $\sqrt{a + bu}$

---

8. $\displaystyle\int u\sqrt{a + bu}\, du = -\frac{2(2a - 3bu)(a + bu)^{3/2}}{15b^2} + C$

9. $\displaystyle\int u^2\sqrt{a + bu}\, du = \frac{2(8a^2 - 12abu + 15b^2u^2)(a + bu)^{3/2}}{105b^3} + C$

10. $\displaystyle\int \frac{u\, du}{\sqrt{a + bu}} = -\frac{2(2a - bu)\sqrt{a + bu}}{3b^2} + C$

11. $\displaystyle\int \frac{u^2 du}{\sqrt{a + bu}} = \frac{2(3b^2u^2 - 4abu + 8a^2)\sqrt{a + bu}}{15b^3} + C$

12. $\displaystyle\int \frac{du}{u\sqrt{a + bu}} = \frac{1}{\sqrt{a}}\ln\left|\frac{\sqrt{a + bu} - \sqrt{a}}{\sqrt{a + bu} + \sqrt{a}}\right| + C, \qquad a > 0$

13. $\displaystyle\int \frac{du}{u\sqrt{a + bu}} = \frac{2}{\sqrt{-a}}\,\text{Arctan}\sqrt{\frac{a + bu}{-a}} + C, \qquad a < 0$

14. $\displaystyle\int \frac{\sqrt{a + bu}\, du}{u} = 2\sqrt{a + bu} + a\int \frac{du}{u\sqrt{a + bu}}$

Forms Containing $a^2 \pm u^2$ and $u^2 \pm a^2$

---

15. $\displaystyle\int \frac{du}{a^2 + u^2} = \frac{1}{a}\,\text{Arctan}\,\frac{u}{a} + C$

16. $\displaystyle\int \frac{du}{a^2 - u^2} = \frac{1}{2a}\ln\left|\frac{a + u}{a - u}\right| + C, \qquad a^2 > u^2$

17. $\displaystyle\int \frac{du}{u^2 - a^2} = \frac{1}{2a}\ln\left|\frac{u - a}{u + a}\right| + C, \qquad a^2 < u^2$

Forms Containing $\sqrt{a^2 - u^2}$

---

18. $\displaystyle\int \sqrt{a^2 - u^2}\, du = \frac{u}{2}\sqrt{a^2 - u^2} + \frac{a^2}{2}\,\text{Arcsin}\,\frac{u}{a} + C$

19. $\displaystyle\int \frac{du}{\sqrt{a^2 - u^2}} = \text{Arcsin}\,\frac{u}{a} + C$

Table 2 *A short table of integrals* (continued)

---

20. $\displaystyle \int \frac{du}{(a^2 - u^2)^{3/2}} = \frac{u}{a^2\sqrt{a^2 - u^2}} + C$

21. $\displaystyle \int \frac{u^2\,du}{\sqrt{a^2 - u^2}} = -\frac{u}{2}\sqrt{a^2 - u^2} + \frac{a^2}{2}\,\text{Arcsin}\,\frac{u}{a} + C$

22. $\displaystyle \int \frac{u^2\,du}{(a^2 - u^2)^{3/2}} = \frac{u}{\sqrt{a^2 - u^2}} - \text{Arcsin}\,\frac{u}{a} + C$

23. $\displaystyle \int \frac{du}{u\sqrt{a^2 - u^2}} = -\frac{1}{a}\ln\left|\frac{a + \sqrt{a^2 - u^2}}{u}\right| + C$

24. $\displaystyle \int \frac{du}{u^2\sqrt{a^2 - u^2}} = -\frac{\sqrt{a^2 - u^2}}{a^2 u} + C$

25. $\displaystyle \int \frac{\sqrt{a^2 - u^2}\,du}{u^2} = -\frac{\sqrt{a^2 - u^2}}{u} - \text{Arcsin}\,\frac{u}{a} + C$

26. $\displaystyle \int \frac{\sqrt{a^2 - u^2}\,du}{u} = \sqrt{a^2 - u^2} - a\ln\left|\frac{a + \sqrt{a^2 - u^2}}{u}\right| + C$

Forms Containing $\sqrt{u^2 \pm a^2}$

---

27. $\displaystyle \int \sqrt{u^2 \pm a^2}\,du = \frac{1}{2}\left[u\sqrt{u^2 \pm a^2} \pm a^2\ln\left|u + \sqrt{u^2 \pm a^2}\right|\right] + C$

28. $\displaystyle \int u^2\sqrt{u^2 \pm a^2}\,du = \frac{1}{8}u(2u^2 \pm a^2)\sqrt{u^2 \pm a^2} - \frac{1}{8}a^4\ln\left|u + \sqrt{u^2 \pm a^2}\right| + C$

29. $\displaystyle \int \frac{\sqrt{u^2 - a^2}}{u}\,du = \sqrt{u^2 - a^2} - a\,\text{Arccos}\,\frac{a}{|u|} + C$

30. $\displaystyle \int \frac{\sqrt{u^2 + a^2}}{u}\,du = \sqrt{u^2 + a^2} - a\ln\left|\frac{a + \sqrt{u^2 + a^2}}{u}\right| + C$

31. $\displaystyle \int \frac{\sqrt{u^2 \pm a^2}}{u^2}\,du = -\frac{\sqrt{u^2 \pm a^2}}{u} + \ln\left|u + \sqrt{u^2 \pm a^2}\right| + C$

32. $\displaystyle \int \frac{du}{\sqrt{u^2 \pm a^2}} = \ln\left|u + \sqrt{u^2 \pm a^2}\right| + C$

33. $\displaystyle \int \frac{du}{u\sqrt{u^2 - a^2}} = \frac{1}{a}\,\text{Arccos}\,\frac{a}{|u|} + C$

34. $\displaystyle \int \frac{du}{u\sqrt{u^2 + a^2}} = \frac{1}{a}\ln\left|\frac{u}{a + \sqrt{u^2 + a^2}}\right| + C$

35. $\displaystyle \int \frac{du}{u^2\sqrt{u^2 \pm a^2}} = -\frac{\left(\pm\sqrt{u^2 \pm a^2}\right)}{a^2 u} + C$

36. $\displaystyle \int \frac{u^2\,du}{\sqrt{u^2 \pm a^2}} = \frac{1}{2}\left(u\sqrt{u^2 \pm a^2} \mp a^2\ln\left|u + \sqrt{u^2 \pm a^2}\right|\right) + C$

37. $\displaystyle \int \frac{du}{(u^2 \pm a^2)^{3/2}} = \frac{\pm u}{a^2\sqrt{u^2 \pm a^2}} + C$

38. $\displaystyle \int \frac{u^2\,du}{(u^2 \pm a^2)^{3/2}} = \frac{-u}{\sqrt{u^2 \pm a^2}} + \ln\left|u + \sqrt{u^2 \pm a^2}\right| + C$

Table 2 *A short table of integrals* (continued)

Exponential and Logarithmic Forms

39. $\displaystyle\int e^u \, du = e^u + C$

40. $\displaystyle\int a^u \, du = \frac{a^u}{\ln a} + C$

41. $\displaystyle\int u e^{au} \, du = \frac{e^{au}}{a^2}(au - 1) + C$

42. $\displaystyle\int u^n e^{au} \, du = \frac{u^n e^{au}}{a} - \frac{n}{a}\int u^{n-1} e^{au} \, du$

43. $\displaystyle\int \frac{e^{au}}{u^n} \, du = -\frac{e^{au}}{(n-1)u^{n-1}} + \frac{a}{n-1}\int \frac{e^{au} \, du}{u^{n-1}}$

44. $\displaystyle\int \ln u \, du = u \ln u - u + C$

45. $\displaystyle\int u^n \ln u \, du = \frac{u^{n+1} \ln u}{n+1} - \frac{u^{n+1}}{(n+1)^2} + C$

46. $\displaystyle\int \frac{du}{u \ln u} = \ln|\ln u| + C$

Trigonometric Forms

47. $\displaystyle\int \sin u \, du = -\cos u + C$

48. $\displaystyle\int \cos u \, du = \sin u + C$

49. $\displaystyle\int \tan u \, du = -\ln|\cos u| + C = \ln|\sec u| + C$

50. $\displaystyle\int \cot u \, du = \ln|\sin u| + C = -\ln|\csc u| + C$

51. $\displaystyle\int \sec u \, du = \ln|\sec u + \tan u| + C$

52. $\displaystyle\int \csc u \, du = \ln|\csc u - \cot u| + C$

53. $\displaystyle\int \sec^2 u \, du = \tan u + C$

54. $\displaystyle\int \csc^2 u \, du = -\cot u + C$

55. $\displaystyle\int \sec u \tan u \, du = \sec u + C$

56. $\displaystyle\int \csc u \cot u \, du = -\csc u + C$

57. $\displaystyle\int \sin^2 u \, du = \frac{1}{2}u - \frac{1}{4}\sin 2u + C$

58. $\displaystyle\int \cos^2 u \, du = \frac{1}{2}u + \frac{1}{4}\sin 2u + C$

59. $\displaystyle\int \sin^n u \cos u \, du = \frac{\sin^{n+1} u}{n+1} + C$

Table 2 *A short table of integrals* (continued)

60. $\displaystyle\int \cos^n u \sin u \, du = -\frac{\cos^{n+1} u}{n+1} + C$

61. $\displaystyle\int \sin mu \sin nu \, du = -\frac{\sin(m+n)u}{2(m+n)} + \frac{\sin(m-n)u}{2(m-n)} + C$

62. $\displaystyle\int \cos mu \cos nu \, du = \frac{\sin(m+n)u}{2(m+n)} + \frac{\sin(m-n)u}{2(m-n)} + C$

63. $\displaystyle\int \sin mu \cos nu \, du = -\frac{\cos(m+n)u}{2(m+n)} - \frac{\cos(m-n)u}{2(m-n)} + C$

64. $\displaystyle\int u \sin u \, du = \sin u - u \cos u + C$

65. $\displaystyle\int u \cos u \, du = \cos u + u \sin u + C$

66. $\displaystyle\int \sin^n u \cos^m u \, du = \frac{\sin^{n+1} u \cos^{m-1} u}{n+m} + \frac{m-1}{n+m} \int \sin^n u \cos^{m-2} u \, du$

67. $\displaystyle\int \sin^n u \, du = -\frac{1}{n} \sin^{n-1} u \cos u + \frac{n-1}{n} \int \sin^{n-2} u \, du$

68. $\displaystyle\int \cos^n u \, du = \frac{1}{n} \cos^{n-1} u \sin u + \frac{n-1}{n} \int \cos^{n-2} u \, du$

69. $\displaystyle\int \tan^n u \, du = \frac{\tan^{n-1} u}{n-1} - \int \tan^{n-2} u \, du$

70. $\displaystyle\int \cot^n u \, du = -\frac{\cot^{n-1} u}{n-1} - \int \cot^{n-2} u \, du$

71. $\displaystyle\int \sec^n u \, du = \frac{\sec^{n-2} u \tan u}{n-1} + \frac{n-2}{n-1} \int \sec^{n-2} u \, du$

72. $\displaystyle\int \csc^n u \, du = -\frac{\csc^{n-2} u \cot u}{n-1} + \frac{n-2}{n-1} \int \csc^{n-2} u \, du$

Other Forms

73. $\displaystyle\int e^{au} \sin bu \, du = \frac{e^{au}(a \sin bu - b \cos bu)}{a^2 + b^2} + C$

74. $\displaystyle\int e^{au} \cos bu \, du = \frac{e^{au}(a \cos bu + b \sin bu)}{a^2 + b^2} + C$

75. $\displaystyle\int \text{Arcsin}\, u \, du = u \,\text{Arcsin}\, u + \sqrt{1 - u^2} + C$

76. $\displaystyle\int \text{Arctan}\, u \, du = u \,\text{Arctan}\, u - \frac{1}{2} \ln(1 + u^2) + C$

B Answers to Selected Exercises

Chapter 1

Section 1.1 (page 3)

1. $\sqrt{13}$ **3.** $4\sqrt{5}$ **5.** $\sqrt{7}$ **7.** $\sqrt{6}$ **9.** $2\sqrt{2}$ **11.** positive in quadrants I and III   **13. a.** y-axis   **b.** x-axis
21. $y^2 - 4x + 4 = 0$ **23.** $(0, 1)$ **25.** $(7, 2)$ **27.** $\left(-\dfrac{9}{2}, 5\right)$

Section 1.2 (page 8)

1. -1 **3.** $\dfrac{3}{2}$ **5.** $\dfrac{6}{5}$ **7.** 2 **9.** undefined   **11.** 0 **13.** $\dfrac{3}{4}$

15. a.

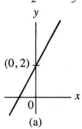

(a)

b.

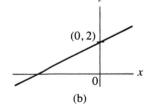

(b)

c.

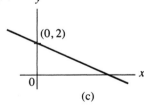

(c)

d.

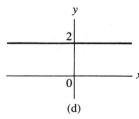

(d)

e.

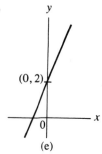

(e)

f.

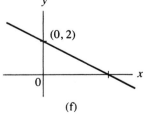

(f)

17. $\dfrac{13}{6}$ **23.** $\dfrac{7}{2}$ **25.** $-\dfrac{1}{5}, -\dfrac{8}{7}, -\dfrac{7}{2}$ **27.** $x = -4$

Section 1.3 (page 12)

1. $x - 2y + 11 = 0$ **3.** $3x - y - 13 = 0$ **5.** $x + 3y = 0$ **7.** $y = 0$ (x-axis)   **9.** $5x + 3y + 3 = 0$ **11.** $x + y - 5 = 0$
13. $x + 8y - 26 = 0$ **15.** $x - y + 10 = 0$ **17.** $x + 3y + 6 = 0$

19. $y = -3x + \dfrac{5}{2}$

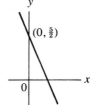

21. $y = \dfrac{2}{3}x + 0$

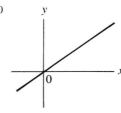

23. $y = 0x + \dfrac{7}{2} = \dfrac{7}{2}$

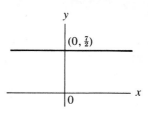

25. parallel   **27.** neither   **29.** perpendicular   **31.** $3x - 4y + 7 = 0$ **33.** $20x + 28y - 27 = 0$

35.

37. $F = 6x$

39. $F = \dfrac{9}{5}C + 32$ **41.** $R = 0.01T + 50$

Section 1.4 (page 21)

In the following answers the intercepts are given first, followed by symmetry, asymptotes, and extent.

1. $y = -1, x = \dfrac{1}{2}$; none; none; all x

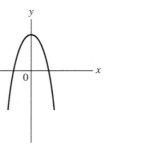

3. $y = -9, x = \pm 3$; y-axis; none; all x

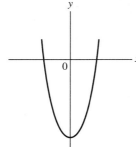

5. $y = 1, x = \pm 1$; y-axis; none; all x

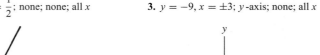

7. $y = 4, x = \pm\sqrt{2}$; y-axis; none; all x

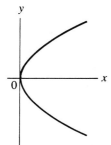

9. origin; x-axis; none; $x \geq 0$

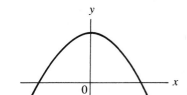

11. $y = \pm 1$, $x = -1$; x-axis; none; $x \geq -1$

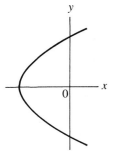

13. $y = -15, x = -5, 3$; none; none; all x

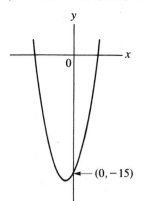

$(0, -15)$

15. $y = 0, x = 0, -3, 2$; none; none; all x

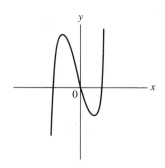

17. $y = 0, x = 0, 1, 2$; none; none; all x

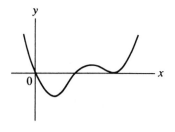

19. $y = 0, x = 0, 1, 2$; none; none; all x

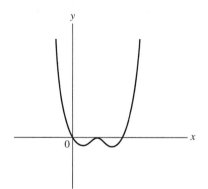

21. $y = 1$; none; $x = -2, y = 0; x \neq -2$

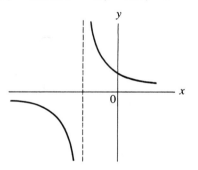

23. $y = 2$; none; $x = 1, y = 0; x \neq 1$

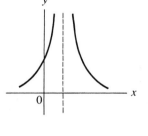

25. $y = 0, x = 0$; none; $x = 1; x \neq 1$

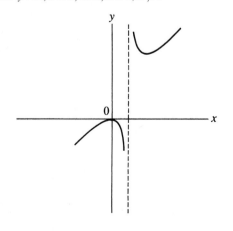

27. $y = -1/2, x = -1$; none; $x = -2, 1, y = 0; x \neq -2, 1$

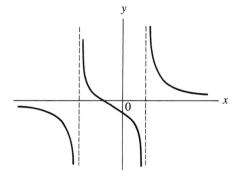

29. $y = 0, x = 0, \pm 1$; origin; none; $-1 \le x \le 1$

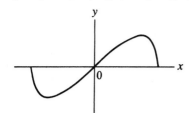

31. $y = 4, x = \pm 2$; y-axis; $x = \pm 1, y = 1$; $x \ne \pm 1$

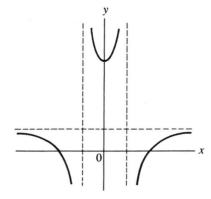

33. $x = -5, 3$; x-axis; none; $x \le -5, x \ge 3$

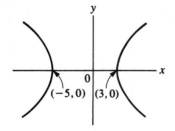

35. $y = 0, x = 0$; x-axis; $x = 2, 3, y = 0$; $0 \le x < 2, x > 3$

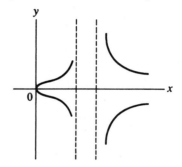

37. $y = \pm 2, x = \pm 2$; both axes; $x = \pm 1,$
$y = \pm 1$; $x \le -2, -1 < x < 1, x \ge 2$

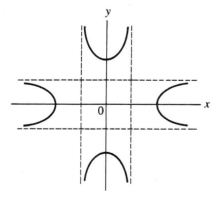

39. $C = 0, C_1 = 0$; none; $C = 10^{-2}$; $C_1 \ge 0$

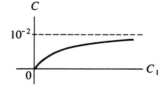

41.

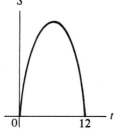

43.

Section 1.5 (page 24)

1. 0, 1, 2

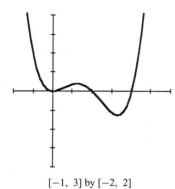

[−1, 3] by [−2, 2]

3. 0, 1, 2

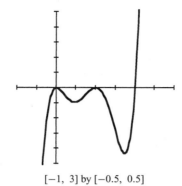

[−1, 3] by [−0.5, 0.5]

5. 0, 2

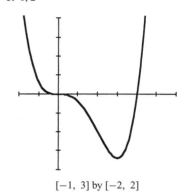

[−1, 3] by [−2, 2]

7. 0, $\sqrt[3]{2}$

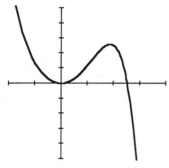

[−1, 2] by [−10, 10]

9. $x \geq 0$

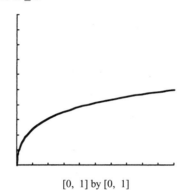

[0, 1] by [0, 1]

11. $x = 0, x > 0$

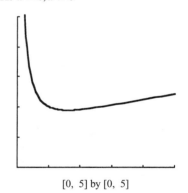

[0, 5] by [0, 5]

13. $x = \pm \dfrac{\sqrt{6}}{2}, x \neq \pm \dfrac{\sqrt{6}}{2}$

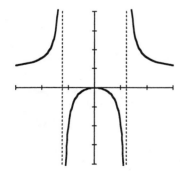

[−3, 3] by [−20, 20]

15. $x = 4, x \geq 0$

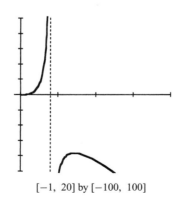

[−1, 20] by [−100, 100]

17. $\left(\dfrac{64}{9}, -1.52 \right)$

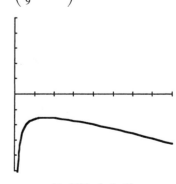

[4, 20] by [−5, 5]

19. 1.80 C

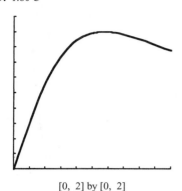

[0, 2] by [0, 2]

21. 2.5 W

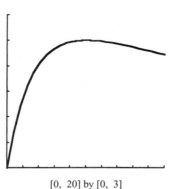

[0, 20] by [0, 3]

23. 0.021 ft

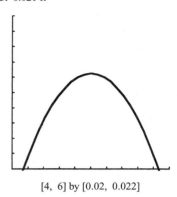

[4, 6] by [0.02, 0.022]

Section 1.7 (page 28)

1. $x^2 + y^2 = 25$ **3.** $x^2 + y^2 = 100$ **5.** $(x + 2)^2 + (y - 5)^2 = 1$ or $x^2 + y^2 + 4x - 10y + 28 = 0$

7. $(x + 1)^2 + (y + 4)^2 = 17$ or $x^2 + y^2 + 2x + 8y = 0$ **9.** $\left(x + \dfrac{1}{2}\right)^2 + \left(y + \dfrac{1}{2}\right)^2 = \dfrac{65}{2}$ or $x^2 + y^2 + x + y - 32 = 0$

11. $(x - 1)^2 + (y - 1)^2 = 4$, $(1, 1)$, $r = 2$ **13.** $(x + 2)^2 + (y - 4)^2 = 16$, $(-2, 4)$, $r = 4$ **15.** $(x + 2)^2 + (y + 1)^2 = 3$, $(-2, -1)$, $r = \sqrt{3}$

17. $(x - 2)^2 + \left(y + \dfrac{1}{2}\right)^2 = 2$, $\left(2, -\dfrac{1}{2}\right)$, $r = \sqrt{2}$ **19.** $(x - 1)^2 + \left(y - \dfrac{3}{2}\right)^2 = 1$, $\left(1, \dfrac{3}{2}\right)$, $r = 1$

21. $(x + 2)^2 + (y - 1)^2 = 9$, $(-2, 1)$, $r = 3$ **23.** $\left(x - \dfrac{1}{2}\right)^2 + (y - 1)^2 = 1$, $\left(\dfrac{1}{2}, 1\right)$, $r = 1$

25. $(x - 2)^2 + \left(y + \dfrac{1}{2}\right)^2 = 2$, $\left(2, -\dfrac{1}{2}\right)$, $r = \sqrt{2}$ **27.** $\left(x + \dfrac{3}{2}\right)^2 + (y + 2)^2 = 5$, $\left(-\dfrac{3}{2}, -2\right)$, $r = \sqrt{5}$

29. point circle, $\left(\dfrac{5}{2}, \dfrac{1}{2}\right)$ **31.** point circle, $(3, -4)$ **33.** imaginary circle   **35.** $x^2 + y^2 = 4.00$, $x^2 + y^2 = 11.6$ **37.** $x^2 + y^2 = 26{,}300^2$

Section 1.8 (page 33)

1. $y^2 = 12x$ **3.** $x^2 = -20y$ **5.** $y^2 = -16x$ **7.** $y^2 = 4x$ **9.** $y^2 = -8x$ **11.** $y^2 = -8x$ **13.** $y^2 = x$, $x^2 = y$
15. $(0, 2)$, $y + 2 = 0$ **17.** $(0, -3,)$, $y - 3 = 0$ **19.** $(4, 0)$, $x + 4 = 0$ **21.** $(-1, 0,)$, $x - 1 = 0$ **23.** $(0, 1)$, $y + 1 = 0$

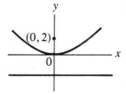
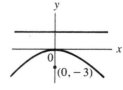
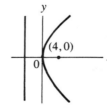
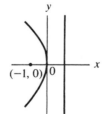
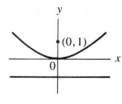

25. $\left(\dfrac{9}{4}, 0\right)$, $x + \dfrac{9}{4} = 0$ **27.** $\left(-\dfrac{1}{4}, 0\right)$, $x - \dfrac{1}{4} = 0$ **29.** $\left(-\dfrac{1}{6}, 0\right)$, $x - \dfrac{1}{6} = 0$ **31.** $x^2 + y^2 - 6y - 27 = 0$

33. $y^2 - 2y - 8x + 17 = 0$ **35.** $x^2 = -y$, $y^2 = -8x$ **37.** 26.3 m   **39.** $200/\sqrt{13} \approx 55.5$ ft   **41.** $2\sqrt{3}$ m   **43.** 4 ft

Section 1.9 (page 39)

1. $(\pm5, 0), (\pm3, 0), 4$

3. $(\pm3, 0), (\pm\sqrt{5}, 0), 2$

5. $(\pm4, 0), (\pm\sqrt{15}, 0), 1$

7. $(0, \pm4), (0, \pm\sqrt{7}), 3$

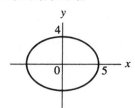
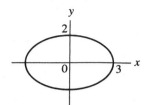
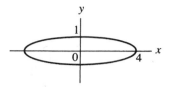
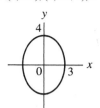

9. $(0, \pm\sqrt{10}), (0, \pm\sqrt{6}), 2$

11. $(0, \pm\sqrt{5}), (0, \pm2), 1$

13. $(\pm\sqrt{6}, 0), (\pm\sqrt{3}, 0), \sqrt{3}$

15. $(0, \pm\sqrt{15}), (0, \pm2\sqrt{2}), \sqrt{7}$

17. $(\pm5, 0), (\pm\sqrt{15}, 0), \sqrt{10}$

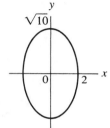
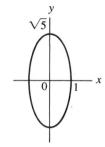
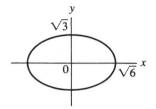

19.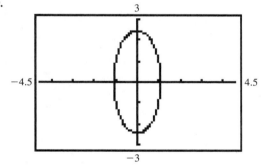

21.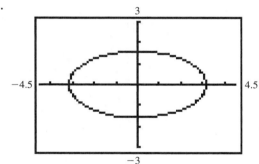

23. $\dfrac{x^2}{16} + \dfrac{y^2}{7} = 1$

25. $4x^2 + 3y^2 = 48$

27. $2x^2 + y^2 = 18$

29. $\dfrac{x^2}{39} + \dfrac{y^2}{64} = 1$

31. $x^2 + 4y^2 = 16$

33. $\dfrac{x^2}{64} + \dfrac{y^2}{28} = 1$

35. $e = 2/3$

37. circle; $x^2 + y^2 - 8x + 12 = 0$

39. $9x^2 + 16y^2 = 36$

41. $\dfrac{12\sqrt{5}}{5} = 5.4$ m

43. $\dfrac{x^2}{16,810,000} + \dfrac{y^2}{16,809,600} = 1$

45. 0.967

Section 1.10 (page 45)

1. $(\pm4, 0), (\pm5, 0)$

3. $(\pm3, 0), (\pm5, 0)$

5. $(0, \pm2), (0, \pm2\sqrt{2})$

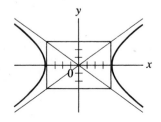
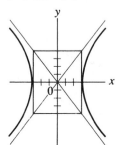
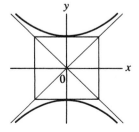

7. $(\pm 1, 0), (\pm\sqrt{6}, 0)$

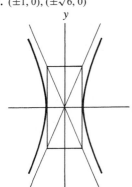

9. $(0, \pm 2\sqrt{3}), (0, \pm 2\sqrt{5})$

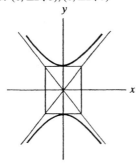

11. $(0, \pm\sqrt{2}), (0, \pm\sqrt{5})$

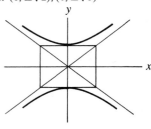

13. $4x^2 - 9y^2 = 36$ **15.** $16y^2 - 9x^2 = 144$ **17.** $\dfrac{y^2}{36} - \dfrac{x^2}{28} = 1$ **19.** $\dfrac{x^2}{16} - \dfrac{y^2}{16} = 1$ **21.** $3x^2 - y^2 = 27$ **23.** $4x^2 - y^2 = 4$

25. $16y^2 - 9x^2 = 144$ **27.** $3x^2 - y^2 + 6x + 4y - 4 = 0$ **29.** $y^2 - 25x^2 = 144$ **31.** $pV = 36$

Section 1.11 (page 50)

1. circle, center at $(1, 2)$, radius $\sqrt{3}$ **3.** parabola, vertex at $(2, -3)$, focus at $(4, -3)$

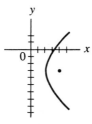

5. $\dfrac{(y+2)^2}{6} - \dfrac{(x+2)^2}{9} = 1$; hyperbola, center at $(-2, -2)$ **7.** $(x+2)^2 + \dfrac{(y - \frac{3}{2})^2}{4} = 1$; ellipse, center at $\left(-2, \dfrac{3}{2}\right)$

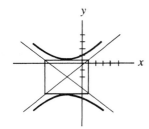

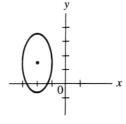

9. $(x+1)^2 + (y-1)^2 = 0$; point: $(-1, 1)$ **11.** $\left(y - \dfrac{5}{2}\right)^2 - \dfrac{x^2}{6} = 1$; hyperbola   **13.** $\left(x - \dfrac{1}{8}\right)^2 + \left(y - \dfrac{3}{4}\right)^2 = 1$; circle, center at $\left(\dfrac{1}{8}, \dfrac{3}{4}\right)$

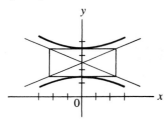

15. $3(x-3)^2 + (y+1)^2 = -1$; imaginary locus

17. $(x + 1)^2 = 12(y - 2)$; parabola, vertex at $(-1, 2)$, focus at $(-1, 5)$

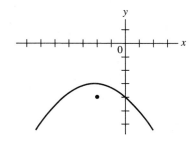

19. $\dfrac{(x + 3)^2}{2} + (y - 1)^2 = 1$; ellipse, center at $(-3, 1)$

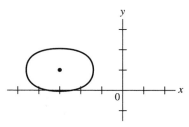

21. $(x + 2)^2 = -4(y + 3)$; parabola, vertex at $(-2, -3)$, focus at $(-2, -4)$

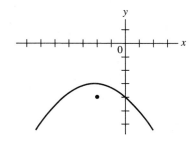

23. $\dfrac{(x - 2)^2}{8} + \dfrac{(y + 3)^2}{4} = 1$; ellipse, center at $(2, -3)$

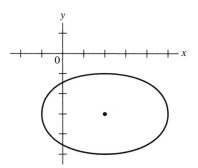

25. $(y - 2)^2 = 16(x + 1)$ **27.** $\dfrac{(x + 3)^2}{9} + \dfrac{y^2}{4} = 1$ **29.** $\dfrac{(x + 3)^2}{16} - \dfrac{(y - 1)^2}{20} = 1$ **31.** $\dfrac{(x - 2)^2}{25} + \dfrac{(y - 3)^2}{4} = 1$

33. $\dfrac{(x - 1)^2}{4} - y^2 = 1$ **35.** $\dfrac{(x + 3)^2}{9} + \dfrac{(y - 1)^2}{25} = 1$ **37.** $(y + 2)^2 = -12(x - 4)$ **39.** $(x + 2)^2 = -16(y + 4)$

41. $\dfrac{(y - 1)^2}{4} - \dfrac{(x + 1)^2}{5} = 1$

Review Exercises for Chapter 1 (page 51)

2. $x + y = 3$ **3.** -40 **6.** $2x - y - 7 = 0$ **7.** $3x + y - 2 = 0$ **9.** $x^2 + y^2 - 2x + 4y = 0$ **10.** $x^2 + y^2 - 4x - 10y + 4 = 0$
11. $(-1, -1), \sqrt{2}$ **13.** ellipse, $(0, \pm 4), (0, \pm\sqrt{7})$ **14.** ellipse, $(\pm 1, 0), (\pm\sqrt{3}/2, 0)$ **15.** hyperbola, $(0, \pm 2), (0, \pm\sqrt{11})$

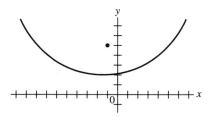

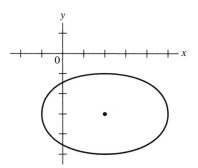

16. hyperbola, $(\pm 3, 0), (\pm 5, 0)$

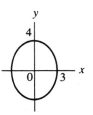

17. parabola, $(0, 0), \left(-\dfrac{3}{4}, 0\right)$

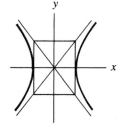

18. parabola, $(0, 0), \left(0, \dfrac{9}{4}\right)$

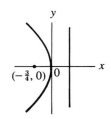

19. $(y+3)^2 = -4(x-2), (2, -3)$, parabola

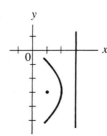

20. $(x-4)^2 + (y+5)^2 = 45$, circle, center at $(4, -5)$, $r = 3\sqrt{5}$

21. $\dfrac{(x-2)^2}{9} + \dfrac{(y+1)^2}{16} = 1, (2, -1)$, ellipse

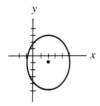

22. $(x+2)^2 = -8(y-3), (-2, 3)$, parabola

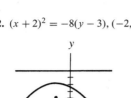

23. $\dfrac{(x-2)^2}{9} - \dfrac{(y-4)^2}{9} = 1, (2, 4)$ hyperbola

24. $y^2 = 8x$ **25.** $(x-1)^2 = 12(y-3)$ **26.** $\dfrac{(x+2)^2}{12} + \dfrac{(y+4)^2}{16} = 1$ **27.** $\dfrac{x^2}{7} + \dfrac{y^2}{16} = 1$ **28.** $9x^2 - 16y^2 = 81$

29. $\dfrac{(y-2)^2}{9} - \dfrac{x^2}{7} = 1$ **30.** $(y-2)^2 = -16(x+4)$ **31.** $x^2 = 4(y-2)$ **32.** $\dfrac{(y+1)^2}{4} - \dfrac{(x-2)^2}{5} = 1$

33. $\dfrac{(x-4)^2}{12} + \dfrac{(y+1)^2}{16} = 1$

34. intercepts: $x = -1, y = 1$; symmetry: none

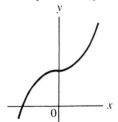

35. intercepts: $x = -1, y = 1$; symmetry: none

36. intercepts: $x = 0, \pm\sqrt{2}, y = 0$; symmetry: origin

37. intercepts: $x = 0, 4, y = 0$; symmetry: x-axis; extent: $x \le 0, x \ge 4$

$(4, 0)$

38. intercept: $(0, 0)$; symmetry: y-axis; asymptote: $y = 1$; extent: all x

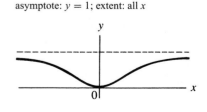

39. intercept: $(0, 0)$; symmetry: origin; asymptotes: $x = \pm 2$, $y = 0$; extent: $x \neq 2, -2$

40. intercepts: $x = 0, -1$, $y = 0$; symmetry: none; extent: $x \geq -1$

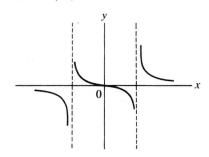

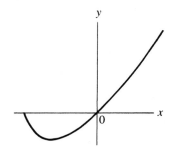

41. 0.10 ft from vertex   **42.** 12.0 ft

43. $V = x(6 - 2x)^2$, $x = 1$ in.   **44.** 144 ft   **45.** 0.94

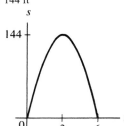

Chapter 2

Section 2.1 (page 58)

1. $A = x^2$ **3.** $V = \frac{2}{3}\pi r^2$ **5.** $R_T = 10 + R$ **7.** $d = 50t$ **9.** $C = 600 + 100(t - 3)$ **11.** $L > 0$ **13.** $(-\infty, \infty)$; $(-\infty, \infty)$

15. $[2, \infty)$; $[0, \infty)$ **17.** $[-2, 2]$; $[0, 2]$ **19.** $x = 0$ and $x \geq 3$; $[0, \infty)$ **21.** $x = 1$ and $x \geq 2$; $[0, \infty)$ **23.** all x except $x = 1$; $y \neq 0$

25. $[-1, 1]$; $[1, 2]$ **27.** $4, 4$ **29.** $0, 12$ **31.** $3, 15$ **33.** $1, 2, -7$ **35.** $\frac{1}{3}, \frac{1}{a}$ **37.** $\sqrt{a^4 - 1}, \sqrt{x^2 - 2x}$

39. $1 - x^2 - 2x\,\Delta x - (\Delta x)^2$, $1 - x^2 + 2x\,\Delta x - (\Delta x)^2$ **41. a.** $x^2 + 2x + 1$ **b.** $x^2 + 1$ **c.** x^4

43. $-1, 1, 1, 1$

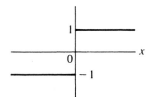

Section 2.2 (page 65)

1. 1   **3.** 1   **5.** 0   **7.** 4; $y = \dfrac{x^2 + 4x}{x}$ and $y = x + 4$ **9.** 5; $y = \dfrac{x^2 - 3x - 4}{x - 4}$ and $y = x + 1$ **11.** 0   **13.** 0   **15.** 4   **17.** -1

19. -8 **21.** 0   **23.** 5   **25.** $\dfrac{3}{2}$ **27.** 0   **29.** -6 **31.** 10   **33.** 2   **35.** $\dfrac{11}{2}$ **37.** $\dfrac{4}{5}$ **39.** $\dfrac{3}{4}$ **41.** 0   **43.** $\dfrac{1}{3}$

45. 2   **47.** 1   **49.** 0   **51. a.** ∞ **b.** 0

Section 2.4 (page 74)

1. $y' = 2$ **3.** $y' = -3$ **5.** $y' = 2x$ **7.** $y' = 4x - 1$ **9.** $y' = 3x^2 - 6x$ **11.** $y' = -\dfrac{1}{(x+1)^2}$ **13.** $y' = -\dfrac{2}{x^3}$

15. $y' = \dfrac{2x}{(1-x^2)^2}$ **17.** $y' = \dfrac{1}{2\sqrt{x}}$ **19.** $y' = -\dfrac{1}{2\sqrt{1-x}}$ **21.** $-2, 4, 6$ **23.** $-\dfrac{1}{16}, -\dfrac{1}{54}$

25. $y = 4x - 5$ **27.** $y = \dfrac{1}{2}(4-x)$

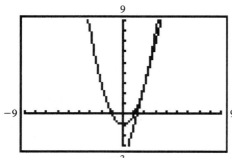

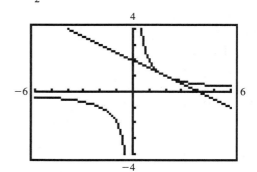

Section 2.5 (page 78)

1. $y' = 1$ **3.** $y' = 2x$ **5.** $y' = 2x + 1$ **7.** $y' = 6x + 4$ **9.** $y' = 15x^2 - 14x$ **11.** $y' = 21x^2 - 2x - 1$ **13.** $y' = x^2 + x + 1$

15. $y' = x - x^2$ **17.** $y' = 200x^9 - 144x^5 + 6x^2$ **19.** $\dfrac{dy}{dt} = \dfrac{7}{5}t^6 - \dfrac{5}{\sqrt{2}}t^4$ **21.** $\dfrac{dy}{dR} = R^5 + \dfrac{4}{5}R^3$

23. $y = -4x + 8, \ y = \dfrac{1}{4}(x-2)$ **25.** $y = 2x - \dfrac{2}{3}, \ y = \dfrac{1}{6}(-3x + 11)$

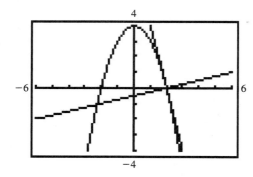

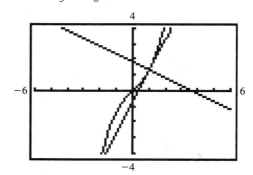

Section 2.6 (page 81)

1. $v = 4t, a = 4, t = 0$ **3.** $v = 2t - 2, a = 2, t = 1$ **5.** $v = 2 - 2t, a = -2, t = 1$ **7.** $v = 12 - 4t, a = -4, t = 3$

9. $v = 9t^2 + 4t, a = 18t + 4, t = 0$ and $t = -\dfrac{4}{9}$ **11.** 1000 m/s   **13.** 33.4 m/s   **15.** 4 cm^2 per cm   **17.** 2.95 $\Omega/°$C   **19.** 63 W/A

21. -0.025 lb/°F   **23.** $\omega = 6$ rad/s, $\alpha = 6$ rad/s^2 **25.** 2.2 A   **27.** 0.20 A   **29.** 22 J/s = 22 W (watts)   **31.** $\beta = \dfrac{1}{2P}$

Section 2.7 (page 88)

The given expressions are equal to dy/dx unless stated otherwise.

Group A   **1.** $16x^3 - 8x$ **3.** $-\dfrac{1}{x^2}$ **5.** $5x^4 + 9x^{-4} - 4x^{-3}$ **7.** $\dfrac{5}{2}x\sqrt{x} - \dfrac{1}{2x\sqrt{x}}$ **9.** $-\dfrac{8}{x^5} + \dfrac{2}{x^{3/2}}$ **11.** $16x(2x^2 - 3)^3$

13. $-24x(4 - 3x^2)^3$ **15.** $100x^9(x^{10} + 1)^9$ **17.** $-\dfrac{3(x^2 - 1)}{(x^3 - 3x)^{3/2}}$ **19.** $\dfrac{dv}{dt} = \dfrac{t^2}{(t^3 - 3)^{2/3}}$ **21.** $x^2(x + 1)(5x + 3)$

23. $4x^3(x+2)(3x+4)$ **25.** $2x(x^2-5)(3x^2-5)$ **27.** $20(x+5)^3(x+1)$ **29.** $12x^2(x-2)^2(x-1)$ **31.** $\dfrac{-1}{(x-1)^2}$

33. $\dfrac{dP}{dt} = \dfrac{-t^2+4t+4}{(t^2+4)^2}$ **35.** $\dfrac{5x^2+4x}{2\sqrt{x+1}}$ **37.** $\dfrac{2(x^2+1)}{\sqrt{x^2+2}}$ **39.** $\dfrac{dR}{ds} = \dfrac{s^2-4s+3}{(s-2)^2}$

Group B   **1.** $-\dfrac{1}{2\sqrt{1-x}}$ **3.** $-\dfrac{2x}{3(1-x^2)^{2/3}}$ **5.** $\dfrac{2x-1}{(x-x^2)^{5/4}}$ **7.** $\dfrac{dn}{dm} = \dfrac{m^4-26m^2-16}{(m^2-8)^2}$ **9.** $\dfrac{2x^2-1}{\sqrt{x^2-1}}$ **11.** $\dfrac{2-3x}{\sqrt{1-x}}$

13. $\dfrac{dT}{d\theta} = \dfrac{7\theta^2(\theta+6)}{2\sqrt{\theta+7}}$ **15.** $-\dfrac{x+4}{2(x-4)^2\sqrt{x}}$ **17.** $\dfrac{3x^2+4x}{2(x+1)\sqrt{x+1}}$ **19.** $\dfrac{2-x^2}{x^3\sqrt{x^2-1}}$ **21.** $\dfrac{x\sqrt{x}(x^2+15)}{2(x^2+3)^2}$ **23.** $\dfrac{3x-5}{2\sqrt{x-2}}$

25. $\dfrac{2x^2+9x-6}{2(2x+3)^2\sqrt{x-1}}$ **27.** $\dfrac{2x^3-30x}{(x+3)^{3/2}(x-5)^{1/2}}$ **29.** 18   **31.** $\dfrac{dZ}{dX} = \dfrac{X}{\sqrt{16+X^2}}$ **33.** $t = 2$ s

Section 2.8 (page 94)

1. $\dfrac{dy}{dx} = -\dfrac{2}{3}$ **3.** $\dfrac{dy}{dx} = \dfrac{x}{y}$ **5.** $\dfrac{dy}{dx} = \dfrac{2x}{3y}$ **7.** $\dfrac{dy}{dx} = -\dfrac{x}{3y^2}$ **9.** $\dfrac{dy}{dx} = \dfrac{9x^2}{16y^3}$ **11.** $\dfrac{dy}{dx} = \dfrac{1-10x}{18y^2}$ **13.** $\dfrac{dy}{dx} = -\dfrac{b^2x}{a^2y}$

15. $\dfrac{dy}{dx} = -\dfrac{y}{x}$ **17.** $\dfrac{dy}{dx} = -\dfrac{2y}{x}$ **19.** $\dfrac{dy}{dx} = -\dfrac{2x+2xy^2+1}{2x^2y}$ **21.** $\dfrac{dy}{dx} = \dfrac{8xy^2-3x^2}{2y-8x^2y}$ **23.** $\dfrac{dy}{dx} = \dfrac{6x^2-10xy^3}{15x^2y^2-4y^3}$

25. $\dfrac{dy}{dx} = -\dfrac{4x^3y^4+5}{4x^4y^3-6y}$

27. $y = \dfrac{1}{4}(x-5), \; y = -4x+3$ **29.** $y = x-1, \; y = -x-3$

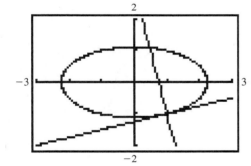

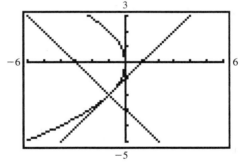

31. $y = -x-1, \; y = x+3$ **33.** $y = \dfrac{1}{3}(-4x-17), \; y = \dfrac{1}{4}(3x-6)$

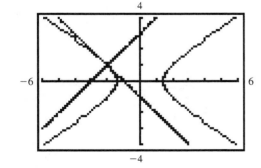

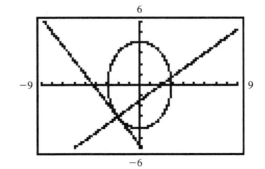

Section 2.9 (page 96)

1. $y'' = 60x^2+30x$ **3.** $y'' = \dfrac{-1}{4(x-1)^{3/2}}$ **5.** $\dfrac{d^3y}{dx^3} = 120x^3-120x^2-24x$ **7.** $f'''(x) = \dfrac{3}{8(5+x)^{5/2}}$ **9.** $\dfrac{d^2y}{dx^2} = \dfrac{48}{(3-2x)^3}$

11. $-\dfrac{4}{y^3}$ **13.** $-\dfrac{4}{y^3}$ **15.** $-\dfrac{4p^2}{y^3}$ **17.** $\dfrac{1}{2x^{3/2}}$

Review Exercises for Chapter 2 (page 96)

1. $-1, 0, 1$ **2.** $\sqrt[3]{x^2 + 2}, x^{2/3} + 2$ **3.** $0, 0, 2$, undefined   **4.** $[0, 1), (1, \infty)$; 0 and 2   **5. a.** $x \geq 1$; $y \geq 0$ **b.** all x and y **6.** 18

7. 8   **8.** 8   **9.** 3   **10.** 2   **11. a.** -3 **b.** -3 **12.** 7   **13.** $\dfrac{1}{2}$ **14.** 2   **15.** 2   **16.** 0   **17.** 0   **18. a.** 2   **b.** 1

19. $1 - 6x$ **20. a.** $3x^2$ **b.** $-\dfrac{2}{x^2}$ **21. a.** $\dfrac{1}{(4-x)^2}$ **b.** $\dfrac{1}{2\sqrt{x}}$ **22.** $-\dfrac{1}{2\sqrt{3-x}}$ **23.** $12x^2(x^3 - 2)^3$ **24.** $-\dfrac{4x^3}{(x^4+3)^2}$

25. $\dfrac{5}{(x+1)^2}$ **26.** $\dfrac{5x^4}{2(7-x^5)^{3/2}}$ **27.** $\dfrac{8x-x^3}{(4-x^2)^{3/2}}$ **28.** $(x^2+1)(5x^2-12x+1)$ **29.** $\dfrac{4-2x^2}{\sqrt{4-x^2}}$ **30.** $\dfrac{dy}{dx} = -\dfrac{2x+3}{2y}$

31. $\dfrac{dy}{dx} = -\dfrac{2xy + y^2}{x^2 + 2xy + 3y^2}$ **32.** $\dfrac{dy}{dx} = \dfrac{1 - 4xy^2 + 4y}{4x^2 y - 4x}$ **33.** $y'' = -\dfrac{2xy}{(3y^2 - x)^3}$ **34.** $-\dfrac{9}{2}$ **35.** $-\dfrac{14}{5}$ **36.** $\dfrac{3}{8}(x+3)^{-5/2}$

37. a. 2   **b.** 1   **38.** 0   **39.** $-\dfrac{2k}{r^3}$ **41.** 0.21 V/s   **42.** $\dfrac{dc}{dT} = 0.65 + 0.000038T$ **43.** 30¢ per widget   **44.** $t = 50$ s

45. -0.0158 dyne/cm

Chapter 3

Section 3.1 (page 102)

1. Increasing on $(-\infty, -1]$, decreasing on $[-1, \infty)$ **3.** Increasing on $(-\infty, -2]$, decreasing on $[-2, 2]$, increasing on $[2, \infty)$
5. min. at $x = 1$ **7.** max. at $x = -1$ **9.** min. at $x = 1$, max. at $x = -2$ **11.** min. at $x = 1$, max. at $x = 3$

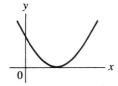

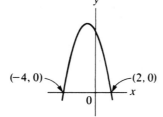

(−4, 0)   (2, 0)

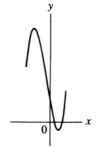

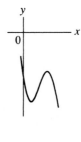

13. min. at $x = \pm 1$, max. at $x = 0$ **15.** critical values: $x = 0, -1$; min. at $x = -1$ **17.** critical values: $x = 0, -1$; max. at $x = -1$

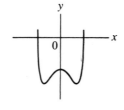

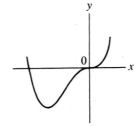

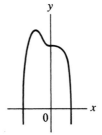

19. no min. or max., vertical tangent at $(0, 0)$ **21.** 320 m   **23.** 2 Ω

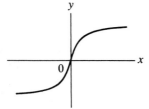

Section 3.2 (page 109)

1. Concave down everywhere   **3.** Concave down on $(-\infty, 0]$, concave up on $[0, \infty)$ **5.** Concave up on $(-\infty, 0]$, concave down on $[0, 2]$, concave up on $[2, \infty)$ **7.** Concave down on $(-\infty, 3]$, concave up on $[3, \infty)$

9. min.: $(1, -2)$; concave up everywhere

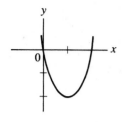

11. max.: $(1/2, 3/2)$; concave down everywhere

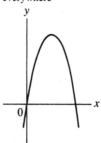

13. max.: $(-3, 1/2)$; concave down everywhere

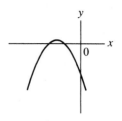

15. max.: $(-1, 5)$, min.: $(1, -3)$; concave up on $[0, \infty)$, concave down on $(-\infty, 0]$; $(0, 1)$ is an inflection point

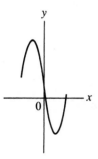

17. max.: $(1, 1)$, min.: $(3, -3)$; concave up on $[2, \infty)$, concave down on $(-\infty, 2]$; $(2, -1)$ is an inflection point

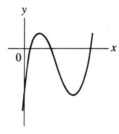

19. max.: $(-1, 16)$, min.: $(3, -16)$; concave up on $[1, \infty)$, concave down on $(-\infty, 1]$; $(1, 0)$ is an inflection point

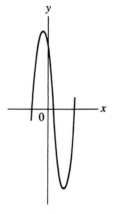

21. max. at $x = 1$, min. at $x = -1$; concave up, $x \le 0$, concave down, $x \ge 0$; inflection point: $(0, 2)$

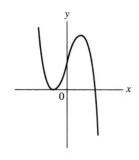

23. min. at $x = 1$; concave up on $(-\infty, 0]$ and $[2/3, \infty)$, concave down on $[0, 2/3]$; inflection points at $x = 0, 2/3$

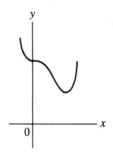

25. min. at $x = -3/4$; concave up on $(-\infty, -1/2]$ and $[0, \infty)$, concave down on $[-1/2, 0]$; inflection points at $x = -1/2, 0$

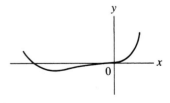

27. min. at $x = 3$; concave up everywhere

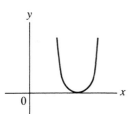

29. intercept $(0, 0)$; vertical asymptote: $x = 3$; horizontal: $y = 1$; no min. or max.; concave down, $x < 3$, concave up, $x > 3$

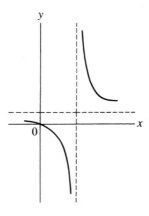

31. vertical asymptote: y-axis; asymptotic curve: $y = x^2$; min. at $x = \sqrt[3]{4}$; concave up on $(-\infty, -2]$, concave down on $[-2, 0)$, concave up on $(0, \infty)$; inflection point at $x = -2$

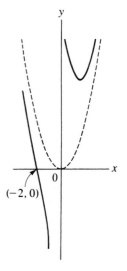

33. vertical asymptote: $x = 2$; horizontal asymptote: $y = 1$; no min. or max.; decreasing everywhere; concave down, $x < 2$, concave up, $x > 2$

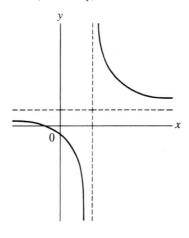

35. domain: $x \geq 0$; vertical tangent at $(0, 0)$; min. at $x = 0$; concave down

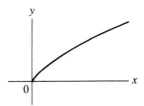

37. no min. or max.; vertical tangent at $(3, 0)$; concave down, $x \geq 3$, concave up, $x \leq 3$; inflection point: $(3, 0)$

39. vertical asymptote: $x = -1$; max. at $x = 1$; concave down on
$(-\infty, -1)$, concave down on $(-1, 2]$, concave up on $[2, \infty)$;
inflection point at $x = 2$

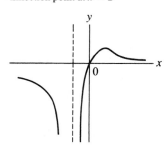

41. max. at $x = \sqrt{3}$, min. at $x = -\sqrt{3}$; inflection points at $x = 0, \pm 3$

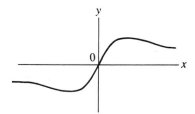

Section 3.3 (page 113)

1. critical point: $(0, 1)$; inflection point: $(0, 1)$

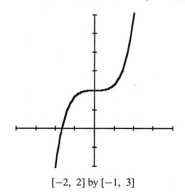

$[-2, 2]$ by $[-1, 3]$

3. min. at $x = 0$, max. at $x = \sqrt[3]{2}$;
inflection point at $x = 1/\sqrt[3]{2}$

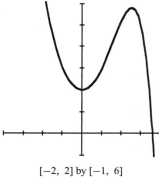

$[-2, 2]$ by $[-1, 6]$

5. min.: $(0, -1.0)$, max.: $(-1, 0.80)$;
inflection point: $(-0.63, 0.071)$

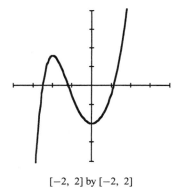

$[-2, 2]$ by $[-2, 2]$

7. min.: $(0, 1.5)$

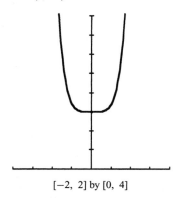

$[-2, 2]$ by $[0, 4]$

9. min.: $(0, 0.20)$, max.: $(\pm 1.41, 2.2)$;
inflection points: $(\pm 1.1, 1.5)$

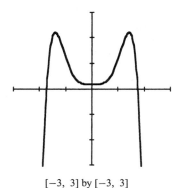

$[-3, 3]$ by $[-3, 3]$

11. max.: $(-\sqrt{2}, 2.26)$, min.: $(\sqrt{2}, -2.26)$;
inflection points: $(-1, 1.4)$,
$(1, -1.4)$, $(0, 0)$

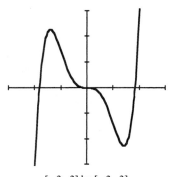

$[-3, 3]$ by $[-3, 3]$

13. max.: $(-1, 2/35)$, min.: $(1, -2/35)$;
inflection points: $(-\sqrt{6}/3, 0.038)$,
$(\sqrt{6}/3, -0.038)$, $(0, 0)$

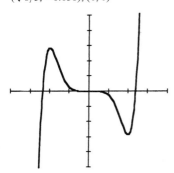

$[-2, 2]$ by $[-0.1, 0.1]$

15. y-axis; min.: $(1.41, 5.06)$

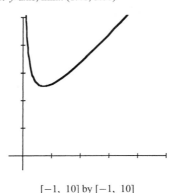

$[-1, 10]$ by $[-1, 10]$

17. $x = \pm 1$; no min. or max.

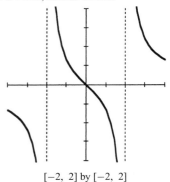

$[-2, 2]$ by $[-2, 2]$

19. $x = \pm\sqrt{3}$; max.: $(-3, -9/2)$,
min.: $(3, 9/2)$

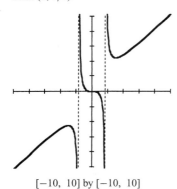

$[-10, 10]$ by $[-10, 10]$

21. y-axis; max.: $(-1.12, -2.79)$,
min.: $(1.12, 2.79)$

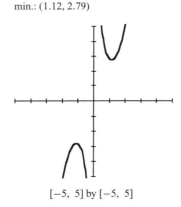

$[-5, 5]$ by $[-5, 5]$

23. y-axis; no min. or max.

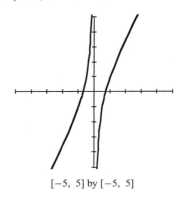

$[-5, 5]$ by $[-5, 5]$

Section 3.4 (page 118)

1. 17.3 W   **3.** 20 in.   **7.** $\frac{1}{2}$ C   **9.** 250 units   **11.** 30, 30   **13.** 20 m × 10 m   **15.** 150 m × 75 m   **17.** 20 ft × 80 ft

19. 2 cm   **21.** width $= \sqrt{3}$ ft, depth $= \sqrt{6}$ ft   **23.** base: 4 in. × 4 in.; height: 2 in.   **25.** $r = \dfrac{5}{4 + \pi}$ m   **27.** $h = r$

29. $(2, 1)$ **31.** $x = \sqrt{3}, y = 6$ **33.** 35   **35.** 6 in. × 18 in. × 4.5 in.   **37.** 18 in. × 18 in. × 36 in.   **39.** 8 in. × 16 in.

41. 8 km from the nearest point   **43.** $\dfrac{1}{3}$ height of cone

Section 3.5 (page 127)

1. 4   **3.** $-\dfrac{9}{4}$ **5.** -2 A/s   **7.** 80 W/s   **9.** Rate of change $= -\dfrac{8}{3}$ m/min   **11.** $\dfrac{1}{20\pi} = 0.0159$ cm/min   **13.** 53 Pa/min

15. 10.1 ft/s   **17.** 210 km/h   **19.** 25 km/h   **21.** $-\dfrac{1}{4}$ unit/min   **23.** $\dfrac{2\sqrt{13}}{3} = 2.4$ m/min   **25.** $\dfrac{3}{16\pi} = 0.060$ ft/s   **27.** 14 ft/s

29. 2.6 m/s   **31.** $\dfrac{4}{\pi} = 1.3$ ft/min   **33.** $\dfrac{4}{75} = 0.053$ m/min   **35.** $\dfrac{6}{25} \dfrac{\text{ft}}{\text{min}}$

Section 3.6 (page 131)

1. $dy = (3x^2 - 1)\, dx$ **3.** $dy = -\dfrac{dx}{(x - 1)^2}$ **5.** $\Delta y = 0.31, dy = 0.3$ **7.** 0.24 cm^2, 0.67%   **9.** ± 0.75 in.^{3}, 0.6%

11. 3%, 2%   **13.** ± 0.07 s, 2.5%   **15.** 4.2 W   **17.** $dA = 2\pi r\, dr$ ⊏━━━⊐ dr
$2\pi r$

Review Exercises for Chapter 3 (page 132)

1. $x - 2y - 1 = 0$; $2x + y - 7 = 0$

3. min. at $x = 2$; concave up everywhere

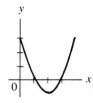

2. $2x + 9y + 23 = 0$; $9x - 2y - 24 = 0$

4. min. at $x = 3$, max. at $x = 1$; concave down on $(-\infty, 2]$, concave up on $[2, \infty)$; inflection point at $x = 2$

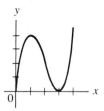

5. min. at $x = -2$, max. at $x = 2$; concave up on $(-\infty, 0]$, concave down on $[0, \infty)$; inflection point at $x = 0$

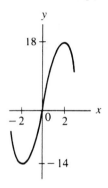

6. min. at $x = 2$; concave up on $(-\infty, 0]$ and $[4/3, \infty)$, concave down on $[0, 4/3]$; inflection points at $x = 0$ and $x = 4/3$

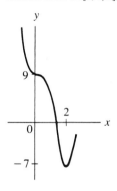

7. min. at $x = 1$; concave up on $(-\infty, 0]$ and $[2/3, \infty)$, concave down on $[0, 2/3]$; inflection points at $x = 0$ and $x = 2/3$

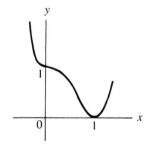

8. max.: $(0, 0)$; concave down on $(-1, 1)$, concave up on $(-\infty, -1)$ and $(1, \infty)$; no inflection points; vertical asymptotes: $x = \pm 1$, horizontal: $y = 1$

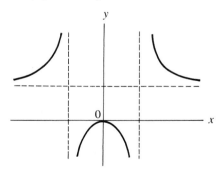

9. min. at $x = -1/\sqrt[3]{2}$; concave up on $(-\infty, 0)$ and $[1, \infty)$, concave down on $(0, 1]$; inflection point: $(1, 0)$; vertical asymptote: $x = 0$, asymptotic curve: $y = x^2$

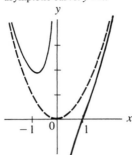

10. max. at $x = \sqrt{3}$, min. at $x = -\sqrt{3}$; concave up on $[-\sqrt{6}, 0)$ and $[\sqrt{6}, \infty)$, concave down on $(-\infty, -\sqrt{6}]$ and $(0, \sqrt{6}]$; inflection points at $x = \pm\sqrt{6}$; vertical asymptote: $x = 0$, horizontal: $y = 0$

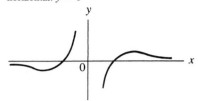

11. $\dfrac{2a}{3}$ **14.** 24 ft $\times$ 60 ft **17.** on land $(50 - 8\sqrt{5})$ m **18.** $2\sqrt{6}$ mi from nearest point **19.** $4\sqrt{2}$ in.

20. 11 cm $\times$ 22 cm $\times$ $(22/3)$ cm **21.** 3.2 Ω/min **22.** $\dfrac{3}{10\pi}\dfrac{\text{mi}}{\text{day}}$ **23.** 0.095 A/s **24.** 3.75 cm/min toward lens **25.** $\dfrac{1}{\pi}$ m/min

26. $6\sqrt{3}$ m^3/min **27.** 840 (lb/in.^{2})/s **28.** $\Delta y = 0.0150062$, $dy = 0.015$ **29.** ± 6 cm^3, 0.6% **30.** 15% **31.** ± 0.63 in.^{2}, 0.8%

Chapter 4

Section 4.1 (page 135)

1. $3x + C$ **3.** $x - x^3 + C$ **5.** $\dfrac{1}{2}x^4 - x^3 + \dfrac{1}{2}x^2 + C$ **7.** $\dfrac{1}{4}x^4 - x^3 + C$ **9.** $\dfrac{1}{6}x^6 - \dfrac{6}{5}x^5 + \dfrac{1}{2}x^4 + 3x + C$ **11.** $-\dfrac{1}{x} - 2x + C$

13. $-\dfrac{3}{x} + 3x^{2/3} + C$ **15.** $-\dfrac{2}{3x} - \dfrac{5}{8x^2} + \dfrac{2}{3}x^{3/2} + C$

Section 4.2 (page 140)

1. $\dfrac{1}{2}$ **3.** $\dfrac{1}{3}$ **5.** 8

Section 4.3 (page 143)

3. 1 **5.** $\dfrac{5}{4}$ **7.** $\dfrac{32}{3}$ **9.** $\dfrac{4}{9}$

Section 4.5 (page 149)

1. $\dfrac{2}{3}x^{3/2} + C$ **3.** $-\dfrac{1}{2x^2} + \dfrac{3}{x} + C$ **5.** $\dfrac{4}{3}x^{3/2} - x^3 + x + C$ **7.** $-\dfrac{1}{3x^3} + 2\sqrt{x} - 4x + C$ **9.** $\dfrac{1}{4}(2x^2 - 3)^4 + C$

11. $-\dfrac{1}{10}(2 - x^2)^5 + C$ **13. a.** $x - \dfrac{1}{2}x^2 + C$ **b.** $-\dfrac{1}{5}(1 - x)^5 + C$ **15.** $-\dfrac{1}{2(x^2 - 1)} + C$ **17.** $\dfrac{1}{4}(2x^2 + x)^4 + C$

19. $-2\sqrt{1 - t} + C$ **21.** $-\sqrt{1 - x^2} + C$ **23.** $\dfrac{5}{12}(x^4 - 2x)^{6/5} + C$ **25.** $\dfrac{1}{5}x^5 + \dfrac{2}{3}x^3 + x + C$ **27.** $-\dfrac{1}{6}(1 - x^2)^3 + C$

29. $x + \dfrac{4}{3}x^{3/2} + \dfrac{1}{2}x^2 + C$ **31.** $-\dfrac{3}{35}(1 - 5s)^{7/3} + C$ **33.** $\dfrac{4}{3}x^{3/4} - \dfrac{2}{\sqrt{x}} - \dfrac{1}{2}x^2 + C$ **35.** $\dfrac{3}{8}x^8 + \dfrac{6}{5}x^5 + \dfrac{3}{2}x^2 + C$

37. $\dfrac{5}{12}(x^3 + 1)^4 + C$ **39.** $24x^8 - \dfrac{96}{5}x^5 + 6x^2 + C$ **41.** $\dfrac{1}{5}(1 + x^3)^5 + C$ **43.** $\dfrac{1}{15}(1 + x^3)^5 + C$ **45.** $3x^8 + \dfrac{24}{5}x^5 + 3x^2 + C$

47. $\dfrac{1}{3}(x^4 + 2)^3 + C$ **49.** $\dfrac{1}{8}x^8 + \dfrac{2}{5}x^5 + \dfrac{1}{2}x^2 + C$ **51.** $-\dfrac{1}{12}(3 - t^4)^3 + C$ **53.** $\dfrac{1}{2}$ **55.** $\dfrac{45}{4}$ **57.** $\dfrac{2}{3}$ **59.** $\dfrac{8}{15}$ **61.** 2

63. $\dfrac{26}{3}$ **65.** 7 **67.** $\sqrt{3}$

Section 4.6 (page 155)

1. 1   **3.** 1   **5.** 1   **7.** $\dfrac{9}{2}$ **9.** $\dfrac{1}{2}$ **11.** $\dfrac{1}{4}$ **13.** $\dfrac{32}{3}$ **15.** $\dfrac{17}{4}$ **17.** $\dfrac{59}{12}$ **19.** 12   **21.** $\dfrac{1}{3}$ **23.** $\dfrac{1}{12}$ **25.** $\dfrac{32}{3}$ **27.** $\dfrac{32}{3}$

29. $\dfrac{9}{2}$ **31.** $\dfrac{16}{3}$ **33. a.** $A = s^2$ **b.** $A = ab$

35. 18

37. $\dfrac{9}{2}$

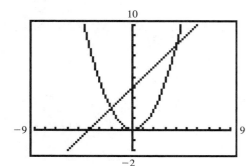

39. $\dfrac{32}{3}$

41. $\dfrac{64}{3}$

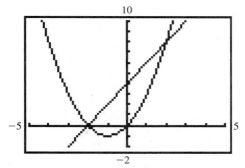

43. 8

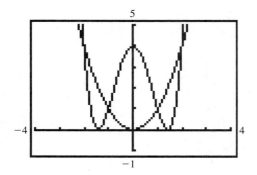

Section 4.7 (page 158)

1. $\dfrac{2}{3}$ **3.** $\dfrac{4}{3}$ **5.** $\dfrac{1}{8}$ **7.** $-\dfrac{1}{36}$ **9.** $\dfrac{2}{3}$ **11.** 1   **13.** 1   **15.** 2   **17.** $\dfrac{1}{36}$ **19.** $\dfrac{1}{7}$ **21.** 4   **23.** 8   **27.** $\dfrac{3}{2}(2^{2/3} - 1) \approx 0.881$

29. does not exist   **31.** $5(1 + 2^{3/5}) \approx 12.58$ **33.** does not exist

Section 4.8 (page 163)

1. $y = \frac{3}{2}x^2 + 1$ **3.** $y = 2x^3 + x + 4$ **5.** $y = x^3 + 2x - 3$ **7.** 3 s   **9.** 45 m   **11.** 5 s   **13.** $t = 2.3$ s, $v = 33$ m/s

15. 5 s, 35 m/s   **17.** 4 s, 30 m/s   **19.** 1 s   **21.** no; he stops in 56 m   **23.** $10\sqrt{3}$ m/s   **25.** $v = \frac{1}{2}\left(21 - \frac{1}{t^2+1}\right)$ **27.** 4.7 C

29. 0.38 C   **31.** 56.7 V   **33.** 56 Ω **35.** 57 J

Section 4.9 (page 169)

1. 8.704, 8.667 $\left(\frac{26}{3}\right)$ **3.** 0.783, 0.785   **5.** 2.793, 2.797   **7.** 3.376   **9.** 1.252   **11.** 2.787   **13.** 8.146   **15.** 3.6535   **17.** 56.7

19. 0.65 C

Review Exercises for Chapter 4 (page 169)

1. 27   **2.** $\frac{124}{5}$ **3.** $2x\sqrt{x} + \frac{1}{3}x^{-3} + x + C$ **4.** $\frac{1}{4}(5x^2 + 4)^4 + C$ **5.** $-\frac{1}{12}(1 - x^2)^6 + C$ **6.** $2\sqrt{x-4} + C$ **7.** $3\sqrt{x^2 - 2} + C$

8. $\frac{4}{9}(x^3 + 1)^3 + C$ **9.** $\frac{3}{8}x^8 + \frac{6}{5}x^5 + \frac{3}{2}x^2 + C$ **10.** $\frac{2}{3}(\sqrt{x} - 1)^3 + C$ **11.** $\frac{1}{2}x^2 - \frac{4}{3}x\sqrt{x} + x + C$ **12.** $-\frac{1}{6}(3 - x^4)^{3/2} + C$

13. $\frac{1}{3}(x^2 - 4x)^{3/2} + C$ **14.** $\frac{16}{3}$ **15.** $\frac{40}{3}$ **16.** $\frac{16}{3}$ **17.** $\frac{32}{3}$ **18.** $\frac{37}{12}$ **19.** $\frac{9}{2}$ **20.** $\frac{9}{2}$ **21.** $-\frac{1}{4}$ **22.** 4   **23.** does not exist

24. does not exist   **25.** $\frac{1}{8}$ **26.** does not exist   **27.** 2   **28.** 2   **29.** 4.75 C   **30.** 46.3 V   **31.** 3.8 s   **32.** 2.1 s, $v = 41$ m/s

33. no; 85 ft/s   **34.** 24 rad   **35.** 0.916

Chapter 5

Section 5.1 (page 174)

1. $\frac{14}{5}$ **3.** 0   **5.** $\frac{\sqrt{2}}{2}$ **7.** $\frac{\sqrt{42}}{6}$ **9.** 30 m/s   **11.** 3.3 A   **13.** 56 W

Section 5.2 (page 177)

1. 84π **3.** $\frac{4\pi}{21}$ **5.** 4π **7.** $\frac{206\pi}{15}$ **9.** $\frac{32\pi}{3}$ **11.** $\frac{2\pi}{3}$ **13.** 4π **15.** $\frac{16\pi}{3}$ **17.** $\frac{32\pi}{3}$ **19.** $\frac{32\pi}{3}$ **21.** $\frac{64\pi}{3}$

23. 2π **25.** $\frac{72\pi}{5}$ **27.** 1880 cm^3 **29.** $99\pi = 310$ ft^3 **31.** π **33.** π

Section 5.3 (page 183)

1. 18π **3.** $\frac{\pi}{6}$ **5.** $\frac{24\pi}{5}$ **7.** $\frac{16\pi}{3}$ **9.** $\frac{8\pi}{3}$ **11.** 3π **13.** $\frac{\pi}{6}$ **15.** $\frac{3\pi}{10}$ **17.** $\frac{\pi}{2}$ **19.** $\frac{8\pi}{3}$ **21.** $\frac{2048\pi}{3}$ **23.** $\frac{5\pi}{3}$

27. 32π **29. a.** 64π **b.** $\frac{512\pi}{7}$ **c.** $\frac{1024\pi}{35}$ **d.** $\frac{704\pi}{5}$ **31. a.** $\frac{32\pi}{15}$ **b.** $\frac{13\pi}{6}$ **35.** $\frac{256\pi}{3}$

37. 18.317006   **39.** 51.294523   **41.** 27π m^3

Section 5.4 (page 193)

1. $\left(\frac{17}{11}, \frac{73}{22}\right)$ **3.** $\left(\frac{17}{3}, \frac{4}{3}\right)$ **5.** $\left(\frac{1}{3}, \frac{1}{3}\right)$ **7.** $\left(\frac{1}{3}, \frac{2}{3}\right)$ **9.** $\left(\frac{2}{3}, \frac{1}{3}\right)$ **11.** $\left(\frac{2}{3}, \frac{4}{3}\right)$ **13.** $\left(0, \frac{8}{5}\right)$

15. $\left(1, -\frac{2}{5}\right)$ **17.** $\left(\frac{62}{15}, \frac{1}{3}\right)$ **19.** $\left(\frac{1}{10}, \frac{1}{2}\right)$ **21.** $\left(\frac{4a}{3\pi}, \frac{4a}{3\pi}\right)$ **23.** $\left(1, \frac{8}{5}\right)$ **25.** $\left(\frac{2}{5}, 1\right)$

27. along the axis, $4r/3\pi$ units from center   **29.** $(a/3, b/3)$, if legs are placed along positive axes

31. a. $\bar{x} = \frac{3}{4}$ **b.** $\bar{y} = \frac{4}{3}$ **33.** $\bar{x} = \frac{5}{6}$ **35.** $\bar{x} = \frac{5}{4}$ **37.** $\bar{x} = \frac{3r}{8}$ **39. a.** $\bar{x} = \frac{5}{8}$ **b.** $\bar{y} = \frac{1}{2}$

41. along the axis, one-fourth of the way from base   **43.** $\left(0, \frac{5}{6}b\right)$ **45.** $\bar{y} = 3.6$ ft

Section 5.5 (page 199)

1. $\dfrac{\rho}{4}, \dfrac{\sqrt{2}}{2}$ **3.** $\dfrac{\sqrt{21}}{7}$ **5.** $I_y = \dfrac{\rho}{6}, R_y = \dfrac{\sqrt{6}}{6}, I_x = \dfrac{2\rho}{3}, R_x = \dfrac{\sqrt{6}}{3}$ **7.** $I_y = \dfrac{64\rho}{15}, R_y = \dfrac{2\sqrt{5}}{5}$ **9.** $I_y = \dfrac{32\rho}{7}, R_y = \dfrac{2\sqrt{42}}{7}$

11. $I_y = \dfrac{243\rho}{20}, R_y = \dfrac{3\sqrt{30}}{10}$ **13.** $I_x = \dfrac{\rho}{20}, R_x = \dfrac{\sqrt{30}}{10}$ **15.** $I_y = \dfrac{192\rho}{5}$ **17.** $I_y = \dfrac{1}{2}\pi r^4 h\rho = \dfrac{1}{2}mr^2; R_y = \dfrac{r\sqrt{2}}{2}$

19. $I_x = \dfrac{2^{11}\pi}{9} \approx 715, R_x = \dfrac{4\sqrt{5}}{3}$ **21.** $\dfrac{8\pi\rho}{5}$ **23.** $\dfrac{2\sqrt{3}}{3}$ **25.** $\dfrac{\sqrt{30}}{9}$ **27.** 7.1 J, $0.38\,\text{kg} \cdot \text{m}^2/\text{s}$

Section 5.6 (page 205)

(Recall that $w = 10{,}000\ \text{N/m}^3$.)   **1.** 96 ft-lb   **3. a.** 12 ft-lb   **b.** 63 ft-lb   **5.** 600 J   **7.** 700 ft-lb   **9.** $54w$ J   **11.** $682.5w$ J

13. $\dfrac{675\pi w}{2}$ J   **15.** $216w$ J   **17.** $\dfrac{75\pi w}{4}$ J   **19.** $128w$ J   **21.** $18\pi w$ J   **23.** $256w$ J   **25.** 6700 ft-lb   **27. a.** 18 ergs   **b.** 20 ergs

29. $8w$ N   **31.** $40w$ N   **33.** $40w$ N   **35.** $\left(\dfrac{9}{2}\right)w$ N   **37.** $\dfrac{8w}{3}$ N   **39.** $42w$ N   **41.** $\dfrac{27w}{2}$ N   **43.** $\dfrac{640w}{3}$ N   **45.** $\dfrac{4w\sqrt{3}}{3}$ N

Review Exercises for Chapter 5 (page 206)

1. 64 ft/s   **2.** 3.2 A   **3.** 30 W   **4.** $\dfrac{4}{3}\pi r^3$ **5. a.** 64π **b.** $\dfrac{1024\pi}{35}$ **c.** $\dfrac{704\pi}{5}$ **d.** $\dfrac{512\pi}{7}$ **6.** $\dfrac{9\pi}{14}$ **7.** $\dfrac{5\pi}{6}$ **8.** $\dfrac{32\pi}{15}$

9. $\dfrac{5\pi}{3}$ **10.** 8π **11.** $\left(\dfrac{3}{4}, \dfrac{8}{5}\right)$ **12.** $\dfrac{64\rho}{15}$ **13.** $\left(\dfrac{3}{5}, \dfrac{2}{35}\right)$ **14.** $\left(-\dfrac{1}{2}, \dfrac{2}{5}\right)$ **15.** $\left(\dfrac{27}{16}, 0\right)$ **16.** $\left(0, \dfrac{7}{22}\right)$ **17.** $\dfrac{\rho}{20}, \dfrac{\sqrt{30}}{10}$

18. $\dfrac{4\sqrt{21}}{7}, \dfrac{2\sqrt{5}}{5}$ **19.** $\dfrac{4\sqrt{5}}{3}$ **20.** $\dfrac{\pi\rho}{26}$ **21. a.** 8 ft-lb   **b.** 16 ft-lb   **22.** 1000 J   **23.** 1.06×10^9 J   **24.** 4.2×10^6 J

25. 585,000 N   **26.** 8.3×10^5 N

Chapter 6

Section 6.1 (page 213)

1. $\dfrac{1}{2}$ **3.** -1 **5.** $-\dfrac{2\sqrt{3}}{3}$ **7.** $-\dfrac{1}{2}$ **9.** -2 **11.** $\dfrac{\sqrt{3}}{3}$ **13.** undefined   **15.** 0   **17.** $\dfrac{\sqrt{2}}{2}$ **19.** $\dfrac{2\sqrt{3}}{3}$ **21.** $-\dfrac{\sqrt{3}}{3}$

23. 1   **25.** $\dfrac{\pi}{3}$ **27.** $\dfrac{5\pi}{6}$ **29.** $\dfrac{3\pi}{4}$ **31.** $\dfrac{4\pi}{5}$ **33.** $\dfrac{\pi}{9}$ **35.** $45°$ **37.** $30°$ **39.** $300°$ **41.** $198°$ **43.** $-100°$

45. $P = \pi, A = \dfrac{1}{3}$

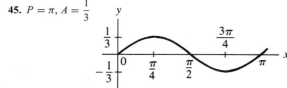

47. $P = 4\pi, A = 3$

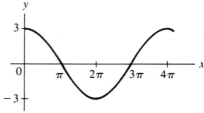

49. $\sin\theta$ **51.** $\dfrac{\sin\theta + 1}{\cos\theta}$ **53.** $-\cos\theta$ **55.** 1   **57.** $\cos^2\theta$ **59.** $\dfrac{1}{\cos^2\theta\,\sin^2\theta}$ **61.** $\dfrac{1}{\sin\theta}$ **63.** $\sin^2 4x$ **65.** $\cos^2 2x$

67. $1 - \sin^2 5x$ **69.** $1 - \cos^2 6x$ **71.** $\sec^2 6x$ **73.** $\tan^2 2x$ **75.** $\sec^2 5x - 1$ **77.** $1 + \tan^2 7x$ **79.** $\csc^2 3x - 1$

85. $\sin 10x$ **87.** $\dfrac{1}{2}\sin x$ **89.** $\cos 6x$ **91.** $\cos 16x$ **93.** $\dfrac{1}{2}(1 - \cos 6x)$ **95.** $\dfrac{1}{2}(1 + \cos 4x)$ **97.** $\dfrac{1}{2}(1 - \cos x)$

Section 6.2 (page 218)

All expressions are equal to dy/dx unless stated otherwise.

1. $-5\sin 5x$ **3.** $-8\sin 4x$ **5.** $2x\cos x^2$ **7.** $\dfrac{ds}{dt} = -9t^2\sin t^3$ **9.** $3\cos 3x$ **11.** $x\cos x + \sin x$

13. $2\sin x\cos x = \sin 2x$ **15.** $\dfrac{dw}{dv} = -8\sin 4v\cos 4v = -4\sin 8v$ **17.** $\dfrac{x\cos x - \sin x}{x^2}$ **19.** $\dfrac{dw}{dv} = -2v\sin(v^2 + 3)$

21. $\cos 2x - 2x \sin 2x$ **23.** $4x \cos(2x+2) + 2\sin(2x+2)$ **25.** $-\dfrac{\cos(1/x)}{x^2}$ **27.** $\dfrac{\cos x + x \sin x}{\cos^2 x}$ **29.** $\dfrac{\sin 4x - 4x \cos 4x}{\sin^2 4x}$

31. $\dfrac{dN}{d\theta} = -\dfrac{2\theta \sin 2\theta + \cos 2\theta}{3\theta^2}$ **33.** $\sqrt{x}\cos x + \dfrac{\sin x}{2\sqrt{x}}$ **35.** $-6x^2 \cos x^3 \sin x^3$ **37.** $\cos^2 x - \sin^2 x = \cos 2x$

39. $\dfrac{\sin^2 x(3x \cos x - \sin x)}{x^2}$ **41.** $\cos 3x(\cos 3x - 6x \sin 3x)$ **47.** slope $= 1$ **49.** 0.28 V **51.** $\dfrac{5\pi}{2}$ cm/s

Section 6.3 (page 222)

All expressions are equal to dy/dx unless stated otherwise.

1. $5 \sec 5x \tan 5x$ **3.** $\dfrac{dy}{dt} = -6 \csc 3t \cot 3t$ **5.** $-12 \csc^2 4x$ **7.** $\dfrac{dz}{dw} = -4w \csc w^2 \cot w^2$ **9.** $\dfrac{ds}{dt} = 2 \sec^2 2t$

11. $\cot 2x - 2x \csc^2 2x$ **13.** $3x^2 \sec(x^3+1) \tan(x^3+1)$ **15.** $\dfrac{dr}{d\theta} = \dfrac{\sec\theta(\theta \tan\theta - 1)}{\theta^2}$ **17.** $-\dfrac{3\csc^2\sqrt{3x}}{2\sqrt{3x}}$ **19.** $\dfrac{\sec^2 2x}{\sqrt{\tan 2x}}$

21. $32 \tan^3 4x \sec^2 4x$ **23.** $\dfrac{dr}{d\omega} = -\omega\sqrt{\csc\omega^2}\cot\omega^2$ **25.** $\dfrac{dT_1}{dT_2} = T_2 \csc T_2(2 - T_2 \cot T_2)$ **27.** $-\cot^2 x - 2\cos^2 x$

29. $\dfrac{\sin x \sec^2 x - \cos x - \sin x}{\sin^2 x}$ **31.** $2x(x-1)\cos(1-x)^2 + \sin(1-x)^2$ **33.** $\dfrac{3x^2(\tan 3x - x \sec^2 3x)}{\tan^2 3x}$ **35.** $\dfrac{1 + 10x \cot 5x}{\csc^2 5x}$

37. $-\dfrac{4x^2 \csc 2x^2 \cot 2x^2 + \csc 2x^2}{4x^2}$ **39.** $\dfrac{-3\sin 3x + 3\sin 3x \cot x^2 - 2x \cos 3x \csc^2 x^2}{(1 - \cot x^2)^2}$ **43.** $16 \sec 4x(\sec^2 4x + \tan^2 4x)$

45. $2\sec^2 x(x \tan x + 1)$ **47.** $\dfrac{\sec^2 x}{2y}$ **49.** $\dfrac{\sec x(x \tan x + 1)}{2y}$ **51.** $\dfrac{\cos(x+y^2)}{2y - 2y\cos(x+y^2)}$ **53.** $\dfrac{\cot y^2}{1 + 2xy \csc^2 y^2}$

55. $\dfrac{2 - 2xy}{x^2 + \sin y}$ **57.** $\dfrac{1}{8}$

Section 6.4 (page 225)

1. $\dfrac{\pi}{4}$ **3.** $-\dfrac{\pi}{6}$ **5.** 0 **7.** $\dfrac{\pi}{6}$ **9.** π **11.** $-\dfrac{\pi}{4}$ **13.** $-\dfrac{\pi}{4}$ **15.** $\dfrac{3\pi}{4}$ **17.** $\dfrac{\pi}{3}$ **19.** $\dfrac{2\sqrt{5}}{5}$

21. $\dfrac{\sqrt{37}}{37}$ **23.** $\dfrac{2\sqrt{5}}{5}$ **25.** $\dfrac{3\sqrt{2}}{4}$ **27.** $-\dfrac{\sqrt{7}}{3}$ **29.** $\sqrt{1-x^2}$ **31.** $\dfrac{1}{x}$ **33.** $\dfrac{1}{\sqrt{1+4x^2}}$ **35.** $\dfrac{\sqrt{1-4x^2}}{2x}$ **37.** $\sqrt{1-9x^2}$

39. $2\tan\dfrac{y}{2}$ **41.** $\text{Arcsin}(y-1)$ **43.** $\dfrac{1}{3}\text{Arccos}(y-1)$ **45.** $x = \dfrac{1}{4}\text{Arctan}\dfrac{y-1}{3}$ **47.** $x = 2\,\text{Arcsin}(y+2)$

Section 6.5 (page 228)

1. $\dfrac{3}{1+9x^2}$ **3.** $\dfrac{-5}{\sqrt{1-25x^2}}$ **5.** $\dfrac{ds}{dt} = \dfrac{4t}{1+4t^4}$ **7.** $\dfrac{du}{dv} = \dfrac{6v}{\sqrt{1-9v^4}}$ **9.** $\dfrac{dy}{dw} = \dfrac{2}{\sqrt{1-4w^2}}$ **11.** $\dfrac{dv_1}{dv_2} = -\dfrac{2v_2}{\sqrt{1-v_2^4}}$

13. $\dfrac{4x}{\sqrt{1-4x^4}}$ **15.** $\dfrac{7}{1+49x^2}$ **17.** $\dfrac{x}{1+x^2} + \text{Arctan}\,x$ **19.** $-\dfrac{2x^2}{\sqrt{1-x^4}} + \text{Arccos}\,x^2$ **21.** $\dfrac{dr}{d\theta} = \dfrac{3\theta}{\sqrt{1-9\theta^2}} + \text{Arcsin}\,3\theta$

23. $\dfrac{dR}{dV} = \dfrac{6V}{1+9V^2} + 2\,\text{Arctan}\,3V$ **25.** $\dfrac{x - \sqrt{1-x^2}\,\text{Arcsin}\,x}{x^2\sqrt{1-x^2}}$ **27.** $\dfrac{1}{2\sqrt{x - x^2}}$ **29.** $\dfrac{\sqrt{x}}{\sqrt{1-x^2}} + \dfrac{\text{Arcsin}\,x}{2\sqrt{x}}$ **31.** $\dfrac{2\,\text{Arcsin}\,x}{\sqrt{1-x^2}}$

33. $-\dfrac{1}{2}[(1-x^2)\text{Arccos}\,x]^{-1/2}$ **35.** $\dfrac{x - (1+x^2)\text{Arctan}\,x}{x^2(1+x^2)}$ **37.** $\dfrac{2\sqrt{1-x^4}\,\text{Arcsin}\,x^2 - 4x^2}{\sqrt{1-x^4}(\text{Arcsin}\,x^2)^2}$ **39.** 0.145 rad/s

Section 6.6 (page 232)

1. $\log_3 27 = 3$ **3.** $\log_9 1 = 0$ **5.** $\log_{32}\dfrac{1}{2} = -\dfrac{1}{5}$ **7.** $3^5 = 243$ **9.** $\left(\dfrac{1}{4}\right)^2 = \dfrac{1}{16}$ **11.** $9^{-1/2} = \dfrac{1}{3}$ **13.** $\dfrac{1}{9}$ **15.** 4 **17.** 27

19. $\dfrac{1}{9}$ **21.** **23.** **25.**

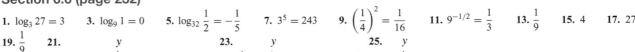

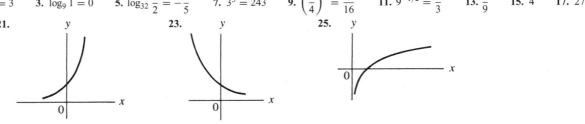

27. $\log_3 24$ **29.** $\log_5 4$ **31.** $\log_b \dfrac{\sqrt{3}}{3} = -\dfrac{1}{2}\log_b 3$ **33.** $\log_3 \dfrac{2y^2}{25}$ **35.** 3   **37.** $\dfrac{1}{2}(1 + \log_6 x)$ **39.** $-2(1 + \log_5 x)$

41. $-\dfrac{1}{3}(1 + \log_3 x)$ **43.** $-\dfrac{1}{2}\log_5(y - 2)$ **45.** $\log_{10} x - \dfrac{1}{2}\log_{10}(x + 2)$ **47.** $\dfrac{1}{2}\log_{10} x - \log_{10}(x + 1)$ **49.** 1.96

51. 0.502   **53.** -0.533

Section 6.7 (page 237)

1. $\dfrac{1}{x}$ **3.** $\dfrac{4}{x}$ **5.** $\dfrac{dR}{ds} = \dfrac{2}{s}$ **7.** $\dfrac{6}{x}$ **9.** $\dfrac{3}{x}\log_{10} e$ **11.** $\dfrac{dR_1}{dR_2} = \cot R_2$ **13.** $1 + \ln x$ **15.** $\dfrac{1}{2(x - 1)}$ **17.** $-\dfrac{1}{2(x + 2)}$

19. $\dfrac{dz}{dt} = \dfrac{2t}{t^2 + 2}$ **21.** $\dfrac{x + 2}{x(x + 1)}$ **23.** $\dfrac{1 - x^2}{x(x^2 + 1)}$ **25.** $\dfrac{2 + 3x^2}{2x(2 - x^2)}$ **27.** $-\dfrac{2x \tan x + 1}{2x}$ **29.** $\dfrac{4(x + 1)\tan x - 1}{2(x + 1)}$

31. $\dfrac{2\ln x}{x}$ **33.** $\dfrac{\sqrt{x + 1}}{x} + \dfrac{\ln x}{2\sqrt{x + 1}}$ **35.** $\dfrac{ds}{d\theta} = \dfrac{\sec\theta}{\theta} + (\sec\theta\tan\theta)\ln\theta$ **37.** $\dfrac{x}{x^2 - 1 + \sqrt{x^2 - 1}}$ **39.** $\dfrac{6\ln^2 x}{x}$

41. $\dfrac{1}{x\ln x}$ **43.** $\dfrac{dv_1}{dv_2} = \dfrac{1 - \ln v_2}{v_2{}^2}$

Section 6.8 (page 240)

1. $4e^{4x}$ **3.** $2xe^{x^2}$ **5.** $\dfrac{dy}{dt} = -4te^{-t^2}$ **7.** $e^{\tan x}\sec^2 x$ **9.** $3x^2(\ln 3)3^{x^3}$ **11.** $\dfrac{2x}{\log_4 e}4^{x^2} = 2x(\ln 4)4^{x^2}$ **13.** $\dfrac{dC}{dr} = 2(r + 1)e^r$

15. $e^{\sin x}\cos x$ **17.** $e^x \cos e^x$ **19.** $\dfrac{dS}{d\omega} = e^{2\omega}(\cos\omega + 2\sin\omega)$ **21.** $\dfrac{\sec^2 x - 2x\tan x}{e^{x^2}}$ **23.** $e^{\sec x}\left(\sec x \tan x \ln x + \dfrac{1}{x}\right)$

25. $2e^{2x}\cot e^{2x}$ **27.** $\dfrac{1 - 2x}{e^{2x}}$ **29.** $\dfrac{2\cos 2x + 2e^x \cos 2x - e^x \sin 2x}{(e^x + 1)^2}$ **33.** $5\cosh 5x$ **35.** $6x^2 \sinh 2x^3$ **37.** $\cosh 3x + 3x \sinh 3x$

39. $x^{\sin x}[(\sin x)/x + \cos x \ln x] = x^{(\sin x - 1)}(\sin x + x \cos x \ln x)$ **41.** $(\ln x)^x [\ln(\ln x) + (1/\ln x)]$ **43.** $(\tan x)^{x-1}(x \sec^2 x + \tan x \ln \tan x)$

45. $i = -1.1$ A   **47.** $\dfrac{dp}{dt} = \dfrac{ake^{-kt}}{(1 + ae^{-kt})^2}$

Section 6.9 (page 242)

1. -4 **3.** $\dfrac{3}{2}$ **5.** 7   **7.** 6   **9.** ∞ **11.** 2   **13.** $-\dfrac{1}{2}$ **15.** 0   **17.** -3 **19.** $\dfrac{1}{2}$ **21.** 0   **23.** 0   **25.** e

Section 6.10 (page 247)

1. max. at $x = 1$; inflection point at $x = 2$ **3.** min. at $x = \dfrac{1}{e}$; concave up everywhere   **5.** min. at $x = 0$; concave up everywhere

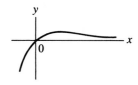

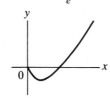

7. $i = \sqrt{2}$ A   **9.** 1.8 V   **11.** $\dfrac{dN}{dt} = -24e^{-2.0t}$ **13.** -0.18 g/min   **15.** 448 bacteria per hour   **17.** $\dfrac{1}{\sqrt{e}}$ **21.** 0.018 rad/s

23. 0.015 rad/s   **25.** $\dfrac{\pi}{6}$ **27.** 9.58 m   **29.** $\dfrac{\pi}{4} + \dfrac{\alpha}{2}$ **31.** $x = n$

Section 6.11 (page 250)

1. -0.739085 **3.** 2.474577   **5.** 0.876726   **7.** 1.326725   **9.** 0.853530   **11.** 4.605551, -2.605551
13. 6.474667, 0.754138, -1.228805 **15.** -0.917988

Review Exercises for Chapter 6 (page 251)

1. $-\dfrac{\pi}{2}$ **2.** $-\dfrac{\pi}{4}$ **3.** $\dfrac{2\pi}{3}$ **4.** 0 **5.** $\dfrac{\sqrt{10}}{10}$ **6.** $\dfrac{2\sqrt{2}}{3}$ **7.** $\dfrac{x}{\sqrt{1+x^2}}$ **8.** $\dfrac{\sqrt{1-x^2}}{x}$ **9.** $3x^2\sec^2 3x + 2x\tan 3x$

10. $\dfrac{4x\cos 4x - \sin 4x}{x^2}$ **11.** $\dfrac{dv}{dt} = \dfrac{e^{2t}(2t-1)}{t^2}$ **12.** $\dfrac{dV_1}{dV_2} = \dfrac{1}{V_2}\cos V_2 - \sin V_2 \ln V_2$ **13.** $-\dfrac{1}{2(x+3)}$ **14.** $\dfrac{1+x}{2x(1-x)}$

15. $-\dfrac{1}{x(2x^2+1)}$ **16.** $-e^{-x}\cos e^{-x}$ **17.** $\dfrac{1}{2x\sqrt{\ln 2x}}$ **18.** $2e^{2\tan x}\sec^2 x$ **19.** $-\dfrac{\sin(\ln x)}{x}$ **20.** $(2x)(\ln 5)5^{x^2}$

21. $e^{2x}(2\cot x - \csc^2 x)$ **22.** $\dfrac{x}{\sqrt{1-x^2}} + \text{Arcsin } x$ **23.** $-\dfrac{3e^{\text{Arccos }3x}}{\sqrt{1-9x^2}}$ **24.** $2\tan x$ **25.** $(\cot x)^x(\ln\cot x - x\csc x\sec x)$

26. $(\sec x)^x(x\tan x + \ln\sec x)$ **27.** $\dfrac{dy}{dx} = e^{-\sin y}\sec y\csc x\cot x$ **28.** $\dfrac{dy}{dx} = -\dfrac{y}{1+y\sec^2 y}$ **29.** 4 **30.** 0 **31.** 1 **32.** $\dfrac{1}{2}$

33. 0 **34.** $x+y-1=0$ **35.** $\dfrac{1}{e}\times 1$ **36.** 20 m/s **37. a.** $22,408.45 **b.** rP **38.** decreases at 27 kg/m² per min

39. 3.00×10^{-15} mm of mercury per min **40.** $\dfrac{dI}{dt} = \dfrac{E}{L}e^{-(Rt)/L}$ **41.** $\theta = \text{Arcsin}\dfrac{1}{\sqrt{3}} \approx 35.3°$ **42.** $\text{Arctan}\dfrac{1}{2} \approx 26.6°$ **43.** 0.415083

Chapter 7

Section 7.1 (page 254)

1. $\dfrac{1}{3}(x^2+1)^{3/2} + C$ **3.** $-2\sqrt{1-x} + C$ **5.** $\dfrac{1}{3}\sin^3 x + C$ **7.** $\dfrac{1}{6}\tan^3 2x + C$ **9.** $\dfrac{1}{12}(1+\tan 3t)^4 + C$ **11.** $\dfrac{1}{16}(1+e^{4r})^4 + C$

13. $\dfrac{1}{20}(1-\cos 5x)^4 + C$ **15.** $-\dfrac{1}{6(\tan 3x+1)^2} + C$ **17.** $\dfrac{1}{2}\ln^2 x + C$ **19.** $2\sqrt{\ln x} + C$ **21.** $-\dfrac{1}{2}(\text{Arccos } x)^2 + C$ **23.** $\dfrac{\sqrt{3}}{8}$

25. $\dfrac{2}{3}$ **27.** $-\dfrac{1}{4}\cot^2 2x + C$ **29.** $\dfrac{2}{3}(\tan x)^{3/2} + C$ **31.** $\dfrac{1}{8}(\text{Arctan } 4R)^2 + C$ **33.** $\dfrac{1}{3}(1+\ln x)^3 + C$

Section 7.2 (page 257)

1. $\ln|x-1| + C$ **3.** $\dfrac{1}{3}\ln|2+3x| + C$ **5.** $-\dfrac{1}{3}\ln|1-3s| + C$ **7.** $\ln\sqrt{2}$ **9.** $-e^{-x} + C$ **11.** $\dfrac{2}{3}(e^6 - 1)$ **13.** $\dfrac{1}{4}e^{4x} + C$

15. $\dfrac{1}{2}e^{r^2} + C$ **17.** $e^{\sin R} + C$ **19.** $\dfrac{1}{2}\ln|1+\tan 2x| + C$ **21.** $e^{\text{Arctan } x} + C$ **23.** $\ln|\ln x| + C$ **25.** $x - e^{-x} + C$

27. $\dfrac{1}{3}\ln|1-e^{-3W}| + C$ **29.** $-\dfrac{1}{1+x^2} + C$ **31.** $x + 2e^x + \dfrac{1}{2}e^{2x} + C$ **33.** $x - \ln|x+2| + C$ **35.** $-\csc x + C$

37. $e^{\sin^2 x} + C$ **39.** $\ln 2$ **41.** $\dfrac{1}{2}\ln(1+\sin 2x) + C$ **43.** $-\dfrac{1}{2(1+\sin 2x)} + C$ **45.** $\dfrac{1}{3}(5^{3x})\log_5 e + C = \dfrac{5^{3x}}{3\ln 5} + C$

47. $-2^{\cot x}\log_2 e + C = \dfrac{-2^{\cot x}}{\ln 2} + C$ **49.** $\dfrac{2}{3}\rho\ln 3$ **51.** $\dfrac{3}{2}(1-e^{-2/3})$ A **53.** $\pi\left(1-\dfrac{1}{e}\right)$ **55.** $8\pi\ln 2$ **57.** $v = \dfrac{10}{k}(1-e^{-kt})$ m/s

59. 18.2 kg

Section 7.3 (page 260)

1. $\dfrac{1}{2}\tan 2x + C$ **3.** $\dfrac{1}{3}\sec 3x + C$ **5.** $-\dfrac{1}{4}\csc 4x + C$ **7.** $2\ln\left|\sec\dfrac{1}{2}x\right| + C$ **9.** $\dfrac{1}{2}\sin 2t + C$ **11.** $\dfrac{1}{2}\ln|\csc y^2 - \cot y^2| + C$

13. $-\ln|1+\cot y| + C$ **15.** $-\dfrac{1}{5}\ln|4-\tan 5t| + C$ **17.** $-\dfrac{1}{2}\cos x^2 + C$ **19.** $\ln\left|\sin\dfrac{1}{2}T^2\right| + C$ **21.** $2\sin\sqrt{x} + c$

23. $\dfrac{1}{16}\tan^4 4x + C$ **25.** $\dfrac{1}{3}\ln|\csc e^{3x} - \cot e^{3x}| + C$ **27.** $-\cot(\ln x) + C$ **29.** $-\cot x + C$ **31.** $-\dfrac{1}{2}\ln|1+2\cos x| + C$ **33.** $\dfrac{1}{2}$

35. $\ln 2$ **37.** 183 V **39.** π **41.** $\dfrac{1}{4}\ln 2$ **43.** $\dfrac{2E}{\pi}$V **45.** 0.66

Section 7.4 (page 266)

1. $\frac{1}{6}\sin^3 2x - \frac{1}{10}\sin^5 2x + C$ **3.** $\frac{1}{3}\cos^3 x - \cos x + C$ **5.** $\frac{1}{7}\cos^7 x - \frac{1}{5}\cos^5 x + C$ **7.** $\frac{1}{4}\sin^4 x - \frac{1}{6}\sin^6 x + C$

9. $\frac{1}{2}x + \frac{1}{16}\sin 8x + C$ **11.** $\frac{1}{8}x - \frac{1}{32}\sin 4x + C$ **13.** $\frac{1}{10}\cos^5 2t - \frac{1}{6}\cos^3 2t + C$ **15.** $\frac{1}{20}\sin^5 4x - \frac{1}{28}\sin^7 4x + C$

17. $\frac{1}{2}\tan^2 x + \ln|\cos x| + C$ **19.** $\frac{1}{3}\tan^3 x + \frac{1}{5}\tan^5 x + C$ **21.** $\frac{1}{3}\sec^3 y + C$ **23.** $\frac{1}{5}\sec^5 x - \frac{1}{3}\sec^3 x + C$

25. $-\frac{1}{14}\cot^7 2x - \frac{1}{18}\cot^9 2x + C$ **27.** $-\cot x - \frac{2}{3}\cot^3 x - \frac{1}{5}\cot^5 x + C$ **29.** $\frac{20}{21}$ **31.** $4 + \frac{3}{2}\pi$ **33.** $\ln 2$

35. $\frac{1}{24}\tan^6 4x + \frac{1}{32}\tan^8 4x + C$ **37.** $\frac{\pi^2}{2}$ **39.** $10\sqrt{2}$ A   **41.** $\frac{15}{4}$ W   **43.** 1.25   **45.** 110 V

Section 7.5 (page 269)

1. $\text{Arcsin } x + C$ **3.** $\frac{1}{6}\text{Arctan }\frac{2}{3}x + C$ **5.** $-\sqrt{1 - x^2} + C$ **7.** $\frac{1}{18}\ln(16 + 9x^2) + C$ **9.** $\frac{1}{4}\text{Arctan }\frac{1}{2}t^2 + C$

11. $-\text{Arcsin}\left(\frac{1}{2}\cot x\right) + C$ **13.** $\frac{\sqrt{3}}{3}\text{Arcsin }\frac{\sqrt{15}}{5}x + C$ **15.** $\ln(2 + \sin y) + C$ **17.** $\frac{\pi}{12}$ **19.** $\frac{1}{\sqrt{3}}\text{Arcsin }\frac{\sqrt{3}}{2}x + C$

21. $-\frac{1}{3}\sqrt{4 - 3x^2} + C$ **23.** $-\frac{1}{x - 3} + C$ **25.** $\text{Arctan}(x - 3) + C$ **27.** $\text{Arcsin }\frac{x + 2}{\sqrt{5}} + C$ **29.** $\frac{2}{\sqrt{3}}\text{Arctan }\frac{2x + 3}{\sqrt{3}} + C$

31. $\frac{1}{2}\ln(x^2 + 16) + \text{Arctan }\frac{1}{4}x + C$ **33.** $\frac{\pi}{4}$ **35.** $\frac{1}{2}\ln(1 + e^{2x}) + C$ **37.** $\text{Arcsin }e^\theta + C$ **39.** $\frac{\pi}{2}$ **41.** $\frac{\pi^2}{4}$

Section 7.6 (page 274)

1. $-\frac{\sqrt{4 - x^2}}{x} - \text{Arcsin }\frac{x}{2} + C$ **3.** $\frac{1}{3}(x^2 + 9)^{3/2} + C$ **5.** $\frac{x}{25\sqrt{x^2 + 25}} + C$ **7.** $\frac{1}{2}\text{Arcsec }\frac{x}{2} + C = \frac{1}{2}\text{Arctan}\left(\frac{1}{2}\sqrt{x^2 - 4}\right) + C$

9. $-\frac{\sqrt{x^2 + 16}}{16x} + C$ **11.** $-\frac{1}{\sqrt{x^2 - 2}} + C$ **13.** $\frac{1}{3}(x^2 + 6)\sqrt{x^2 - 3} + C$ **15.** $-\frac{\sqrt{x^2 + 1}}{x} + \ln\left|x + \sqrt{x^2 + 1}\right| + C$ **17.** $\frac{\sqrt{2}}{32}$ **19.** $\frac{9\pi}{4}$

21. $\left(40\pi + \frac{16}{3}\right)w$ newtons, where $w = 10{,}000$ N/m³   **23.** $\frac{8a^5\pi\rho}{15} = \frac{2}{5}ma^2$ **25.** $(108\pi + 144)w$ J $(w = 10{,}000$ N/m³$)$

27. $2\,\text{Arcsin }\frac{x}{2} + \frac{1}{2}x\sqrt{4 - x^2} + C$ **29.** $\frac{1}{15}(x^2 - 4)^{3/2}(3x^2 + 8) + C$

Section 7.7 (page 278)

1. $xe^x - e^x + C$ **3.** $\frac{1}{4}\sin 2x - \frac{1}{2}x\cos 2x + C$ **5.** $x\tan x + \ln|\cos x| + C$ **7.** $\frac{1}{4}x^2(2\ln x - 1) + C$ **9.** $x\text{Arcsin }x + \sqrt{1 - x^2} + C$

11. $\frac{1}{3}x\sin 3x + \frac{1}{9}\cos 3x + C$ **13.** $-e^{-x}(x^2 + 2x + 2) + C$ **15.** $x\,\text{Arccot }x + \frac{1}{2}\ln(1 + x^2) + C$ **17.** $-\frac{1}{2}\cos x^2 + C$

19. $\frac{1}{2}e^x(\sin x - \cos x) + C$ **21.** $\frac{1}{\pi^2 + 1}[e^{-x}(\pi\sin\pi x - \cos\pi x)] + C$ **23.** $\frac{2\ln 2 - 1}{(\ln 2)^2}$ **25.** π **27.** $\left(\frac{\pi^2}{4} - 2\right)\rho$ **29.** 482 V

31. $\left(-\frac{1}{4}x^3 - \frac{3}{16}x^2 - \frac{3}{32}x - \frac{3}{128}\right)e^{-4x} + C$ **33.** $4x(x^2 - 6)\cos x + (x^4 - 12x^2 + 24)\sin x + C$ **35.** $\frac{2}{5}e^{-x}\sin 2x - \frac{1}{5}e^{-x}\cos 2x + C$

Section 7.8 (page 287)

1. $\frac{1}{4}\ln\left|\frac{x - 2}{x + 2}\right| + C$ **3.** $\ln\left|\frac{x}{1 - x}\right| + C$ **5.** $\ln\left|(x - 2)^2(x + 1)^3\right| + C$ **7.** $\frac{1}{2}x^2 + 2x + \frac{27}{4}\ln|x - 3| + \frac{1}{4}\ln|x + 1| + C$

9. $\ln\left|\frac{(x - 4)^2(x - 1)}{(x + 2)^2}\right| + C$ **11.** $-\frac{2}{x + 1} + 3\ln|x + 1| + C$ **13.** $-\frac{1}{x - 2} - 2\ln|x + 1| + C$ **15.** $2\ln|x - 2| - \frac{8}{x - 2} - \frac{9}{2(x - 2)^2} + C$

17. $\ln|x + 2| - \frac{3}{2}\text{Arctan }\frac{x}{2} + C$ **19.** $\ln\left|(x + 1)(x^2 + 1)\right| + 2\text{Arctan }x + C$ **21.** $\frac{1}{2}x^2 - 4\ln(x^2 + 4) - \frac{8}{x^2 + 4} + C$

23. $\ln|x| - \frac{1}{2}\text{Arctan }\frac{1}{2}(x - 1) + C$ **25.** $\frac{1}{2}\ln|x| - \frac{1}{4}\ln|x^2 + 2x + 2| - \frac{1}{2}\text{Arctan}(x + 1) + C$

Section 7.9 (page 290)

1. $\dfrac{1}{2}\ln\left|\dfrac{x}{2+x}\right| + C$ **3.** $\dfrac{1}{2}\left(x\sqrt{x^2-7} - 7\ln\left|x + \sqrt{x^2-7}\right|\right) + C$ **5.** $\dfrac{1}{2\sqrt{5}}\ln\left|\dfrac{\sqrt{5}+x}{\sqrt{5}-x}\right| + C$ **7.** $-\dfrac{\sqrt{5x^2+4}}{4x} + C$

9. $-\dfrac{1}{6}\sin 3x + \dfrac{1}{2}\sin x + C$ **11.** $\dfrac{1}{\sqrt{3}}\ln\left|\sqrt{3}x + \sqrt{3x^2+5}\right| + C$ **13.** $\dfrac{1}{2}x^2 e^{2x} - \dfrac{1}{2}xe^{2x} + \dfrac{1}{4}e^{2x} + C$

15. $\dfrac{1}{5}\tan^5 x - \dfrac{1}{3}\tan^3 x + \tan x - x + C$ **17.** $\dfrac{1}{12}\ln\left|\dfrac{2x-3}{2x+3}\right| + C$ **19.** $\dfrac{1}{\sqrt{3}}\ln\left|\dfrac{\sqrt{3+x}-\sqrt{3}}{\sqrt{3+x}+\sqrt{3}}\right| + C$ **21.** $\dfrac{1}{2\sqrt{15}}\ln\left|\dfrac{\sqrt{3}x-\sqrt{5}}{\sqrt{3}x+\sqrt{5}}\right| + C$

23. $\sqrt{x^2-10} - \sqrt{10}\,\text{Arccos}\dfrac{\sqrt{10}}{x} + C$ **25.** $-\dfrac{1}{10}\cos 5x - \dfrac{1}{2}\cos x + C$ **27.** $\dfrac{x}{5\sqrt{4x^2+5}} + C$

Review Exercises for Chapter 7 (page 291)

1. $\dfrac{1}{2}\ln(x^2+1) + C$ **2.** $-\dfrac{1}{2(x^2+1)} + C$ **3.** $2\,\text{Arctan}\,x + C$ **4.** $e^{\text{Arctan}\,x} + C$ **5.** $-\sqrt{9-t^2} + C$ **6.** $\text{Arcsin}\dfrac{\theta}{3} + C$

7. $\dfrac{1}{4}\sin 2x^2 + C$ **8.** $\dfrac{1}{4}\cos 2x + \dfrac{1}{2}x\sin 2x + C$ **9.** $\dfrac{1}{2}\text{Arctan}\dfrac{1}{2}e^x + C$ **10.** $\ln(4 + e^x) + C$ **11.** $\dfrac{-1}{4+e^x} + C$

12. $\dfrac{x^2}{4}(2\ln x - 1) + C$ **13.** $\dfrac{1}{2}\ln^2 x + C$ **14.** $\sin(\ln x) + C$ **15.** $\dfrac{1}{2}\ln|x^2 + 4x + 5| + C$ **16.** $\text{Arctan}(x+2) + C$ **17.** $-\dfrac{1}{x+2} + C$

18. $\text{Arctan}(\ln x) + C$ **19.** $x\ln^2 x - 2x\ln x + 2x + C$ **20.** $\dfrac{1}{\sqrt{5}}\text{Arcsin}\dfrac{\sqrt{5}}{2}x + C$ **21.** $\dfrac{1}{2}\ln\left|\dfrac{2-\sqrt{4-x^2}}{x}\right| + C$

22. $-\dfrac{1}{3}(4-x^2)^{3/2} + C$ **23.** $\dfrac{1}{6}\ln|1 + \sin^3 2x| + C$ **24.** $\dfrac{1}{10}\cos^5 2x - \dfrac{1}{6}\cos^3 2x + C$ **25.** $\dfrac{1}{2}x - \dfrac{1}{8}\sin 4x + C$ **26.** $\dfrac{1}{4}e^{2x^2} + C$

27. $\dfrac{1}{4}e^{2x}(2x-1) + C$ **28.** $\dfrac{1}{4}(\text{Arctan}\,2x)^2 + C$ **29.** $x\,\text{Arctan}\,2x - \dfrac{1}{4}\ln(1+4x^2) + C$ **30.** $-e^{-\sin^2 x} + C$ **31.** $e^{\tan x} + C$

32. $\cos\dfrac{1}{x} + C$ **33.** $x - \dfrac{4}{\sqrt{3}}\text{Arctan}\dfrac{x}{\sqrt{3}} + C$ **34.** $\ln|1 - \cos\omega| + C$ **35.** $\ln|\csc\omega(\csc\omega - \cot\omega)| + C = -\ln|1 + \cos\omega| + C$

36. $\dfrac{3}{2}(\ln x)^{2/3} + C$ **37.** $-\dfrac{1}{5}\cot^5 x - \dfrac{1}{7}\cot^7 x + C$ **38.** $\text{Arcsin}\dfrac{x-2}{\sqrt{2}} + C$ **39.** $\dfrac{1}{10}\sec^5 2x - \dfrac{1}{6}\sec^3 2x + C$ **40.** $e^{\tan x}(\tan x - 1) + C$

41. $\dfrac{1}{3}\tan^3 x + C$ **42.** $-\ln|2 - \ln x| + C$ **43.** $\dfrac{1}{3}\tan 3x + \dfrac{1}{9}\tan^3 3x + C$ **44.** $\dfrac{1}{12}\csc^3 4x - \dfrac{1}{20}\csc^5 4x + C$

45. $\dfrac{1}{17}e^x(4\sin 4x + \cos 4x) + C$ **46.** $-\dfrac{1}{4}e^{\cos 4x} + C$ **47.** $\dfrac{1}{2\sqrt{5}}\text{Arctan}\dfrac{\sqrt{5}}{2}x + C$ **48.** $\dfrac{3}{10}\ln(5x^2 + 4) + C$

49. $\dfrac{1}{2}\ln|x^2 + 2x - 8| + C$ **50.** $\dfrac{1}{3}\ln|(x+4)(x-2)^2| + C$ **51.** $\ln|(x-2)(x^2+9)| + C$

Chapter 8

Section 8.1 (page 297)

1. $x - 3y + 3 = 0$

3. $y^2 = x - 1$

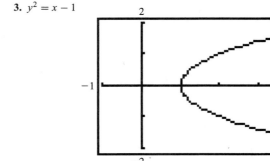

5. $y = 1 - x^2$

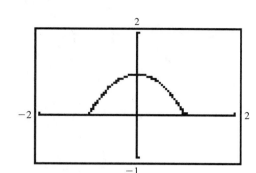

7. $y = 1 + \dfrac{1}{9}x^2$

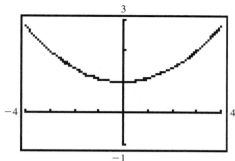

9. $y = 2 + e^x$

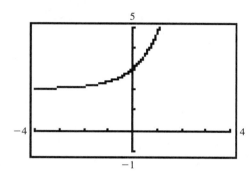

11. $\mathbf{v} = 4\mathbf{i} - \mathbf{j}, (\sqrt{17}, 346.0°)$ **13.** $\mathbf{v} = -2\mathbf{i} + 3\mathbf{j}, (\sqrt{13}, 123.7°)$ **15.** $\mathbf{v} = -2\mathbf{i} + 2\mathbf{j}, (2\sqrt{2}, 135°)$ **17.** $\mathbf{v} = \mathbf{i} - \mathbf{j}, (\sqrt{2}, 315°)$
19. $\mathbf{v} = \dfrac{2}{3}\mathbf{i} + \dfrac{8}{3}\mathbf{j}, \left(\dfrac{2}{3}\sqrt{17}, 76.0°\right)$ **21.** $(\sqrt{17}, 166.0°), (2\sqrt{5}, 333.4°)$ **23.** $(\sqrt{2}, 315°), (4, 180°)$ **25.** $(4, 225°), (4, 315°)$
27. $(13.87 \text{ m/s}, -2.75°)$ **29. b.** $(10\sqrt{13} \text{ m/s}, 16.1°), (10 \text{ m/s}^2, -90°)$

Section 8.2 (page 301)

1. $\dfrac{14}{3}$ **3.** 1.18 **5.** $\dfrac{14}{3}$ **7.** 19 **9.** $\dfrac{3}{2}$ **11.** $\sqrt{2}(e - 1)$ **13.** $6a$ **15.** 18.4599 **17.** 3.3428 **19.** 4.0333

Section 8.3 (page 306)

3. $(-1, \sqrt{3})$ **5.** $(-3, -3\sqrt{3})$ **7.** $(1.97, -0.35)$ **9.** $\left(\sqrt{2}, \dfrac{7\pi}{4}\right)$ **11.** $(5, 0.93)$ **13.** $r = 2\sec\theta$ **15.** $r = \sqrt{2}$

17. $2r\cos\theta - 4r\sin\theta = 5$ **19.** $r^2\cos^2\theta - 2r\cos\theta + r\sin\theta = 2$ **21.** $r^2 = \sec 2\theta$ **23.** $r^2 = \dfrac{1}{2(1 + \sin^2\theta)}$ **25.** $y = 2$ **27.** $y = x$
29. $x^2 + y^2 = x$ **31.** $(x^2 + y^2 - x)^2 = x^2 + y^2$ **33.** $(x^2 + y^2 + x)^2 = x^2 + y^2$ **35.** $(x^2 + y^2 + 2y)^2 = x^2 + y^2$
37. $(x^2 + y^2 - 4y)^2 = 4(x^2 + y^2)$ **39.** $(x^2 + y^2)^2 = 2xy$ **41.** $y^2 = 4(x + 1)$ **43.** $x^2 = 8(y + 2)$ **45.** $(x^2 + y^2)^2 = 3(x^2 - y^2)$
47. $(x^2 + y^2)^2 = a(3x^2y - y^3)$ **49.** $(x^2 + y^2 + 2y)^2 = x^2 + y^2$

Section 8.4 (page 312)

1.

3.

5.

7.

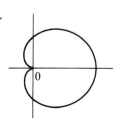

9.

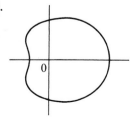

11.

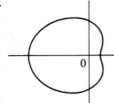

13.
15.

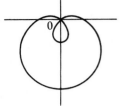

17.

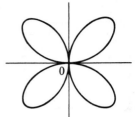

19.

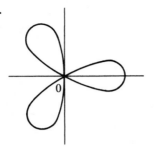

21.

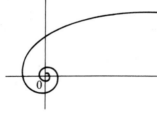

23.

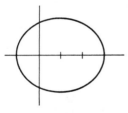

25.
27.

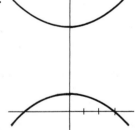

29.

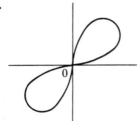

31.

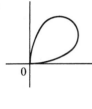

33.

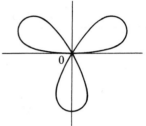

35.

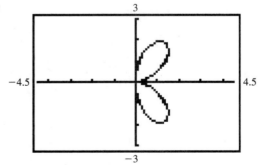

37.

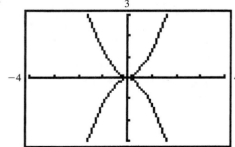

39.

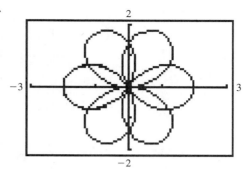

41.

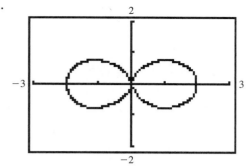

43.

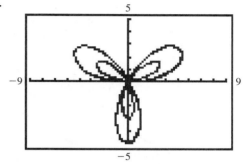

Section 8.5 (page 316)

1. $\dfrac{\pi}{3}$ **3.** $\dfrac{1}{2}$ **5.** $\dfrac{4}{3}$ **7.** π **9.** 4π **11.** $\dfrac{9\pi}{2}$ **13.** 9π **15.** 1 **17.** $\dfrac{9\pi}{2}$ **19.** 6π **21.** $\dfrac{9\pi}{2}$ **23.** $\dfrac{27\pi}{2}$ **25.** 4

27. 2 **29.** 6 **31.** $\dfrac{1}{2}\left(2\pi - 3\sqrt{3}\right)$ **33.** $\dfrac{16 + \pi}{4}$ **35.** $4\left(\dfrac{\pi}{3} + \dfrac{\sqrt{3}}{2}\right)$ **37.** $1 - \dfrac{\pi}{4}$ **39.** 8π **41.** $\dfrac{5\pi}{4}$

Review Exercises for Chapter 8 (page 317)

1. $4x + y^2 = 16$ **2.** $4x^2 - y^2 - 8x + 8 = 0$ **3. a.** $\mathbf{v} = -\dfrac{5}{\sqrt{2}}\mathbf{i} + \dfrac{5}{\sqrt{2}}\mathbf{j}$; (5 m/s, 135°); $\mathbf{a} = -\dfrac{5}{\sqrt{2}}\mathbf{i} - \dfrac{5}{\sqrt{2}}\mathbf{j}$; (5 m/s^2, 225°)

b. $\mathbf{v} = \dfrac{5}{\sqrt{2}}\mathbf{i} - \dfrac{5}{\sqrt{2}}\mathbf{j}$; (5 m/s, 315°); acceleration vector: same as in part **(a)**   **4. a.** $\mathbf{a} = -\mathbf{i} - \dfrac{\sqrt{3}}{2}\mathbf{j}$; $\left(\dfrac{\sqrt{7}}{2}\ \text{m/s}^2,\ 221°\right)$ **5.** $\ln\left(1 + \sqrt{2}\right)$

6. $\dfrac{1}{27}(13\sqrt{13} - 8)$ **7.** $r = 3\cot\theta\csc\theta$ **8.** $y = 4$ **9.** $x(x^2 + y^2) = 4y$ **10.** $(x^2 + y^2 + 3x)^2 = 4(x^2 + y^2)$

11. circle

12. cardioid

13. four-leaf rose

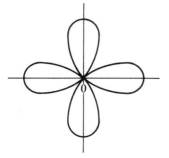

14. limaçon with loop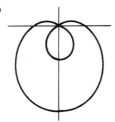

15. lemniscate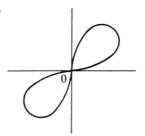

16. three-leaf rose

17. $\dfrac{1}{4} - \dfrac{\pi}{16}$ **18.** 6π **19.** a^2 **20.** $\dfrac{19\pi}{2}$ **21.** $\dfrac{1}{2}(2\pi - 3\sqrt{3})$ **22.** $\dfrac{9\sqrt{3}}{2} - \pi$ **23.** $4\left(\sqrt{3} - \dfrac{\pi}{3}\right)$ **24.** $\dfrac{\pi}{8}$ **26.** -1 **27.** $-\dfrac{2}{\pi}$

Chapter 9

Section 9.1 (page 324)

1.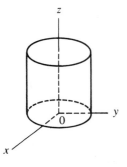

3.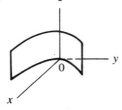

5.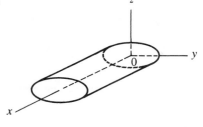

7.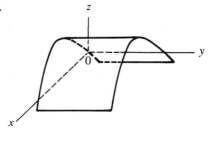

9.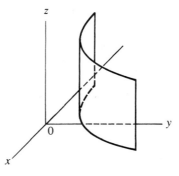

11.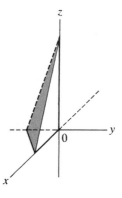

13.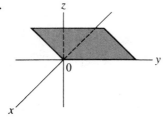

15.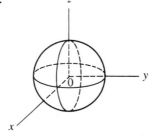

17.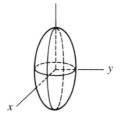

19.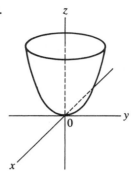

21.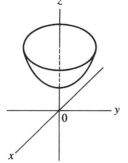

23.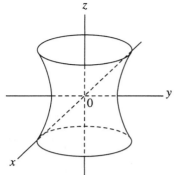

25.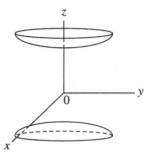

27.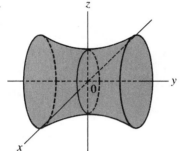

29.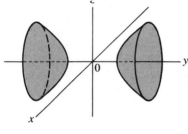

31.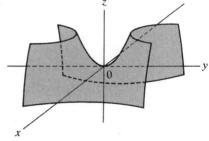

Section 9.2 (page 329)

1. a. $4x$ **b.** $10y$ **3. a.** 1   **b.** $2\cos 2y$ **5. a.** x^2 **b.** $3\sin y$ **7. a.** $2e^{2x}$ **b.** $\dfrac{1}{y}$ **9. a.** $2x\ln y$ **b.** $\dfrac{x^2}{y}$

11. a. $6x\tan 2y$ **b.** $6x^2\sec^2 2y$ **13. a.** $\dfrac{3x+2y}{2\sqrt{x+y}}$ **b.** $\dfrac{x}{2\sqrt{x+y}}$ **15. a.** $y^3\sec^2 xy$ **b.** $xy^2\sec^2 xy + 2y\tan xy$

17. a. $-2x\sin x^2 y$ **b.** $\dfrac{-x^2 y\sin x^2 y - \cos x^2 y}{y^2}$ **19. a.** $\dfrac{-x-2y^2}{2x^2\sqrt{x+y^2}}$ **b.** $\dfrac{y}{x\sqrt{x+y^2}}$ **21. a.** 5   **b.** -2 **c.** 0   **d.** 0   **e.** 0

23. a. $4x+6xy^2+5y$ **b.** $6y+6x^2 y+5x$ **c.** $4+6y^2$ **d.** $6+6x^2$ **e.** $12xy+5$ **25. a.** $2\cos(2x+y)$ **b.** $\cos(2x+y)$

c. $-4\sin(2x+y)$ **d.** $-\sin(2x+y)$ **e.** $-2\sin(2x+y)$ **27. a.** $\dfrac{2}{x}$ **b.** $\dfrac{1}{y}+\sec^2 y$ **c.** $-\dfrac{2}{x^2}$ **d.** $-\dfrac{1}{y^2}+2\sec^2 y\tan y$ **e.** 0

29. a. $-\dfrac{y}{x^2+y^2}$ **b.** $\dfrac{x}{x^2+y^2}$ **c.** $\dfrac{2xy}{(x^2+y^2)^2}$ **d.** $-\dfrac{2xy}{(x^2+y^2)^2}$ **e.** $\dfrac{-x^2+y^2}{(x^2+y^2)^2}$ **31. a.** $\dfrac{\sqrt{xyz}}{2x}$ **b.** $\dfrac{\sqrt{xyz}}{2y}$

c. $-\dfrac{\sqrt{xyz}}{4x^2}$ **d.** $-\dfrac{\sqrt{xyz}}{4y^2}$ **e.** $\dfrac{\sqrt{xyz}}{4xy}$ **33. a.** $-e$ **b.** e **35. a.** $\sqrt{2}$ **b.** 1   **37.** $\dfrac{25}{169}=0.15$ **39.** $\dfrac{4}{9}=0.444$

41. $0.67(v_g+0.1v_p)^{1/3}\ \Omega^{-1}$ **43.** $12°\text{C/m}$

Section 9.3 (page 335)

1. $dP = \dfrac{2}{V^2}\,dL - \dfrac{4L}{V^3}\,dV$ **3.** $dL = -\dfrac{X\,dX}{(X^2+Y^2)^{3/2}} - \dfrac{Y\,dY}{(X^2+Y^2)^{3/2}}$ **5.** $dM = \dfrac{\cos\theta_1}{\sin\theta_2}\,d\theta_1 - \dfrac{\sin\theta_1}{\sin\theta_2\tan\theta_2}\,d\theta_2$

7. $df = (1 + \omega\sec^2 r\omega)\,dr + (r\sec^2 r\omega)\,d\omega$ **9.** ± 0.17 **11.** $33\ \text{cm}^3$ **13.** ± 0.017 A, 2.1% **15.** ± 0.0076 s, 0.98%

17. $\dfrac{500\pi}{3} = 520\ \text{cm}^3/\text{min}$ **19.** 2300 W/s

21. min.: $\left(-\dfrac{2}{3}, 1, -\dfrac{13}{3}\right)$

23. max.: $\left(3, \dfrac{3}{2}, \dfrac{19}{2}\right)$

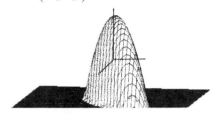

25. saddle point: $(-1, 1, -2)$

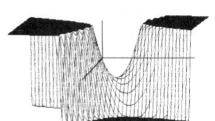

27. min.: $\left(\dfrac{1}{2}, \sqrt{2}, -15.6\right)$, saddle point: $\left(\dfrac{1}{2}, -\sqrt{2}, 7.1\right)$

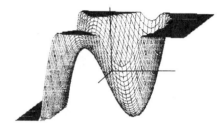

29. saddle points: $(0, 1, 0)$, $(0, -1, 0)$

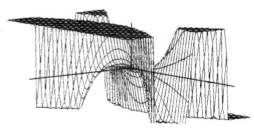

31. saddle point: $(0, 0, 0)$; max.: $(3, 3, 27)$

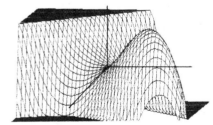

33. $20, 20, 20$ **39.** $\dfrac{dy}{dx} = -\dfrac{6x^5 + 10x^4 y^2 - 18x^2 y^3}{4x^5 y - 18x^3 y^2 - 16y^3}$ **41.** $\dfrac{dy}{dx} = -\dfrac{10x^4 + 12x^3 y - 12x^2 y^2 + 14xy^2}{3x^4 - 8x^3 y + 14x^2 y}$

43. $\dfrac{dy}{dx} = -\dfrac{28x^3 y^8 + 48x^2 y^5 + 50x}{56x^4 y^7 + 80x^3 y^4 - 14y}$ **45.** $\dfrac{dy}{dx} = \dfrac{y\sin x}{2y + \cos x}$ **47.** $\dfrac{dy}{dx} = \dfrac{2x\csc^2(x^2 - y^2) + 3y}{2y\csc^2(x^2 - y^2) - 3x}$ **49.** $\dfrac{dy}{dx} = \dfrac{y\sin y}{1 - xy\cos y + 6y^2}$

Section 9.4 (page 340)

1. $y = 9.9x - 4.8$ **3.** $y = 0.724x + 0.057$ **5.** $R = 0.0047T^2 + 4.2933$ **7.** $y = 1.988\ln x - 2.986$

Section 9.5 (page 346)

1. $\dfrac{1}{18}$ **3.** $\dfrac{1}{4}$ **5.** $\dfrac{1}{3}$ **7.** $\dfrac{16\sqrt{2}}{3}$ **9.** $\dfrac{1}{4}$ **11.** 1 **13.** $2e^3 + 1$ **15.** $\dfrac{397}{5}$ **17.** 1 **19.** 2 **21.** 16 **23.** $\dfrac{16}{3}$ **25.** $\dfrac{1}{6}$

27. 36 **29.** $\dfrac{4}{3}$ **31.** 36 **33.** $\dfrac{8}{3}$ **35.** $\dfrac{1}{6}$ **37.** 3

Section 9.6 (page 350)

1. $\dfrac{1}{2}$ **3.** $\dfrac{1}{4}$ **5.** $\dfrac{8}{3}$ **7.** $\dfrac{1}{6}$ **9.** 9   **11.** $\dfrac{32}{3}$ **13.** $\dfrac{48}{35}$ **15.** $8\displaystyle\int_0^r\int_0^{\sqrt{r^2-y^2}}\sqrt{r^2-x^2-y^2}\,dx\,dy = 8\int_0^r\int_0^{\sqrt{r^2-x^2}}\sqrt{r^2-x^2-y^2}\,dy\,dx$

17. $4\displaystyle\int_0^2\int_0^{\sqrt{4-x^2}}(9-x^2-y^2)\,dy\,dx = 4\int_0^2\int_0^{\sqrt{4-y^2}}(9-x^2-y^2)\,dx\,dy$ **19.** $\displaystyle\int_0^1\int_{\sqrt{y}}^{2-y}\sqrt{8-2x^2-y^2}\,dx\,dy$ **21.** $\displaystyle\int_1^2\int_0^{y^2+1}xy\,dx\,dy$

23. $\displaystyle\int_0^1\int_y^1 F(x,y)\,dx\,dy$ **25.** $\displaystyle\int_0^2\int_{y^2}^4 F(x,y)\,dx\,dy$ **27.** $\displaystyle\int_0^4\int_{-\sqrt{x}}^{\sqrt{x}} F(x,y)\,dy\,dx$ **29.** $\displaystyle\int_0^{16}\int_0^{\sqrt{x}/2} F(x,y)\,dy\,dx$

31. $\displaystyle\int_0^1\int_{x^2}^{\sqrt{x}} F(x,y)\,dy\,dx$ **33.** $\dfrac{32}{3}$ **35.** 2

Section 9.7 (page 357)

1. $4,\dfrac{8}{3}$ **3.** $\dfrac{8}{3},\dfrac{16}{3}$ **5.** $\dfrac{32}{3},\dfrac{128}{15}$ **7.** $\left(\dfrac{3}{2},\dfrac{6}{5}\right)$ **9.** $\dfrac{\sqrt{6}}{3}$ **11.** $\dfrac{3\rho}{35}$ **13.** $\sqrt{\dfrac{e^2-1}{2}}$ **15.** $e-1$ **17.** $\dfrac{1}{3}\rho$ **19.** $\dfrac{16\rho}{5}$

21. $\left(\dfrac{4-e}{2},\dfrac{e^2+1}{4e}\right)$ **23.** $\dfrac{1}{4}(3e^4+1)$ **25.** $\left(\dfrac{5}{12},\dfrac{2}{7},\dfrac{5}{14}\right)$ **27.** $\dfrac{1}{3}\rho$ **29.** $\dfrac{7\rho}{288}$ **31.** $\dfrac{62\rho}{63}$ **33.** $\left(\dfrac{1}{3},\dfrac{1}{3},1\right)$

35. $\displaystyle\int_0^2\int_0^3\int_{4-2x}^{4-x^2}(1+x)\,dz\,dy\,dx$

Section 9.8 (page 360)

1. 18π **3.** $\dfrac{81\pi}{2}$ **5.** $\dfrac{7\pi}{8}$ **7.** 16   **9.** $\dfrac{81\pi}{16}$ **11.** $\dfrac{1}{8}$ **13.** $\dfrac{64}{3}$ **15.** $\dfrac{\pi}{10}$ **17.** 0

Review Exercises for Chapter 9 (page 360)

1. plane
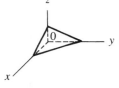

2. plane
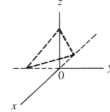

3. cylinder
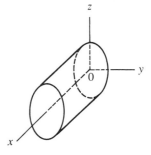

4. ellipsoid
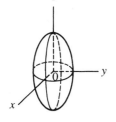

5. cylinder
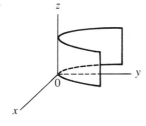

6. paraboloid
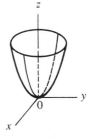

7. hyperboloid of two sheets

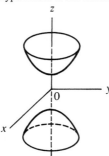

8. sphere

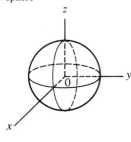

9. hyperboloid of one sheet

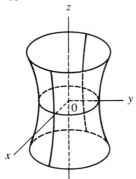

10. a. $\dfrac{-y}{2\sqrt{y^2 - x}}$ **b.** $\dfrac{2y^2 - x}{\sqrt{y^2 - x}}$ **c.** $\dfrac{-y}{4(y^2 - x)^{3/2}}$ **d.** $\dfrac{2y^3 - 3xy}{(y^2 - x)^{3/2}}$ **e.** $\dfrac{x}{2(y^2 - x)^{3/2}}$ **11. a.** $\dfrac{x}{x^2 + y^2} + y\cos xy$

b. $\dfrac{y}{x^2 + y^2} + x\cos xy$ **c.** $\dfrac{y^2 - x^2}{(x^2 + y^2)^2} - y^2 \sin xy$ **d.** $\dfrac{x^2 - y^2}{(x^2 + y^2)^2} - x^2 \sin xy$ **e.** $-\dfrac{2xy}{(x^2 + y^2)^2} - xy\sin xy + \cos xy$

13. $9.5°C$ **14.** $T = -\dfrac{5}{4}°C$ **15.** $11°C/cm$ **16.** 7.29 ft/s **17.** $\dfrac{8}{3}$ **18.** $\dfrac{512}{15}$ **19.** 1 **20.** $\dfrac{1}{6}$ **21.** $\dfrac{64\rho}{15}$ **22.** $\left(\dfrac{3}{4}, \dfrac{8}{5}\right)$

23. $\left(-\dfrac{1}{2}, \dfrac{2}{5}\right)$ **24.** $\dfrac{63\rho}{20}, \dfrac{\sqrt{70}}{10}$ **25.** $\displaystyle\int_0^2 \int_0^{x^2} F(x, y)\, dy\, dx$ **26.** $\displaystyle\int_{-4}^4 \int_0^{16-x^2} F(x, y)\, dy\, dx$ **27.** $\pm 11.0\,\text{cm}^3, 0.23\%$ **28.** $\pm 1.2\,\text{ft}^3$

29. $\dfrac{384\rho}{5}$ **30.** $\dfrac{324}{5}$ **31.** $\displaystyle\int_{-1}^1 \int_0^{\sqrt{1-y^2}} \int_x^{2x} \rho(y^2 + z^2)\, dz\, dx\, dy$ **32.** $\dfrac{35}{96}$ **33.** 128π

Chapter 10

Section 10.1 (page 365)

1. $\dfrac{3}{2}$ **3.** 1 **5.** 4 **7.** $\dfrac{1}{2}$ **9.** $\dfrac{2}{5}$ **11.** $\dfrac{4}{7}$ **13.** $\dfrac{7}{9}$ **15.** $\dfrac{7}{33}$ **17.** $\dfrac{25}{33}$ **19.** $\dfrac{1}{999}$ **21.** $\dfrac{251}{495}$

Section 10.2 (page 371)

5. conv. **7.** div. **9.** conv. **11.** div. **13.** conv. **15.** conv. **17.** conv. **19.** div. **21.** conv. **23.** div. **25.** conv.
27. conv. **29.** conv. **31.** conv. **33.** div. **35.** conv. **37.** conv. **39.** div. **41.** conv. **43.** conv. **45.** conv. **47.** conv.
49. conv. **51.** div. **53.** div. **55.** conv. **57.** div. **59.** div.

Section 10.4 (page 380)

1. $3x - \dfrac{3^3 x^3}{3!} + \dfrac{3^5 x^5}{5!} - \cdots$ **3.** $1 - x + \dfrac{x^2}{2!} - \dfrac{x^3}{3!} + \dfrac{x^4}{4!} - \cdots$ **5.** $1 - \dfrac{x}{2!} + \dfrac{x^2}{4!} - \cdots$ **7.** $x - \dfrac{x^3}{2!} + \dfrac{x^5}{4!} - \dfrac{x^7}{6!} + \cdots$

9. $x^2 - \dfrac{x^4}{2} + \dfrac{x^6}{3} - \dfrac{x^8}{4} + \cdots$ **11.** $1 - \dfrac{x^2}{3} + \dfrac{x^4}{5} - \cdots$ **13.** $1 - \dfrac{x}{2} + \dfrac{x^2}{3} - \dfrac{x^3}{4} + \cdots$ **19.** $1 - x + \dfrac{1}{3}x^3 - \dfrac{1}{6}x^4 + \cdots$

21. $3\left(x^2 - \dfrac{x^4}{2} + \dfrac{x^6}{3} - \dfrac{x^8}{4} + \cdots\right)$ **23.** $2x - \dfrac{x^2}{2} - \dfrac{x^4}{4} + \dfrac{2x^5}{5} - \dfrac{x^6}{6} - \dfrac{x^8}{8} + \dfrac{2x^9}{9} - \cdots$ **25.** $2e^{5\pi j/6}$ **27.** $3e^{\pi j/2}$

29. $2\sqrt{2}e^{3\pi j/4}$

Section 10.5 (page 386)

1. 0.644234, max. error: -0.000016, 0.6442 **3.** 0.98477, max. error: 0.00004, 0.9848 **5.** 0.818667, max. error: 0.000067, 0.8187
7. 0.36225, max. error: 0.0001, 0.362 **9.** 0.095333, max. error: -0.000025, 0.0953 **11.** 3.14 **13.** 0.06185 **15.** 0.76355

17. 0.29124 **19.** 0.4848 **21.** 0.85717 **23.** 0.5299 **31.** 1.96040 A **33.** $x(t) = t - \dfrac{t^5}{5\cdot 2!} + \dfrac{t^9}{9\cdot 4!} - \cdots$ **35.** 0.3413
37. -0.33 A **39.** 1.0005

Section 10.6 (page 397)

1. a. $a_0 = \dfrac{1}{2}, a_1 = 0, a_2 = 0$ **b.** $b_1 = \dfrac{2}{\pi}, b_2 = 0, b_3 = \dfrac{2}{3\pi}$ $\dfrac{1}{2} + \dfrac{2}{\pi}\left(\sin \pi t + \dfrac{1}{3}\sin 3\pi t + \dfrac{1}{5}\sin 5\pi t + \cdots\right)$

3. a. $a_0 = \dfrac{1}{2}, a_1 = 0, a_2 = 0$ **b.** $b_1 = \dfrac{2}{\pi}, b_2 = 0, b_3 = \dfrac{2}{3\pi}$ $\dfrac{1}{2} + \dfrac{2}{\pi}\left(\sin \dfrac{\pi t}{5} + \dfrac{1}{3}\sin \dfrac{3\pi t}{5} + \dfrac{1}{5}\sin \dfrac{5\pi t}{5} + \cdots\right)$

5. a. $a_0 = 0, a_1 = 0, a_2 = 0$ **b.** $b_1 = \dfrac{2a}{\pi}, b_2 = -\dfrac{2a}{2\pi}, b_3 = \dfrac{2a}{3\pi}$ $\dfrac{2a}{\pi}\left(\sin \dfrac{\pi t}{a} - \dfrac{1}{2}\sin \dfrac{2\pi t}{a} + \dfrac{1}{3}\sin \dfrac{3\pi t}{a} - \cdots\right)$

7. $\dfrac{a}{4} - \dfrac{2a}{\pi^2}\left(\cos \dfrac{\pi t}{a} + \dfrac{1}{3^2}\cos \dfrac{3\pi t}{a} + \dfrac{1}{5^2}\cos \dfrac{5\pi t}{a} + \cdots\right) + \dfrac{a}{\pi}\left(\sin \dfrac{\pi t}{a} - \dfrac{1}{2}\sin \dfrac{2\pi t}{a} + \dfrac{1}{3}\sin \dfrac{3\pi t}{a} - \cdots\right)$

9. $\dfrac{1}{\pi} + \dfrac{1}{2}\sin t - \dfrac{2}{\pi}\left(\dfrac{1}{3}\cos 2t + \dfrac{1}{15}\cos 4t + \dfrac{1}{35}\cos 6t + \dfrac{1}{63}\cos 8t + \cdots\right)$

Review Exercises for Chapter 10 (page 397)

1. $\dfrac{3}{4}$ **2.** $\dfrac{131}{990}$ **3.** div.   **4.** conv.   **5.** conv.   **6.** conv.   **7.** div.   **8.** conv.   **9.** conv.   **10.** div.

13. $\sin x^2 = x^2 - \dfrac{x^6}{3!} + \dfrac{x^{10}}{5!} - \cdots$ **14.** $\cos \sqrt{x} = 1 - \dfrac{x}{2!} + \dfrac{x^2}{4!} - \cdots$ **15.** $-1 - \dfrac{x}{2!} - \dfrac{x^2}{3!} - \cdots$ **16.** $\sinh x = x + \dfrac{x^3}{3!} + \dfrac{x^5}{5!} + \cdots$

17. 0.877604, max. error: -0.000022, 0.8776   **18.** 0.94608   **19.** 0.8666   **20.** 0.100   **21.** 0.10 C   **22.** 0.8387   **23.** 0.69466

Chapter 11

Section 11.1 (page 402)

17. $y = x^3 - 3$ **19.** $y = \tan x$ **21.** $y = e^x + (2-e)x - 1$

Section 11.2 (page 407)

1. $2x^3 + 3y^2 = c$ **3.** $6x + 2x^3 = 9y^2 + c$ **5.** $\ln(1 + x^2) + y = c$ **7.** $\ln|y| + \operatorname{Arctan} x = c$ **9.** $x^2 y^3 = c$ **11.** $xy^2 - 2xy - 2 = cx$
13. $x = c(1-y)$ **15.** $PV = c$ **17.** $2\tan x + 4y - y^2 = c$ **19.** $3y^2 - 2\cos^3 t = c$ **21.** $t\sqrt{v^2 + 1} - 1 = ct$
23. $2\sin T_1 - 2T_1 \cos T_1 + 2T_2 + T_2{}^2 = c$ **25.** $(y^2 - 1)\sin x = c$ **27.** $xe^x - e^x - e^{-y} = c$ **29.** $(e^x + 1)\sec y = c$
31. $y \sin y + \cos y + x^2 = c$ **33.** $y = 2x$ **35.** $(y+2)(x-3) = -7$ **37.** $x = \cos y$

Section 11.3 (page 412)

1. $y = 1 + ce^{-x}$ **3.** $y = e^{3x} + ce^{2x}$ **5.** $y = \left(\dfrac{1}{2}x + c\right)e^{2x^2}$ **7.** $y = 1 + ce^{-2\sin x}$ **9.** $y = \dfrac{1}{2}x + \dfrac{c}{x}$ **11.** $xy = xe^x - e^x + c$

13. $y = x^2 \tan x + cx^2$ **15.** $y = x^3 \sin x - x^4 \cos x + cx^3$ **17.** $y = x^2 e^x + cx^2$ **19.** $y = 1 + ce^{\cos x}$ **21.** $4y\cos x = 2x + \sin 2x + c$

23. $y(x+1) = x + 2\ln|x-1| + c$ **25.** $y = -\dfrac{1}{2}x\cos x^2 + cx$ **27.** $y\sec t = \tan t + c$ **29.** $r = \dfrac{1}{2}t\ln t - \dfrac{1}{4}t + \dfrac{c}{t}$

31. $r = \dfrac{1}{3}s^2 e^{3s} - \dfrac{1}{9}se^{3s} + cs$ **33.** $y = 2e^{-x}(3x + 1)$ **35.** $y = 2 + e^{(-1/3)x^3}$ **37.** $i = \dfrac{E}{R}(1 - e^{-Rt/L})$

Section 11.4 (page 419)

1. $i = 1 - e^{-25t}$ **3.** $N = 100e^{-0.0223t}$ **5.** 12.2 years   **7.** 1380 years   **9.** 11,200 years   **11. a.** \$2170.59   **b.** 6.9%   **c.** \$335.16
13. $T = 70 - 50e^{-0.1783t}$; 21.9 min   **15.** 42.2 min   **17.** 91 m/s; 500 m/s   **19.** $\ln[(5+v)/(5-v)] = 4t$ or $v = 5(1 - e^{-4t})/(1 + e^{-4t})$;
$v = 5$ m/s   **21.** 80 s   **23.** $y = kx$ **25.** $2x^2 + y^2 = k$ **27.** $y^2 = x + k$ **29.** $y^2 - x^2 = k$
31. $N = \dfrac{10{,}000}{1 + 9e^{-[\ln(9/4)]t}}$ **33.** $x(t) = 40 - 20e^{-t/50}$; 29 lb   **35.** $x(t) = 20e^{-t/50}$; 11.0 lb

Section 11.5 (page 424)

Answers for selected values of x and y

1. a.

x	0	0.3	0.6	1.0	1.3	1.6	2.0
y	2	1.7351	1.5404	1.3585	1.2635	1.1937	1.1285

b.

x	0	0.3	0.6	1.0	1.3	1.6	2.0
y	2	1.7380	1.5446	1.3632	1.2681	1.1978	1.1319

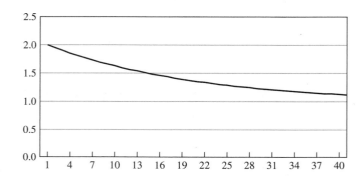

3. a.

x	0	0.1	0.2	0.3	0.4
y	3	2.7175	2.4716	2.2588	2.0759

b.

x	0	0.1	0.2	0.3	0.4
y	3	2.7190	2.4743	2.2624	2.0802

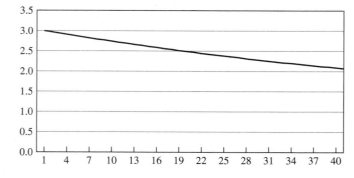

5.

x	0	0.1	0.2	0.3	0.4	0.5
y	1	1.1095	1.2438	1.4111	1.6254	1.9112

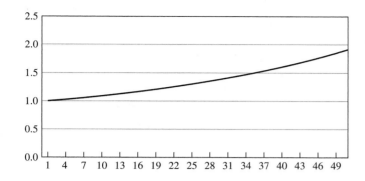

7. a.

x	0	0.2	0.4	0.6	0.8	1.0
y	0	0.1983	0.3835	0.5451	0.6787	0.7852

b.

x	0	0.2	0.4	0.6	0.8	1.0
y	0	0.1976	0.3809	0.5400	0.6714	0.7761

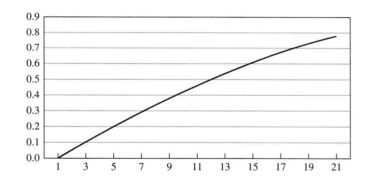

9.

x	0	0.3	0.6	1.0	1.3	1.6	2.0
y	2	1.7408	1.5488	1.3679	1.2725	1.2019	1.1353

11.

x	0	0.1	0.2	0.3	0.4
y	3	2.7194	2.4749	2.2633	2.0813

13.

x	0	0.1	0.2	0.3	0.4	0.5
y	1	1.1108	1.2470	1.4175	1.6373	1.9337

Review Exercises for Chapter 11 (page 424)

1. $y = x - 1 + ce^{-x}$ **2.** $3x - \dfrac{3}{x} + y^3 = c$ **3.** $y = \dfrac{1}{2} + ce^{-x^2}$ **4.** $y = -1 + ce^{x^2}$ **5.** $2\operatorname{Arctan} x + \ln(1 + y^2) = c$

6. $y\sin^2 x = -\cos x + c$ **7.** $x^2 y = c$ **8.** $x^3 + \ln|y - \tan y| = c$ **9.** $y = \dfrac{1}{2}x^4 + cx^2$ **10.** $y(\sec x + \tan x) = \cos x - x + c$

11. $x^2 + \csc^2 y = c$ **12.** $e^{2x} - 2ye^{-y} - 2e^{-y} = c$ **13.** $y(\sec x + \tan x) = 2(\sqrt{2} + 1)$ **14.** $y\ln y - y + \ln|\ln x| = c$ **15.** 3.2 h

16. 1590 years   **17.** $-29°F$ **18.** 14.4 min   **19.** $v = 60(1 - e^{-(1/6)t})$ **20.** $y - 4 = k(x + 3)$ **21.** $y = \dfrac{k}{x^2}$

22. a.

x	0	0.08	0.12	0.16	0.2
y	0	0.00486	0.01231	0.02328	0.03788

b.

x	0	0.08	0.12	0.16	0.2
y	0	0.00657	0.01498	0.02697	0.04267

Chapter 12

Section 12.1 (page 431)

1. $y = c_1 e^{6x} + c_2 e^{7x}$ **3.** $y = c_1 e^{2x/3} + c_2 e^{-x/2}$ **5.** $y = c_1 e^{x/4} + c_2 e^{-2x}$ **7.** $y = e^{x/2}\left(c_1 e^{\sqrt{5}x/2} + c_2 e^{-\sqrt{5}x/2}\right)$

9. $y = c_1 e^x + c_2 e^{x/2}$ **11.** $y = e^{3x} - e^{-3x}$ **13.** $y = \dfrac{e}{1 - e^3}(e^{-x} - e^{2x})$ **15.** $y = c_1 e^x + c_2 e^{2x} + c_3 e^{-3x}$

17. $y = c_1 e^{-x} + c_2 e^{(1+\sqrt{3})x} + c_3 e^{(1-\sqrt{3})x}$ **19.** $y = e^{-x/2}\left(c_1 e^{\sqrt{5}x/2} + c_2 e^{-\sqrt{5}x/2}\right)$ **21.** $y = e^x\left(c_1 e^{\sqrt{3}x} + c_2 e^{-\sqrt{3}x}\right)$

23. $y = e^{2x}\left(c_1 e^{\sqrt{6}x} + c_2 e^{-\sqrt{6}x}\right)$ **25.** $y = e^{-3x}\left(c_1 e^{\sqrt{15}x} + c_2 e^{-\sqrt{15}x}\right)$ **27.** $y = e^{-x}\left(c_1 e^{(\sqrt{2}/2)x} + c_2 e^{-(\sqrt{2}/2)x}\right)$ **29.** $y = c_1 e^x + c_2 e^{-(2/3)x}$

Section 12.2 (page 434)

1. $y = c_1 e^{-3x} + c_2 x e^{-3x}$ **3.** $y = c_1 e^{(1/2)x} + c_2 x e^{(1/2)x}$ **5.** $y = c_1 e^{-(2/3)x} + c_2 x e^{-(2/3)x}$ **7.** $y = c_1 e^{(5/2)x} + c_2 x e^{(5/2)x}$

9. $y = e^{2x}(c_1 \cos x + c_2 \sin x)$ **11.** $y = e^{-2x}(c_1 \cos 2x + c_2 \sin 2x)$ **13.** $y = e^{(1/2)x}\left(c_1 \cos \dfrac{1}{2}x + c_2 \sin \dfrac{1}{2}x\right)$ **15.** $y = c_1 \cos 5x + c_2 \sin 5x$

17. $y = c_1 e^{3x} + c_2 x e^{3x}$ **19.** $y = c_1 + c_2 x + c_3 x^2 + c_4 e^{-2x}$ **21.** $y = e^{-x/2}\left(c_1 \cos \dfrac{\sqrt{7}}{2}x + c_2 \sin \dfrac{\sqrt{7}}{2}x\right)$

23. $y = e^{3x/2}\left(c_1 \cos \dfrac{\sqrt{11}}{2}x + c_2 \sin \dfrac{\sqrt{11}}{2}x\right)$ **25.** $y = e^x\left(c_1 \cos \dfrac{\sqrt{6}}{2}x + c_2 \sin \dfrac{\sqrt{6}}{2}x\right)$ **27.** $y = e^{-x}\left(c_1 e^{(\sqrt{6}/2)x} + c_2 e^{-(\sqrt{6}/2)x}\right)$

29. $y = c_1 e^{10x} + c_2 e^{-10x}$ **31.** $y = c_1 \cos 10x + c_2 \sin 10x$ **33.** $y = c_1 + e^{x/3}\left(c_2 \cos \dfrac{\sqrt{2}}{3}x + c_3 \sin \dfrac{\sqrt{2}}{3}x\right)$

35. $y = e^{2x}\left(c_1 e^{\sqrt{2}x} + c_2 e^{-\sqrt{2}x}\right)$ **37.** $y = c_1 \cos 2x + c_2 \sin 2x$ **39.** $y = c_1 + c_2 e^{2x} + c_3 x e^{2x}$ **41.** $y = c_1 + c_2 e^{-3x} + c_3 e^{3x}$

43. $y = c_1 e^{2x} + c_2 x e^{2x} + c_3 x^2 e^{2x} + c_4 x^3 e^{2x}$ **45.** $y = \sin x$ **47.** $y = -e^x \sin x$ **49.** $y = c_1 \cos 3x + c_2 \sin 3x + c_3 x \cos 3x + c_4 x \sin 3x$

51. $y = c_1 e^{-x} + e^{x/2}\left(c_2 \cos \dfrac{\sqrt{2}}{2}x + c_3 \sin \dfrac{\sqrt{2}}{2}x\right)$ **53.** $y = c_1 e^{6x} + c_2 e^{-4x} + c_3 x e^{-4x}$

55. $y = c_1 e^{-x} + c_2 x e^{-x} + e^x\left(c_3 \cos \dfrac{\sqrt{5}}{2}x + c_4 \sin \dfrac{\sqrt{5}}{2}x\right)$

Section 12.3 (page 440)

1. $y = c_1 e^{3x} + c_2 x e^{3x} + \dfrac{1}{4} e^x$ **3.** $y = c_1 e^{3x} + c_2 x e^{3x} + x + \dfrac{2}{3}$ **5.** $y = c_1 e^{2x} + c_2 e^{-x} - x^2 + x - \dfrac{3}{2}$

7. $y = e^{(1/2)x}\left(c_1 \cos \dfrac{\sqrt{3}}{2}x + c_2 \sin \dfrac{\sqrt{3}}{2}x\right) - x^2 - 2x + 1$ **9.** $y = e^{x/2}\left(c_1 \cos \dfrac{\sqrt{7}}{2}x + c_2 \sin \dfrac{\sqrt{7}}{2}x\right) + \dfrac{1}{2} e^{3x}$

11. $y = c_1 e^{3x} + c_2 x e^{3x} - \dfrac{1}{2} \sin 3x$ **13.** $y = c_1 e^{-x} + c_2 e^{-3x} + \dfrac{1}{8} e^x + 2$ **15.** $y = c_1 \cos x + c_2 \sin x - 2 \sin 2x$

17. $y = c_1 e^{-2x} + c_2 e^{-3x} - \dfrac{1}{5} \cos x + \sin x$ **19.** $y = c_1 e^x + c_2 x e^x + 3 e^{2x}$ **21.** $y = c_1 e^x + c_2 e^{-x} + c_3 e^{2x} + e^{3x}$

23. $y = e^x(c_1 \cos 2x + c_2 \sin 2x) + x e^x$ **25.** $y = \dfrac{1}{2}(\cos 3x + e^{3x})$ **27.** $y = 5 \cos x + \sin x - 2 \cos 2x$

29. $y = e^{-x}(3 \cos 2x + 4 \sin 2x) + 2 \cos x + \sin x$ **31.** $y_p = -2 \sin 3x$ **33.** $y_p = \dfrac{1}{2} \cos x - \dfrac{1}{2} \sin x$

35. $y_p = -x^2 - x - 2$ **37.** $y_p = \dfrac{1}{3} e^{2x} + 1$ **39.** $y_p = -\dfrac{1}{4} x + \dfrac{1}{16} + e^{3x}$ **41.** $y_p = \dfrac{2}{3} x + \dfrac{1}{2} \cos x$

43. $y_p = -2x e^x - \dfrac{2}{3} e^x$ **45.** $y = c_1 e^{-2x} + c_2 e^{2x} + 2x e^{2x}$ **47.** $y = c_1 e^{3x} + c_2 e^{-2x} - 2x e^{-2x}$ **49.** $y = c_1 e^{-7x} + c_2 e^{4x} + x e^{4x}$

51. $y = c_1 e^{-x} + (c_2 + x)e^x$ **53.** $y = c_1 e^{-2x} + c_2 e^{2x} + \dfrac{1}{4} x e^{2x} - 1$ **55.** $y = c_1 \cos x + c_2 \sin x - x \cos x$

Section 12.4 (page 449)

1. $x(t) = \dfrac{1}{4} \cos 8t$

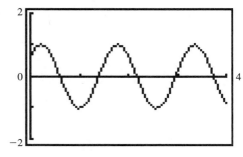

3. $x(t) = \dfrac{5}{28} \cos 8t + \dfrac{1}{14} \cos 6t$

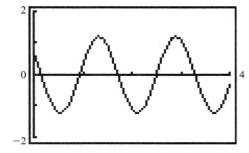

5. $x(t) = \dfrac{2}{3} \cos 4t + \dfrac{3}{4} \sin 4t$

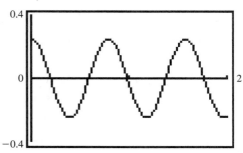

7. $x(t) = \dfrac{2}{3} \cos 4t - \sin 4t$

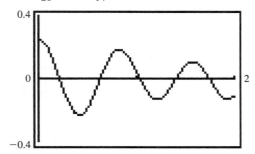

9. $x(t) = e^{-t/2}\left(\frac{1}{2}\cos\frac{3}{2}\sqrt{7}t + \frac{1}{6\sqrt{7}}\sin\frac{3}{2}\sqrt{7}t\right)$

11. $x(t) = e^{-t/2}\left(-\frac{1}{4}\cos\frac{\sqrt{47}}{2}t - \frac{1}{4\sqrt{47}}\sin\frac{\sqrt{47}}{2}t\right)$

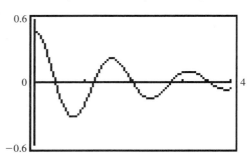

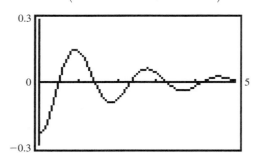

13. $x(t) = e^{-t}\left(\frac{1}{2}\cos\sqrt{31}t - \frac{5}{2\sqrt{31}}\sin\sqrt{31}t\right)$

15. $x(t) = \frac{1}{2}e^{-8t} + 4te^{-8t} - \frac{1}{4}\cos 8t$

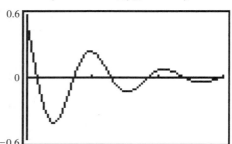

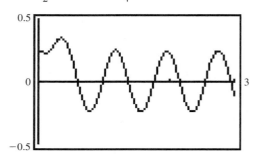

17. $x(t) = 0.25\cos 10t$ **19.** $x(t) = e^{-t}\left(0.25\cos\sqrt{99}t + \frac{0.25}{\sqrt{99}}\sin\sqrt{99}t\right)$ or $x(t) = e^{-1.0t}(0.25\cos 9.9t + 0.025\sin 9.9t)$

21. $x_c = e^{-1.0t}(0.27\cos 9.9t - 0.039\sin 9.9t)$ **23.** $q = 0.1\sin 100t, i = 10\cos 100t$
$x_p = -0.017\cos 5t + 0.13\sin 5t$

25. $q = \frac{2}{75}\sin 50t - \frac{1}{75}\sin 100t$ **27.** $x_p = 0.20\sin 5t - 0.029\cos 5t$ **29.** $q_p = \frac{1}{2}\sin 10t, i_p = 5\cos 10t$

Review Exercises for Chapter 12 (page 450)

1. $y = c_1 + c_2x + c_3x^2 + c_4x^3$ **2.** $y = c_1e^{2x} + c_2e^{-2x}$ **3.** $y = c_1\cos 2x + c_2\sin 2x$ **4.** $y = c_1e^{4x} + c_2xe^{4x}$

5. $y = e^x\left(c_1e^{\sqrt{3}x} + c_2e^{-\sqrt{3}x}\right)$ **6.** $y = e^x(c_1\cos x + c_2\sin x)$ **7.** $y = e^{(1/6)x}\left(c_1\cos\frac{\sqrt{11}}{6}x + c_2\sin\frac{\sqrt{11}}{6}x\right)$

8. $y = e^{-x/4}\left(c_1\cos\frac{1}{4}\sqrt{15}x + c_2\sin\frac{1}{4}\sqrt{15}x\right)$ **9.** $y = c_1e^{2x} + c_2xe^{2x} + c_3\cos x + c_4\sin x$ **10.** $y = c_1 + c_2x + c_3x^2 + c_4\cos 3x + c_5\sin 3x$

11. $y = e^{(1/4)x}\left(c_1\cos\frac{\sqrt{7}}{4}x + c_2\sin\frac{\sqrt{7}}{4}x\right)$ **12.** $y = e^{(1/2)x}\left(c_1e^{(\sqrt{3}/2)x} + c_2e^{-(\sqrt{3}/2)x}\right)$ **13.** $y = e^{(1/3)x}\left(c_1e^{(\sqrt{7}/3)x} + c_2e^{-(\sqrt{7}/3)x}\right)$

14. $y = e^x(c_1\cos x + c_2\sin x)$

In Exercises 15–20, $y_c = c_1e^{4x} + c_2e^{-x}$

15. $y = y_c - e^x$ **16.** $y = y_c - 2x + \frac{3}{2}$ **17.** $y = y_c + \frac{3}{17}\cos x - \frac{5}{17}\sin x$ **18.** $y = y_c - 4xe^{-x}$ **19.** $y = y_c - \frac{2}{5}xe^{-x} - x^2e^{-x}$

20. $y = y_c + \frac{1}{5}xe^{4x}$ **21.** $y = e^{-x} - e^{3x}$ **22.** $y = 2\cos x$ **23.** $i(t) = -1.08\sin 20.0t + 1.88\cos 20.0t$

25. $x(t) = \frac{7}{30}\cos 8t - \frac{5}{8}\sin 8t + \frac{1}{60}\cos 4t$

Chapter 13

Sections 13.1–13.3 (page 459)

3. $\dfrac{5s+2}{s(s+1)}$ **5.** $\dfrac{s^3+s^2+4}{s^2(s^2+4)}$ **7.** $\dfrac{5}{(s-2)^2+25}$ **9.** $\dfrac{6}{(s+4)^4}$ **11.** $\dfrac{48}{(s+1)^5}$ **13.** $\dfrac{4}{s}-\dfrac{10}{s^2+4}$ **15.** $5\sin 2t$ **17.** $\dfrac{1}{2}t^2 e^{-2t}$

19. $\dfrac{1}{4}(\sin 2t - 2t\cos 2t)$ **21.** $\dfrac{1}{3}t\sin 3t$ **23.** $5\cos\sqrt{6}t$ **25.** $\dfrac{1}{2}t-\dfrac{1}{4}\sin 2t$ **27.** $2e^{-2t}\cos 2t$ **29.** $\dfrac{1}{\sqrt{5}}e^{-3t}\sin\sqrt{5}t$

31. $e^t\sin\sqrt{10}t$ **33.** $e^{3t}(\cos t+3\sin t)$ **35.** $e^t(\cos\sqrt{5}t+\dfrac{1}{\sqrt{5}}\sin\sqrt{5}t)$ **37.** $1-e^{-t}$ **39.** $e^{2t}+e^{-3t}$ **41.** $\dfrac{1}{6}e^{-2t}-\dfrac{1}{2}e^{2t}+\dfrac{4}{3}e^{4t}$

43. $\dfrac{8}{3}e^{-2t}-4te^{-2t}+\dfrac{1}{3}e^t$ **45.** $\dfrac{1}{2}e^{-t}-\dfrac{1}{2}\cos t+\dfrac{1}{2}\sin t$ **47.** $-\dfrac{1}{2}e^{-t}+\dfrac{1}{2}\cos t+\dfrac{1}{2}\sin t$ **49.** $6te^{-2t}-e^{-2t}+e^t$ **51.** $e^{-t}-te^{-t}$

53. $e^{-t}\cos t+e^t$ **55.** $\dfrac{1}{7}e^{-4t}-\dfrac{1}{7}e^{-2t}\cos\sqrt{3}t+\dfrac{2\sqrt{3}}{21}e^{-2t}\sin\sqrt{3}t$

Section 13.4 (page 464)

1. e^t **3.** $2e^{2t}-2$ **5.** te^{2t} **7.** $\dfrac{1}{2}t^2 e^{-4t}+3e^{-4t}$ **9.** $\cos 3t-\dfrac{2}{3}\sin 3t$ **11.** $\sin t-t\cos t$ **13.** $-2\cos 2t+\dfrac{1}{2}t\sin 2t$

15. $\dfrac{1}{5}e^t-\dfrac{1}{5}\cos 2t+\dfrac{2}{5}\sin 2t$ **17.** $-3te^{2t}-e^{2t}+e^{3t}$ **19.** $t^4 e^{3t}$ **21.** $e^{2t}\left(-3\cos\sqrt{6}t+\sqrt{6}\sin\sqrt{6}t\right)$ **23.** $2(e^t-\cos t-\sin t)$

25. $\dfrac{3}{10}e^{2t}+\dfrac{3}{10}e^{-2t}-\dfrac{3}{5}\cos t$ **27.** $e^t-e^{-t}\cos 2t-e^{-t}\sin 2t$ **29.** $q(t)=0.03e^{-50t}-0.15e^{-10t}+0.12$ **33.** $\cos 2t-\cos 3t$

35. $\dfrac{2}{5}\sin 3t+\dfrac{2}{5}\sin 2t$ **37.** $-2e^{-2t}\cos\sqrt{2}t-\dfrac{19}{4\sqrt{2}}e^{-2t}\sin\sqrt{2}t+\dfrac{3}{4\sqrt{6}}\sin\sqrt{6}t$ **39.** $x(t)=\cos 4t+\dfrac{32}{15}\sin t-\dfrac{8}{15}\sin 4t$

41. $q(t)=0.00208e^{-50t}\cos 50\sqrt{3}t-0.00110e^{-50t}\sin 50\sqrt{3}t-\dfrac{5}{2404}\cos 20.0t+\dfrac{6}{601}\sin 20.0t$

Review Exercises for Chapter 13 (page 465)

1. $\dfrac{2}{s+3}$ **2.** $\dfrac{4-s^2}{s(s^2+4)}$ **3.** $\dfrac{12}{s^4}+\dfrac{3}{s^2+9}$ **4.** $\dfrac{s+3}{(s+3)^2+36}$ **5.** $\dfrac{8}{s^2(s^2+4)}$ **6.** $\dfrac{120}{(s-3)^5}$ **7.** $e^{2t}\cos\sqrt{5}t$

8. $\dfrac{1}{6}t^3 e^{-3t}$ **9.** $e^t\left(\cos 2t+\dfrac{1}{2}\sin 2t\right)$ **10.** $\dfrac{1}{16}(\sin 2t-2t\cos 2t)$ **11.** $\dfrac{1}{3}e^{-t}+\dfrac{2}{3}e^{2t}$ **12.** $\dfrac{4}{9}e^{-t}-\dfrac{1}{3}te^{-t}-\dfrac{4}{9}e^{-4t}$

13. $\dfrac{1}{30}e^{-2t}-\dfrac{1}{5}e^{3t}+\dfrac{1}{6}e^{4t}$ **14.** $e^{-2t}+2\sin t-\cos t$ **15.** e^{-2t} **16.** te^t **17.** $e^{-2t}\left(\dfrac{1}{2}t^2-1\right)$ **18.** $\sqrt{2}\sin\sqrt{2}t$

19. $e^{-t}-e^{3t}$ **20.** $\dfrac{1}{2}(e^{-t}-\cos t+\sin t)$ **21.** $e^{-t}\left(\cos 2t+\dfrac{1}{2}\sin 2t\right)$ **22.** $\dfrac{1}{2}e^t\sin 2t$ **23.** $\dfrac{3}{5}e^{-2t}+\dfrac{2}{5}e^{-t}\cos 2t+\dfrac{13}{10}e^{-t}\sin 2t$

24. $\dfrac{8}{25}e^{-2t}+\dfrac{2}{5}te^{-2t}-\dfrac{8}{25}\cos t+\dfrac{6}{25}\sin t$ **25.** $x(t)=-4te^{-8t}$ **26.** $x(t)=\dfrac{1}{4}t\sin 4t-\dfrac{1}{2}\cos 4t$

Index